Advances in 21st Century Human Settlements

Indexed by SCOPUS

This Series focuses on the entire spectrum of human settlements – from rural to urban, in different regions of the world, with questions such as: What factors cause and guide the process of change in human settlements from rural to urban in character, from hamlets and villages to towns, cities and megacities? Is this process different across time and space, how and why? Is there a future for rural life? Is it possible or not to have industrial development in rural settlements, and how? Why does 'urban shrinkage' occur? Are the rural areas urbanizing or is that urban areas are undergoing 'ruralisation' (in form of underserviced slums)? What are the challenges faced by 'mega urban regions', and how they can be/are being addressed? What drives economic dynamism in human settlements? Is the urban-based economic growth paradigm the only answer to the quest for sustainable development, or is there an urgent need to balance between economic growth on one hand and ecosystem restoration and conservation on the other – for the future sustainability of human habitats? How and what new technology is helping to achieve sustainable development in human settlements? What sort of changes in the current planning, management and governance of human settlements are needed to face the changing environment including the climate and increasing disaster risks? What is the uniqueness of the new 'socio-cultural spaces' that emerge in human settlements, and how they change over time? As rural settlements become urban, are the new 'urban spaces' resulting in the loss of rural life and 'socio-cultural spaces'? What is leading the preservation of rural 'socio-cultural spaces' within the urbanizing world, and how? What is the emerging nature of the rural-urban interface, and what factors influence it? What are the emerging perspectives that help understand the human-environment-culture complex through the study of human settlements and the related ecosystems, and how do they transform our understanding of cultural landscapes and 'waterscapes' in the 21st Century? What else is and/or likely to be new vis-à-vis human settlements – now and in the future? The Series, therefore, welcomes contributions with fresh cognitive perspectives to understand the new and emerging realities of the 21st Century human settlements. Such perspectives will include a multidisciplinary analysis, constituting of the demographic, spatio-economic, environmental, technological, and planning, management and governance lenses.

If you are interested in submitting a proposal for this series, please contact the Series Editor, or the Publishing Editor:

Bharat Dahiya (bharatdahiya@gmail.com) or
Loyola D'Silva (loyola.dsilva@springer.com)

T. M. Vinod Kumar
Editor

Indo-Pacific Smart Megacity System

Emerging Architecture and Megacity Studies

Springer

Editor
T. M. Vinod Kumar
School of Planning and Architecture
New Delhi, Delhi, India

ISSN 2198-2546 ISSN 2198-2554 (electronic)
Advances in 21st Century Human Settlements
ISBN 978-981-99-6217-4 ISBN 978-981-99-6218-1 (eBook)
https://doi.org/10.1007/978-981-99-6218-1

This Springer imprint is published by the registered company Springer Nature Singapore Pte Ltd.
The registered company address is: 152 Beach Road, #21-01/04 Gateway East, Singapore 189721, Singapore

Paper in this product is recyclable.

Dedicated to Shinzo Abe

Contents

Introduction

The Smart Megacity System of Indo-Pacific: Emerging Architecture and Megacities Studies

T. M. Vinod Kumar

Abstract In 2016 as per UN-HABITAT, there were about thirty-one gigantic cities/urban agglomerations/conurbations/megalopolises of population size ten million and above-called megacities in the universe which is likely to be 41 in 2030. Twenty million population and above cities are called meta cities which were eight in 2016 and will have eleven in 2030. Most of the megacities and meta cities are in the Indo-Pacific, amounting to a total of twenty-one out of 41 in 2030. They all form a close-knit and integrated system manifesting the spatial system of the global economy. Most newly emerging megacities built with newer industrial revolutions and technological basis are in the Indo-Pacific region. These megacities are functionally, spatially, and economically interconnected by sea and air routes and fiber-optic internet networks, generating daily economic interactions of great significance. The individual megacities in the system can be called wise to various degrees. Indo-Pacific region envelopes the boundary of the Indian and the Pacific Oceans, which are contiguous and most extensive of all oceans of the universe. The region can be called a combination of the homogeneous, program, and polarized regions for conducting various Indo-Pacific-specific designed economic activities and functions. These cities are genuinely global, work based on the ever-changing dynamics of global politics based on international diplomacy against national bureaucracy faithfully local the set of policies, programs, regulatory laws, and budgets of cities in the sub-national and below following a democratic pattern of governance with mixed economy or capitalist economy of various proportions, by largely local self-government in a Federal Governing setup such as in India, the USA or even to some extent Russia and China with autocracies in position. This local self-governance must be transformed uniquely to use the global megacity system's international and regional economic development. The Indo-Pacific megacity system is defined as an interconnected urban network functionally and economically by twenty-one megacities in the Indo-Pacific, servicing sub-national regions and servicing countries some thirty-one and above outside sharing the same megacity gateway physical infrastructure and related economic, institutional, and physical infrastructure of megacities of the Indo-Pacific. In this region, these megacities have a disproportionately

T. M. Vinod Kumar (✉)
School of Planning and Architecture, New Delhi, India
e-mail: tmvinod@gmail.com

T. M. Vinod Kumar (ed.), *Indo-Pacific Smart Megacity System*, Advances in 21st Century Human Settlements, https://doi.org/10.1007/978-981-99-6218-1_1

higher NDP share as well as disproportionately higher average household income; they have distinct income and expenditure patterns than more miniature cities and villages of the respective Nations and assume a significant role in regional and international economic development and politics of the Indo-Pacific. In the Indo-Pacific region, all leading economies of the world are situated, such as the US, the People's Republic of China, India, Russia, Japan, Australia, South Korea, and ASEAN countries as well as SARC countries, and many other countries groupings. Others situated outside the Indo-Pacific, like France, the Netherlands, Germany, the UK, and the European Economic Community have substantial economic and security concerns in this region and long-term historical connections to the legacy of colonialization that connects them to the Indo-Pacific and have their independent National Strategies for the Indo-Pacific. Ships carrying goods and aero planes carrying Indo-Pacific persons of rare servicing capabilities and entrepreneurial force crisscross the Indo-Pacific every instant which also uses sea routes carrying many Trillian dollars' worth of manufactured goods annually a large world share and home for major fiber-optic internet cable networks of the world. These gigantic habitats are significant because it has all the potential to configure themselves into a system of intelligent and global megacities for their sustainability and rapid economic development, as well as collectively participating in major concerns like a war against the COVID-19 pandemic, other disaster management such as Tsunami, earthquake, forest fires, and floods, climate change issues and other security challenges like piracies and aggressions of certain countries for territorial expansion. This creative configuration of megacities into intelligent and global megacities resulted from two books, Smart Global Megacities, published by Springer Nature in 2021, through city case studies and edited by the author, articulating how such cities function. The vast population, cultural and ecosystem diversity, diverse institutional endowments, diverse political systems such as capitalist and socialist, and autocratic governing systems, supply chains connectivity, global linkages, and size of income and expenditure in these megacities create many opportunities for configuring an intelligent global megacity and then a system of smart global megacities which is the subject matter of this book. These two books mentioned above tried to understand the smart global nature of megacities and surveyed their growth, development, distribution, and distribution across geographic regions. Now this book explores further a single system of megacities in the Indo-Pacific. These cities are significant contributors to the Gross Domestic Product of their home nation. Theories of global cities are briefly studied and finally end up with broad approaches to configure these megacities into smart and global megacities systems sustainably based on the dynamics of international politics, which drives the dynamics of the worldwide economy. The second part of this book uses megacity case studies that specifically study the role of a few individual megacity cases in the megacity system of the Indo-Pacific. This chapter serves as a background for the Regional Development of Indo-Pacific area and several megacity case studies across countries in this book. This book is a continuation of the above two books, which looks at how these megacities, a sizable number in the universe located in the Indo-Pacific, work together as a system to promote an open, accessible, and prosperous region in the world for the next century.

Keywords Megacities · Smart Cities · Global Cities · Smart city configuration · Global city configuration · Megacity time series growth and distribution · Theories on global cities · Sustainable approach · Methodology for configuring megacities to smart and international cities · Megacity system

1 Introduction

This chapter introduces two parts, "Indo Pacific System of Megacities: Emerging Architecture" and "Indo Pacific System of Megacities: Megacity Studies," which are the outcome of international collaborative research to explore the art and science of Indo-Pacific megacity system-based Regional Development. The Indo-Pacific region is a diverse multi-countries region culturally rich with an ancient civilizational long history, willing to collaborate for their welfare and economic development as well as collectively attempt to ensure their security against aggression, terrorism, piracy, corruption as well as natural disasters like Tsunami, floods and earthquake and pandemic like the COVID-19 as a collective effort. The Indo-Pacific Economic Framework has already brought together thirteen countries among forty of these regions on the first day it was introduced by the US President on May 23, 2022, in Tokyo during the fourth QUAD Summit. It will be signed after due discussion, deliberation, and dialogue in the next year. This region, undoubtedly global, will develop the path of prosperity. This country's grouping is only next to the United Nations Assembly and much larger than the European Union and G20 Nations. Countries outside the Indo-Pacific, France, Germany, the Netherlands, and the European Union have their Indo-Pacific country policy, strategies, and guidelines to pursue engagement on a long-term basis with the Indo-Pacific region to collaborate and engage actively. India, the US, Australia, Japan, and even those reluctant to join Indo-Pacific Russia conducted two by two dialogues with foreign and defense ministers of India between these countries. Comprehensive trade deals recently enacted exist between India, Australia, and India UAE, with India and the European Union and India and India US under-speed track. On behalf of Indo-Pacific, the US, Japan, Australia, and India have met twice virtually during COVID-19 days and twice in person in Washington and Tokyo for the summit and will again meet in the next summit in Sydney in 2023. These meetings resulted in the formulation of joint actions with individual responsibilities of each of the four countries identified and now moving toward program implementation. In this digital age, the QUAD with the US, Japan, Australia, and India collectively, as an effort of diplomats and many think tanks working groups, formulated Indo-Pacific programs and Indo-Pacific Economic Framework, with no brick-and-mortar office, Governing institution, and their Indo-Pacific bureaucracy deliberately. A look at global and world cities theories, summarized later in this chapter, shows no idea that the global megacity system region exists. The benefit of Indo-Pacific regional global city system-based development is directed to people in the collaborating countries and outside since the region is externally linked with the rest of the world. For example, STEM Educational fellowship in the US universities for one hundred students programs Indo-Pacific is open to individuals. The two billion Vaccination against COVID-19 can be implemented in collaboration with the QUAD leading countries and some selected offices of the public healthcare system as deemed helpful by QUAD as the most suitable to individuals in person in the Indo-Pacific region.

In contrast, QUAD countries have sent Vaccines to many countries outside, as well as the maritime surveillance program and anti-corruption drives. A closer study of citizen rights and responsibilities in the constitution of Indo-Pacific region as against existing institutional, terms of reference, citizens individually and organized as a community spatially or in cyberspace as districts has been given more rights than institutions to participate in the march of the global region toward prosperity which will be dealt with in the conclusion of this chapter. The underlying thrusts of the Indo-Pacific are to disseminate emerging technologies with supply chains and standards such as 5G and 6G, Industry 4, related space technologies, renewable energy technologies, and a green economy to benefit humanity, save from climate change, and more resilient supply chains for any eventuality such as war and climatic calamities and to enlist the participation of Indo-Pacific region on the path of prosperity. Climate change, disasters such as tsunamis, floods, pandemics, and security arrangements for aggression, piracy, unauthorized fishing, and corruption are essential to Indo-Pacific and programs that address marine and space and cyber surveillance. There are also military alliances, such as AUKUS, which is beyond the scope of this book.

In all countries, urban economies have a maximum share of NDP, and cities of larger sizes trigger the growth of NDP since they are natural leaders of the urban system. These cities sustain clean air, water, physical, social, and economic infrastructure, and by-products and services provided by people in adjacent cities, villages, and countries, manifesting international cooperation as a daily reality. The larger the cities, the more collaboration with more countries, cities, and villages, and if such large cities exist in one area, all these cities also interact with each other through communication, travel, goods movement, and money flow. Most Start-ups and Unicorns are located selectively in certain megacities such as Bengaluru in Karnataka in India, Los Angeles in the US, and Shanghai in China. At the same time, the US, China, and India have unicorns, the maximum number in the world, and all of them in Indo-Pacific. These cities have high per capita income and good purchasing power, making them the first market for innovative products and services. Access to megacities by air, road, sea, and railways is better than other cities in the world, and these megacities also have better.

Interconnectivity within themselves, expanding the market for goods and services, and triggering the economic multiplier is also significant if the strategy of using these megacities as a system is in totality.

These megacities are also connected with other megacities outside the region serving all countries with independent Indo-Pacific strategies, such as the European Union, France, Germany, the UK, and the Netherlands. To sustain the existing reality between Indo-Pacific and the outside, some leading countries with high GDPs want to develop a relationship with potentially high-growth areas in the Indo-Pacific.

In 2004, Australia, India, Japan, and the United States came together to manage the humanitarian assistance and disaster relief efforts in the aftermath of the 2004 Indian Ocean earthquake and tsunami. This group focused on this narrow mission in 2004 and ceased Quadrilateral engagement after the Tsunami response but continued to persist as an idea among strategists. By 2006, then-Prime Minister of India, Manmohan Singh, on a state visit to Japan with his newly elected counterpart, Prime Minister Shinzo Abe, noted, in a joint statement, the "usefulness of having dialogue among India, Japan and other like-minded countries in the Asia–Pacific region on themes of mutual interest." With the United States and Australia making up these "other like-minded countries in the Asia–Pacific," the QUAD was born.

By 2007, the QUAD had two components. The first was diplomatic—an informal meeting of officials from all four countries on the sidelines of the East Asia Summit in Manila. The second was military—a joint exercise involving all four countries, as well as Singapore, under the aegis of the then-bilateral US-India Malabar Exercise (Japan, Australia, and Singapore were invited as "non-permanent partners" in 2007. MALABAR was officially upgraded to a trilateral exercise in 2015 when Japan became a "permanent partner." However, by 2008, the QUAD was dormant, partly because domestic political compulsions forced the countries to take a step back from the QUAD. Prime Minister Singh faced backlash from his domestic political alliance over growing US-India ties and the QUAD. However, the QUAD was put down after Australian officials, sensitive to China's vocal concerns over the QUAD, declared that they "would not be proposing to have a dialogue of that nature" again. In the years since, the idea of resurrecting the QUAD has continued to linger, gaining speed in recent years through various bilateral and trilateral mechanisms. The rise of China and its increasingly assertive behavior, which has threatened the critical interests of all four countries, has also given additional impetus to restart the group. On Nov. 12, 2017, an entire decade after the first informal meeting of the QUAD, the four countries announced that officials from all four countries had met to discuss "issues of common interest in the Indo-Pacific region."

1.1 Peace and Prosperity in the Indo-Pacific: [1–4]

Indo-Pacific is a vast region forming two oceans and is under the self-appointed leadership of QUAD. The fundamental concept behind the Indo-Pacific System of Megacities requires elaboration (Fig. 1).

The term 'Indo-Pacific' is being used increasingly in geopolitical discourse. Many analysts sought to describe the 'geo-economic' connection between the Indian and Pacific Oceans. However, in the contemporary context, in the 2000s, analysts began to observe the 'security' linkage between the two Oceans. From about 2011 onwards, the term has been used frequently by strategic analysts and high-level government/ military leadership in Australia, Japan, and the US to denote said region. The term 'Indo-Pacific' refers to the maritime space stretching from the littorals of East Africa and West Asia, across the Indian Ocean and the Western Pacific Ocean, to the littoral

Fig. 1 Oceans of the World, and the Asia–Pacific and Indo-Pacific Regions. *Source* [5]

of East Asia. It is much bigger than the Asia–Pacific, which leaves out the African east coast as the economic leader of the next century. The spirit of the term was picked up by Japan's Prime Minister Shinzo Abe, as reflected in his speech to the Indian Parliament in August 2007 that talked about the "Confluence of the Indian and Pacific Oceans" as "the dynamic coupling as seas of freedom and prosperity" in the "broader Asia." From 2010 onwards, the term Indo-Pacific acquired salience within the Indian government and has since been used often by India's apex political leadership. It is also "symbiotically linked" with the Quadrilateral Security Dialogue—an informal

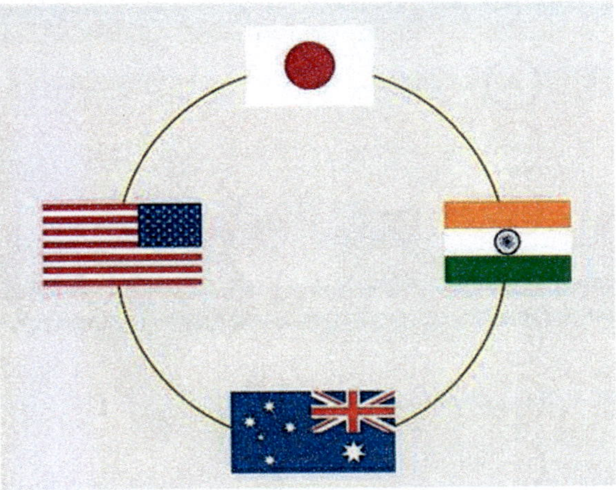

Fig. 2 The QUAD core countries

grouping of like-minded democracies in the region, comprising Australia, Japan, India, and the US (Fig. 2).

In 2019, the US State Department published a document formalizing the concept of a "Free and Open Indo-Pacific," to be sustained among members of "the QUAD," a partnership of four Indo-Pacific democracies led by the United States, in concert with Australia, India, and Japan. 'Indo-Pacific' has also featured prominently in top-level US strategic documents such as the 2017 National Security Strategy, the 2018 Nuclear Posture Review, and the 2018 National Defense Strategy. It has been argued that the concept of the Indo-Pacific may lead to a change in popular "mental maps" of how the world is understood in strategic terms. In 2013, US officials began using the word "Indo-Asia Pacific," which enabled America to maintain its geographic inclusiveness in the new Indo-Pacific coinage (Fig. 3).

The economic aspect of Indo-Pacific is summarized in the following Map.

It has its share of natural disasters, as shown in the figure below. India, the USA, Japan, and Australia came together in 2004 to aid people in the region most affected by the Tsunami, which was the beginning of these four countries coming together (Fig. 4).

The term's importance was raised when it found mentioned in the joint statement issued by Indian Prime Minister Narendra Modi and US President Donald Trump after the former's state visit to the White House on June 26, 2017. "As responsible stewards in the Indo-Pacific region, President Trump and Prime Minister Modi agreed that a close partnership between the United States and India is central to peace and stability. In marking 70 years of diplomatic relations between India and the United States, the leaders resolved to expand and deepen the strategic partnership between the countries and advance common objectives. These objectives include combatting terrorist threats, promoting stability across the Indo-Pacific region, increasing free and fair trade, and strengthening energy linkages." However, President Trump's November 2017 articulation on Indo-Pacific was widely seen as something that would

Fig. 3 Indo-Pacific status of selected attributes concerning the world. *Source* [6]

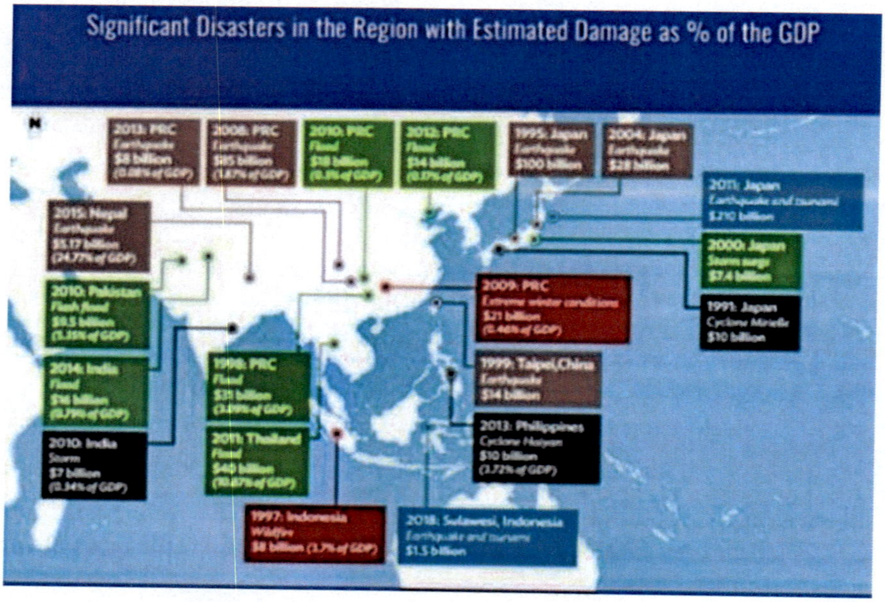

Fig. 4 Disasters in the Indo-Pacific. *Source* [6]

usher in a new (US-China) Cold War. This led to the Indian Prime Minister spelling out the Indian vision of Indo-Pacific as an enabler for "a common pursuit of progress and prosperity… not directed against any country… (albeit based on) our principled commitment to the rule of law."

As usual, there have been different perceptions of the geographic extent of the Indo-Pacific. They are summarized in the figure below (Fig. 5).

Major countries in the Indo-Pacific are given below.

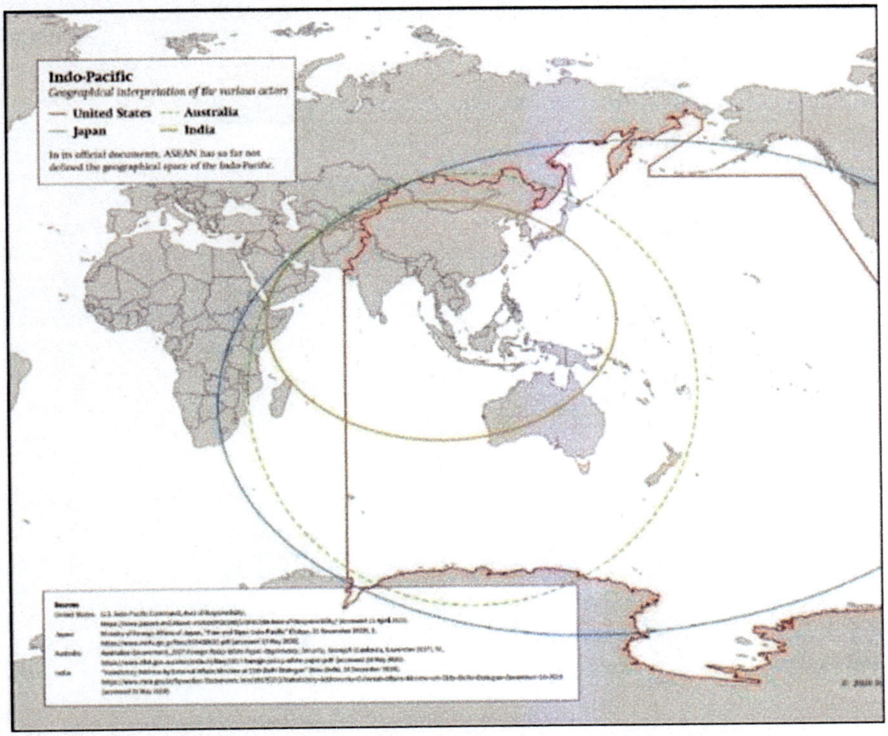

Fig. 5 Geographic interpretations of Indo-Pacific

Major Milestones of Indo-Pacific Regional Development are summarized in the Table 1.

Indo-Pacific is a connected region with several connectivity types in the figure below. This region is interconnected and accounts for a significant part of global trade. Connectivity can be of two kinds. One is physical connectivity, such as seaport, airport, railway, and highway connectivity. Another connectivity creates additional access within the landlocked servicing countries and subregions of individual megacities, some twenty-one subregions (Table 1).

However, the overall connectivity of the Indo-Pacific is not satisfactory now and requires additional infrastructure augmentation as per emerging practical other demands. The creation of overcapacity and the capacity of infrastructure are equally bad. Creating overpower, as in the case of many existing ghost towns in China and Sri Lanka, not so well-planned new towns, airports, and harbors can lead to the economic

Table 1 Indo-Pacific key milestones

Year	Major milestones of Indo-Pacific regional development
2007	Japanese Prime Minister Shinzo Abe calls the Indian Ocean and the Pacific Ocean shared space in his speech "Confluence of the two seas" in New Delhi
2011	The US secretary of State Hillary Clinton speaks of the Indo-Pacific in the context of the pivot of Asia
2012	The Indo-Pacific appears for the first time officially in the White paper Australia in the Asian century
2015	The document ensuring secure seas Indian maritime security strategy makes use of the word Indo-Pacific
2016	Shinzo Abe declares his "free and open Indo-Pacific Strategy to be the core of the Japan's foreign policy doctrine."
2017	At the APEC summit in Da Nang US, President Trump calls the "Free and Open Indo pacific" the focal point of US Asian Policy
	The national security strategy of the United States dedicated a special section to the "Free and Open Indo Pacific"
2018	French President Emmanuel Macron speaks about a French Strategy for the Indo-Pacific
	Indian Prime Minister Narendra Modi presents his country's view on Indo-Pacific Concept in his speech at the IISS Shangri-La Dialogue in Singapore
	Japan, from now on, calls "Free and Indo-Pacific" a vision rather than a strategy
2019	The French Defense Ministry publishes the strategy paper France and Security in Indo-Pacific
	At the summit in Bangkok, the ASEAN states agree on a joint document ASEAN. Outlook on Indo-Pacific
	The US Department of defense publishes the Indo-Pacific Strategy Report. Preparedness, Partnership, and Promoting a Networked region
	The US Department of State publishes the strategy paper A free and Open Indo. Pacific Advancing Shared Values
2020	Germany announces its Indo-Pacific strategy
	The Netherlands issues "Indo-Pacific Guidelines for strengthening Dutch and EU cooperation with Partners in Asia
2021	In April 2021, the European Union (EU) released the 'EU Strategy for Cooperation in the Indo-Pacific,' a document outlining a cooperation plan with the Indo-Pacific
	Prime Minister Shri Narendra Modi attended the first in-person QUAD Leaders' Summit in Washington D.C., USA, on 24 September 2021, along with Prime Minister Scott Morrison of Australia, Prime Minister Yoshihide Suga of Japan, and President. Joseph Biden of the United States of America
2022	On February 11, the White House announced its new strategy for a "free and open. Indo-Pacific" region, which pledges support for regional connectivity, trade, and investment, and deepening bilateral and multilateral partnerships
	Prime Minister participated in a virtual summit of QUAD leaders, along with US President Joe Biden, Australian Prime Minister Scott Morrison, and Japanese Prime Minister Fumio Kishida

(continued)

Table 1 (continued)

Year	Major milestones of Indo-Pacific regional development
	The second in-person and the fourth summit of the Quadrilateral Security Dialogue, or QUAD, hosted by Japan as early as 23–24 May 2022 against the backdrop of pledges by the leadership of India, Australia, Japan, and the US to forge ahead with efforts to ensure a free and open Indo-Pacific and announced Indo-Pacific Maritime Domain Awareness, among QUAD fellowship, Advanced technical standards, cyber security, critical emerging technologies, infrastructure with $ 50 billion, and Disaster. relief
	With thirteen initial member countries, Indo-Pacific Economic Framework came up for discussion by Indo-Pacific democracies
	The US Indo-Pacific Strategy February 2022
	Joint Statement of the Leaders of India, Israel, United Arab Emirates, and the United States (I2U2)., July 14, 2022
	Indo-Pacific Strategy of Canada announced in November 2022
2023	The fourth summit of the Quadrilateral Security Dialogue Tokyo
	The fifth summit Quad Leaders' Joint Statement 20 May 2023, Hiroshima

Source Author

collapse of Nations. A policy of timely creation of additional infrastructure capacity to be an adequate and marginally higher degree in tune with growing demand is necessary to avoid the cost of congestion (Fig. 7).

The other type of important connectivity is institutional connectivity, in which nations are connected for common goals and objectives by treaties, covenants, and agreements. Such institutional connectivity for the Integrated Megacity System for the spatial, social, and economic development of the Indo-Pacific does not exist for collective implementation anywhere. Institutions require bureaucracy, and complex bureaucracies with complex administrative procedures often tend to be a drag for the speedy performance of regional development. It is best to be done by an open and free-market mechanism and implementation agency directly plan and implement avoiding lengthy multi-administrative procedures creating more corruption, and less efficiency and cost escalation including an administrative expense. As an example, India-centric institutional connectivity is shown below in the diagram. An example of the creation of a market-oriented and crowd-funded international airport in the megacity of Kochi is the one well-functioning and profit-making and innovating cum solar energy run airport of Kochi in Kerala, India as against airports under government such as Kozhikode and Thiruvananthapuram airports in Kerala, its growth, performance, and profitability despite increasing demands.

There exists similar connectivity among all other countries of QUAD. Some are for collective security through defense forces, and others for economic development and social welfare (Fig. 8).

Peace and prosperity are the goals of the Indo-Pacific. These two goals do not require any lobbying to make them acceptable to all countries of the Indo-Pacific region. However, the four countries, Japan, the USA, Australia, and India, made it

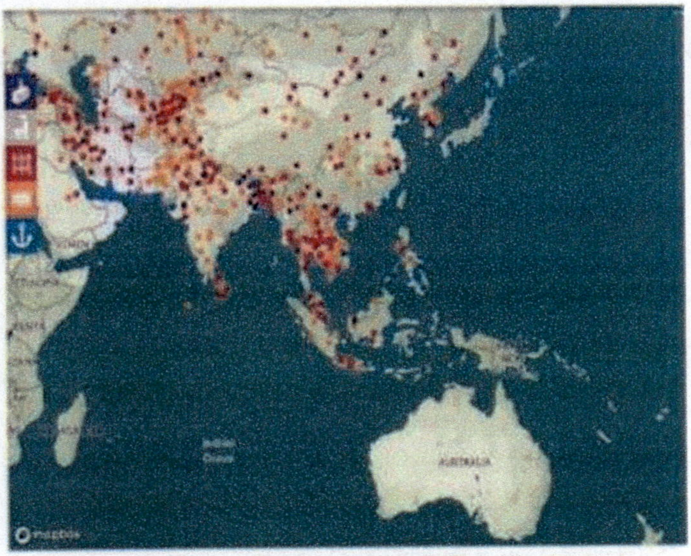

<https://csis-website-prod.s3.amazonaws.com/s3fs-public/190409_figure1_web.jpg>

CSIS Reconnecting Asia Project, https://reconnectingasia.csis.org/map

Fig. 7 Spatial connectivity of Indo-Pacific

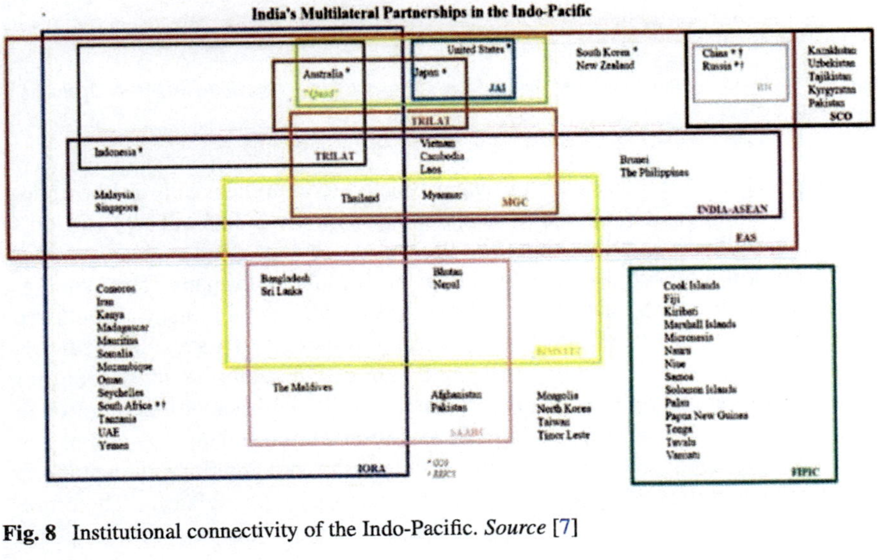

Fig. 8 Institutional connectivity of the Indo-Pacific. *Source* [7]

Fig. 9 Vasudhaiva Kutumbakam the Indian Concept of the Universe. *Sources* Maha Upanishad and Government of India

their task to follow the peace and prosperity goals comprehensively in the Indo-Pacific. The Indio Pacific regional concept is based on the Indian concept of God's family or Vasudhaiva Kutumbakam (Fig. 9).

Indian origin religions such as Hinduism and Buddhism in many forms such as Saivism, Vaishnavism, Mahayana, and Theravada Buddhism with many schools and Vedanta and Vashist Advaita and Dvaita the duality school, and the resulting culture, adaptive arts, and way of living is the foundation of the culture of most countries in this region overlaid with layers of Christianity a colonial legacy or Islam through merchants later. It changed very marginally the initial culture and the way of living of the majority of nations in the Indo-Pacific. As the author did, it will be evident if one lives in many countries. They all believe in the concept of a global family. Therefore, there was no difficulty in evolving the Indo-Pacific Regional Development for peace and prosperity through this concept.

As a result, the Indo-Pacific is a diverse area around the Indian and Pacific oceans, the largest of many seas in the world. Peace in this region is a prerequisite to continuing the path of prosperity.

Peace is an absence of war. The security of peace in the Indo-Pacific is impossible without complete preparation.

For war from now onwards in the best workable way by countries of Indo-Pacific individually and collectively.

Outer space, cyberspace, sea, and land, with their unique topography, climate, and choking points, are domains of war. War is not mere fighting of soldiers but multi-faceted, such as cyber war, AI-induced war, economic sanctions, and war with physical obstruction to ships, airways, and the internet. Finding strategies and means that ensure massive deterrence is considered sufficient to ensure peace. These need to be formulated by countries individually and collectively, emphasizing the individual. Stability is also created by the ability to have diplomatic dialogues. The best practical way is to develop collective defense intelligence of the sea such as five eyes, land, outer space, and cyberspace in the Indo-Pacific, which is being started by QUAD countries, alliances, and partners of the Indo-Pacific, for example, NATO alliance of Japan, South Korea and AUKUS the US, the UK, and Australia. History has shown that large countries such as the USA were defeated by small countries like Vietnam or Afghanistan in wars, which gives hope of winning a war by small countries even against superpowers. The war was against the Taliban and Vietnamese, respectively, and to defeat them, but in the end, the Taliban or Vietnamese came into power in the respective countries. This shows that small and less wealthy countries with less economic and defense capability can also win wars. Prolonging the war, as we see in Ukraine, is resulting in large-scale destruction of cities, deaths, disability, and injury to many people, and displacement of the large-scale population by outmigration, temporary or permanent, without giving diplomacy and dialogue a chance. These culminate in a worldwide recession and scarcity of essentials (Fig. 10).

Two large armies supported by Airforce and the Navy are face to face and eyeballed in the Western Himalayas for two years and above in the worst climate, but the war has not occurred on a full scale. India and China are alone, with no armies stationed on the border other than these two. Here diplomacy and talk between defense forces are given a chance with an unpredicted outcome. This is far better

Fig. 10 The five eyes

than the prolonged warlike in Ukraine. The Ukraine war is not a good example to follow for the Indo-Pacific.

The book's central theme is not the collective security of the Indo-Pacific but the prosperity of all in the Indo-Pacific and sharing a common future. This book focuses on people, their life, work, welfare, and wealth, and any war that causes hardship to people is not acceptable to Indo-Pacific. Other means, discussed briefly above, such as diplomacy and deterrence, are advocated. This book aims not to elaborate on the security aspect but to mention that Indo-Pacific security is the uniting factor for countries in the Indo-Pacific and their city system-based Regional Development.

1.2 A Free and Open Indo-Pacific

1. A target of **Free and Open** Indo-Pacific is a goal of QUAD for achievement. Freedom of movement of goods through container ships accounts for a significant share of the international trade in the Indo-Pacific, which is the lifeblood of the global economy. Some belligerent countries in this region have been aggressive by salami-slicing other countries' sovereign territories and almost prevented such international goods traffic and trade by building artificial islands with defense infrastructure. Preventing such barriers to international trade is the Indo-Pacific's primary security goal, which will be elaborated on in another chapter of this book. This behavior requires capable self-defense and massive deterrence giving way to lasting peace in the Indo-Pacific. Many countries interested in the Indo-Pacific are nuclear-armed, such as the US, China, Russia, India, Pakistan, North Korea, the UK, and France. Some have a sizeable army, navy, air force, and experience, having fought many battles in the last century. Some of them have hypersonic and space warfare capabilities and cyber warfare capabilities. Some others, even today, keep some 100,000-army face-to-face and eyeball to eyeball preventing salami-slicing of Himalayas. In the Indo-Pacific, a group of countries' defense forces can use other countries' facilities and services. Some groups of countries have executed periodic war game exercises. These countries have also conducted innumerable dialogues and diplomatic interventions in the past, avoiding full-scale war. Some counties have formed AUKUS Australia, the UK, and the US defense treaty by transferring nuclear-powered submarine technology and trying to place submarine forces with lethal weapons in the Indo-Pacific to ensure freedom of movement. Indo-Pacific is not a group of countries like NATO with defense cooperation that says any aggression toward any one country will be aggression to NATO, and all these countries will come into a war if attacked. There are US allies like South Korea, Australia, Philippines, Japan, and collaborators like India and other countries who do not have alliance treaties with NATO and have no such future intentions.
2. Free and Open Indo-Pacific also have a developmental connotation. They are given below.

Countries in the Indo-Pacific are free to hold independent foreign policies and military and commercial relationships. For example, despite the two armies of India and China standing face to face in the Himalayas, there has been a rise in the trade percentage from one hundred billion to 130 billion in the year 2022. However, the US-China economic relationship is facing sanctions and restrictions by tariffs. There have been dialogues between some of these countries and implementing certain sanctions that do not allow foreign direct investment in India or buying products and services such as 5G and telecommunication infrastructure from China for security reasons. India and Russia have long defense cooperation, including joint manufacturing of defense products such as Brahmos supersonic missiles. On the other hand, some countries in the Indo-Pacific have sanctions imposed because of the Ukraine war.

3. Freedom of easy movement of people for commerce, education, health care, and tourism of various kinds is needed for regional economic development. This can be eased by eVisa and less complicated visa rules that quickly help the movement of people within the region. However, some countries have such facilities many countries of the Indo-Pacific need to embark on these initiatives.
4. Freedom of easy transfer of money. Indian Rupee is an accepted currency in certain countries such as Bhutan, Nepal, and Singapore. In India, the digital transfer of funds by the Unified Payment Interface (UPI) has replaced significant cash transactions. UPI is.

   Gradually spreading in a few countries in the Indo-Pacific, such as the UAE, Australia, and Nepal, and Russians are considering its acceptability. SWIFT is another US dollar-centered mode of dollar transfer through messaging. There needs to be a liberalized regime of all types of electronic money transfers using smartphones in all countries of the Indo-Pacific.
5. Freedom to easily migrate to the Indo-Pacific countries for studies, work, and business. A start-up in one country requires a partner in another country who wants to migrate into a third country that can better nurture and evolve the start-up into a Unicorn, which should not be bogged down by migration-related bureaucracy. They should experience the least amount of time in the administrative process, and most of the process could be executed on the website or smartphone without visiting the respective embassies.
6. Freedom to get skills in demand and advanced degrees from universities free of cost to students of the Indo-Pacific. The Indo-Pacific Regional Development is based on advanced technologies such as ICT, IoT, 5 G, 6 G, AI, Block Chains, and Industry 4. This requires all countries to have a sizeable population with relevant skills from universities that impart high-quality education such that advanced knowledge on science, technology, and mathematics are available as per demand to drive emerging technology-based Indo-Pacific regional economic development. Although the US, India, Japan, and Australia have limited scholarship programs, they are not widespread and highly limited in scope to all countries in the Indo-Pacific. The best way to make high-quality university education imparted in Indo-Pacific is to make it accessible so that all students in the

Indo-Pacific, regardless of nationality, income, religion, caste, or class, have easy access to free education.

7. Of the twenty-four countries that provide free education, sixteen are based in Europe, and one is in North America. It is rare in Asia. Three countries in South America and three in Africa also have this provision.

Europe

Norway: Tuition is not only reduced or free for citizens but also for international students. Taxpayers' money covers attendance at state universities, but higher living expenses are a negative issue for international students.
Sweden: Although previously offered to all students, their tuition is now free to citizens and European students. There was an increase in scholarship programs to compensate, but they were not remarkably effective.
Germany: Some universities have recently begun charging some tuition fees, but most still offer free education to all students, even international students. They may ask for a small contribution each semester, but it is nowhere near the total tuition price.
Denmark: All Danish citizens are offered scholarships and aid, but most colleges are free. They offer their programs to citizens, European students, and students with certain visas.
Finland: Finland is a country that offers completely free education and only charges fees to non-European students that wish to take classes in English. Living expenses are not covered.
Austria: Not a tuition-free country, but remarkably close. Tuition and school fees are meager for citizens and European students, with a slight increase for non-European students.
Greece: Free education is available for citizens of Greece and European students; international fees are meager compared to regular tuition fees in other countries.
France: Not free, but exceptionally low. If one is native to France or Europe, one will only pay a few hundred euros. International fees do go up to thousands per year.

South America

Uruguay: Education is accessible to all Uruguay citizens. It does not extend to European or non-European students.
Brazil: University-level education is accessible even to international students. All classes are in Portuguese as they do not offer English.
Argentina: Free education is only offered to Argentinian students and citizens; it is not available for international students.
North America Panama: Accessible to all students, with no concern for nationality. That includes citizens, Europeans, and all international students.

Asia

Malaysia: Free tuition is uncommon within Asia; free education is only available to Malaysian citizens.

Africa Morocco: Free tuition to all citizens but does not outreach to international students. Egypt:

Also, only free tuition for all citizens, but not for international students. Their programs are meant to decrease poverty within their country.

Kenya: Unlike other African countries, they allow free education for citizens and international students. International students are limited to public tuition if they are high-scoring secondary school students.

Countries in the Indo-Pacific shall develop self-sustaining ecology of high-quality university education at a near-zero marginal cost to students. Greenfield or Brownfield, such as an Indian Institute of Technology, is free education. This requires a system to make the marginal cost per student near zero through various possible strategies.

1. Zero healthcare costs to the population shall be implemented in the Indo-Pacific. It is crucial to have a healthy population for regional economic development. Since all countries of the Indo-Pacific are not high- or middle-income countries, it is essential to make health care universally accessible and free to all. Marginal cost-based country-specific design can be researched.

    Forty-three countries in the world offer free or universal healthcare. However, the standards among these countries can vary widely. The list includes everyone from Norway, one of the healthiest countries in the world and the first country to introduce free healthcare in 1912, to another extreme case in Yemen, where raging conflict has made health care from international relief agencies impossible to access.

List of Countries Offering Universal Health Care Systems

Not all citizens or residents receive free healthcare in all these countries. In many of these countries, employers and individuals share in the cost of healthcare through contributions, cost-share arrangements, co-pays, and other related fees. However, these programs aim to make health care as affordable and accessible as possible for the most significant number of people—"Universal Care."

Countries With Universal Healthcare are Australia (Healthcare Systems in Australia and Austria). Bahrain, Belgium, Brunei, Canada (Canadian Health Care System), Cyprus, Denmark, Finland, France (French Healthcare System), Germany (German Healthcare System), Greece, Hong Kong, Iceland, Ireland (Healthcare System in Ireland), Israel, Italy (Healthcare System in Italy), Japan (Japanese Healthcare System), Kuwait, Luxembourg, Netherlands, New Zealand (New Zealand's Healthcare System), Norway, Portugal (Healthcare System in Portugal), Singapore, Slovenia, South Korea, Spain (Healthcare System in Spain), Sweden (Sweden's Healthcare System), Switzerland, United Arab Emirates, and UK (Healthcare System in the UK).

1.3 *Indo-Pacific Subregions*

Before we look at the megacity system of Indo-Pacific, we need to consider the Indo-Pacific region, which is the spatial home of the megacity system of the Indo-Pacific. Chapter 2 of emerging architecture will spell out and identify this region, and Chapter 3 will go into more detail about the components of this region. Without going into the specific outcomes in that chapter, the regional concept, the basis of their work, identifies first all megacities in the Indo-Pacific and then investigates the influence region of megacities. Megacities have distances between them. When one is spatially partitioning these individual megacities, then one gets the subregions of the megacities. Suppose one studies the landlocked and other adjacent countries that use these megacities as their gateways to the Indo-Pacific for international trade. In that case, one gets the servicing countries of the Indo-Pacific Megacity System. The subregion of individual megacities and the servicing countries constitute the Indo-Pacific subregion, the geographical space for integrated development. If one draws a boundary of these spatial components of the Indo-Pacific, we get the Indo-Pacific region.

1.3.1 Indo-Pacific Homogenous Subregions

The region is homogeneous such that the area has common developmental issues. If development issues and their solutions are expected, then the answer can be applied to these homogeneous regions without looking at each case as exceptional (Fig. 11).

After defining the region, the next question is how to enact regional development of the Indo-Pacific and how to integrate it. For example, the Indo-Pacific is home to innumerable islands, as shown in the figure. Islands are susceptible ecologically and culturally, and developmental activities, including the building up essential infrastructure, require great care keeping in view the impacts, which is not the case in other areas. For example, removing sand from the island beach is a prescription for water to invade the very scarce island land. Pumping more water using a high-power pump can draw saltwater instead of potable water. The expertise in mindful ecology for regional development is more important than mindless economics. Another issue for island communities is access to megacity gateways. There is a need to break into the transport system using multiple types of ships requiring birthing facilities. Even though the islands of the Indo-Pacific look the same geographically and ecologically, they fall into different climatic zones. Agriculture changes with climatic zones. The agricultural technology requirements in these regions can also change. Therefore, connecting these agro-climatic zone-based islands to specialized and relevant farm research and extension centers in different countries may be the best first step to embark on regional development using island communities' high-tech and advanced intelligent agriculture.

There can be many more homogeneous subregions, such as low-income, middle-income, and high-income regions. They can have a different program or project

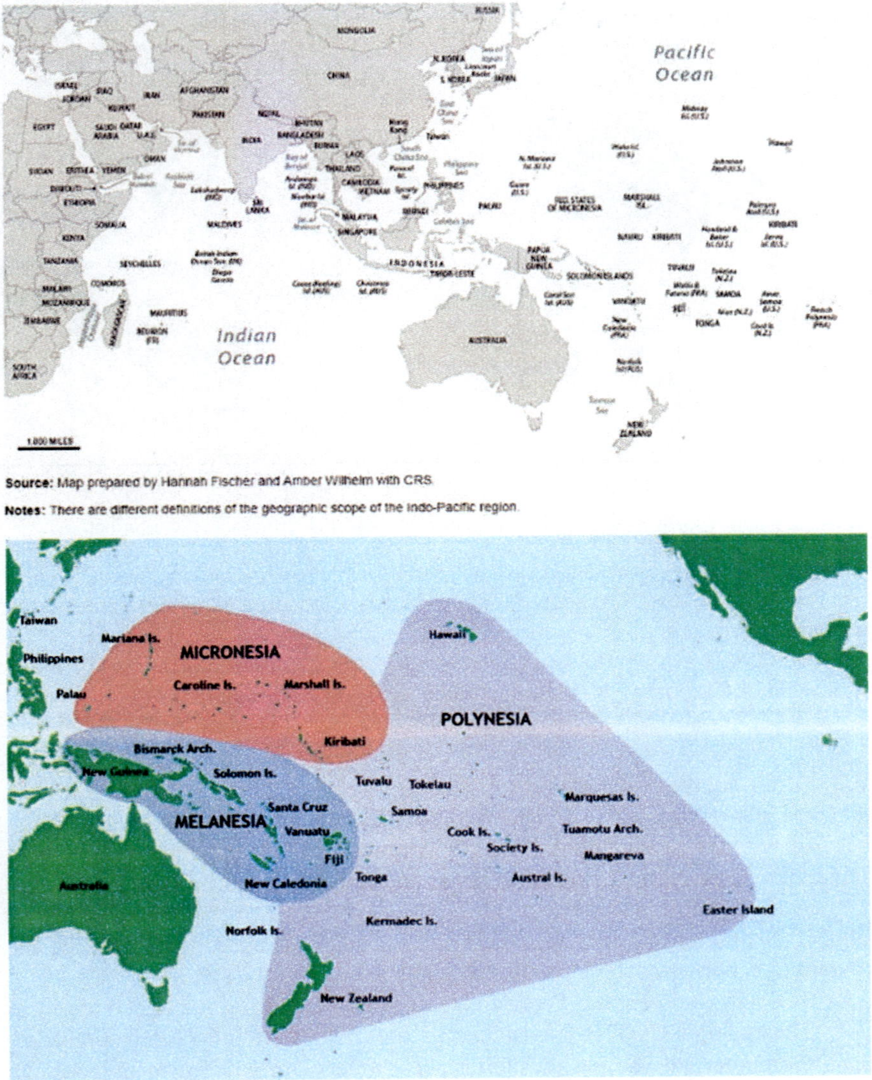

Fig. 11 Indo-Pacific region with small island details

regime in the Indo-Pacific. There can also be regions worst affected by disasters, such as tsunamis, and middle and least affected by disasters. They can be classified, mapped, and then evolved appropriate projects/programs of action suited to them. As we know, the effort required by programs Indo-Pacific is specific and related to the characteristics of the subregions. Therefore, the agenda of the Indo-Pacific needs

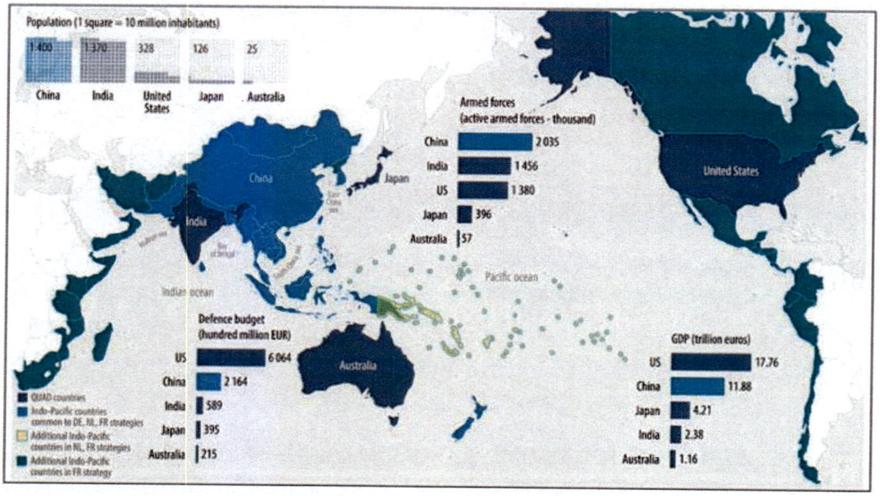

Fig. 12 The Indo-Pacific as defined in strategies from France, Germany, and the Netherlands

to be formulated to the homogeneous and region-specific to make them minimal risk and workable.

In addition, it is essential to present the economic and defense might of major countries of the Indo-Pacific documented in the following map (Fig. 12).

1.3.2 Indo-Pacific Program Subregions

These homogeneous subregions can be the appropriate program regions of the Indo-Pacific, first based on the homogeneous regions' specific needs, say for earthquakes or other disasters. Management or economic development needs based on Gross National Income or and second, where such need-based appropriate Indo-Pacific Program of action protocol is feasible to implement (Figs. 13 and 14).

As we know, Indo-Pacific Programs are discussed and formulated comprehensively. The example COVID-19 vaccination program in the Indo-Pacific considers selecting appropriate vaccine technology, it is financing, patents, manufacturing, cold-chain logistics, last-mile implementation, and many other components by the working groups of QUAD leading countries, the US, Japan, Australia, and India through dialogue, and discussion and implemented in the Indo-Pacific with no constitutionally sanctioned institution such as like the European Community Parliament or Administration which is installed above hierarchically many countries in the Indo-Pacific.

Barring a few, many of the countries in the Indo-Pacific are democracies that follow a rule-based system. Others are autocracies like North Korea or China, where the final word of all things lies with its supreme leader. Here the top leader supports any program he feels appropriate, and then the program can be implemented.

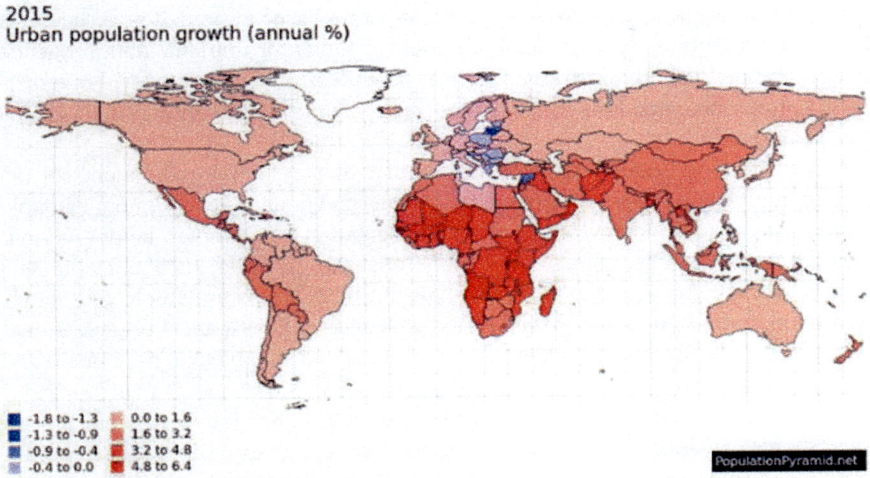

Fig. 13 Annual Urban Population Growth (%) 2015. *Source* [8]

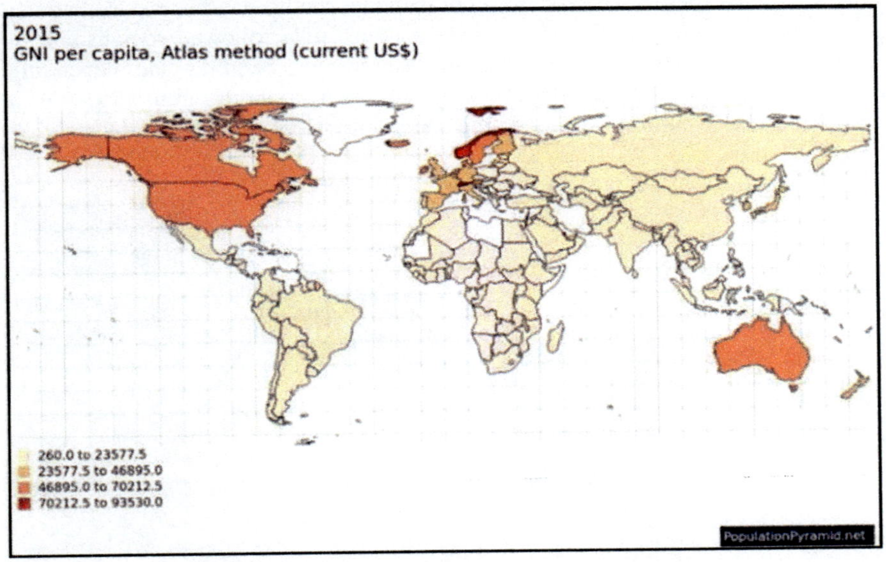

Fig. 14 GNI Per capita 2015. *Source* [8]

In democracies, there are periodically elected legislators to frame policies and administrative rules for issues such as, say, disaster management enacted in parliaments and assemblies and regulators and organizational mechanisms to implement the regulations, and the judiciary to see legislations and rules according to the principles laid down in the constitution, and finance commissions at national and state level which allocate various taxes collected for a rule-based system of governance as above. Although there is a provision to mobilize funds for implementation outside the tax revenues, the administrator's legislative leadership with a low level of expertise in entrepreneurship fails to do so, and such income for governing bodies tends to be near zero.

A case study of India is presented to explain two essential components of the program region legislative, administrative bodies, and finance commission, and how it functions in a democracy. A simplified presentation is attempted below without considering the modifications in the Union Territories of India, such as Delhi. The diagram below represents the area-wise-constitutionally constituted elected bodies and administrative machinery with a periodic election and permanent organizational structure tasked to execute the rules and regulations laid out above in parliament and legislative assemblies (Fig. 15).

The above diagram represents the general structure, but it can vary since state governments can adopt or modify the overhead system given in the way they want as per the constitution. Further delineation of these areas is democratically done at the state and union level.

The multi-level planning, administrative, taxing, and tax revenue sharing system is tabulated below. There has been some modification recently with the central Planning Commission, which abolishes the Five-year Plan. India followed socialism with a mixed economy from 1947 to 1991, fully supporting, weighing, and depending heavily on state-owned enterprises like any communist country until 1991, when India faced near bankruptcy changed the course from socialism to Capitalism with

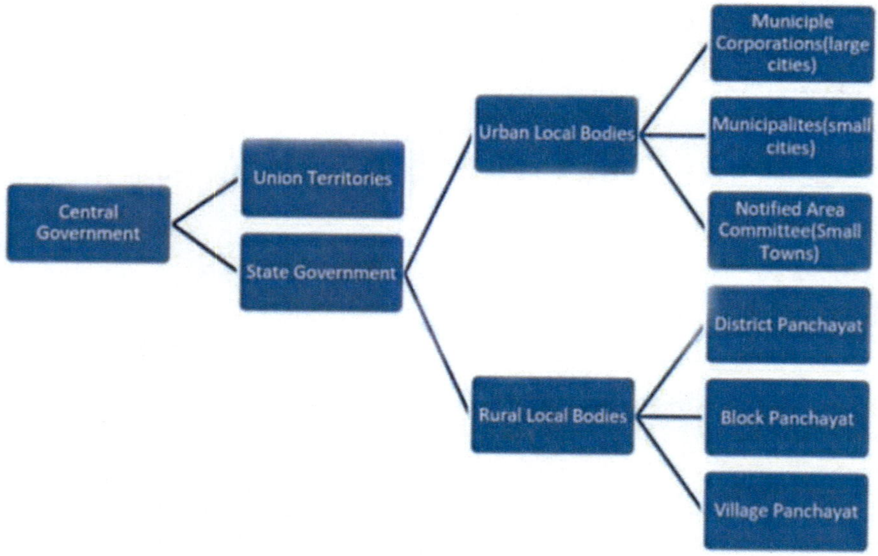

Fig. 15 Legislative and administrative structure of India. *Source* Author

disinvestment programs of public sector enterprises, such as Air India. Then it moved to a capitalist economy where private entrepreneurship is accepted, and state-owned enterprises are not. The above diagram and table below National Five-year Plan do not exist. However, some states with the Communist Party still ruling worship state planning board and the five years plan the state planning board in the respective state makes, following the yesteryears communist country protocol for public investment. At the same time, the Chinese Communist Party follows capitalism, and the Russian Communist Party, except for Cuba and Venezuela, is in poverty (Table 2).

The multi-level governing system is tabulated below.

The Finance Commission of India is a constitutional body that has survived for over 50 years in India. Formed under Article 280 of the Indian Constitution, the Finance Commission of India functions as a quasi-judicial body. The purpose behind the formation of this Commission is to determine the methods and formulas necessary for distributing the tax proceeds between the Centre and states as well as among the states as per the arrangement provided by the Constitution of India and current requirements. Along with this, the taxes and grants that are to be provided to the local bodies in states for their functioning are also determined by the Finance Commission of India.

Table 2 Multilevel political, administrative, and taxing and sharing of the tax system of India

Planning level	Political/ Administrative level	Planning approach	Power to Tax	Share of Tax
National not existing in India	Union Government	Central Policy Sectoral Policies and Plans	Yes	Yes
Federal	State Government	State Policy Sectoral/ Regional/River Valley/ Hill area planning	Yes	Yes
Sub Federal 1	District Panchayat /Metropolitan/ Administration/ /Committee	Metropolitan /District/Planning/ Budget ng	Yes	Yes
Sub Federal 2	Block Panchayat/ Administration/ Urban Zones	Block/Zonal/Planning/ Budgeting	No	Yes
Sub Federal 3	City/Gram Panchayat/Ward Committee/ Village	Annual Planning/ Budgeting	No	Yes
Sub Federal 4	Community/ Resident Welfare Association	No	No	No

Source Author

Article 281 of the Constitution provides that the President of India is required to lay the Finance Commission report before each House of Parliament along with a note that explains the actions taken by the government based on the recommendations the Commission gave. It was the 73rd Constitutional Amendment Act, 1992 that facilitated the constitution of a Finance Commission at a 5-year interval by the state governments to decide the division of resources between the state government and the Panchayat institutions at all levels.

The functions of the Finance Commission are;

1. It is the responsibility of the Finance Commission to recommend the distribution of the net proceeds of taxes that are supposed to be shared between the Union, and the states, along with the inter-state distribution.
2. The Finance Commission recommends the principles that are applied to govern the grants-in-aid to the states and the Union Territories by the Union from the Consolidated Fund of India.
3. The Commission recommends the measures that need to be adopted to augment the consolidated fund of a state to facilitate supplying the required resources to the panchayats and the local bodies of the state to avoid hindrance in their functioning. The Commission must carry out this function based on the recommendations made by the state finance commissions as per their requirement.
4. As it is the President of India who carries out all the necessary formalities with the Finance Commission of India, any matter which the President feels needs to be considered by the Finance Commission from time to time will be taken up by the Commission as a function only. The Commission submits a report to the President after delivering the necessary parts allotted to it. The President presents this report before the Houses of Parliament and accompanies a memorandum explaining the Commission's essential actions to fulfill its functions.

Given the above multi-level planning, administrative, and financing structure of one democracy, India in the Indo-Pacific, it is necessary to investigate how the Indo-Pacific programs can function. We must understand the concept of the Indo-Pacific sector scheme or program.

For example, suppose a community sub-Federal 4-level entity can independently plan, finance, and implement a project or program for the community with no outside help. In that case, it is called the Community Sector scheme.

Suppose a watershed management program cutting across two adjacent villages must be taken up in a community development block. In that case, it becomes the next level sector scheme, precisely the sub-Federal 3-level gram panchayat/ward committee level. This is called a gram panchayat/ward committee sector scheme or program.

In the case of Indo-Pacific, we all know that only a few countries have vaccine technology for COVID-19. Further, many countries have no manufacturing facility for vaccines, and the vaccine-specific cold-chain infrastructure and logistics are not available for many countries. Vaccine availability is limited, and rich countries have more access than poor, low-income countries; foreign exchange for vaccine procurement is not open, and then it becomes a QUAD Sector project or Indo-Pacific Sector program since the US, Japan, Australia, and India can jointly execute all in collaboration with National Governments. They have patents, technology, manufacturing facility, logistics and cold-chain expertise, and finance capabilities. The only limitation is that QUAD has no Governing status like the European Union, which can be overlooked. This is more of an advantage than a limitation.

Given this limitation, how can QUAD execute all its regional programs? There are two possibilities. The mistakes made by the Union Government of India can be repeated with less.

Application of mind by propagating it as a program outside the Governing structure and functioning. JNNURM, an urban development program, had a City Development Plan as the basis of implementation, bypassing the legal Master and Zonal Plans. This weakened the existing lawful provision on Urban Development and Regulations. Again, in the one hundred smart city programs of the Union Government, they had a particular purpose vehicle that effectively bypassed the local self-government of the Municipal Corporation in the Indian Constitution of an ignorant bureaucracy and poor level of the knowledge base of an elected politician. This weakened the Municipal Corporation and did not allow them to experience urban development projects, which is their constitutional right.

First, divide all QUAD programs or projects into welfare or economic development projects. Make this welfare program an external collaborative project or program using all types of machinery, both legislative, technical, and administrative, available following all existing protocols strictly based on cost-effective efficiency.

Since most of the democracies in the Indo-Pacific follow a capitalist economy where the private sector conducts industry and services, QUAD, through the diplomatic channel, can enable such collaboration.

In addition to collaborating with start-ups for the advanced technology-based economic development of QUAD, they can also collaborate with levels and institutions marked "No" in the above table if it is a community-based program.

1.3.3 Indo-Pacific Polarized Region

The Indo-Pacific is a polarized region. This region is the creation of the QUAD, the US, Japan, Australia, and India. The QUAD advocates for a multipolar world in Indo-Pacific since it is too big for a unipolar world. When we consider the multipolar world, we see countries here as units of discussion and not villages, towns, or people.

People matter, and for Indo-Pacific smart people matter more for economic and social development. Intelligent people have their habitats, villages, and towns. They can form a spatial community for converting their work and living into a smart community habitat that creates an innovative community economy. They can form a community of like-minded professionals, such as a community of chartered accountants, and a non-spatial community in cyberspace to extend professional services such as filing income tax returns and all other accounting services of other distant countries using broadband internet.

The following data presentation speaks about the ability of Indo-Pacific into the path of the digital economy much faster than the other parts of the world based on the statistics given below on broadband usage (Fig. 16 and Table 3).

Generating a large pool of intelligent people in the Indo-Pacific is required for an innovative community. It means creating a competent, professional class with expertise in Science, Technology, Engineering, Mathematics (STEM), and entrepreneurship through an enlightened educational policy. India has a brand new National

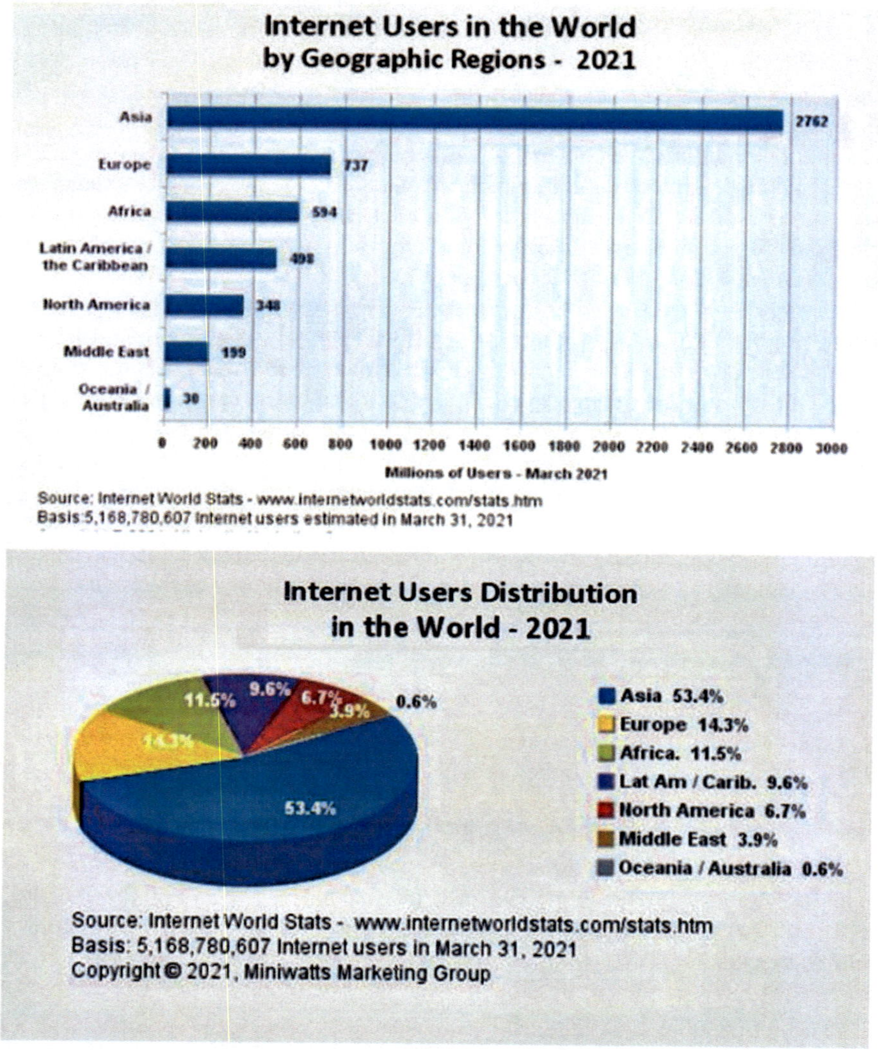

Fig. 16 Internet usage

Table 3 World internet usages by population statistics 2021 Year Q1 Estimate

World region's	Population estimate 2001	Population % of World	Internet users March 31, 2021	Penetration rate *% pop	Growth 2000 to 2021	Internet World %
Asia	4,327,333,821	54.9	2,762,187,516	63.8	2316.5	53.5
Europe	835,817,920	10.6	736,995,638	88.2	601.3	14.3
Africa	1,373,486,514	17.4	594,008,009	43.2	13.058	11.5
Latin America/ Caribbean	659,743,522	8.4	498,437,116	75.6	2658.5	9.6
North America	370,322,393	4.7	347,916,627	93.9	221.9	6.7
Middle East	265,587,661	3.4	198,850,130	74.9	5953.6	3.9
Oceania/ Australia	43,473,756	0.6	30,385,571	69.9	298.7	0.6
World Total	7,875,765,587	100	5,168,780,607	65.6	1331.9	100

Note Population-based on UN Population Division, Internet data published by Nielson Online by the International Telecommunication Union (www.internetworldstats.com)

Educational Policy, and other countries in the Indo-Pacific need to review their policy. This professional education is long-drawn and at present costly for many people. A free university education, recommended earlier in the chapter, ensures the sustained supply of smart people to the Indo-Pacific. Not all graduates can be the generator of Unicorns, a one-billion-dollar industry that is a tiny part of the university.

Graduate population. This tiny population grows in numbers and can only be available if there is a large pool of STEM graduates in the Indo-Pacific. Free education removes the burden of the debt-ridden university-educated population, which the West has presented as the capitalist model of university education. Free education makes the STEM graduate happier to live without debts, improving their quality of living and income since only a reasonably high quality of living brings about more Unicorns. The university shall be greenfield, freshly designed, or brownfield retrofitted and unbundled to make the marginal cost of education near zero to be sustainable. This calls for international collaborative research to study countries' experiences with free or partially free education and formulate a generalized design that all regional universities can adopt. However, there are limitations in several countries in the Indo-Pacific where the demographic future is not promising, with the share of the working population decreasing. The aged and non-working population is disproportionately increasing, resulting in the depopulation of megacities like Tokyo and soon many megacities in the People's Republic of China unless there are enlightened migration policies that remedy the situation, such as in the USA. This book recommends open and enlightened migration policies which need to be studied by countries that practice it, such as Switzerland, and evaluating and adopting it (Figs. 17, 18 and 19).

In many countries in the Indo-Pacific, the GDP share of the urban area exceeds rural. In India, with a comparatively low level of urbanization as in countries with a high urban population, such as the US, it is estimated to be much higher than 70% urban GDP share.

Despite that, India recorded a high GDP growth rate compared to other countries, including China (Table 4).

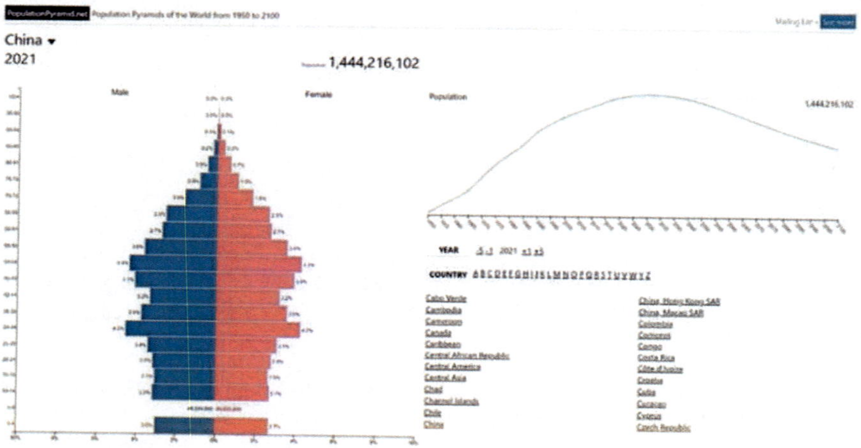

Fig. 17 Population pyramid and growth China

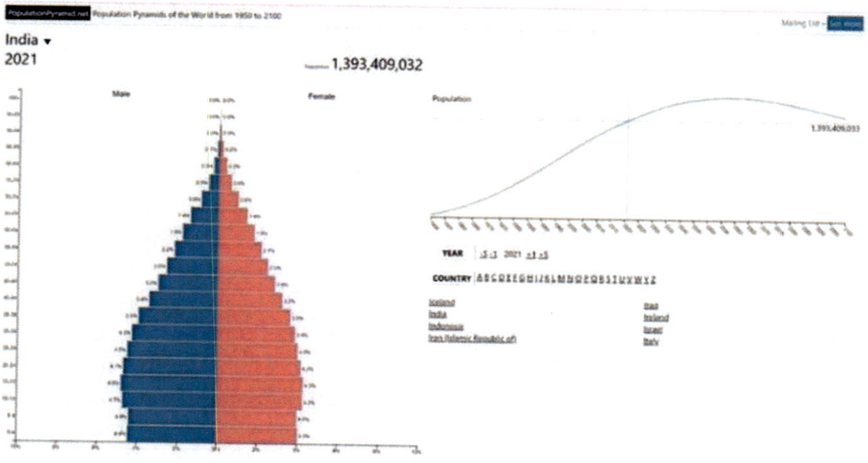

Fig. 18 Population pyramid and growth India. *Source* [8]

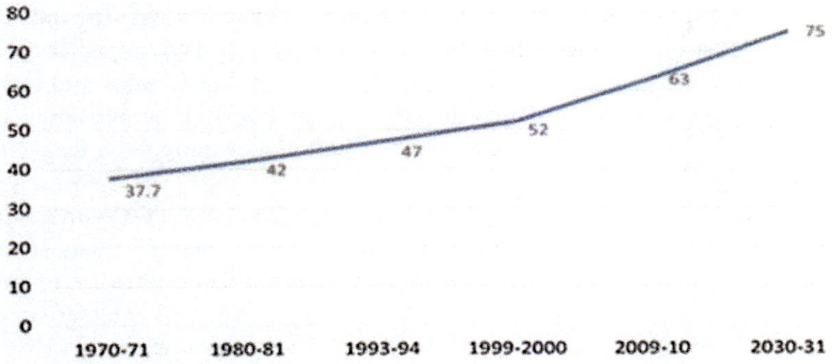

Fig. 19 Urban Share of GDP in India 1970–71 to 2030–31. *Source* [9]

Table 4 Trends in India's Urbanization 1961–2011

Census year	Urban population (in million)	Proportion urban	Annual exponential growth rate (%)
1961	78.64	17.97	–
1971	209.11	19.91	3.23
1981	159.46	23.34	3.79
1991	217.18	25.72	3.09
2001	286.12	27.86	2.75
2011	377.10	31.16	2.76

Source Government of India (2011) Census of India 2011, Provisional Population Total, Urban Agglomeration and Cities, New Delhi

The economic relationship between urban and rural areas sustains the functional region through sustainable, healthy, and robust economic linkages.

This book advocates a model of the System of Indo-Pacific Megacities for Indo-Pacific. To sustain and prosper a megacity of more than ten million population, it requires a well-established functional region comprising a system of villages, towns, and other neighboring countries that sustain it by supplying the megacities with all goods and services to survive daily. In addition, these megacities also cater to the economic activities of the adjoined countries by providing physical and institutional gateway infrastructure such as harbors, sea and dry ports and airports, and related institutions and services. Imagine wrong monetary policies cutting off this functional region in Sri Lanka or Pakistan. Then there will not be long-range survival of the megacity. Imagine a prolonged war like in Ukraine with no diplomatic dialogue to end the war; then, the megacity is destroyed physically and economically.

The megacity system model simplifies the complex reality of innumerable villages, cities, metropolises, and megacities in some countries of the Indo-Pacific and gives an operational framework for all programs in the Indo-Pacific. Megacities are smart and global, with robust functional linkages as an operating region.

So, suppose one initiates any Indo-Pacific program in megacities with the intention of its spread in the immediate subregion use. In that case, it can spread fast with existing linkages, and there is no need to work directly in the villages and towns. This reality can be empirically verified by the daily movement of people and goods from megacities and the flow of communications and money among the megacities of the Indo-Pacific.

The worst thing one can do on a map of a city or region is to draw its boundary around a city or area as if a wall exists around it. This conceptually has limited the growth of the international megacity system. Statistically, the census of India considers a metropolitan area as a contiguous spatial area with urban characteristics of one million population or more within a district, even if it spreads beyond the district as if there is a wall. Again, suppose many districts have a spatially interconnected metropolitan urban area, like five districts from Kannur to Kochi. In that case, it is a megacity when it crosses a population of ten million. Still, the census of India considers it as five metropolitan areas and not a megacity, an excellent example of statistical forces or enemies against megacity numbers in India.

Institutionally, there is a Metropolitan Area Committee for legislative and administrative purposes. Still, no megacity institutional setup appropriate exists in India, although the biggest meta city in the world surpasses Tokyo Delhi in the making soon.

The model we put forward for the megacity system of Indo-Pacific system of cities consists of all megacities within the region and the serving subregions of the megacities and the adjoining countries which makes use of the megacity for its gateway to the Indo-Pacific Ocean. Here the Author considers the adjacent landlocked country as a part of the megacity family, like Bhutan and Nepal as part of the family of Kolkata megacity.

Our previous works have established that all megacities are innovative and global. In this book, Author states that megacities are functionally interconnected as a system of 21 megacities in the Indo-Pacific. This interconnection is already manifested by the daily movement of goods, services, money, and communications between these megacities. Global politics determines the change of intensity of comprehensive interactions between megacities and related supply chain movements and affects the international economics of people who do not love peace. The imminent wars India faces with China and amassing of soldiers on the country border create economic barriers for megacity functional economic relationships for security reasons with the neighboring country, which affect them considerably, for example, India and China today and European Union and Russia. India is now open to all countries in the world to participate in the 5G introduction barring China for security reasons, and so also the Belt and Road Initiative is locked in India based on sovereignty issues, and Foreign Direct Investment from China is very limited or collaboration with them by Indian industry and commercial establishment. Chinese industries and commercial firms in India are scrutinized very minutely by intelligence agencies of India for Indian security breaches consequent to the determinant of Chinese commercial and industrial activities in India, and evidence of their shrinking is manifested by

Chinese firms in Indian megacities nowadays. So, China's commercial relationship will decline with India in the future, and the megacity system that connects India with China gets reduced or even broken. In the coming years, with India and no Industries from China can be established in India in comparison with other countries, and goods and services from China will be highly restricted with the policy that China can like policies such as selling only smartphones for less than Rs 12,000 in India. The development of ICT, IoT, M2M, and Industry 4 used in smart cities extensively opens all possibilities for the integrated creation of products and services in multi-countries in multi-megacities and subregion in a seamless fashion barring China through the megacity system of Indo-Pacific.

Such products and services constitute the supply chains that sustain the needs of the megacities and their hinterlands and adjoint serviced countries in the most resilient manner. Megacity resilience has three sets of factors global politics, the resulting Global Economics, and social, cultural, and ecological resilience.

The smart global megacity system operated by smart people is highly responsive to likely changes in supply chains in the market economy. It acts with less ideological hesitation to overcome shortages such as food and energy and possible economic downturns due to gradual movement from a frozen state to a deep frieze storage state, such as economic relationships with India and China. Smart people acts by replacing the supply chains with feasible and cost-effective means to overcome this in which part of the megacity system within China is in a deep freeze stage for other megacities while others are active. China has a problem of a shortage of working-age groups in the coming decades, as indicated in the figure on the age pyramid projection given above, which may result in a gradual reduction of population in their megacities as experienced by Japanese megacities and even South Korea from their demographic and migration profiles. This can be remedied by immigration, but China is not a famous immigration country for many countries, even if policy changes occur in China, unlike the United States. This phenomenon narrated above was evident after sanctions due to the Ukraine's war for food and energy commodities was felt in distant Africa, South Asia, and Europe. This became near normal with certain countries affected by the recession and others not based on ideology superimposed over them but by smart people who know how to act.

1.3.4 Countries in Indo-Pacific

Thirty-six nations are 50% of the world's population, 3,000 different languages, several of the world's largest militaries, and five nations allied with the US through mutual defense treaties that are part of the Indo-Pacific. Australia, Bangladesh, Bhutan, Brunei, Cambodia, Fiji, India, Indonesia, Japan, Laos, Malaysia, Maldives, Myanmar, Nepal, New Zealand, Papua New Guinea, Philippines, Singapore, Sri Lanka, Taiwan, Thailand, Timor-Leste, United States, and Vietnam. They are divided into the following.

Central Indo-Pacific
The Central Indo-Pacific includes the numerous seas and straits connecting the Indian and Pacific oceans, including the waters surrounding the Indonesian archipelago except for Sumatra's northwest coast, which is part of the Western Indo-Pacific, the South China Sea, the Philippine Sea, the north coast of Australia, and the seas surrounding New Guinea, western and central Micronesia, New Caledonia, the Solomon Islands, Vanuatu, Fiji, and Tonga. The Central Indo-Pacific, due in part to its central location at the meeting of two oceans, has the most incredible diversity of corals and mangroves.

Eastern Indo-Pacific
The Eastern Indo-Pacific surrounds the mostly volcanic islands of the central Pacific Ocean, extending from the Marshall Islands through central and southeastern Polynesia to Easter Island and Hawaii.

Western Indo-Pacific
The Western Indo-Pacific covers the western and middle portion of the Indian Ocean, including Africa's east coast, the Red Sea, the Gulf of Aden, the Persian Gulf, the Arabian Sea, the Bay of Bengal, and the Andaman Sea, as well as the coastal waters surrounding Madagascar, Seychelles, Comoros, Mascarene Islands, Maldives, and the Chagos Archipelago.

2 Megacity System

The megacity system is highly complex spatially and very expensive for systematic governance, as practiced by existing structured government administration. Still, systemically much more accessible, this social ecology is very natural for all person-to-person or community-to-community or institution-to-institution value-added economic interactions transcending country boundaries in the Indo-Pacific if administrators do not interfere "legitimately" for doubting corruption or valid reasons in the name of Governance. This can be replaced for mutual benefit by platform governance as practiced by the Indian Income Tax Department of the Government of India or GST using AI and blockchain technology to replace corruption possibilities footprint of governance and all participants. Platform governance will create job opportunities for start-ups and STEM graduates. The unemployed waiting to migrate to other countries are available in many countries in this region.

Understanding it and deciphering how countries can benefit from the Indo-Pacific megacity system requires modeling. Modeling is a simplification of the reality of the Indo-Pacific megacity system.

Based on the reality of the Indo-Pacific, this region is conceived as the first megacity system transcending countries physically well interconnected by internet, sea, air, and railway and road connection. Megacities are part of different countries, having dissimilar central, federal, and local self-governance systems. The immense amount of money flows in a limited geographic space of megacity for daily expenditure and income of households, commercial establishments, industrial, service sector institutions, and other sectors of the economy ensures lasting sustenance. An approach of a self-generating megacity economic system developed in cooperation with other megacities based on experience is the only way for accelerated development out of the experience.

These households and institutions, both the government and private sector, have immense potential to interact economically and politically with the system of megacities. The countries where these megacities are located are mostly not bound by any treaty of any one country to another, but few are. They have immense opportunity to work together without a formal international governance system of the Indo-Pacific system like the European Community.

While international laws, court judgments, and intercountry treaties are being broken, like in the case of the Philippines and China border disputes, international bodies have no further action for global peace. There is a weakening of international bodies like WHO in the COVID-19 pandemic, which is unable to find out the origin of COVID-19 and intervene scientifically based on facts emerging in the source of the Pandemic and ineffective in wars such as in Ukraine. Increasingly these bodies' effort is effectively complemented or even replaced by mainly by intercountry institutions such as NATO, G20, BRICS, SCO, or individual countries.

These intercountry collaborative activities further strengthen the Indo-Pacific, although countries can cooperate or reject the cooperation even for disaster relief or free COVID-19 vaccination.

There is unlikely to be a similar European Union Governing structure in the Indo-Pacific.

2.1 The Regional Megacity System of the Indo-Pacific.

The megacity system of the Indo-Pacific is part of the international. Furthermore, regional human settlement systems open to mutually beneficial economic outcomes through economic interactions such as joint production as part of the supply chain, a rich market, and a provider of superior services which only megacities can provide to other megacities and their subregions. This collaboration can happen in the multi-megacity, between governments and private sectors, government to government, from individual to individual, community to community, or institution to an institution if the country in question is willing to collaborate and prosper with another megacity in the megacity system of Indo-Pacific with governments support without corruption. This collaboration of human settlement is creating its influence felt on the subregion of each megacity and system of megacities forming a powerful engine of economic growth and development in the Indo-Pacific region unknown until Indo-Pacific. A deliberate push and initiative are required from the participants here. It can never evolve as a part of incremental regional economic development due to the allocation of tax revenues discussed above. These initiatives can be classified as national and international. The nation's bureaucrats administer the policies, programs, and tax allocation.

Revenues within the Nation. Outside the nation, it is the diplomats of the countries in the Indo-Pacific who facilitate innumerable groups in their own countries and abroad to collaborate with foreign megacities, which include businesses, institutions, universities, international agencies, and countries in the Indo-Pacific. The subregion of the megacity with a lower-level urban settlement system also benefits from this economic development. Many landlocked countries depend on the megacity in other countries for their gateway functions for international financial activities. These countries are the international part of the megacity sub-system. For example, Kolkata has its subregion for economic activity in Eastern India. It is servicing landlocked countries such as Bhutan, Nepal, and others and other megacities in the Indo-Pacific region, such as Bangkok, Dhakka, and Jakarta. Furthermore, a few countries can interact with Kolkata because of the adjacency of Thailand, Bangladesh, Indonesia, Myanmar, and other SARC countries and a few ASEAN countries to Kolkata.

2.2 *The Indo-Pacific Region*

The Indo-Pacific region is shown below in Fig. 1. This covers a substantial area of earth and sea. At the same time, the western part of the USA faces Russia and China, the Indian Ocean envelopes East Africa, ASEAN countries, and Australia and New Zealand, as shown in the map below (Fig. 20).

While the allies may follow the leadership of the US, other collaborators of the Indo-Pacific may disagree with the narrative of the US allies and the US, for example, the war in Ukraine and as shown in the G 20 countries' finance ministers meeting in 2022 in which Russian speech was boycotted by only 10 out of 20 of the US allies and voting in the United Nations assembly and security council's took place in 2022. This is an expression of a robust multipolar Indo-Pacific which articulates independent foreign policies that fully account for countries' self-interest and long historical relationships and **does only agree** with United Nations sanctions and not any one individual country or a grouping of countries such as NATO country sanctions. Whether sanctions help the cause of the sanction and the results intended is debatable.

With the rising involvement of the US, UK, France, and the Netherlands in the new growth areas of Indo-Pacific, the idea of the Indo-Pacific Economic Corridor was conceptualized during the US-India Strategic Dialogue of 2013, when Secretary of State John Kerry referred to the potential of the Indo-Pacific Economic Corridor, in transforming the prospects for development and investments as well as for trade and transit between the economies of South and Southeast Asia Indo-Pacific economic corridor [3].

K. Y. Home, in his scholarly study, has mapped out the potential for various emerging trans-regional corridors in Asia along with the challenges of linking IPEC

Fig. 20 Countries in Indo-Pacific. *Source* www. Free Mapofworld.com

into the more extensive web of regional economic integration initiatives taking shape in the region in 2017 [4].

The following summarizes significant countries in the Indo-Pacific (Table 5).

Table 5 Major countries of Indo-Pacific region population and GDP

S No	Country	Land Area (sq. km)	Population 2017	GDP Million US$ (2017) GDP per Capita US$ 2017	GDP per capita US$ 2017	Capital
1	Australia	7,692,024	23,731,000	1,482,539	49,928	Canberra
2	Bangladesh	147,570	156,594,962	205,715	1359	Dhaka
3	Bhutan	38,394	742,737	6384	2804	Thimphu
4	Brunei	5765	415,717	17,426	26,939	Bandar Ser Begawan
5	Cambodia	181,035	13,205,539	16,899	1270	Phnom Penh
6	Fiji	18,274	909,389	8330	5153	Suva
7	India	3,287,263	1,324,171,354	1,690,000	1939	New Delhi
8	Indonesia	1,904,569	252,164,800	856,066	3570	Jakarta
9	Japan	377,944	126,434,969	4,769,804	38,884	Tokyo
10	Laos	236,800	6,320,000	11,707	2353	Vientiane
11	Malaysia	329,847	30,185,787	336,913	9503	Kuala Lumpur
12	Maldives	298	341,356	4920	8602	Male
13	Myanmar	676,578	50,496,000	65,291	1275	Naypyidaw
14	Nepal	147,181	26,494,503	62,384	730	Kathmandu
15	New Zealand	268,021	4,357,437	201,028	39,427	Wellington
16	Papua New Guinea	462,840	6,732,000	16,096	2183	Port Moresby
17	Philippines	343,448	107,242,000	289,686	2951	Manila
18	Singapore	710	5,183,700	307,085	52,961	Singapore
19	Sri Lanka	65,610	20,277,597	233,637	3835	Colombo
20	Taiwan	36,191	23,119,772	505,452	31,900	Taipei
21	Thailand	513,120	67,764,000	380,491	5908	Bangkok
	Timor-Leste					
22	United States	983,520	328,271,859	19,386,800	59,531	Washington
23	Vietnam	331,210	88,069,000	187,848	2186	Hanoi

Source [10]

2.3 Geopolitical Context of Indo-Pacific

In 2019, the US State Department published a document formalizing the concept of a "Free and Open Indo-Pacific," to be sustained among members of "the QUAD," a partnership of four Indo-Pacific democracies led by the United States, in concert with Japan, Australia, and India [11], and Japan.

As stated earlier, 'Indo-Pacific' has also featured prominently in top-level US strategic documents such as the 2017 National Security Strategy [12], the 2018 Nuclear Posture Review, and the 2018 National Defense Strategy [13]. It has been argued that the concept of the Indo-Pacific may lead to a change in popular "mental maps" of how the world is understood in strategic terms. In 2013, US officials began.

It was using the term "Indo-Asia Pacific." This enabled America to maintain its geographic inclusiveness in the new Indo-Pacific coinage.

The term's profile was mentioned in the joint statement issued by Indian Prime Minister Narendra Modi and US President Donald Trump after the former's state visit to the White House on June 26, 2017. "As responsible stewards in the Indo-Pacific region, President Trump and Prime Minister Modi agreed that a close partnership between the United States and India is central to peace and stability. In marking 70 years of diplomatic relations between India and the United States, the leaders resolved to expand and deepen the strategic partnership between the countries and advance common objectives. These objectives include combatting terrorist threats, promoting stability across the Indo-Pacific region, increasing free and fair trade, and strengthening energy linkages." However, President Trump's November 2017 articulation on Indo-Pacific was widely seen as something that would usher in a new (US-China) Cold War. This led to the Indian Prime Minister spelling out the Indian vision of Indo-Pacific as an enabler for "a common pursuit of progress and prosperity... not directed against any country. (Albeit based on) our moral commitment to the rule of law."

The Pacific Ocean is vast, covering North and South America, but the Indo-Pacific is confined to the west coast of the USA and not South America. The exclusion of South American megacities is due to the gravitational pull of the population concentration of India and China as well as the attention of the megacities in the North than in the south of the Pacific Ocean and the south-east the ASEAN countries linking to the north of Pacific. It is possible to have a like UAE, the USA, Israel, and India QUAD2 as QUAD 3 to South America connecting and opening to the US and India if there is a national will. The export of such economic activities from developing countries to developed countries is likely to narrow down to a city with megacities with a better endowment of production and marketing factors. The research's subject is the landmass, governments, and megacities interconnected to the Indo-Pacific. This region also accounts for numerous island countries, as stated earlier. Some islands are as big as Java, but others are very tiny. The megacities system needs to service these island communities for rapid economic development.

3 Urbanization, Economic Development, the Growth of Megacities in the Indo-Pacific

The world population is progressing toward more of the urban than the rural economy. While in 2005, urban and rural people were equal in percentage, as shown in the figure below, ever since the urban population surpassed the rural. Although Africa lags behind Asia in urbanization, the Indo-Pacific is less urban than Europe but more urban than Africa, as stated in the table below on the share of the urban population in the world for major regions.

The share of the urban population in the world for major regions is shown in the table below (Table 6).

Among the countries in the Indo-Pacific, the growth of the working population is an essential indication of sustaining economic growth. If the worker population is inadequate, a need is there to attract more workers, and the government must have a highly welcoming immigration policy or promote economic development in other countries where these endowments are high. The graph below shows the growth of the working population in different countries. It is almost flat for South Korea and the US. It is declining in China and increasing in India (Fig. 21).

With increasing urbanization, there has been a shift in the pattern of the economy. These shifts and economic changes may vary from country to country as urbanization progresses. The time series change in India is as shown (Tables 7 and 8).

It is essential to consider the changes in demography, especially the working-age group, immigration patterns, and trends of a shift in the economy and develop

Table 6 Urban population (in million) and its proportion to the total population by major regions, 1950–2050

Major region	1950	1975	2000	2025	2050
Africa	32 *14.0*	103 *24.7*	279 *34.5*	659 *44.9*	1339 *55.9*
Asia	245 *17.5*	597 *25.0*	1,393 *37.5*	2,561 *53.9*	3313 *64.2*
Europe	283 *51.5*	443 *65.4*	517 *70.9*	562 *75.8*	581 *82.0*
Latin America and the Caribbean	69 *41.3*	197 *60.7*	396 *75.3*	567 *82.1*	674 *86.2*
North America	110 *63.9*	179 *73.8*	249 *79.1*	325 *83.4*	390 *87.4*
Oceania	8 *62.4*	15 *71.9*	22 *70.5*	32 *71.1*	42 *73.5*
World	764 *29.6*	1,535 *37.7*	2,856 *46.6*	4,706 *58.2*	6,339 *66.4*

Note Urban population is shown in millions. Proportion (%) is shown in *italics* *Source* United Nations, Department of Economic and Social Affairs, Population Division (2014). World Urbanization Prospects: The 2014 Revision, CD-ROM Edition

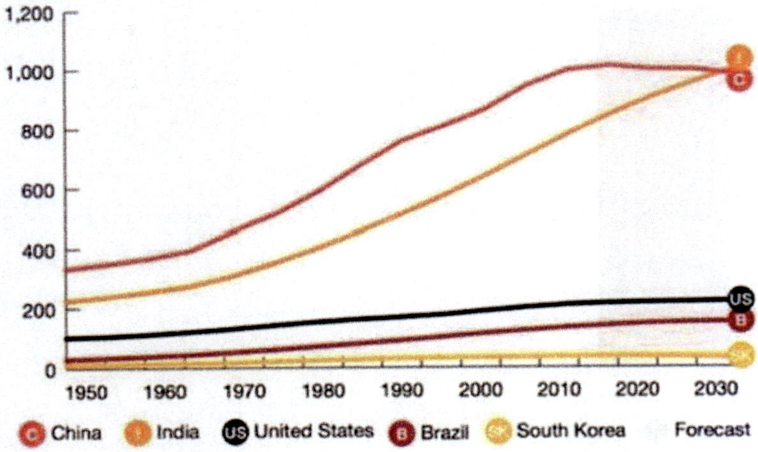

Fig. 21 Growth of working-age population of countries 1950 to 2030. *Source* PwC: [14]

scenarios for regional development. One such alternate design is presented below for India.

With increased urbanization, cities generate higher GDP than rural areas. The bigger the city size, the more the share of GDP. For example, the megacity Dhaka accounts for a sixty percent share of the GDP of Bangladesh, while the Indian megacities' percentage is not skewed to anyone's megacity. A megacity economy drives Bangladesh's economic and social indicators far ahead of India's.

The following four maps show the Growth rate of Urban Agglomerations by size class and percent urban and rural in 1979, 1999, 2018, and 2030.

From 1990 to 2018, many metropolises and megacities started growing in the Indo-Pacific, and this is projected to accelerate from 2018 to 2030. This shows a more significant shift in increased GDP growth in the Indo-Pacific in the coming decades.

It can also be seen that there is no appreciable change in the settlement pattern in Europe, while the USA shows some marginal growth in large cities (Figs. 22 and 23).

Megacity in a democracy with a market economy means people, and they are intelligent people primarily who smartly drive the international economy, foreseeing all possibilities. The institutions and infrastructure in megacities are intended to create smart people required for following the above strategy. These are available in the world's megacities and not in small cities or villages.

The figure showed the Projected World Population Distribution by the size of settlements in 2016: it can be seen that the share of the population living in megacities is only 8.2% and other lesser measures, as shown in the figure.

The time series growth of the world's megacities is presented in the table below.

The growth of megacities in the world is shown below for the following benchmark years 1950, 1975, 2000, 2005, and 2015 (Table 9).

Table 7 Size of national domestic product (NDP) by sectors and per capita NDP in India

Year	Urba n NDP as % of total	Value at the current price					Value at (1999–2003 thousand price)				
		(Rs in a million)			(In Rs)		(Rs in a million)			(In Rs)	
		Agric culture e	Indus try	Service	Tota l GDP	Per capita GDP	Agriculture	Industry	Service	Total NDP	Per capita GDP
1970–71	37.65	6.5 (4.7)	51.74 (37.4)	80.26 (57.9)	138.5 (100)	1294	74.34 (4.9)	594.2 (39.4)	839.8 (55.7)	1508.3 (100)	14,142
1980–81	41.09	22.5 (5.0)	170.9 (37.7)	259.9 (57.3)	453.4 (100)	2888	114.8 (5.4)	801 (37.4)	1225.7 (57.2)	2141.6 (100)	13,951
1993–94	45.73	139.4(4.4)	1046.9 (32.8)	2005.7 (62.8)	3192 (100)	13,525	222.4 (4.6)	1583.9 (32.5)	3069.7 (63.0)	4875.9 (100)	20,997
1999–00	51.7	291.4 (3.5)	2097.9 (25.3)	5911.1 (71.2)	8300.4 (100)	30,183	291.4 (3.5)	2097.9 (25.3)	5911 (71.2)	8300.4 (100)	30,183
2004–05	52.02	308.7 (2.2)	3649.7 (26.5)	9808.2 (71.2)	13,766.5 (100)	44,223	279.7 (2.4)	2942.1 (25.6)	8258.3 (71.9)	11,480 (100)	37,245

Note Figures in the parentheses are shares in percent
Source National Accounts Statistics (various years)

Table 8 Three economic growth scenarios for India, 2014–2034

Scenario one	Scenario two	Scenario three
Pushing old ways faster outlines a focus on investment in education, health, and other dimensions related to human capital. Our analysis suggests that in this scenario, India's GDP could see a 6.6% compound annual growth rate (CAGR) between now and 2034	*Turbocharging investment* outlines the impact of rapid and significant investment in physical infrastructure and envisions a 7 trillion GDP leading up to 2034	*The Winning Leap* includes investment in human and physical capital (as in the previous two scenarios) but also focuses on investment in R&D and innovation and envisions a 9.0% CAGR for GDP between now and 2034. This scenario forecasts the most aggressive growth and is the only scenario that will generate 240 million new jobs India's growing population needs over the next 20 years

Source [15]

In 1950, there were only two megacities Tokyo in the Indo-Pacific and New York outside. It became 3 in 1975, including Mexico and New York outside the Indo-Pacific. In 2005, fifteen megacities were added, and then in 2015, another nineteen megacities were added, which shows a rapid addition rate of megacities. Most of these came up in the Indo-Pacific, as shown in the (Table 10).

Megacities distribution among continents is as shown in the map below. Here one can see that it is more in the Indo-Pacific in the eastern part than in the west or south (Figs. 24 and 25).

The population growth rate megacities are shown with Lagos leading and Osaka and Tokyo showing harmful growths. Again, the left half of the high-growth megacities are from the Indo-Pacific.

The Indo-Pacific's megacity system research focuses on discovering emerging architecture. It has an issue-based sectoral dimension and a city-based system of city-based megacity dimensions, the subject matter of many chapters in the two parts. This Indo-Pacific in recent decades has shown rapid economic development in several countries and reduced the number of people below the poverty line in any earlier decades. There are about eighteen megacities in 2016 which will increase to twenty-one gigantic megacities of population size ten million and above-called megacities in the Indo-Pacific in 2030, as given in one table below. In 2016 as per UN-Habitat, there were thirty-one megacities in the universe which are likely to be 41 in 2030. Some of the twenty million and above cities are meta cities like Tokyo and soon Delhi, which will overtake Tokyo. The megacity system is defined as functionally and economically interconnected twenty-one or more megacities, servicing sub-national regions and servicing countries outside, sharing the same megacity gateway infrastructure and related economic, institutional, and physical infrastructure of megacities of the Indo-Pacific (Figs. 26, 27 and 28).

These megacities have shown a higher GDP growth rate as shown below.

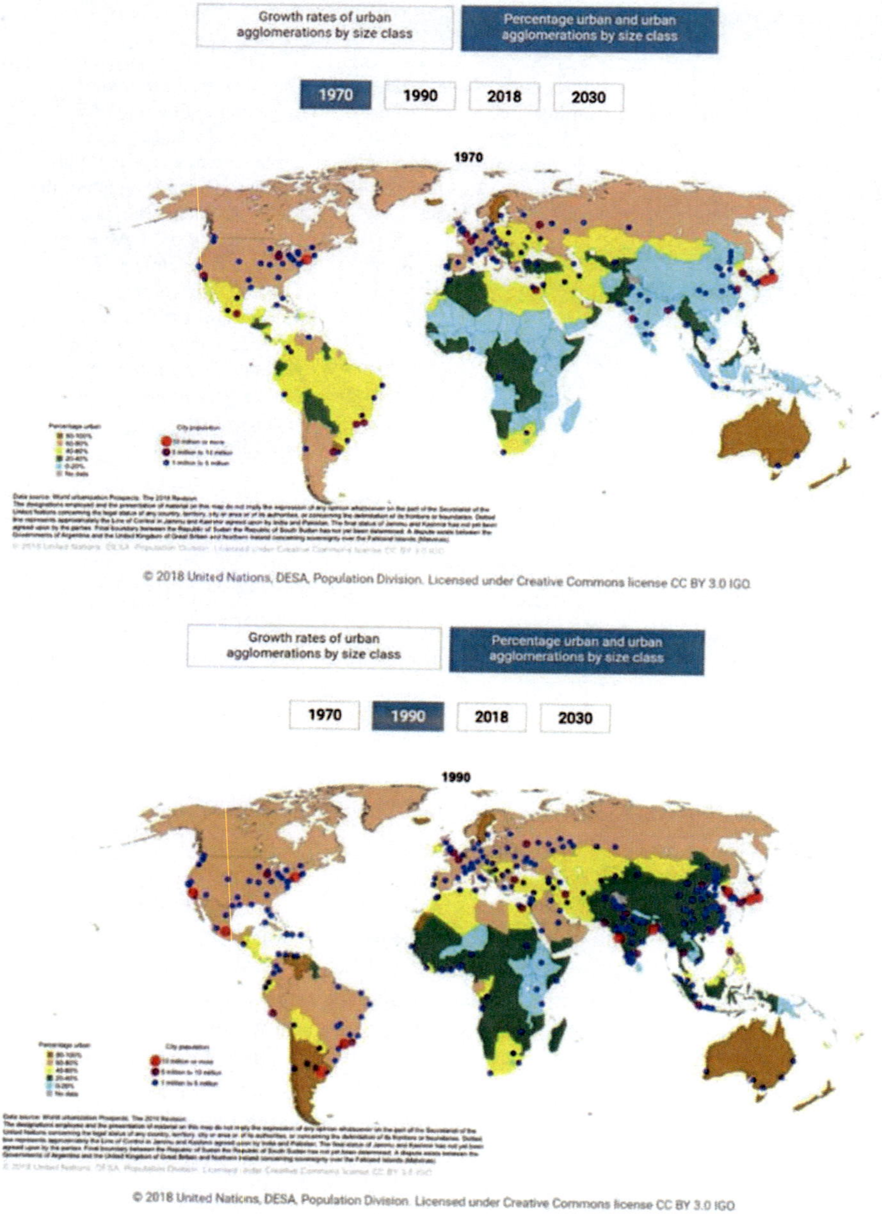

Fig. 22 Growth of Urban Agglomeration by Size 1970 to 2030

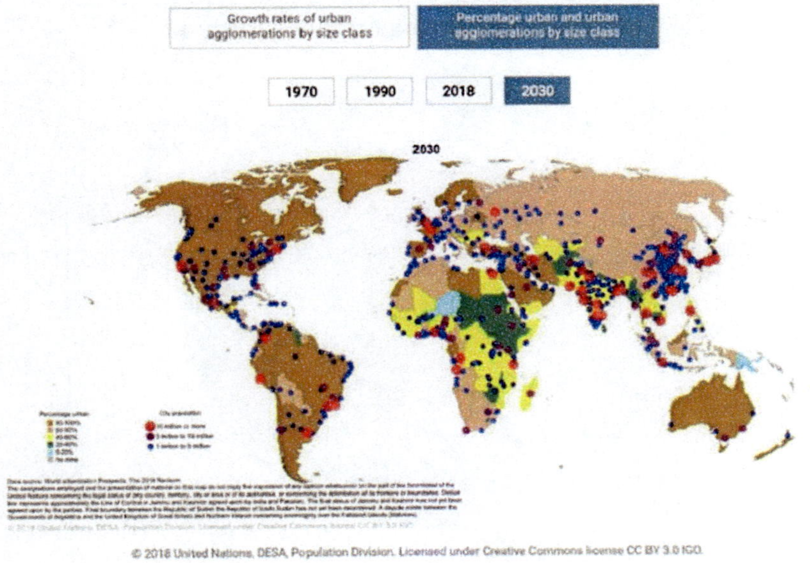

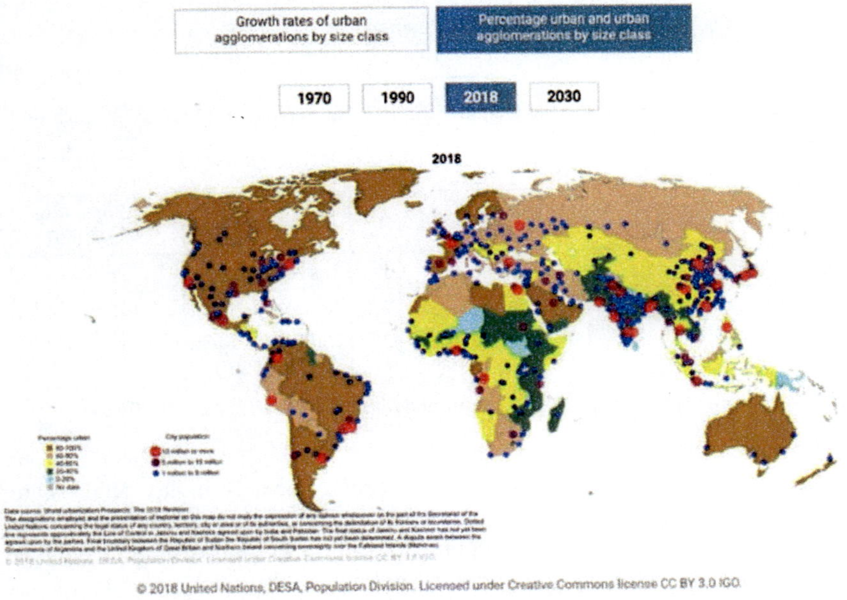

Fig. 22 (continued)

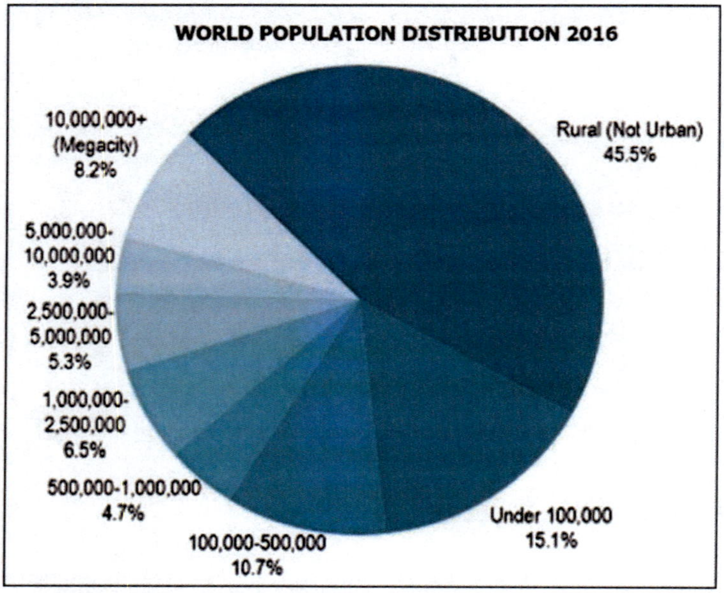

Fig. 23 Population Distribution 2016. *Source* UN-HABITAT

Table 9 Meta cities and megacities population 1950, 1975 the population of cities with ten million inhabitants or more, 1950–75

	1950			1975	
	City	Population (million)		City	Population (million)
1	New York-New ark	12.3	1	Tokyo	26.6
2	Tokyo	11.3	2	New York-New ark	15.9
			3	Cludad DE México (Mexico City)	10.7

Source United Nations, Department of Economic and Social Affairs, No. ESA/P/WP/200

In this region, these megacities have a disproportionately higher NDP share, average household income, and expenditure than more miniature cities and villages of the respective Nations assume a significant role in the Indo-Pacific's regional and international economic development and politics. In the Indo-Pacific region, all leading economies of the world are situated, such as the US, the Peoples's Republic of China, India, Japan, Australia, South Korea, and ASEAN countries. Others like the UK, France, the Netherlands, and Germany have substantial economic and security concerns and connections in Indo-Pacific and have their National Strategies for the Indo-Pacific.

Table 10 Meta cities and megacities population 1950, 1975, 2000, 2005, 2015 the population of cities with ten million inhabitants or more, 1950, 1975, 2000, 2005, 2015

	2000			2005			2015	
	City	Population (million)		City	Population n (million)		City	Population (millionn)
1	Tokyo	14.4	1	Tokyo	35.4	1	Tokyo	35.5
2	Cludad de Mexico (Mexico City)	18.1	2	Mumbai (Bombay)	21.9	2	Mumbai (Bomba y)	21.9
3	New York- New Ark	17.8	3	Cludad de Mexico (Mexico City)	21.6	3	Cludad de Mexico (Mexico City)	21.6
4	Sao Paulo	17.1	4	Sao Paulo	20.5	4	Sao Paulo	20.5
5	Mumbai (Bomb ay)	16.1	5	New York-New Ark	19.9	5	New York-New Ark	19.9
6	Shanghai	13.2	6	Delhi	18.6	6	Delhi	18.6
7	Kolkata a (Calicutta)	13.1	7	Shanghai	17.2	7	Shanghai	17.2
8	Delhi	12.4	8	Kolkata (Calcutta)	17.0	8	Kolkata (Calcutta)	17.0
9	Butanes Aires	11.8	9	Jakarta	13.2	9	Dhaka	16.8
10	Los Angles Long Beach Sanat Ana	11.8	10	Buenos Aires	12.6	10	Jakarta	16.8
11	Osaka-Kobe	11.2	11	Dhaka	12.4	11	Lagos	16.1
12	Jakarta	11.1	12	Jakarta	11.1	12	Karachi	15.2
13	Rio Dejan rio	10.8	13	Rio de Janeiro	10.8	13	Buenos Aires	13.4
14	Al-Qahira Cairo	10.4	14	Al-Qahirah (Cairo)	10.4	14	Al-Qahirah (Cairo)	13.1

(continued)

Table 10 (continued)

	2000			2005			2015	
	City	Population (million)		City	Population n (million)		City	Population (millionn)
15	Dhaka	10.2	15	Dhaka	10-0.2	15	Los Angles, Long Beach-Santa Ana	13.1
16	Moskva (Moscow)	10.1	16	Moskva (Moscow)	10.1	16	Manila	12.9
17	Karachi	10.0	17	Karachi	10.0	17	Beijing	12.9
18	Manila	10.0	18	Manila	10..0	18	Rio de Janario	12.8
						19	Osaka-Kobe	11.3
						20	Istambu l	11.2
						21	Moskva (Mosco w)	11.0
						22	Guangz hou Guangd ong	10.4

Source United Nations, Department of Economic and Social Affairs, No. ESA/P/WP/200

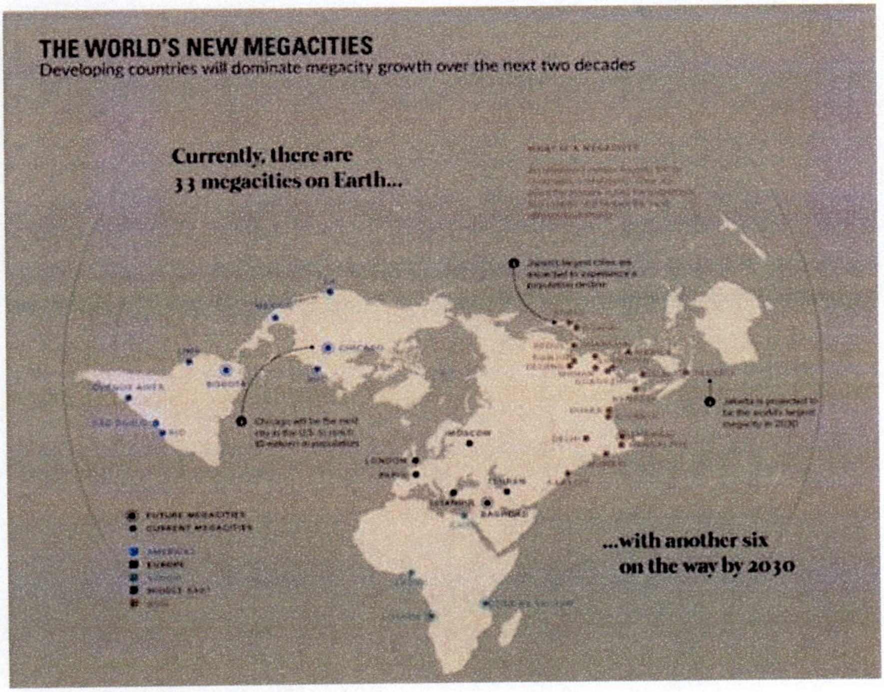

Fig. 24 Megacity Distribution by Continents and population growth of Megacities. *Sources* https://visualcapitalist.com

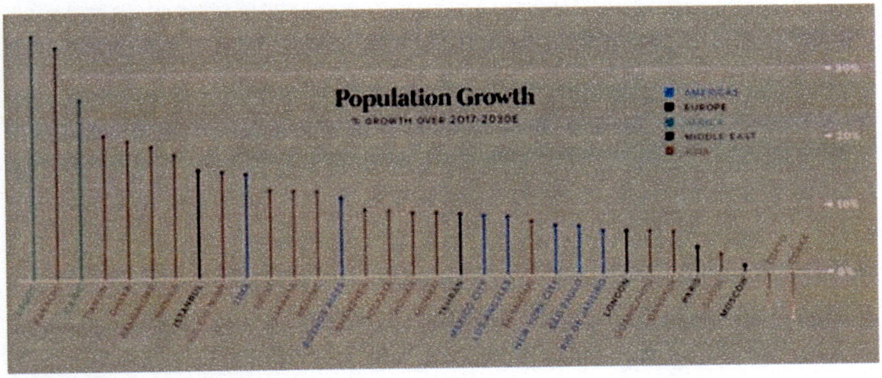

Fig. 25 Population growth rate of megacities in the Indo-Pacific. *Sources* https://visualcapitalist.com/

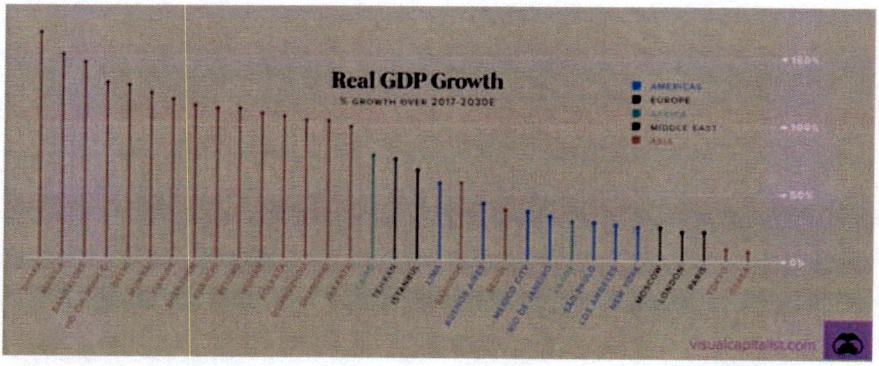

Fig. 26 GDP Growth rate of megacities of the Indo-Pacific. *Source* https://visualcapitalist.com/

These gigantic habitats are significant because it has all the potential to convert themselves into a system of intelligent and global megacities if configured for their sustainability and rapid economic development articulating major concerns like a war against the COVID-19 pandemic, other disaster management such as Tsunami, forest fires, and floods, climate change issues and other security challenges. The vast population, cultural and ecosystem diversity, diverse institutional endowments, supply chain connectivity, global linkages, and size of income and expenditure in these megacities create opportunities for configuring an intelligent global megacity and then progressing toward a system of smart global megacities, which is the subject matter of this book (Tables 11 and 12).

"World Cities Report 2016," New York

From the above, what exists as megacities in Indo-Pacific is shown.

There are three types of megacities population decelerating, accelerating, and slow growth. As shown in the figure above, Tokyo and other cities are decelerating. The megacity strategies vary consequently (Fig. 29)

4 How Megacity System Works in the Indo-Pacific

Theories of megacities are scholarly ideas to explain how megacities systems function in the Indo-Pacific. Megacities are global and functionally interconnected with each other, and their functioning has been given by many. This is summarized below.

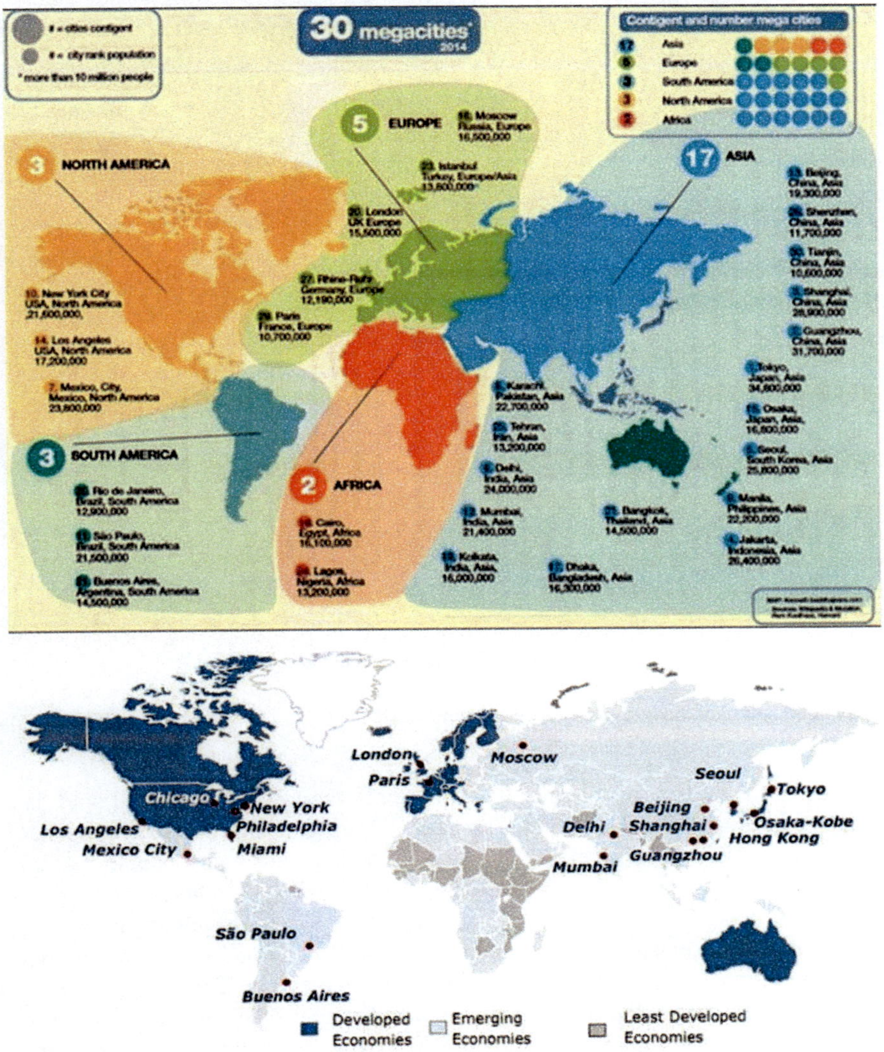

Fig. 27 Megacities in 2014 and Megacities in 2025 by country classification. *Source* [16]

4.1 *Megacity Systems the Connected Global Cities [17]*

Global cities existed in ancient times in port cities or cities en route to the silk road where people.

They have assembled from different countries for various purposes. They learned each other's languages and culture and returned to their country the best they felt, including commercial products. The global population who participated in related

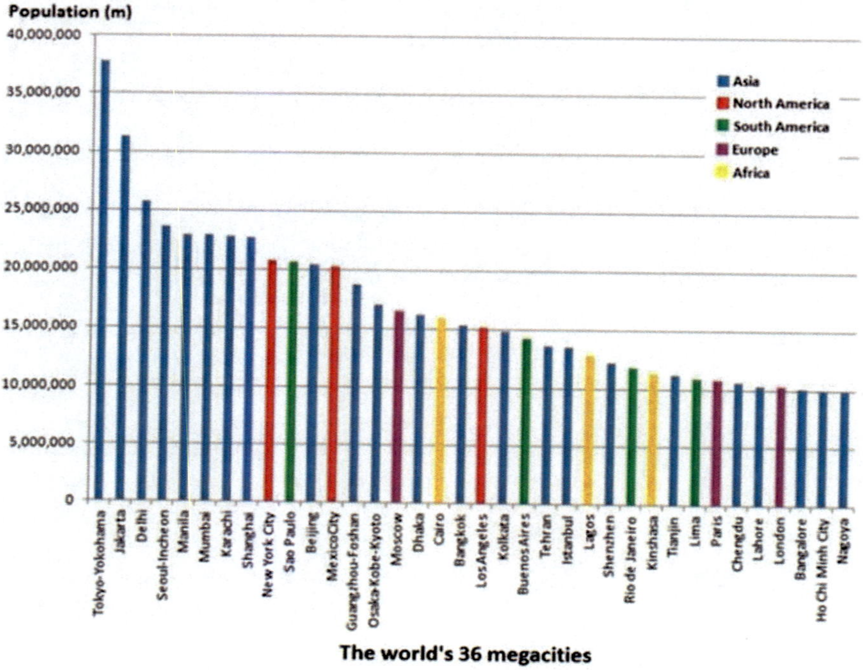

Fig. 28 Population of thirty-five megacities. *Source* [16]

activities of the ancient historical cities was small, the towns were small, and the intensity of workouts was also small. Things changed drastically with the emergence of megacities, where global economic relations occur with many people participating—communication technologies, such as ICT. IoT and Industries 4 created global cities functionally connected as a system of megacities. This also forms supply chains for high-end technological products. In terms of economic, social, and cultural development, megacities gained due to massive globalization. This interconnected production and services indirectly provided world peace since any war like that of Ukraine would disrupt supply chains and people's lives and work in cities worldwide. Therefore, the dispute shall be settled through diplomacy and dialogue as soon as possible with the help of world bodies or bodies of country grouping and moved toward a rule-based system. The trade war between the US and China, directly and indirectly, affects the rest of the world and sanctions on Russia, China, and Iran. Still, the result is devastating for the global economy per emerging indicators. In the past, few people interacted, but the intensity of people-to-people interactions in megacities, meta cities, and often metro cities is immense. This is a key to globalization, a state of interconnectedness that transcends and ignores national boundaries. With the emergence of global cities, a dual city region emerged for each global city. One the city region is within the country where the city is located, and the second is outside

Table 11 Megacities and Meta cities 2016–2030

Rank	City, country	Population in 2016 (thousands)	City, country	Population in 2030 (thousands)
1	Tokyo, Japan	38,140	Tokyo, Japan	37,190
2	Delhi, India	26,454	Delhi, India	36,060
3	Shanghai, China	24,484	Shanghai, China	30,751
4	Mumbai (Bombay), India	21,357	Mumbai (Bombay), India	27,797
5	São Paulo, Brazil	21,297	Beijing, China	27,706
6	Beijing, China	21,240	Dhaka, Bangladesh	27,374
7	Ciudad de México (Mexico City), Mexico	21,157	Karachi, Pakistan	24,838
8	Kinki M.M.A.(Osaka), Japan	20,337	Al-Qahirah (Cairo), Egypt	24,502
9	Al-Qahirah (Cairo), Egypt	19,128	Lagos, Nigeria	24,239
10	New York, Newark, USA	18,604	Ciudad de México (Mexico City), Mexico	23,865
11	Dhaka, Bangladesh	18,237	São Paulo, Brazil	23,444
12	Karachi, Pakistan	17,121	Kinshasa, Democratic Republic of the Congo	19,996
13	Buenos Aires, Argentina	15,334	Kinki M.M.A.(Osaka), Japan	19,976
14	Kolkata (Calcutta), India	14,980	New York-Newark, USA	19,885
15	Istanbul, Turkey	14,365	Kolkata (Calcutta), India	19,092
16	Chongqing, China	13,744	Guangzhou, Guangdong, China	17,574
17	Lagos, Nigeria	13,661	Chongqing, China	17,380
18	Manila, Philippines	13,131	Buenos Aires, Argentina	16,956
19	Guangzhou, Guangdong, China	13,070	Manila, Philippines	16,756
20	Rio de Janeiro, Brazil	12,981	Istanbul, Turkey	16,694
21	Los Angeles-Long Beach-Santa Ana, USA	12,317	Bengaluru, India	14,762
22	Moskva (Moscow), Russian Federation	12,260	Tianjin, China	14,655
23	Kinshasa, Democratic Republic of the Congo	12,071	Rio de Janeiro, Brazil	14,174
24	Tianjin, China	11,558	Chennai (Madras), India	13,921
25	Paris, France	10,925	Jakarta, Indonesia	13,812

(continued)

Table 11 (continued)

Rank	City, country	Population in 2016 (thousands)	City, country	Population in 2030 (thousands)
26	Shenzhen, China	10,828	Los Angeles-Long Beach-Santa Ana, USA	13,257
27	Jakarta, Indonesia	10,483	Lahore, Pakistan	13,033
28	Bengaluru, India	10,456	Hyderabad, India	12,774
29	London, UK	10,434	Shenzhen, China	12,673
30	Chennai (Madras), India	10,163	Lima, Peru	12,221
31	Lima, Peru	10,072	Moskva (Moscow), Russian Federation	12,200
32			Bogotá, Colombia	11,966
33			Paris, France	11,803
34			Johannesburg, South Africa	11,573
35			Krung Thep (Bangkok) Thailand	11,528
36			London, UK	11,467
37			Dar es Salaam, United Republic of Tanzania	10,760
38			Ahmadabad, India	10,527
39			Luanda, Angola	10,429
40			Thành Pho Ho Chí Minh (Ho Chi Minacity), Viet Nam	10,200
41			Chengdu, China	10,104

Source United Nations, 2016 "World Cities in 2016 Data Booklet," New York. United Nations, 2016

the country interacting with global megacities, systems of global megacities, or even international subregions. The economic interaction from outside the city region will often be more dominant and dynamic in terms of money transacted than inside the city region. This economic interaction that leads to manufacturing and trading in the highest quality and globally standardized goods and services at the lowest cost to consumers who have the immense choice to shape their quality of living was the result of globalization. The country-specific regulations that restrict the easy flow of goods and services are discussed and solved mutually by the countries involved, as seen in the current negotiation with China and India, and the US or deliberations in the World Trade Organization. Such arrangements create a division of labor across city regions transcending country boundaries and sharing the economic responsibilities as per capabilities at lower costs. Global urban economies benefited greatly from advanced and standardized producer services such as finance, banking, insurance,

Table 12 Megacities and meta cities in the Indo-Pacific 2016–2030

Rank	City, country	Population in 2016 (thousands)	City, country2	Population in 2030 (thousands)
1	Tokyo, Japan	38,140	Tokyo, Japan	37,190
2	Delhi, India	26,454	Delhi, India	36,060
3	Shanghai, China	24,484	Shanghai, China	30,751
4	Mumbai (Bombay), India	21,357	Mumbai (Bombay), India	27,797
5	Beijing, China	21,240	Beijing, China	27,706
6	Kinki M.M.A.(Osaka)Japan	20,337	Dhaka, Bangladesh	27,374
7	Dhaka, Bangladesh	18,237	Karachi, Pakistan	24,838
8	Karachi, Pakistan	17,121	Kinki M.M.A.(Osaka)Japan	19,976
9	Kolkata (Calcutta), India	14,980	Kolkata (Calcutta), India	19,092
10	Chongqing, China	13,744	Guangzhou, Guangdong, China	17,574
11	Manila, Philippines	13,131	Guangdong, China	17,574
12	Guangzhou, Guangdong, China	13,070	Chongqing, China	17,380
13	Los Angeles-Long Beach-Santa Ana, USA	12,317	Manila, Philippines	16,756
14	Tianjin, China	11,558	Bengaluru, India	14,762
15	Shenzhen, China	10,828	Tianjin, China	14,655
16	Jakarta, Indonesia	10,483	Chennai (Madras), India	13,921
17	Bengaluru, India	10,456	Jakarta, Indonesia	13,812
18	Chennai (Madras), India	10,163	Los Angeles-Long Beach, Santa Ana USA	13,257
19			Lahore, Pakistan	13,033
20			Hyderabad, India	12,774
21			Shenzhen, China	12,673

Source United Nations, 2016 "World Cities in 2016 Data Booklet," New York. United Nations, 2016

logistics, law, management consultancy, advertising, and other services, generating various local employment opportunities that never existed before globalization.

There was also a demand related to skilling. It further triggers technological development and a higher level of professional service capability. The intranational flow of capital becomes much easier with research in Fintech based on the internet, such

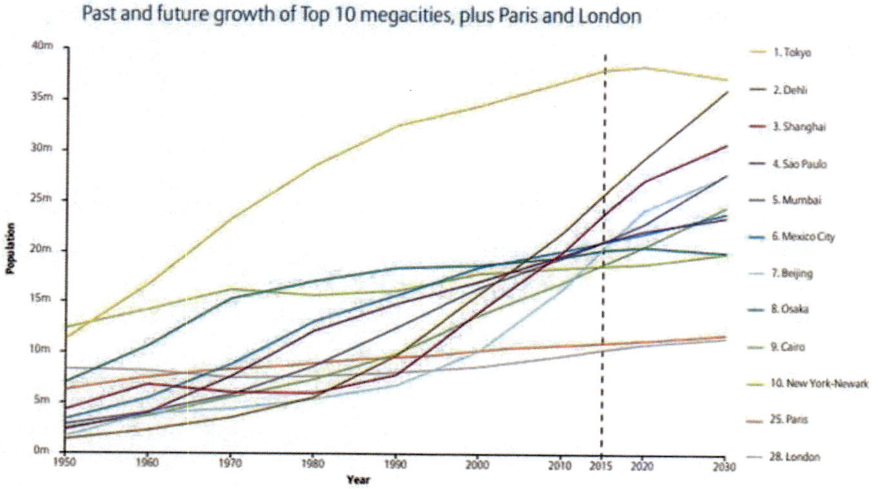

Fig. 29 Population growth patterns of selected megacities. *Source* United Nations Population Division: World Urbanization Prospects 2014 edition 24-4-2022

as SWIFT and UPI. The technology revolution and uniformity of standard practices have made it possible for business enterprises to hire these services anywhere in the world. The intense use of ICT in smart cities has a crucial role. Towns in the Indo-Pacific have taken advantage of this to provide customer services at highly competitive rates by drawing on a large labor force at low pay rates. Unfortunately, this trend contributes to the creation of divided cities, with this professional class earning a higher income with superior skill compared to locally employed labor forces from a local source of traditional economic activities. This can create a high standard of living in a gated community of the rich adjacent to slums with inadequate housing and infrastructure provision, another symptom of divided cities. Protectionism can emerge as an opposing force against globalization and irrational migration policies, eventually leading to the country's lower economic development and isolation of such countries. Such a situation can be avoided if legislators and policymakers can evolve acceptable solutions acceptable to all beyond vote bank solutions. This big political challenge is thrown at legislators and policymakers who must be more knowledgeable about globalization.

Examples of cities that have become hubs of global economic activity are.

1. Finance—Frankfurt, Hong Kong, Amsterdam, Singapore, Sao Paulo, Shanghai.
2. Transport—Dubai, Rotterdam.
3. Information technology—Bengaluru, Seattle, Silicon Valley.

Once the sole domain of Europe and North America, world cities are emerging in the Indo-Pacific and outside to compete with long-established capitals of commerce. Istanbul and Mumbai are culturally influential within their regions through Bollywood and Turkish films, literature, satellite TV networks, and other forms of entertainment. Nairobi, Addis Ababa, and Bangkok, among others, host regional headquarters for international agencies and development partners. Megacities have become the seat of education and research, bringing out innovations attracting students and Professors from across the country.

4.2 Global Functions of Megacities [18–24]

Megacities influence globally with economic, financial, physical, communication, and cultural linkages. Urban growth with severe problems, such as congestion, pollution, social segregation, Terrorism, or high crime rates, threatens the achievement of sustainable urban development. Megacities across the globe are engaging in an intense search for strategies to address these issues. Megacities need immediate solutions for better public transportation, public realm development, energy, solid waste disposal, air and water pollution, or water supply. The critical role megacities play in this context is two-fold: they represent the centers where economic and social change occurs and function as crucial 'actors' behind these changes. Globalization and intense deployment of ICT in cities together result in the increasing importance of megacities at the very top of the hierarchy, the so-called world cities or global cities.

4.3 Five Phenomena of Global Megacities [25, 26]

1. Innovation capacity and the potential for economic development reside in a large part of a megacity. Research-intensive industries and knowledge-based services are increasingly concentrated in megacities since there are reputed Universities and Research Institutions there.
2. Megacities enable innovative companies to exploit opportunities and establish contacts while allowing them to access information and reduce risks. In addition, megacities also provide access to specialized resources and employees and specific routines, traditions, values, and other local institutions. Megacities are characterized by a high degree of economic, social, and cultural complexity, functionally interconnected and embedded in the global division of labor. On the other hand, they constitute the nodes of overlapping financial flows as well as trade, production, political, cultural, and social networks cutting across national boundaries. Innovation capacity and the potential for economic development reside predominantly in megacities.
3. The national hierarchy of cities and the division of labor within the national economy has a counterpart in a global division of labor transcending country boundaries. Megacities become part of an emerging international hierarchy based on a competitive division of work worldwide by international connections that affect financial flows in the knowledge-intensive service sector. It leads to the

emergence of a vertical hierarchy of a globalized system of megacities. The relationship between spatial decentralization and territorial concentration alters the position of peripheral areas in this emerging spatial structure. Unbalanced spatial development reinforces the regional character through expansion, stagnation, and negative growth. Highly skilled employment, high-value infrastructure, and investment co-exist in large megacities.

4. Megacities align their specialization with the global division of labor and forge links with other dynamic megacities. Interconnections between megacities, their hinterlands, and immediate surrounding (structurally weak) areas may decrease and become increasingly disconnected from economic development. This adverse development calls for institutional structures that extend beyond the megacity level.
5. Megacities as economic development hubs generate changing spatial patterns impacting institutional action within the state hierarchy. Devolution of State powers to transnational levels leads to a newly constituted tier of the megacity and megacity regional governance.

   Considering the growing importance of megacity regions, the need for adequate organizational structures ('for megacity governance') emerge in megacity and megacity regions.

4.4 World Cities and Global Cities [27–30]

In his book, Cities in Evolution, Patrick Geddes coined the term "world city" in 1915. Geddes taught biology at the University of Dundee and advised India's rulers on how to run their cities.

World cities are characterized by a sum of political power (both national and international) and organizations related to government; national and international trade, whereby cities function as a gateway for their own and sometimes also neighboring countries; providing superior banking, insurance, and related financial services; advanced professional activities of all kinds; information gathering and diffusion. The form and extent of a city's integration with the world economy and the functions assigned to the town in the new spatial division of labor will be decisive for any structural changes within it. Key cities throughout the world are used by global capital as bases—points in the spatial organization and articulation of production and markets. The resulting linkages make arranging world cities into a complex spatial hierarchy possible. The global control functions of world cities are directly reflected in the structure and dynamics of their production sectors and employment. They are important sites for the concentration and accumulation of international capital.

They are points of destination for many domestic and international migrants. World city formation brings into focus the significant contradictions of industrial capitalism, including spatial and class polarization. World city growth generates social costs at rates that tend to exceed the state's fiscal capacity—conspicuous consumption: arts, culture, and entertainment and the ancillary activities that cater to them. Three main functions of world cities are headquarters function, financial center function, and articulator cities that link a national or regional economy to the global economy.

The world cities hypothesis of Freidman is enumerated below.

1. The nature of a city's integration with the world economy is decisive for any structural changes. It is occurring within it.
2. Key cities throughout the world are used by global capital as 'basing points for the organization of production and markets.'
3. The global control functions of world cities manifest in their industrial structure and job markets.
4. World cities are significant sites for the concentration and accumulation of capital.
5. World cities are destinations for large numbers of migrants—both domestic and international.
6. World city formation exposes the significant contradictions of industrial capitalism, particularly spatial class polarization.
7. World city growth generates social costs that exceed the state's fiscal capacity.

Modern technologies, telecommunications, and information technology have led to decentralization and agglomeration of economic activities. This combination of spatial dispersal and global integration has created a new strategic role for major cities, thus leading to a new type of city (different from historical banking and trade centers).

4.5 *Global Cities by Saskia Sassen [31, 33, 34, 43]*

Global cities, for Sassan, form a 'virtual economic cycle' and function in four new ways: the demand for control creates cities as 'command points'; this leads to a request for finance and business services, whereby cities become the 'key locations' for leading economic sectors; cities become sites of production and innovation for these leading economic sectors; cities constitute markets for primary economic sector production.

Saskia Sassen's Global City Hypothesis

1. First, the geographic dispersal of economic activities that marks globalization and the simultaneous integration of such geographically dispersed activities is a critical factor in feeding the growth and importance of central corporate functions. The more distributed a firm's operations across different countries, the more complex and strategic its main parts are—managing, coordinating, servicing, and financing a firm's network of operations.
2. These central functions become so complex that increasingly the headquarters of large global firms outsource them: they buy a share of their primary tasks from highly specialized service firms—accounting, legal, public relations, programming, telecommunications, and other such services.

3. Third, those specialized service firms engaged in the most complex and globalized markets are subject to agglomeration economies.
4. The more headquarters outsource their most complex, unstandardized functions, particularly those subject to uncertain and changing markets, the more accessible they are to opt for any location because less work done in the headquarters is subject to agglomeration economies.
5. These specialized service firms need to provide a global service which means a worldwide network of affiliates or some other form of partnership. As a result, we have seen a strengthening of cross-border city-to-city transactions and networks. At the limit, this may well be the beginning of the formation of transnational urban systems. The growth of global markets for finance and specialized services, the need for transnational servicing networks due to sharp increases in international investment, the reduced role of the government in the regulation of international economic activity, and the corresponding ascendancy of other institutional arenas—notably global markets and corporate headquarters—all point to the existence of a series of transnational networks of cities.
6. A sixth hypothesis is that the growing numbers of high-level professionals and high profit-making specialized service firms raise the spatial and socio-economic inequality evident in these cities. These technical services' strategic role as inputs increases top-level professionals' value and numbers. Further, the fact that talent can matter enormously for the quality of these strategic outputs and given the importance of speed, proven talent is an added value, the structure of rewards is likely to experience rapid increases.
7. A seventh hypothesis is that one result of the dynamics described in hypothesis six is the growing informalization of a range of economic activities which find their effective demand in these cities yet have profit rates that do not allow them to compete for various resources with the high profit-making firms at the top of the system. Informalizing part of or all production and distribution activities, including services, is one way.

Saskia Sassen never considered the potential of a megacity or megacity system as an extensive reservoir of multitudes of specialized supplies of products and services required for global business operations under one single megacity or a system of megacities but was looking at a much smaller scatter of several cities including metropolitan cities, which was what existed before the 1990s.

4.6 Seven Types of Global Cities [35]

The report, "Redefining global cities: The seven types of global metro economies," by Brookings Fellow Joseph Parrilla and former Brookings Research Analyst and Associate Fellow Jesus Leal Trujillo, uses a first-of-its-kind database of dozens of indicators to examine the global city economy.

Characteristics, industrial structure, and critical competitiveness factors: tradable clusters, innovation, talent, and infrastructure connectivity. Their analysis focuses on the 123 largest metropolitan economies in the world, which collectively account for one-third of global GDP despite containing only 13 percent of the worldwide population. Analysis of these data reveals seven types of global cities—the Global Giants, Asian Anchors, Emerging Gateways, Factory China, Knowledge Capitals, American Middleweights, and International Middleweights—each with distinctive assets, challenges, and growth trajectories. "This report provides fresh evidence that there is no one way to be a global city," Parrilla said. "Cities that understand their unique niche in the global economy are better positioned to create economic strategies that will be successful in the long-term." The accompanying online interactive report uses the typology to help megacity decision-makers understand the local assets that drive economic competitiveness, benchmark their performance against peer cities, and identify the most relevant global innovations for regional growth and prosperity (Fig. 30).

1. Global Giants: six large, wealthy hubs with concentrations of corporate headquarters; they serve as the command-and-control centers for the world's largest advanced economies.
2. Asian Anchors: five large business and financial nodes anchoring inward investment into the Asia–Pacific and Russia.
3. Emerging Gateways: twenty-eight large businesses and transportation entry points for major national and regional emerging markets in Africa, Asia, Eastern Europe, and Latin America.
4. Factory China: twenty-two second and third-tier Chinese cities distinctly reliant on export-intensive manufacturing to power economic growth and global engagement.
5. Knowledge Capitals: nineteen mid-sized, highly productive knowledge creation centers in the United States and Europe with talented workforces and elite research universities.
6. American Middleweights: sixteen mid-sized US metro areas were striving for a post-recession niche in the global economy.
7. International Middleweights: twenty-six mid-sized cities in Australia, Canada, and Europe are globally connected by people and investment flows, but growth has lagged after the fiscal crisis.

4.7 Connected Global Cities [36, 37]

We are all living in a globalized world. The world is interconnected, and the policies of one country, rational or irrational, affect others and megacities. With America, the First Policy of the USA under President Donald Trump and Brexit in the UK, which divorces the UK from the European Union, migration policy, trade policies, custom union issues, and tariffs as an economic weapon against saying China for the

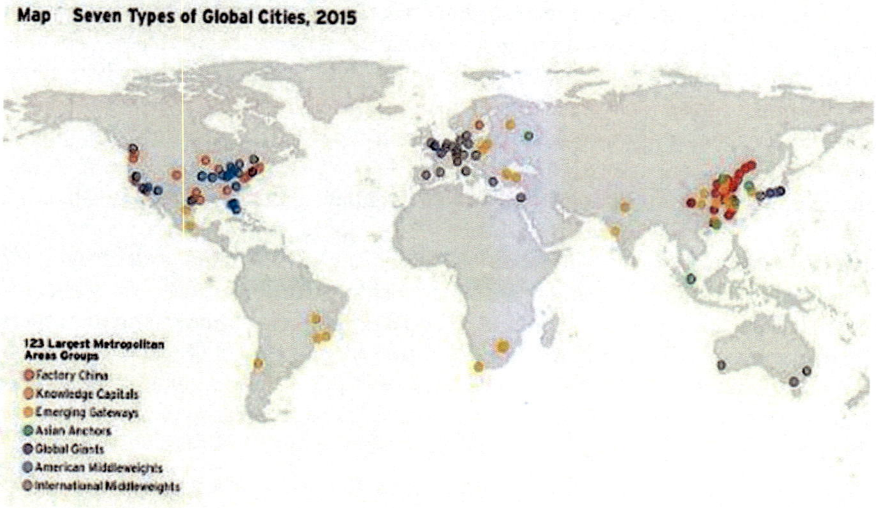

Fig. 30 Typology of Global Cities 2015. *Source* [35]

balance of payment issues have gained considerable attention. Democrat opposes what Republican advocates in the US and vice versa. Germany is more tolerant than some other European Union countries. Switzerland remains a country with a better immigration policy than others. Undoubtedly, there will be retardation in economic development for irrational policies, which will not be limited to a few countries but all countries in a very connected world.

"Since 1998, the Globalization and World Cities (GaWC) Research Network has extensively studied the impact of word cities on globalization. GaWC provides a geographic and economic-based overview of the world and is evolving configuration. GaWC is a research network based in the Geography Department at Loughborough University (UK), founded by Peter J. Taylor in 1998.

Particularly relevant in this discourse is its categorization of world cities into α, β, and γ tiers based on their international connectedness. Suppose the world is observed from the point of view of the connectivity of the world cities. In that case, a new image emerges, where each city is oriented to other cities of the same level of interconnectivity. National or continental maps give way to a new world configuration intended as an archipelago, where each town appears utterly separated from its surroundings and closer to other cities of the same level. The leading parameters for the new configuration are based on mutual connections, primarily in the global economic system. The atlas shows the world's cities according to the GaWC categorization 2010. A new configuration emerges from this, where expected national and local boundaries are presented in their natural disposition based upon their international connectedness" (Fig. 31).

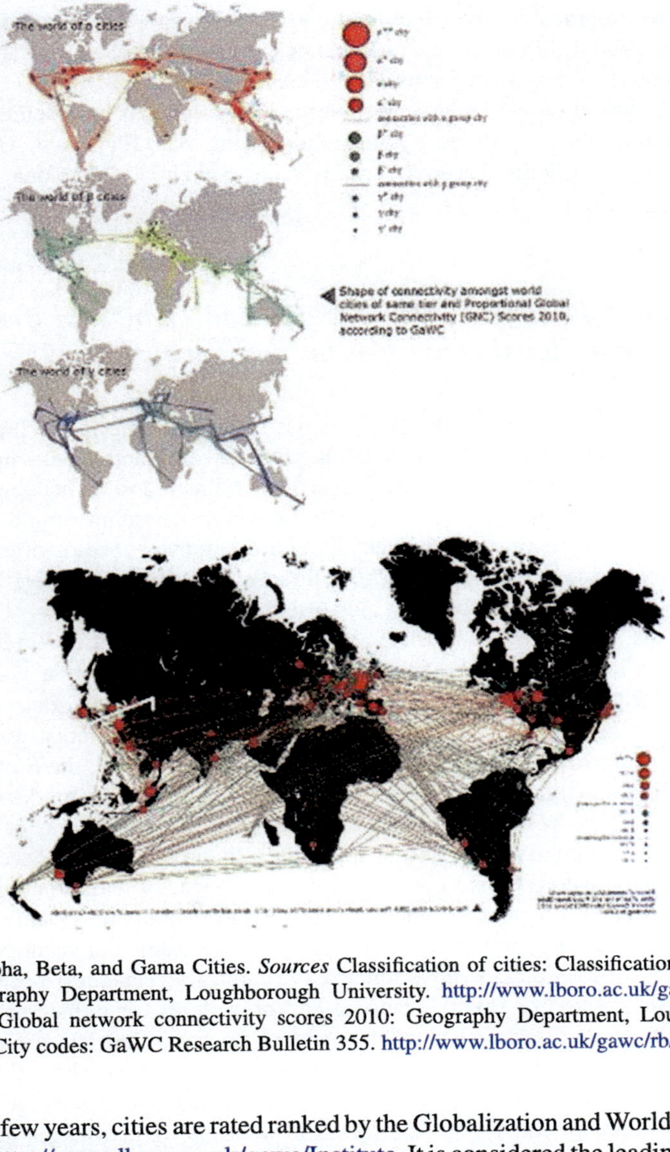

Fig. 31 Alpha, Beta, and Gama Cities. *Sources* Classification of cities: Classification of GaWC 2010. Geography Department, Loughborough University. http://www.lboro.ac.uk/gawc/world2010t.html—Global network connectivity scores 2010: Geography Department, Loughborough University. City codes: GaWC Research Bulletin 355. http://www.lboro.ac.uk/gawc/rb/rb355.html

"Every few years, cities are rated ranked by the Globalization and World Rankings Research http://www.lboro.ac.uk/gawc/Institute. It is considered the leading institute ranking in world cities. Cities are ranked into Alpha, Beta, and Gamma cities by considering many factors; economic factors are deemed more important than cultural or political factors though in this ranking. Here is a list with the latest data from 2016 (GAWC has not done a study in 2017 or 2018). Spotted by Locals cities (most of "our" 71 cities are listed) are linked. Alpha + + world cities: London, New York.

Alpha + world cities: Singapore, Hong Kong, Paris, Beijing, Tokyo, Dubai, Shanghai.

Alpha world cities: Sydney, Sao Paulo, Milan, Chicago, Mexico City, Mumbai, Moscow, Frankfurt, Madrid, Warsaw, Johannesburg, Toronto, Seoul, Istanbul, Kuala Lumpur, Jakarta, Amsterdam, Brussels, Los Angeles.

Alpha – world cities: Dublin, Melbourne, Washington, New Delhi, Bangkok, Zurich, Vienna, Taipei, Buenos Aires, Stockholm, San Francisco, Guangzhou, Manila, Bogotá, Miami, Luxembourg, Riyadh, Santiago, Barcelona, Tel Aviv, Lisbon."

4.8 Thomas Friedman's World is Flat and Thanks for Being Late and Global Megacities in Indo-Pacific [38, 39]

Thomas Friedman's idea that the world is flat is applied here in the Indo-Pacific. Globalization. He visualizes as the flattening of the world, and we attempt to show how it can happen in the system of megacities in the Indo-Pacific. Smart global megacities are ideal places in the Indo-Pacific where flattening should occur. He says the world's flattening is achieved through the convergence of often political events, innovations, and companies. The political events are the initiatives of QUAD in the Indo-Pacific and other country Indo-Pacific Policies.

Flatteners 1: 11/9/89 when the Walls Came Down, and the Windows went up. Friedman refers to the fall of the Berlin Wall, which is merely a symbol from the point of view of the Indo-Pacific global city. Before this date, there were two ways of planning a global megacity, namely the socialist/communist way and the democratic way with the market economy. Soon after that date, the Soviet Union disintegrated, and China moved closer to following the capitalist market economy with its bible as the WTO rule book. In the Socialist/Communist way of megacity planning, the Communist Party, staffed mainly by a single party system, decides what to produce and what services to be developed in a megacity with no consideration about what happens in a market economy within or outside the country and everything conforming the ideology of socialism or Marxism or Maoism or any autocrats "isms by his name" of the socialist government. The targets are made as postulated by Planning Commission or Board. Planning and procedures for implementation are formulated for administrators subjectively; more tax from the rich was collected, and the bureaucracy implemented it using tax collected and using a licensing system with highly restrictive controls, which resulted in inefficiencies out of ignorance of markets and human behavior, encourages and often giving patronages to massive corruption of political and administrative persons and resulting in slow economic growth and much slower poverty reduction. These happened in India before 1991 when India tried to copy the Soviet Union with a socialism tag which they learned from their colonial masters, the British, when their leaders were university students in the UK. The biggest problem of the command economy is that they are prisoners of an ancient theory of Marxism, a product of the ills of the first industrial revolution in the UK. Still, the British never allowed the industrial revolution to happen in India as a British colony, not considering what was happening globally in colonies of past empires and post-colonial countries in an ever-changing world. There have been many industrial revolutions, and Marxism is not a satisfactory answer to the fourth industrial revolution that follows today. As a result, the Marxist party in India has less membership trending toward disappearance in the Indian Parliament, and only one state with less than 3% of the population of India is ruled by them at the rate of once in ten years. Indians have written off Marxism since 1990, but Marxist leaders who misdirect uneducated people have yet to realize that (Fig. 32).

This has resulted in the symbolic breaking of the Berlin Wall by people who protested the slow progress of economic development in countries under socialism, threatening monitory situations toward bankruptcy and more incidence of abject poverty. This increased democratic power for people to decide what they want, how their jobs should evolve, and how they will integrate with the world. In a democratic global city, people and communities determine what is good for them based on emerging facts, and they interact with the local and international community; they invest their money and run the production of goods and services as per their perception and initiative within the rights available under the constitution fully prepared to face any risks. In such a setup, licensing and control of socialist bureaucracy were reduced to minimal regulations to protect all. India practiced the socialist way strictly until 1991 and found poverty never came down, and the growth rate never went up more than 3% India was going to be a bankrupt Nation for the first time in 1991. In 1991 the existing Indian government moved to the democratic way of governance and found growth rates going up to 8 and 9%. Poverty was reduced faster by removing license raj and related bureaucratic heavy-handed controls and planning board-designed subjective procedures called administrative rules. Subsequently, the Planning Commission was dissolved by the Union Government of India and replaced by a policy research think tank called Niti Aayog. This is more useful and productive than subjective administrative rules and targets set by the Planning Commission based on some

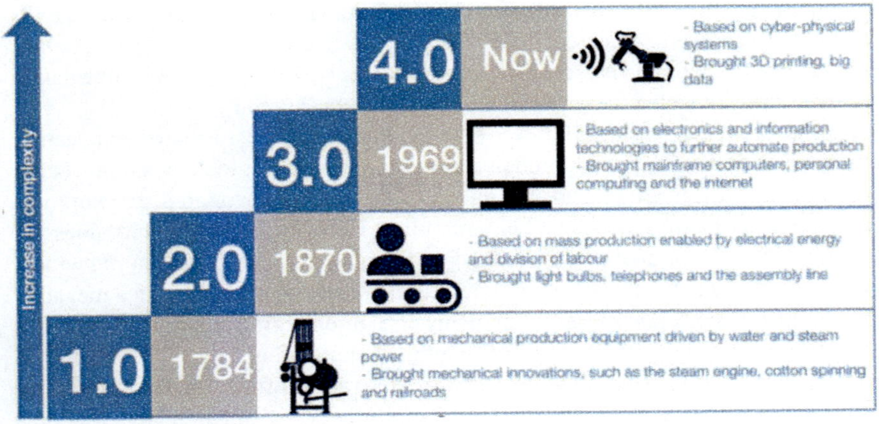

Fig. 32 Industry 1, 2, 3, and 4. *Source* Author

abstract ideas they got from the UK and USSR not supported by empirical research in 24 h and seven days week framework. This means all households in the intelligent global megacity should be empowered, so also local face-to-face communities for economic, social, spatial, environmental, and cultural development with few administrative restrictions and licensing under the constitution of rights enshrined there. The Government, Chamber of Commerce and Industries, and Academia can lead in supporting households and the local community. Under socialism and communism, global-city-information dissemination was restricted and not freely available to all but to a privileged few. With the rise of the internet, information was open freely to people leading them to decide on actions and soon to adjust themselves to global production and services, aligning them to common standards globally acceptable. With the advent of TV, journalism, and mass media, people and households have started knowing what was happening. They started communicating their views and finding opportunities without dictation from parties, planning boards, commissions, and administrators.

Flattener #2 8/9/95 When Netscape Went Public.
The concept of the World Wide Web as a system for creating, organizing, and linking all types of documents so they could be easily browsed was first created by British Computer Scientist Tim Berners-Lee in 1991. It became a reality for public use when Netscape was produced, and the company tried to market it as a commercial web browser by Netscape. The immense possibility of WWW triggered development in fiber-optic networks worldwide at an unimaginable speed that can carry a massive amount of data connecting the whole universe. Then PC and Mac's computers started their accelerated growth path by increasing capabilities and reducing costs by incorporating web browsers independent of the operating system. These were free and triggered by the thought that consequent increasing users meant higher profitability in commercial returns, and entrepreneurs invested in the network infrastructure of fiber-optic cables by entrepreneurs that created the WWW infrastructure meant for increasing business activities globally. Every office needed to interact with workers through the computer using a local area network. With the internet, this office expands globally to the universe transferring all types of data, voice, music, images, movies, and messages at almost zero cost. It was widely used for emails and information sharing across the world for a multitude of businesses (Fig. 33).

This development was significant to smart global megacities. The megacity can be developed, managed, and monitored through the internet. All global economic activities scattered worldwide can use the internet for the global economic development of the megacity by all citizens and communities. Here the size of the city does not matter whether it is a megacity or meta city; the amount of information movement does not matter, and what matters is global economic development using a megacity which is also a digital city intensively using ICT in emerging smart or digital city. This flattening gave rise to the foundation of the smart global megacity. This movement was done fast because all required was free or almost free. The Berlin Wall we discussed above needs to be broken down before the advent of free Netscape, and both cannot co-exist.

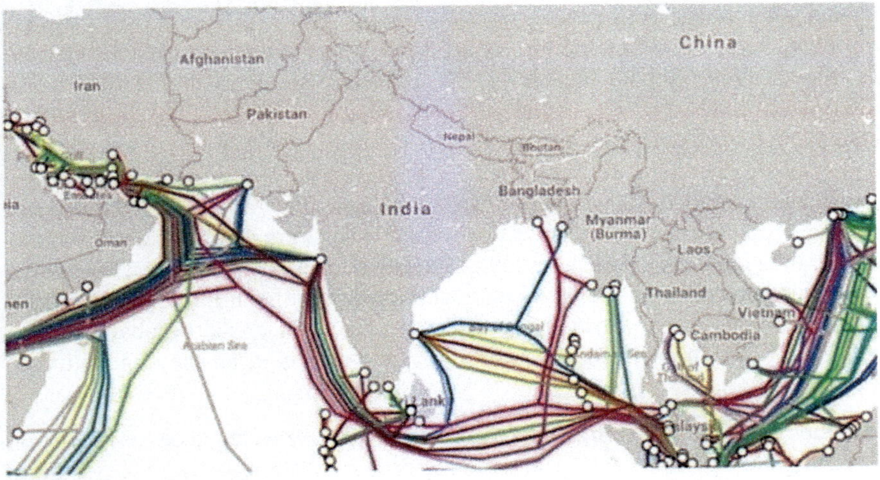

Submarinecablemap.com

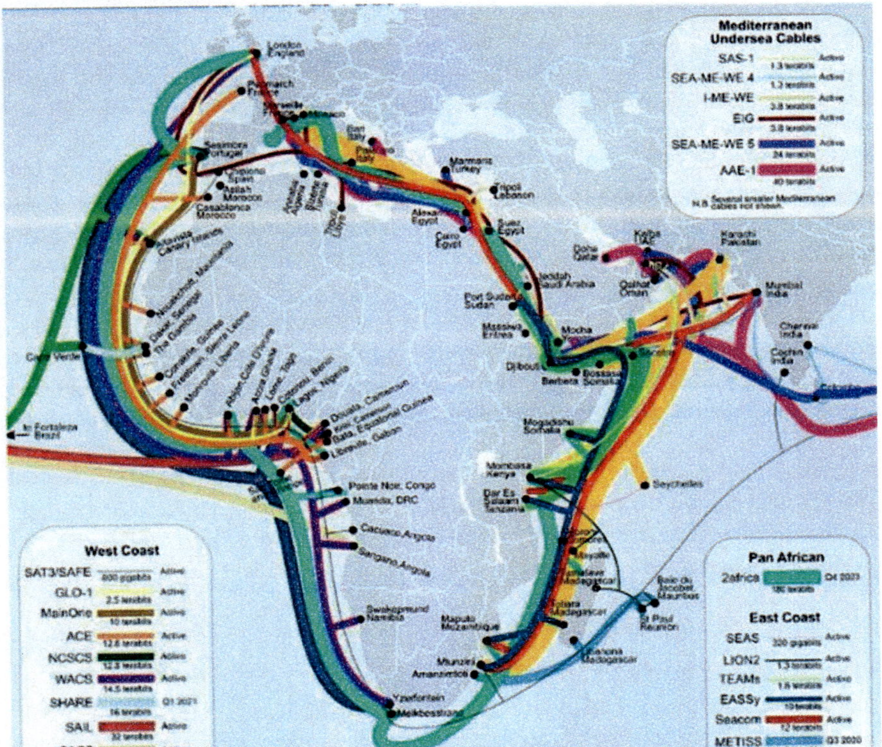

Fig. 33 Fiber Network India and Africa

Flattener #3 Workflow Software.
Workflow software is internet-based software where work can be shared from multiple locations globally. GIS and CAD are used in megacity digitization, monitoring, and production of a spatial decision support system that can be used and implemented from multiple locations for many purposes, including commercial and industrial. Products and services can also be designed, and prototypes generated at various locations using web-based workflow programs. Many GIS Wenders and open-source Linux have their brands of such free software. This became the backbone of the global supply chain in fields like culture, visual art, and movie-making. Workflow software soon moved to automation using artificial intelligence. Worldwide collaboration and sharing economy, two pillars of the smart global megacity system, got institutionalized with workflow software based on the digitization of everything.

Flattener #4 Open-Sourcing Self-Organizing Collaborative Communities
A self-organizing, smart, and collaborative economic community is the building block of smart global megacities. The open-source software movement used the World Wide Web and professionals who wanted to freely contribute their capabilities to make software and web browsers that can be freely used and whose source code is available to all for modification and improvement. This self-organizing smart community had its management system to organize.It produced the most outstanding software for a variety of tasks. Using Fab Lab and freeware CAD-CAM software, any smart community can deliver exceptional, innovative products that may not exist today. Often private companies shared their software and joined with the open-source movement accepting their code of conduct. Thus, Samsung Smart Things, which provides a hub for smart thing IoT smart home automation, which is very much used in the smart global city, is progressing much faster in software development than other commercial software not using Open sourcing. Amazon's Alexa is also another example whose SDK is readily available for one to design and use Alexa for one use. The result is cheaper products and services and more accessible to people. Often the same company gives out such software free of cost. The marginal cost of production near zero is one of the approaches to an intelligent city economy, and the open-source movement is the first practitioner of this philosophy. It is the first post-capitalist and post-socialist model of shared community products and services. The Governing structure of the open-source movement is self-governance, and it is a model for megacity governance. For smart global megacities, Open sourcing and self-organizing communities are the foundation for smart global megacity economic development and smart living.

Flattener #5 Outsourcing Y2K
According to Thomas Friedman India, the second buyer buys a hotel, a mall, a golf course, or an industry at 10% of the cost when the first owner grows bankrupt. Here the first owner is those fiber-optic cable companies that produced a giant global network undersea to carry digital information of unimaginable quantity, assuming that Netscape going commercial can reap huge profits unimaginable before. Still, all the web browsers that subsequently came up wanted it to be free so that it could

attract more information traffic for fiber-optic that helps other commercial ventures, and this created a situation where the whole WWW industry became free, which necessitated selling the fiber-optic network cheap or for at no cost since there is no possibility of web browser going commercial. Indians did not purchase the share of fiber-optic companies, but they and their American clients benefited from the over-capacity creation and used it almost free. They could use these resources because of the English-educated Science, technology, engineering, and mathematical graduates they produced in abundance and who found unemployed under socialism and the militant labor force they created for the vote bank. So, America, in turn, became the second buyer of Indian capability at a much lower cost of university education than available in their own country, which could not find suitable jobs under the "Berlin Wall" socialist mixed-up economy. In addition to fiber-optic, the Y2K crisis that the computer industry faced was a blessing for India to start in the knowledge economy. The early start of many companies gave Indians educated an alternative future in their own country, which the Soviet Russia model of command economy practiced before 1991 could not provide. The alternative was a migration to another country which they did but is now being opposed by USA First and movements like Brexit and other nationalistic and right-wing migration policies. The start-up culture and shared infrastructure owned by the community became another building block because some of them became one-billion-dollar Unicorns in Bengaluru of the intelligent economy of a smart global megacity from this flattener. This also gave the smart global megacity the art of becoming a knowledge-based industry entrepreneur breaking free from the Berlin Wall, socialism, or old and outdated capitalism.

Flattener #5 Offshoring. Running with Gazelles. Eating with Lions

On December 11, 2001, China joined the World Trading Organization, which means China accepted world trade as a rule-based activity and eventually became the world's factory. The spirit behind such transformation was summarized in a saying written in a Chinese factory, as reported by Friedman.

> "Every morning in Africa, a gazelle wakes up.
> It knows it must run faster than the fastest lion, or it will be killed. Every morning a lion wakes up.
> It knows it must outrun the slowest gazelle, or it will starve to death. It does not matter whether One is a lion or a gazelle.
> When the sun comes up, One better start running."

This is the spirit behind China becoming a factory of the world. One cannot see the Berlin Wall discussed above and Marxism, Leninism, and Maoism. There is no Communist Party or Planning Commission but the people's Republic of China's simple, unadulterated autocratic capitalism. We also saw in Russia an Autocratic Oligarchical Capitalism replacing socialism after the breakup of the USSR but adding a democracy tag with periodic elections during Deng's era.

Offshoring is a process in which a foreign company moves stock and barrel the entire factory production to another country to take advantage of lower factors of production such as labor, water, energy, space, and infrastructure cost, for example, and produce the product much cheaper than made at home and reap the benefit in the world market. China came forward as a destination for offshoring. It made itself suitable when it joined the World Trade Organization and accepted a rule-based system from a communist system where the party decides what is good, often in an arbitrary manner with so-called superior knowledge of the supreme party leader. It had about 160 million-plus population in metro and megacities cities on the east coast near harbors to accommodate offshoring with necessary infrastructure. The leadership of Deng, who proclaimed it does not matter the color of the cat and what matters is the capacity to catch a mouse, made China more accommodative and adaptable to offshoring by removing bureaucratic controls characteristic of the socialist economy but supporting most efficiently all such attempts in offshoring. These attracted significant offshoring from the USA, Europe, Taiwan, and Japan to China. Instead of China making itself statuesque, it became more dynamic. The vast Chinese population became students in Western universities and became innovators themselves. The existing managerial class, which was passive in the past under the communist target-keeping system, became more innovative in management and started thinking of out-of-the-box solutions and made themselves more than efficient Japanese, South Korean, European, and American managers. Quality of output, non-failing infrastructure, and efficiency moved China to become a factory for the world. We also found a role reversal here to make it more confusing to know who a lion is and who is a gazelle in Offshoring.

Flattener #7 Supply Chaining. Eating Sushi in Arkansas

The world has said goodbye to a situation where every part of a product is produced under one roof in one location or country and sold from there. The concept of self-reliance is no longer valid and is replaced by the least cost, reliability, and efficiency of supply chains from selected competent countries. The frequency of market demand by location in the world marketplaces determines the assembly location of final products in any country. At the same time, parts and components are produced in many countries and areas based on their patents and efficiency of production and cost factors. The final assembly destination has a series of supply warehouses around it. When an order, for example, on a website, is made, the e-commerce firm sends information about the purchase with the time required to fulfill the order to—the e-commerce seller. A computer algorithm for reduced cost and efficiency allows the message to be transmitted to the best location for the product assembly. The assembly orders all components required from different warehouses and then assembles and couriers them to the consumer. Another algorithm based on real-time supply statistics allows the warehouse to stock an optimum number of parts so that there is no loss due to excess inventory being unsold. In this situation, we have a global system of megacity operating a business where the nationality of the participant supply chain does not matter, but what matters is cost reduction and reduced time for delivery to the customer. All these activities can be synchronized using artificial intelligence, which guides the production, assembly, dispatch, and logistics using facts and algorithms, not Marxism, Socialism, or Maoism. Indirectly, it helps in world peace since war is likely to disrupt the economic activity of many nations involved because of the supply chain disruption, which nobody wants. The supply chain resulted in a series of technological innovations to strengthen the supply chains further and created a complete set of new jobs that were not there earlier.

Sushi in Arkansas expresses the global co-existence of an interconnected intelligent mega-global system of cities.

They Were Flattening # 8 Insourcing. What Are the Guys in Funny Brown Shorts Doing?

Outsourcing is an activity in which a particular corporate function sends out to another country based on efficiency, cost, and another factor of production and services. Here, for example, an accounting firm hired to file income tax outsources his job from the USA to an Indian firm at 5 pm, and the Indian firm works on it, giving all precautions of secrecy of data, and then sends it back to the accounting.

Firm before 9 am when the firm opens its shop. There is no change in the working hours of US and Indian firms since when it is night in the US, it is a day in India. This is an example of outsourcing. Insourcing is another form of outsourcing in which a job remains within the county, but formal office space for this activity is no longer required. The position is given to a stay-at-home spouse and retired personnel who are competent to do the work at home or in shared workspaces, and they may be scattered in many states within the country.

A good example is ticketing airway travel involving receiving instructions, discussing, and finally marking, using electronic payment, ticket preparation, and email. Further, the traveler prints his boarding pass based on his education and then travels to the airport, where face recognition admits him into his aircraft. However, an imaginary corporate office exists based on a telecommunication or web connection. When One phone for an air ticket or apply on the website, one of the in-shoring households receives the call, answers, and tickets it. An algorithm that fixes who should receive the call and execute the job directs the phone call to this individual. The company no longer requires a brick motor office to function here. He can hire the activity dispersed throughout the country at a much cheaper cost. I have used another example here, but Thomas Friedman gave an example of how a courier company UPS doubled up as a computer service company for Toshiba. Imagine what happens to a physical city when we are inshore all Administration of the Government of India and all State Governments in India to the brick and motor collectorates and secretariat; these buildings transform into recreation areas for the local population. They are not required, and so also most government servants, but administrative services are performed much more efficiently and cost-effectively.

Flattener # 9 In-Forming Google, Yahoo, and MSN Web Search

According to Friedman, in-forming through web search is a form of collaboration that is a foundation of globalization and global cities. In-forming is the individual personal analog to open-sourcing, outsourcing, supply chaining, and offshoring. In-forming is the ability to build and deploy Oner supply chains, a chain of information, knowledge, and entertainment that will make One part of the global economy. In-forming is about self-collaboration, an Oner researcher, and the editor, without going to the library. It is about seeking like-minded people and communicating using social media. It has more functions than One can imagine. Web search can also act as a personal and free detective.

Flattener # 10 the Steroid Digital, Mobile, Personal and Virtual

The tenth Flattener is a piece of digital information and its access to all through smartphones in One's pocket. The world is digitizing all information continuously and available in WWW. One can access it through Oner's smartphone even if one travels on a bullet train at 150 miles per hour. One can do all nine flatteners discussed above through the smartphone. The smartphone is One the most extensive library in the world; it is the biggest theater in the world with the choicest cinema, the enormous music concert hall in the world. Many of these digital resources are free of cost. So, the intelligent mobile phone should be part of smart global megacities.

The many study cities were global before the colonial period, like ancient cities in China, India, and many other countries. The town was international but not megacities then and has even unique traditions such as the "City of Truth for Kozhikode Kerala India," coined by travelers before the colonial era a few hundred years before. The rediscovery of the ancient cultural system of the global city is the first step in creating global megacity. With more than 100% smartphone ownership in the study area of Indo-Pacific, the world is already interconnected to megacity, but more actions are required to make it an accurate and most cost-effective global megacity leveraging opportunities in the global workplaces are scattered in the study megacity. Globally operable workflow applications are already available but not yet learned and widely used in the study area for global interconnectedness and collaborative economic development. The Study City is yet to develop a self-organizing collaborative community for international services, trade, app development, and production by designing appropriate intelligent communities. Study City is yet to practice as part of outsourcing. Study City is yet to practice and reap the benefit of offshoring foreign manufacturing and services by making factors competitive and cheaper elsewhere. Study City is yet to practice and reap the advantage of being part of global supply chains. Study City is yet to practice and reap the benefit of being part of insourcing. Study City is yet to practice and reap the advantage of being part of In-forming. Study City is yet to practice and reap the benefit of being part of personal and virtual computing.

Here Thomas Friedman's book explains how a megacity system of Indo-Pacific can become a global city system without mentioning the word megacity.

4.9 New Town Megacities in the Indo-Pacific as Charter Cities [40, 45, 46]

The megacity endowment of a country depends on the demography, with countries having large populations, such as China and India having a more significant number of megacities. In contrast, countries such as Russia, with vast geographic areas and fewer people, have fewer megacities, and Australia, with less population with a large geographic area, does not have even one megacity. It is a fact that a megacity like Tokyo performs as a more robust economic engine none of any village and a small city, say 500,000 plus population, can accomplish. Further, vaster the land area calls for megacity creation for equitable incidence of socio-economic development benefits using more robust economic growth engines.

Since the economy of scale matters in the Indo-Pacific, the new town megacity must be studied for feasibility; for example, Shenzhen is a megacity new Town in China specializing as an electronic manufacturing hub. Since Shenzhen is functionally integrated with the financial capital of Hong Kong, its growth of a higher order is well assured with formidable export figures. These proposed new town megacities can be in strategic locations of the Indo-Pacific and can be a well-thought-out economic engine with complementary industrialization. I suggest two new town megacities; one of their feasibilities will be evaluated in the part Megacity Case Studies.

Vladivostok in the Indo-Pacific is now being connected by a new sea route as part of Indo-Russian collaboration, which can be navigated in 24 days instead of the current 60 days through gateways ports of many other megacities. Russia manufactures high-end and less costly defense equipment for India. Their S 400 system and recently tested multiple nuclear-headed hypersonic missiles make them one of the leaders in defense industries in the world. India is dependent on Russian weapons. I recommend that if there is a multi-country defense equipment manufacturing and service megacity in Vladivostok as a collaboration of Russia, India, and many Indo-Pacific countries participating can be an ideal proposition since the security environment of Indo-Pacific countries demands each country in the Indo-Pacific acquire defense equipment to fight against any aggression can be solved. The need for such a New Town megacity is further substantiated by Military spending of the world in 2021 at $ 2.1 trillion, and the first five largest spenders published by the Stockholm Peace Research Institute were the US, China, India, UK, and Russia all of them part of Indo-Pacific. India's defense spending was $ 76.6 billion, while the US and China were much higher, accounting for 52% of the total $ 2.1 trillion. While in economic activities, many countries like to collaborate; in war, each country is alone and must fight their war alone, just like Ukraine fought a war alone with the Russian invasion and India and China with a massive build-up of defense forces facing each other in the cold.

And inhospitable western Himalayas. Since defense technology uses high-end education and research, we can also have an education hub specializing in STEM education in Vladivostok. This can be free.

Another possibility of Indo-Pacific a second megacity is in the strategic location of Andamans. This lowers the distance between two megacities Bangkok in Thailand and Jakarta in Indonesia, with the Indian subcontinent and many other ASEAN countries nearby. This can be a multi-functional and diverse Tourist cum Industrial and commercial megacity with an international education hub for STEM education and health tourism again accessible. This can be the gateway of India to ASEAN within 60 to 90 min by flight.

The chartered cities concept is an institutional setup for the proposed megasocieties. A charter city functions based on the city's charter document, which fully considers interests of all stakeholders of the proposed city, irrespective foreigners or locals, who want to invest and live in these cities and protect all their interests. It will not follow the general law based on the country's constitution. In this public law, the Central Government, State Government, and City Governments share the Governing and management functions based on what the state, central, and city functions should perform. They will have a unique schedule called state, mid, and city schemes with specific powers and responsibilities. The charter can be designed to function as a compelling global city, meeting the needs and fears of foreign and local participants. It is designed to attract international capital and institutions to cooperate with this charter and participate in global providers of products and services at the least cost. Unlike city laws under the country's constitution, this charter provides for autonomy, with less independence and limited power delegated to the city by the state. Many countries had special economic zones with a limited delegation of freedom to participants. "A charter city is a new type of special zone that can serve as an incubator for reform. In partnership with credible allies, which can be from as many countries as possible, a developing country can pursue reforms in a special zone large enough to one day accommodate a city with millions of residents originally to ten million in these cities." It is a greenfield project and not a brownfield. The formal rules in a charter city and the norms that these rules encourage can differ markedly from those prevailing elsewhere in the country. These rules can nevertheless be legitimate in the eyes of the migrants to the zone, just as the regulations in high-income countries are legitimate in the eyes of the few immigrants that manage to move from less developed countries.

Economist Paul Romer proposes founding many new charter cities in developing countries, especially South America, which supplies many migrants to the USA. This can stop or control the limited extent of immigration to the USA. Romer suggests that a developing country pass a law that sets aside a tract of land for a new charter city. A developed third-party guarantor government would administer this charter city, and citizens from the host country (and other countries) could move in and out as they please. The point of the charter cities idea is to give citizens a choice about where they want to live and to provide the basic rules and amenities required for economic growth. Ideally, by establishing a city with highly developed regulations and governance in an underdeveloped region, living and working in a charter city may provide a closer and more attractive alternative to moving far away to more developed countries.

A city charter is a unique document that, in many ways, acts as a constitution for the city adopting it. It can only be adopted, amended, or repealed by a majority vote of a city's voters. The primary advantage of a charter is that it allows greater authority for a city's governance than that provided by state law.

A charter transfers the power to adopt municipal legislation from the state legislature to the city adopting it. A municipality operating under a charter can acquire complete control over its Municipal affairs. These affairs are unaffected by the general laws passed by the state legislature on the same subject matter. This, in effect, gives the local voters more control over their local government and the city's affairs. However, a city operating under a charter is still subject to the general laws, as passed by the state legislature, on matters that are not municipal and are of state-wide concern.

All indices of easy-to-do business are fully honored in a chartered city since it is open to reforms and the town decides on it, not the legislative assembly of the state where the city is located or the Central Government. The threshold issue is whether there is a conflict between state law and a charter city enactment. The court examines the rules and regulations laid down by the Charter to see whether it conforms to the constitution, and if not, the Charter laws are discarded.

Some rules monopolize urban services like electric generation and supply as a sole public sector domain where private enterprises are not allowed. "As a result, the citizens of less developed countries often end up paying the most for basic goods such as electricity, if they have access to it at all. The places where rules are weak and inefficient are also where job-creating investment could offer the highest returns. Nevertheless, firms build new factories, not in the parts of the world where people are most eager for formal-sector jobs but where electricity is inexpensive, people and property are safe, and the rules for doing business are straightforward, predictable, and efficient. Though better rules should be easy to replicate, experience shows that social groups at all scales, from firms to cities to nations, have great difficulty achieving the consensus required to change their rules and norms. In this setting, competition between different jurisdictions –each trying to attract new members by implementing better rules – can be a powerful source of human progress. This competition is compelling if new entrants, new start-up jurisdictions, can challenge incumbents."

Structure of a Charter City.

A charter city's governance structure could vary significantly following what is laid down in the country's constitution for the city depending on where it is established, "but all charter cities would share four common elements. The first is an undeveloped piece of land large enough to host an entire city eventually. The good target size is 1,000 square kilometers or multiple; the size of Hong Kong and Singapore is higher density than existing and more compact development with adequate public spaces. The second common element is a charter that pre-specifies the general rules that would apply there. The third element is a commitment to choose, backed by both voluntary entry and free exit for all residents, employers, and investors. The fourth is a commitment to the equal treatment of all residents under the law."

"The broad commitment to choose means that only a country that wants to create a charter city will offer the land. It also means that only people and firms who make an affirmative decision to move to the new town will live under its rules – staying if those rules are as good as those offered by competing cities. A charter or constitution should outline general principles and describe the process of establishing and enforcing a city's detailed rules and regulations. It should provide a foundation for a legal system that will let the city grow and prosper. This legal system, which a partner country might give instant enforcement credibility, will be essential for attracting foreign direct investment in long-lived urban infrastructure. Of course, a charter should not specify a centralized economic plan. It should not contain rules for a detailed spatial plan for the city, which determines what people can do in every location. The regulations encouraging economic opportunities and vibrant city life will leave broad scope for experiments and let competition and choice determine which experiments persist. Moreover, in.

In a city where population and income per capita will increase, the rules should allow much more flexibility and change required in a less dynamic environment. Participating nations have three distinct roles: host, source, and partner. The host country provides the land. The source country supplies the people who move to the new city. The partner country helps to ensure that the charter will be respected and enforced for decades into the future. Because these roles can be played by a single nation or by several countries working together as partners, there are many potential charter city arrangements. One possibility is for one country to assume all three roles, much as China did in establishing the special economic zone where the new city of Shenzhen emerged, with the Central Government acting as a "partner" to the local authorities. India is considering such a path, using innovative governance structures and public–private partnerships to create new cities on greenfield sites. The key to going it alone is the ability of the Central Government to credibly commit to would-be residents and investors that the special rules in a new reform zone will be upheld. In Romer's conception, three main factors exist in creating a charter city. First, there is a developing host country. The host country provides the land and designates that land as a special reform zone, subject to the foundational set of rules. Second, the developed guarantor country administers the region with a board of governors and an appointed chairperson, like the Federal Reserve System in the United States. Third, the source country will be. Where the charter city's residents come, this may be from the host country, but they may also be many source countries."

"In practice, some countries have been receptive to Romer's idea. After a meeting of Romer with President Marc Ravalomanana, Madagascar considered creating two charter cities. Still, the plan was scrapped when the political leadership that supported the idea was removed from power. The government of Honduras has considered creating a charter city, though without the oversight of a third-party government. In 2011 Honduras made the necessary legal changes. Romer served as chair of a "transparency committee" but resigned in September 2012 when the Honduran Government agency responsible for the project signed agreements with international developers without the committee's knowledge. In October 2012, the Honduran Supreme Court declared charter cities unconstitutional because the laws of Honduras would not be applicable here."

4.10 Smart Global Megacities [47, 49]

Smart global megacities will have 100% smartphone ownership with higher broadband connectivity in every building, whether a house, factory, hospital, or airport. There shall be a bandwidth below which is not allowed to operate in a megacity.

Economic, social, political, and cultural activities and environmental resources management functions will be conducted by a strategically located system of functional IOTs connected to an ICT system. Smartphone apps articulate this general connectivity by ICT and IoT systems with voice interfaces like Alexa, Google Assistant, or Hub. Local universities will be involved in producing relevant E-Democracy and E-Governance Apps for use by inhabitants for various day-to-day activities and functions. Spatial Decision Support Systems using GIS in smartphone apps will be used extensively in E-Governance and E-Democracy. An attempt will be made to use every available network cost-effectively and efficiently. Smart communities, universities located in the megacities, and the Chamber of Commerce and Industry will be given all critical roles in the Megacity for its future.

A Smart City System comprises six fundamental building blocks: (I) intelligent people, (ii) innovative city economy, (iii) smart mobility, (iv) smart environment, (v) smart living, and (vi) innovative governance. These six building blocks are closely interlinked and contribute to the 'Smart City The system.' Some authors treat the six elements of a smart city system equally [50–55]. However, I give prominence to smart people because, without their active participation and involvement, a Smart City System would not function in the first place. A Smart City System will risk its efficient functioning without Smart People.

We define intelligent people of the global megacity as a part of a smart community of people. The sum of smart households, which promotes global/international city activities by relating and connecting local economic and cultural activities to the world, is a smart community that can interact face to face in a walkable space of a 1.5 to 2 km radius around a major activity center. Here smart households cannot be an isolated entity but an integral part of a smart local society. Local institutions such as academic and business institutions participate with the regulative government to make people smart.

5 Evolution of QUAD and the Program Region of Indo-Pacific

QUAD is originally a Latin word in English, meaning the number FOUR. In Geographic Information System, QUAD can be the Quadrilateral object; in cartography, Quadrilateral can be how four-point objects, namely capital cities, the seat of governance of four distant countries covering Indo-Pacific, are interconnected as in the map below. In architecture, QUAD can be a courtyard; in urban design, it can be a square or public realm where people meet. For scholars of international security and military sciences, it can be Quadrilateral Security Dialogue, and this book deals with the emerging regional development architecture through the prism of megacity system architecture; it is all about how life and work of the people of the Indio Pacific can be changed as a collaborative effort led by the US, Japan, Australia, and India.

In 2004, Australia, India, Japan, and the United States came together spontaneously out of compassion to manage the humanitarian assistance and disaster relief efforts in the aftermath of the 2004 Indian Ocean tsunami. They used all the appropriate capacity they had in this joint and synchronized effort for the Tsunami response mission in 2004 and successfully executed the first Quadrilateral engagement for the Indo-Pacific. It was so well executed as if these four countries were from one family of blood relatives working for a common cause of their big family of the Indo-Pacific irrespective of the fact that the society there is multi-racial and diverse politically, ecologically, economically, and culturally. This mental image retained was so satisfying to these four countries that it persisted as an idea among strategists of the four countries.

By 2006, then-Prime Minister of India, Manmohan Singh, on a state visit to Japan with his newly elected counterpart, Prime Minister Shinzo Abe, noted, in a joint statement, the "usefulness of having dialogue among India, Japan and other like-minded countries in the Asia–Pacific region on themes of mutual interest." With the United States and Australia making up these "other like-minded countries in the Asia–Pacific," the QUAD was born that year.

By 2007, the QUAD had two components. This was diplomatic—an informal meeting of officials from all four countries on the sidelines of the East Asia Summit in Manila or other places. The second was military—a joint exercise involving all four countries and Singapore under the aegis of the then-bilateral US-India MALABAR Exercise. Japan, Australia, and Singapore were invited as "non-permanent partners" in 2007; MALABAR was officially upgraded to a trilateral exercise in 2015 when Japan became a "permanent partner." This military and security aspect of such collaboration is not part of this book, including development like AUKUS.

There was a setback for QUAD by 2008, and it became dormant due to domestic political compulsions. It forced the countries to take a step back from the QUAD. Prime Minister Abe—seen as the driving force behind the grouping—resigned from his position in September 2007. Prime Minister Singh faced backlash from the coalition political alliance over growing US-India ties and the QUAD. However, the QUAD was put down after Australian officials, sensitive to China's vocal concerns over the QUAD, declared that they "would not be proposing to have a dialogue of that nature" again.

In the years since, the idea of resurrecting the QUAD has continued to linger and gain speed. Since 2008, India has deepened its ties with all three countries through various bilateral and trilateral mechanisms. The rise of China and its increasingly assertive behavior, which has threatened the critical interests of all four countries, has also given additional impetus to restart the group. On November 12, 2017, an entire decade after the informal meeting of the QUAD, the four countries announced that officials from all four countries had met to discuss "issues of common interest in the Indo-Pacific region." Since its resurrection, the QUAD has steadily grown in its engagements.

Since 2017, the QUAD has risen steadily from the government's low-level official level to the ministerial and leader's level and expanded its cooperation in scope and substance. During the leader-level meeting due to the COVID-19 aftermath, the QUAD announced the creation of three working groups—on COVID-19 vaccines, climate change, and critical technologies—to underscore its evolution and utility to the Indo-Pacific region. The last summits at the country leadership level further expanded its substantive cooperation, announcing new initiatives among the existing workings groups and creating new workings groups as below.

The Indo-Pacific is a well-connected region with good infrastructure and economies like China and India, with the fastest GDP growth rate globally. The following maps give the existing and proposed connectivity of the Indo-Pacific region. These connections are toward twenty-one megacities, as shown in the map below (Figs. 34, 35, 36, 37 and 38).

The only thing happening is that the trade between India and Pakistan under political compulsion has been reduced directly but increased via Dubai port. This should be read concerning ambitious Belt and Road Initiative as shown below in Pakistan.

Despite two big armies standing eyeball estimate in the eastern Himalayas, the trade between these two countries has not stopped but increased. The details are given below. This is a manifestation of the megacity system-based economies.

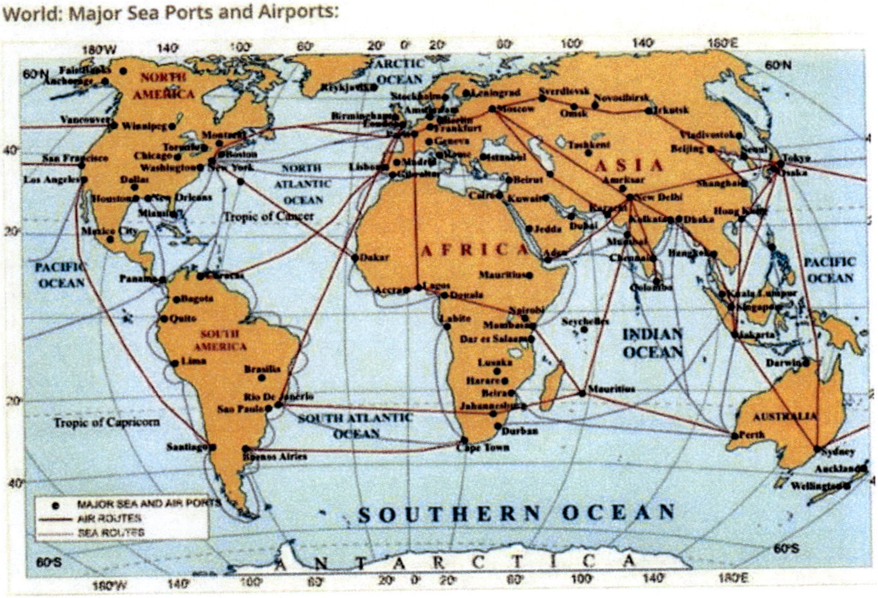

Fig. 34 Major seaports and airports

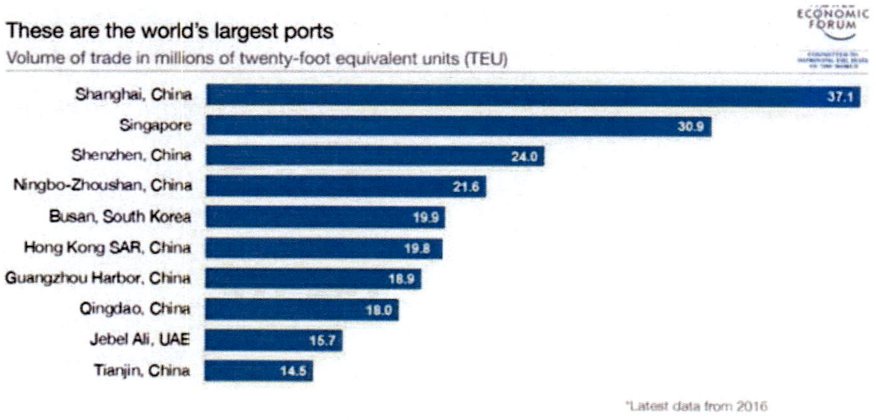

Fig. 35 Ports by importance

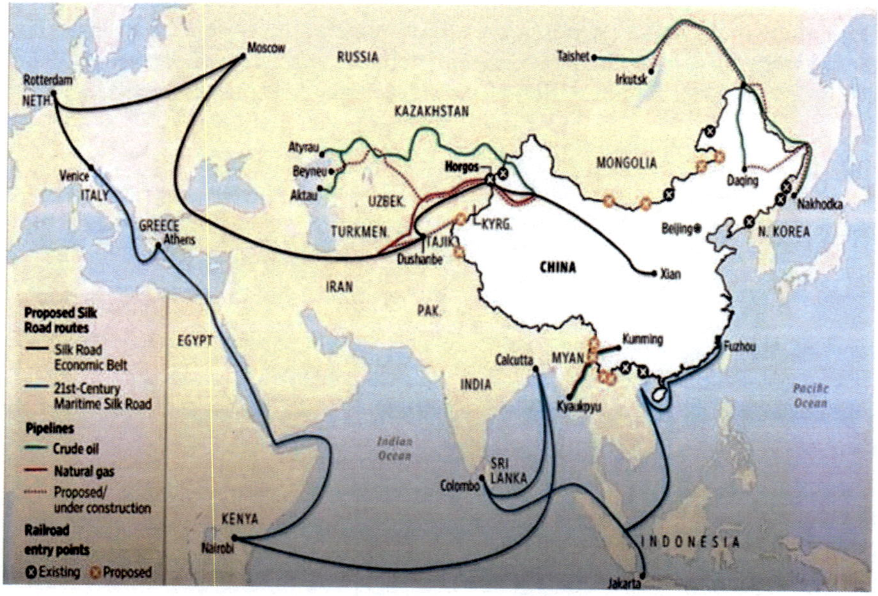

Fig. 36 Silk road ancient and current. *Source* https://www.Onetube.com/watch?v=pUf17M3GLaM

5.1 QUAD Working Groups

QUAD Leaders' policies, strategies, financing, and action programs are based on the constituted working groups. A brief introduction to these workgroups below enlightens the scope of these functional groups.

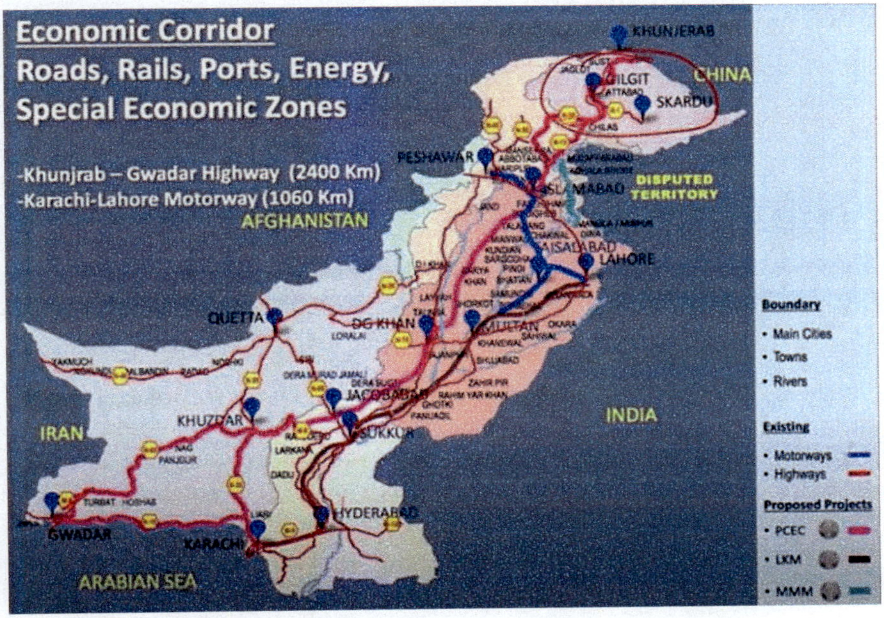

Fig. 37 Economic Corridor Pakistan. *Source* https://nha.gov.pk/projects

1. QUAD Vaccines Experts Group.

This workgroup is to fulfill the target of one billion COVID-19 vaccine doses distributed by the end of 2022 in the Indo-Pacific, with investments of $100 million for COVID-19 vaccines, logistics expertise from Japan, and a management system focusing on last-mile connectivity from Australia and vaccine manufacture from India and vaccine patent from the USA. QUAD countries will jointly build and conduct a pandemic preparedness exercise in 2022.

The United States, through the DFC, will work with Biological E Ltd. to finance increased capacity to support Biological E's effort to produce at least one billion doses of COVID-19 vaccines by the end of 2022 with Stringent Regulatory Authorization (SRA) of the World Health Organization (WHO) Emergency Use Listing (EUL), that includes the Johnson & Johnson vaccine.

Japan, through JICA, is in discussions to provide concessional yen loans for the Government of India to expand manufacturing for COVID-19 vaccines for export, with a priority on producing vaccines that have received authorization from WHO Emergency Use Listing (EUL) or Stringent Regulatory Authorities.

QUAD partners will ensure expanded manufacturing will be exported for the global benefit and procured through key multilateral initiatives, such as COVAX, that provide life-saving vaccines for low-income countries and countries in need.

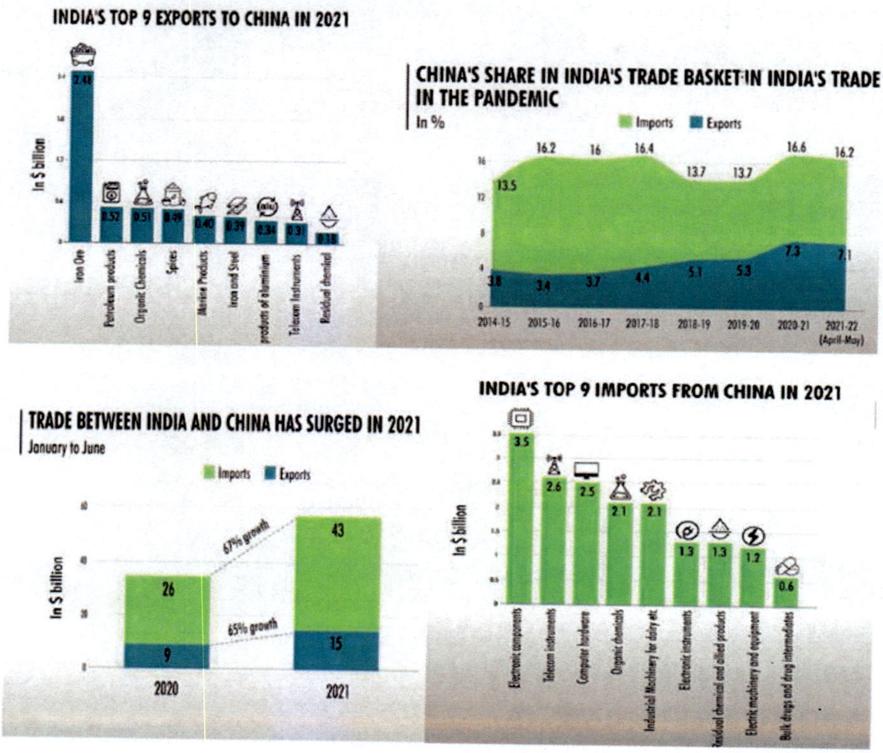

Fig. 38 India Trade to China 2021

QUAD partners will also cooperate to strengthen "last-mile" vaccination, building on existing health security and development programs, and across our governments to coordinate and improve our programs in the Indo-Pacific.

- This includes supporting countries with vaccine readiness and delivery, vaccine procurement, health workforce preparedness, responses to vaccine misinformation, community engagement, immunization capacity, and more.
- Australia will contribute US$77 million for the provision of vaccines and "last-mile" delivery support with a focus on Southeast Asia, in addition to its existing commitment of US$407 million for regional vaccine access and health security, which will provide full vaccine coverage to nine Pacific Island countries and Timor-Leste, and support procurement, prepare for vaccine delivery and strengthen health systems in Southeast Asia.
- Japan will assist vaccination programs in developing countries such as the purchase of vaccines and cold-chain support, including through the provision of grant aid of $41 million and new concessional yen loans, ensuring alignment with and support of COVAX.

- The United States will leverage existing programs to boost vaccination capability further, drawing on at least $100 million in regional efforts focused on immunization.
- The commitment of QUAD will be implemented by the launch of a senior-level QUAD Vaccine Experts Group, comprised of top scientists and officials from our governments. This group will support QUAD cooperation in the long term and use science and evidence to:
- Design and implementation plan for the QUAD COVID-19 vaccine effort; identify hurdles impeding vaccine administration in the region.
- Work with financers and production facilities to monitor timely and sufficient. Capacity expansion will lead to broader distribution of safe and effective vaccines.
- Share governmental plans to support Indo-Pacific health security and COVID-19 response and identify practical cooperation on "last-mile" delivery for hard-to-reach communities in need.
- Strengthen and support the life-saving work of international organizations, including the WHO, COVAX, Gavi, CEPI, UNICEF, the G7, ASEAN, and governments. Furthermore, calls on other countries to do the same.
- Make additional concrete recommendations before the end of the year.

2. Climate Change Working Group.

This program envisages the formation of a Green Shipping Network. The region's many shipping ports will establish a network dedicated to greening and decarbonizing the shipping value chain. QUAD will establish a Clean Hydrogen Partnership to strengthen and reduce costs across all elements of the clean hydrogen value chain. It includes techno development and efficiently scaling up clean hydrogen production, developing regional delivery infrastructure, and boosting demand for pure Hydrogen. It will create a Climate and Information Services Task Force on Climate change.

- Cooperation, both among QUAD leader countries and with other countries, to strengthen implementation of the Paris Agreement, including keeping a Paris-aligned temperature limit within reach.
- Working with other countries to support, strengthen, and enhance climatic actions globally.
- We are committing to advancing low-emissions technology solutions to support emissions reduction.
- Cooperation on climate mitigation, adaptation, resilience, technology, capacity-building, and climate finance.

3. Disaster-resilient Infrastructure Working Group:

Military disaster relief involving humanitarian Assistance and disaster relief in QUAD countries has evolved to considerable maturity in the Indo-Pacific in its many past operations. These defense forces have started using shared infrastructure and protocol with the Malabar exercise. Military-to-military cooperation in disaster relief was not initially a strategy for regional engagement, but with QUAD, in place, it is changing. The central rationale is alleviating suffering, preserving human dignity, and, most importantly, saving lives.

- It is strengthening international humanitarian action, notably in the Indo-Pacific region.
- They are reducing disaster risk through the implementation of long-term strategies and investment in capabilities.
- supporting other countries to better respond to disasters through effective Humanitarian Aid/Disaster Relief; and
- supporting rapid recovery and redevelopment following disasters

4. Critical and Emerging Technology Working Group:

Since its establishment in March 2021 using virtual meetings, this working group has remained the primary mechanism for technology cooperation in the QUAD. QUAD Critical Technologies Working Group encourages cooperation on telecommunications deployment, diversification of equipment suppliers, and future telecommunications through close partnerships with private sectors and industry. It facilitates collaboration to monitor trends and opportunities related to critical and emerging technology developments, including biotechnology. Convene dialogues on essential technology supply. Partner countries remain committed to developing necessary and emerging technology that yields tangible benefits for society, developed through a multi-stakeholder approach that is adaptive, dynamic, and aligns with universal values, including respect for freedom of expression and privacy.

- Develop a statement of principles on technology design, development, and use.
- Facilitate coordination on technology standards development, including between our national technology standards bodies and working with a broad range of partners.
- Encourage cooperation on telecommunications deployment, diversification of equipment suppliers, and future telecommunications, including through close collaboration with private sectors and industry.
- Facilitate cooperation to monitor trends and opportunities related to critical and emerging technology developments, including biotechnology.
- Convene dialogues on critical technology supply chains.

The QUAD members support technological design, development, governance, and using approaches that promote shared values, including individuals' autonomy, agency, and dignity. The statement highlighted that technology design, development, governance, and use should be an equitable and inclusive process that neither involves nor results in unfair discriminatory action. Misuse of Technology or abuse for malicious activities such as authoritarian surveillance and oppression, for terrorist purposes, or disseminating disinformation shall be excluded and are committed to building trust, integrity, and resilience.

According to the official release, QUAD members acknowledge that technology ecosystems should be built upon trust, integrity, and resilience to foster innovation and highlight the need to support openness and interoperability to enable collaboration and to welcome innovators who are both diverse in their gender, race, and ethnicity, as well as in the size, structure, and age of their organizations.

QUAD members expect technology suppliers, vendors, and distributors to produce and maintain secure systems and to be trustworthy, transparent, and accountable in their practices. In contrast, technology developers should build safety and security-by-design approaches so that robust safety and security practices are a part of the technology development process. The statement highlighted that Illicit transfer or theft of technology is a common challenge that undermines the foundation of global technological development and should be addressed.

The QUAD supports resilient, diverse, and secure technology supply chains—for hardware, software, and services—that are vital to the shared national interests of the member states. Close cooperation on supply chains with allies and partners who share the combined values of the QUAD members will enhance our security and prosperity and strengthen allies' capacity to respond to international disasters and emergencies. The members also encouraged healthy competition and international collaboration to advance the frontier of science and technology.

5. QUAD principles of the fair and open marketplace are a cornerstone of innovation and inclusive prosperity.

The QUAD principles suggested that free and fair market competition in which the best technical solutions succeed, including with the help of transparently awarded state incentives, would be prioritized, and highlighted the following points:

The development of competitive technology ecosystems that welcome new market entrants, including start-ups, and enhance innovation, resilience, and sustainability would be encouraged. The importance of industry-led, consensus-based, multi-stakeholder approaches to international standards development that foster interoperability, compatibility, and inclusiveness would be affirmed. The QUAD would remain committed to reducing barriers to data and knowledge sharing for research projects and more significant innovation while protecting research security. The partners would also intensify efforts to facilitate the exchange of researchers and the movement of highly skilled personnel to enhance science and technology collaboration. Remain committed to developing shared research and development agendas, including joint projects and joint capacity-building where possible, that align with foundational scientific principles, further our shared values, and promote workforce diversity and education. "Together, we will continue to keep pace with changes in the global economy and innovation processes by monitoring future trends and harnessing the opportunities to advance technologies to address many of our shared objectives: security and resilience, sustainability, economic inclusion, and health and wellbeing," stated the White House. The White House also highlighted the commitments to foster technology development that accelerates economic progress and serves the needs of all citizens and welcomed all nations to join the QUAD members in pursuit of a shared vision for technologies guided by these principles.

6. QUAD Statement of Principles on Technology Design, Development, Governance, and Use:

This Statement of Principles is a collaborative effort to firm that any technological progress in the QUAD countries is shaped by their shared democratic values and respect for universal human rights. Technology should make the lives of its citizens more secure, prosperous, and rewarding. The Observer Research Foundation, India, the National Graduate Institute for Policy Studies, Japan, and the United States Centre for a New American Security, US, have commissioned papers on the region's critical issues. These papers offer analysis and recommendations on shared challenges facing Indo-Pacific partners in the cyber and technology environment.

7. Infrastructure Working Group:

Building on the G7's announcement of Build Back Better World (B3W)—an infrastructure partnership focused on digital connectivity, climate, health and health security, and gender equality infrastructure—the QUAD will rally expertise, capacity, and inference to strengthen ongoing infrastructure initiatives in the region and identify new opportunities to meet the needs there.

The QUAD will launch the QUAD Infrastructure Coordination Group: Building on existing leadership from QUAD partners on high-standards infrastructure, a senior QUAD infrastructure coordination group will meet regularly to share assessments of regional infrastructure needs and coordinate respective approaches to deliver transparent, high-standards infrastructure. The group will also coordinate technical assistance and capacity-building efforts, including with regional partners, to ensure these efforts are mutually reinforcing and complementary in meeting the significant infrastructure demand in the Indo-Pacific.

Lead on High-Standards Infrastructure: QUAD partners are leaders in building quality infrastructure in the Indo-Pacific region. These complementary approaches leverage both public and private resources to achieve maximum impact. Since 2015, QUAD partners have provided more than $48 billion in official enhancements for infrastructure in the region. This represents thousands of projects, including capacity-building, across over thirty countries supporting rural development, health infrastructure, water supply and sanitation, renewable power generation (e.g., wind, solar, hydro), telecommunications, road transportation, and more. Our infrastructure partnership will amplify these contributions and further catalyze private sector investment in the region.

8. Establish Technical Standards Contact Groups:

The QUAD will establish contact groups on Advanced Communications and Artificial Intelligence, focusing on standards-development activities and foundational pre-standardization research. Technical Standards Contact Groups in Advanced Communication and Artificial Intelligence, Technical standards, 5G diversification, deployment, horizon-scanning, and technology supply chains for critical and emerging technologies will be addressed.

9. Publish a QUAD Statement of Principles Consult on Norms and Guidelines:

QUAD will also consult on norms, guidelines, principles, and rules for ensuring the long-term sustainability of the outer space environment. After months of collaboration, QUAD will launch a statement of principles on technology design, development, governance, and use that QUAD hopes will guide the region and the world toward responsible, open, high-standards innovation. A statement of principles on technology design, development, governance, and use regarding lofty standards of innovation.

10. Support 5G Deployment and Diversification:

Supporting the critical role of QUAD governments in fostering and promoting a diverse, resilient, and secure telecommunications ecosystem, the QUAD has launched a Track 1.5 industry dialogue on Open RAN deployment and adoption, coordinated by the Open RAN Policy Coalition. QUAD partners will jointly facilitate enabling environments for 5G diversification, including efforts related to testing and test facilities.

11. Monitor Biotechnology Scanning:

The QUAD will monitor trends in critical and emerging technologies, starting with advanced biotechnologies, including synthetic biology, genome sequencing, and biomanufacturing. In the process, they will identify related opportunities for cooperation.

12. Semiconductor Supply Chain Initiative:

QUAD partners will launch a joint initiative to map capacity, identify vulnerabilities, and bolster supply chain security for semiconductors and their vital components. This initiative will help ensure QUAD partners support a diverse and competitive market that produces the secure critical technologies essential for AI economies globally—the launch of a Semiconductor.

Supply Chain Initiative Launch a Semiconductor Supply Chain Initiative: QUAD partners will launch a joint initiative to map capacity, identify vulnerabilities, and bolster supply chain security for semiconductors and their vital components. This initiative will help ensure QUAD partners support a diverse, competitive market that produces the secure critical technologies essential for global digital economies.

13. Space Cooperation Working Group:

QUAD will also begin space cooperation for the first time. The four countries will exchange satellite data on monitoring and adapting to climate change and disaster preparedness. Capacity-building in space-related domains in other Indo-Pacific countries will be facilitated and encouraged to drive continuous improvement in areas including the adoption and implementation of shared cyber standards, development of secure software, building workforce and talent, and promoting the scalability and cybersecurity of safe and trustworthy digital infrastructure QUAD countries are among the world's scientific leaders, including in space.

The QUAD will: Share Satellite Data to Protect the Earth and its Waters: Our four countries will start discussions to exchange Earth observation satellite data and analysis on climate change risks and the sustainable use of oceans and marine resources. Sharing this data will help QUAD countries to better adapt to climate change and to build capacity in other Indo-Pacific states that are at grave climate risk in coordination with the QUAD Climate Working Group. Enable Capacity-Building for Sustainable Development: The QUAD countries will also enable capacity-building in space-related domains in other Indo-Pacific countries to manage risks and challenges. The QUAD countries will work together to support, strengthen, and enhance space applications and technologies of mutual interest.

14. Launch a QUAD Senior Cyber Group:

Leader-level experts will meet regularly to advance work between government and industry on driving continuous improvements, including adopting and implementing shared cyber standards; developing secure software; building workforce and talent; and promoting the scalability and cybersecurity of safe and trustworthy digital infrastructure.

15. People-to-People Exchange and Education:

Today's students will be tomorrow's leaders, innovators, and pioneers. To build ties among the next generation of scientists and technologists, QUAD partners announced the QUAD Fellowship: a first-of-its-kind scholarship program operated and administered by a philanthropic initiative and in consultation with a non-governmental task force comprised of leaders from each QUAD country. This program will bring together exceptional American, Japanese, Australian, and Indian master's and doctoral students in science, technology, engineering, and mathematics to study in the United States. This new fellowship will develop a network of science and technology experts committed to advancing innovation and collaboration in the private, public, and academic sectors, in their nations and among QUAD countries. The program will build a foundational understanding among QUAD Scholars of one another's societies and cultures through cohort-wide trips to each QUAD country and robust programming with each country to scientists, technologists, and politicians.

The QUAD will: Launch the QUAD Fellowship: The Fellowship will sponsor one hundred students per year—25 from each QUAD country—to pursue master's and doctoral degrees at leading STEM graduate universities in the United States. It will serve as one of the world's teaching graduate fellowships. Still, uniquely, the QUAD Fellowship will focus on STEM and unite the top minds of Australia, India, Japan, and the United States. Schmidt Futures, a philanthropic initiative, will operate and administer the fellowship program in consultation with a non-governmental task force comprising academic, foreign policy, and private sector leaders from each QUAD country. Founding sponsors of the fellowship program include Accenture Blackstone, Boeing, Google, Mastercard, and Western Digital, and the program welcomes additional sponsors interested in supporting the Fellowship.

This growing cooperation between the QUAD countries presents three significant pieces of misinformation, namely the notion that the QUAD is a "do-nothing talk shop," "a clique of selected countries, or a new cold war defense coalition against China for" for example, the new AUKUS security agreement between the United States, Australia, and the UK and "lack of utility" which undoubtedly is a misrepresentation in light of the above 15 initiatives.

The QUAD has quickly become a coalition collaborating across security areas such as AUKUS, non-security regions, and places that may overlap across both domains (such as critical technologies).

The QUAD's growth and evolution have been driven by growing and shared concerns among the four countries on China's rising assertiveness, territorial expansions, and the actions it has taken to undermine the core interests of all four members of the QUAD. However, the QUAD is not an organization against China if the Chinese system of megacity is still a producer and supplier of Industrial goods on which the world still depends. The Chinese megacity system is also part of the Indo-Pacific megacity system. QUAD intends to address the key challenges facing the Indo-Pacific region, which is not China, and provide ways and means to countries willing to collaborate. The implicit goal of QUAD is to highlight a vision shared by these four countries about a rules-based order and how it may serve the region's interests.

There were country policies for Indo-Pacific by the four QUAD countries: Australia, India, the US, and Japan. Since the commitments of these countries in their summit override all their policies based on consensus, I have not spared to point out the thinking within each country. There are many documents available on that. As a sample, I present one of four countries, namely Japan, diagrammatically below (Fig. 39).

5.2 *The First QUAD Summit on March 12, 2021 [56]*

President of the USA Joseph R. Biden, Prime Minister of India Narendra Modi, Prime Minister of Australia Scott Morrison, and Prime Minister of Japan Yoshihide Suga participated in the first Leaders' Summit of the Quadrilateral Framework, held on March 12, 2021. They discussed regional and global issues of shared interest and exchanged views on practical areas of cooperation toward maintaining a free, open, and inclusive Indo-Pacific region. They discussed contemporary challenges such as resilient supply chains, emerging and critical technologies, maritime security, and climate change leading to the discussions on ongoing efforts to combat the COVID-19 pandemic. They explored opportunities for collaboration in ensuring safe, equitable, and affordable vaccines in the Indo-Pacific region.

1. They reaffirmed their joint commitment to Quadrilateral cooperation between Australia, India, Japan, and the United States. They shared a vision for the free and open Indo-Pacific: inclusive, healthy, anchored by democratic values, and unconstrained by coercion. Recalling their successful joint efforts toward international tragedy, the Tsunami of 2004, they renewed their commitment to fighting jointly against global devastation wrought by COVID-19, the threat of climate

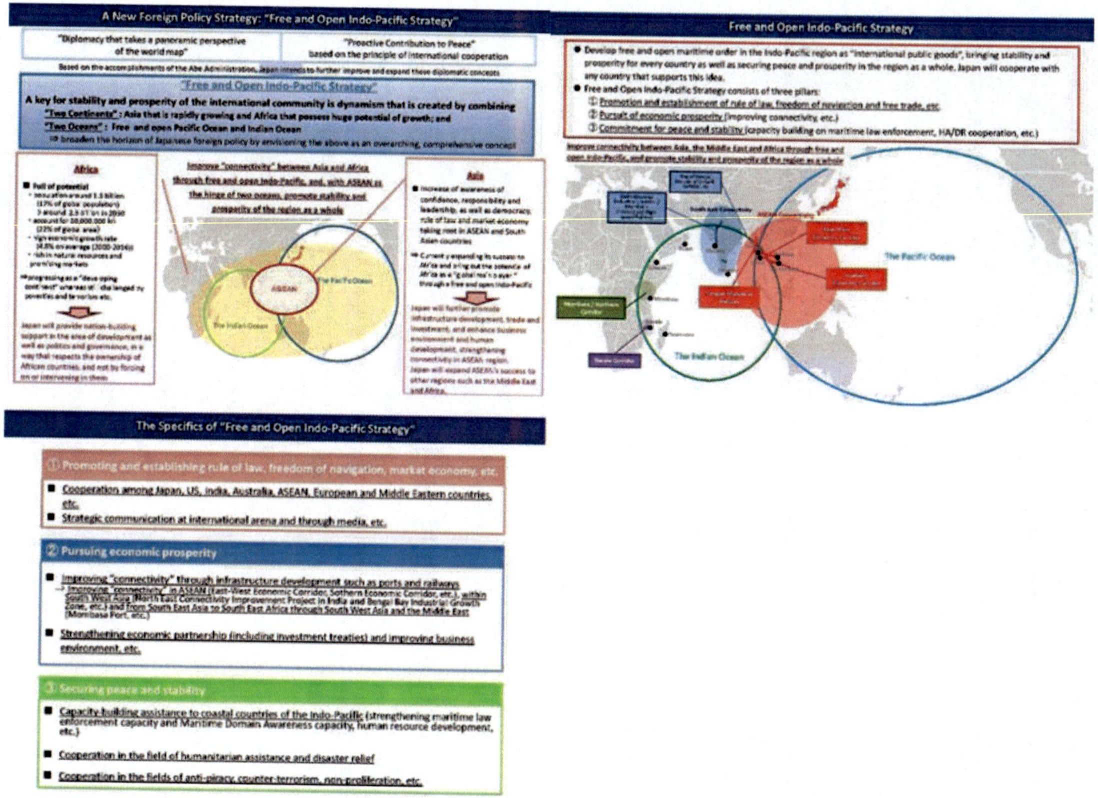

Fig. 39 Free and Open Indo-Pacific of Japanese Vision. *Source* The Government of Japan (November 2019), Toward Free and Open Indo-Pacific Sailing on Indo-Pacific Ocean Finding Opportunity for Prosperity, Tokyo

change, as well security challenges facing the region. On this historic occasion of March 12, 2021, the first-ever leader-level summit of the QUAD, they jointly pledged to strengthen their cooperation on the defining challenges of our time.

2. Together, they committed to promoting a free, open, rules-based order rooted in international law to advance security and prosperity and counter threats to Indo-Pacific and beyond. They supported the rule of law, freedom of navigation and overflight, and peaceful resolution of disputes, democratic values, and territorial integrity. They committed to working together and with a range of partners. They reaffirmed their dedicated support for ASEAN's unity and centrality and the ASEAN Outlook on the Indo-Pacific. Full of potential, the QUAD looks forward to the future; it seeks to uphold peace and prosperity and strengthen democratic resilience based on universal values.
3. Their common goals required them to reckon with the most urgent global challenges. They pledged to respond to the economic and health impacts of COVID-19, combat climate change, and address shared challenges, including in cyberspace, critical technologies, counterterrorism, quality infrastructure investment, humanitarian assistance, disaster relief, and maritime domains.
4. Building on the progress these four countries have achieved in health security, they will join forces to expand safe, affordable, and effective vaccine production and equitable access to speed economic recovery and benefit global health. With a steadfast commitment to the health and safety of their people, they also recognize that none of them can be safe if the pandemic continues to spread. They will, therefore, collaborate to strengthen equitable vaccine access for the Indo-Pacific, with close coordination with multilateral organizations, including the World Health Organization and COVAX. They are united in recognizing that climate change is a global priority and will work to strengthen the climate actions of all nations, including keeping a Paris-aligned temperature limit within reach. They will begin cooperation on the critical technologies of the future to ensure that innovation is consistent with a free, open, inclusive, and resilient Indo-Pacific. They will continue to prioritize the role of international law in the maritime domain, mainly as reflected in the United Nations Convention on the Law of the Sea (UNCLOS) and facilitate collaboration, including in maritime security, to meet challenges to the rules-based maritime order in the East and South China Seas. They reaffirmed their commitment to the complete denuclearization of North Korea following United Nations Security Council resolutions and confirmed the necessity of immediate answers to Japanese abductees. As long-standing supporters of Myanmar and its people, they emphasized the urgent need to restore democracy and the priority of strengthening democratic resilience.
5. They will redouble their commitment to QUAD engagement to advance these goals and others. They will combine their nations' medical, scientific, financing, manufacturing and delivery, and development capabilities and establish a vaccine expert working group to implement their path-breaking commitment to safe and effective vaccine distribution; they will launch a critical and emerging technology working group to facilitate cooperation on international standards and innovative technologies of the future; and they will establish a climate working group to

strengthen climate actions globally on mitigation, adaptation, resilience, technology, capacity-building, and climate finance. Their experts and senior officials will continue to meet regularly; their foreign ministers will converse often and meet at least once a year. At the leader level, they will hold an in-person summit by the end of 2021. The ambition of these engagements is fit to the moment; they are committed to leveraging their partnership to help the world's most dynamic region respond to the historic crisis so that it may be free, open, accessible, diverse, and thriving Indo-Pacific they all seek.

5.3 The Second QUAD Summit on September 24, 2021 [57]

The leaders of Australia, India, Japan, and the United States convened in person as "the QUAD" for the first time. On this important event, they recommitted to their partnership and to a region that is a bedrock of their shared security and prosperity—a free and open Indo-Pacific, which is also.

Inclusive and resilient. Just six months have passed since their last meeting. Since March, the COVID-19 pandemic has caused continued global suffering; the climate crisis has accelerated; and regional security has become ever more complex, assessing all our countries individually and together. This cooperation, however, remains unflinching.

The occasion of the QUAD Summit is an opportunity to refocus themselves and the world on the Indo-Pacific and on their vision for what they hope to achieve. Together, they recommend promoting the free, open, rules-based order rooted in international law and undaunted by coercion to bolster security and prosperity in the Indo-Pacific and beyond. They stand for the rule of law, freedom of navigation and overflight, peaceful resolution of disputes, democratic values, and territorial integrity of states. They commit to working together and with a range of partners. They reaffirm their dedicated support for ASEAN's unity and centrality and ASEAN's Outlook on the Indo-Pacific, and they underscore their dedication to working with ASEAN and its member states—the heart of the Indo-Pacific region—in practical and inclusive ways. They also welcomed the September 2021 EU Strategy for Cooperation in the Indo-Pacific. Since their first meeting, they have tackled the world's most pressing challenges: the COVID-19 pandemic, the climate crisis, and critical and emerging technologies.

Their partnership on COVID-19 response and relief marked a historic new focus for the QUAD. They launched the QUAD Vaccine Experts Group, comprised of top experts from their respective governments, charged with building strong ties and better aligning their plans to support Indo-Pacific health security and COVID-19 response. In doing so, they shared assessments of the state of the pandemic. They aligned their efforts to combat it, reinforced shared diplomatic principles for mitigating COVID-19 in the region, and actively improved coordination of our efforts to support safe, effective, quality-assured vaccine production and equitable access in close collaborations with multilateral efforts, including the COVAX Facility. In addition to doses financed through COVAX, Australia, India, Japan, and the United States have pledged to donate more than 1.2 billion safe and effective COVID-19 vaccines globally. Moreover, to date, they have delivered seventy-nine million safe, effective, and quality-assured vaccine doses to countries in the Indo-Pacific as part of those commitments. Thanks to the QUAD Vaccine Partnership's financing of increased manufacturing capacity at Biological E LTD, additional production in India will come online later this year. In line with our March announcement and recognizing the continuing global supply gap, they will ensure this expanded manufacturing is exported to the Indo-Pacific and the world, and they will coordinate with vital multilateral initiatives, such as the COVAX Facility, to procure proven safe, effective, and quality-assured COVID-19 vaccines for low- and middle-income countries. We also recognize the importance of open and secure supply chains for vaccine production.

Despite months of pandemic hardship throughout the region and world, they have accomplished much. The QUAD leaders welcome Biological E LTD's production, including through their QUAD investments, of at least one billion safe and effective COVID-19 vaccines by the end of 2022. Today, they are proud to announce an initial step toward that supply that will immediately help the Indo- Pacific and the world to end the pandemic. The QUAD also welcomes India's announcement to resume exports of safe and effective COVID-19 vaccines, including COVAX, beginning in October 2021. Japan will continue to help regional partners purchase vaccines through a \$3.3 billion of COVID-19 Crisis Response Emergency Support Loan. Australia will deliver \$212 million in grant aid to purchase vaccines for Southeast Asia and the Pacific. In addition, Australia will allocate \$219 million to support last-mile vaccine rollouts and lead in coordinating QUAD's last-mile delivery efforts in those regions.

They will also strengthen their Science and Technology (S&T) cooperation in the areas of clinical trials and genomic surveillance so that they can accelerate their efforts to end this pandemic and build better health security. They are committed to aligning around shared global targets to help vaccinate the world, save lives now, and build back better by strengthening international health security financing and political leadership. Their countries will also conduct a joint pandemic preparedness tabletop or exercise in 2022.

They had joined forces to tackle the climate crisis, which must be addressed with the urgency it demands. QUAD countries will work together to keep the Paris-aligned temperature limits within reach and will pursue efforts to limit it to 1.5 °C above pre-industrial levels. To this end, QUAD countries intend to update or communicate ambitious NDCs by COP26 and welcome those who have already done so. QUAD countries will also coordinate their diplomacy to raise global ambition, including reaching out to key stakeholders in the Indo-Pacific region.

Their work is organized across three thematic areas: climate ambition, clean energy innovation and deployment, and climate adaptation, resilience, and preparedness, with the intent to pursue enhanced actions during the 2020s, contributing to the aim of achieving global net-zero emissions preferably by 2050, and considering national circumstances. They are pursuing nationally appropriate sectoral decarbonization efforts, including those aimed at decarbonizing shipping and port operations and deploying clean hydrogen technology. They will cooperate to establish responsible and resilient clean energy supply chains and will strengthen the Coalition for Disaster Resilient Infrastructure and climate information systems. QUAD countries will work together for successful outcomes at COP26 and G20 that uphold the climate ambition and innovation this moment requires.

They established cooperation on critical and emerging technologies to ensure that their shared values and respect for universal human rights shape how technology is designed, developed, governed, and used. In partnership with the industry, they are advancing the deployment of secure, open, and transparent 5G and beyond-5G networks and working with various partners to foster innovation and promote trustworthy vendors and approaches such as OpenVPN. Acknowledging the role of governments in fostering an enabling environment for 5G diversification, they will work together to facilitate public–private cooperation and demonstrate in 2022 the scalability and cybersecurity of open, standards-based technology. Concerning the development of technical standards, they will establish sector-specific contact groups to promote an open, inclusive, private sector-led, multi-stakeholder, and consensus-based approach.

They will also coordinate and cooperate in multilateral standardization organizations such as the International Telecommunication Union. They are mapping the supply chain of critical technologies and materials, including semiconductors, and affirm their positive commitment to resilient, diverse, and secure supply chains of essential technologies, recognizing the importance of government support measures and policies that are transparent and market-oriented.

They are monitoring trends in the critical and emerging technologies of the future, beginning with biotechnology and identifying related opportunities for cooperation. They are also launching QUAD Principles on Technology Design, Development, Governance, and Use that they hope will guide the region and the world toward responsible, open, high-standards innovation. In the future, they will deepen their cooperation in these critical areas and broaden it to new ones.

Building upon each of their regional infrastructure efforts, separately and together, they are launching a new QUAD infrastructure partnership. As QUAD, they will meet regularly to coordinate their efforts, map the region's infrastructure needs, and coordinate regional needs and opportunities. They will cooperate in providing technical assistance and empowering regional partners with evaluative tools will promote sustainable infrastructure development. They support the G7.

Infrastructure efforts and look forward to cooperating with like-minded partners, including the EU.

They reconfirmed the G20 Quality Infrastructure Investment Principles and will re-energize their efforts to provide high-standard infrastructure in the Indo-Pacific. They reaffirmed their interest in continuing their engagement with the Blue Dot Network. They emphasized the importance of supporting open, fair, and transparent lending practices aligned with international rules and standards for significant creditor countries, including debt sustainability and accountability. They call on all creditors to adhere to these rules and standards.

In September 2021, they began new cyberspace cooperation and pledged to combat cyber threats, promote resilience, and secure their critical infrastructure. In space, they will identify new collaboration opportunities and share satellite data for peaceful purposes such as monitoring climate change, disaster response and preparedness, sustainable uses of oceans and marine resources, and responding to challenges in shared domains. They will also consult on rules, norms, guidelines, and principles for ensuring the sustainable use of outer space.

They are proud to begin a new educational and people-to-people cooperation chapter as they inaugurate the QUAD Fellowship. Stewarded by Schmidt Futures, a philanthropic initiative, and with generous support from Accenture, Blackstone, Boeing, Google, Mastercard, and Western Digital, this pilot fellowship program will provide one hundred graduate fellowships to leading science, technology, engineering, and mathematics graduate students across our four countries. Through the QUAD Fellowship, the next generation of STEM talent will be prepared to lead the QUAD and other like-minded partners toward the innovations that will shape their shared future.

In South Asia, they will closely coordinate their diplomatic, economic, and human rights policies toward Afghanistan and will deepen their counterterrorism and humanitarian cooperation in the months ahead following UNSCR 2593. They reaffirmed that Afghan territory should not be used to threaten or attack any country or shelter or train terrorists, plan or finance terrorist acts. They reiterated the importance of combating terrorism in Afghanistan. They denounced terrorist proxies and emphasized the importance of denying any logistical, financial, or military support to terrorist groups that could be used to launch or plan terror attacks, including cross-border attacks. They support Afghan nationals and call on the Taliban to provide safe passage to anyone wishing to leave Afghanistan and ensure that all Afghans' human rights, including women, children, and minorities, are respected.

They also recognize that their shared futures will be written in the Indo-Pacific, and they will redouble their efforts to ensure that the QUAD is a force for regional peace, stability, security, and prosperity. Toward that end, they will continue to champion adherence to international law, mainly as reflected in the UN Convention on the Law of the Sea (UNCLOS), to meet challenges to the maritime rules-based order, including in the East and South China Seas.

They affirm their support to small island states, especially those in the Pacific, to enhance their economic and environmental resilience. They will continue their assistance with Pacific Island countries on responses to the health and economic impacts of COVID-19 and on quality, sustainable infrastructure, as well as a partner to mitigate and adapt to the effects of climate change, which poses severe challenges for the Pacific.

They reaffirmed their commitment to the complete denuclearization of North Korea under United Nations Security Council resolutions and confirmed the necessity of immediate solutions for Japanese abductees. They urged North Korea to abide by its UN obligations and refrain from Provocations. They also call on North

Korea to engage in substantive dialogue. They are committed to building democratic resilience in the Indo-Pacific and beyond. They continue to call for ending violence in Myanmar, releasing all political detainees, including foreigners, engaging in constructive dialogue, and the early restoration of democracy. They further call for the urgent implementation of the ASEAN Five-Point Consensus. They will deepen their cooperation in multilateral institutions, including the United Nations, where reinforcing their shared priorities enhances the resilience of the multilateral system itself. Individually and together, they will respond to the challenges of the present time, ensuring that the region remains inclusive, open, and governed by universal rules and norms.

They will continue to build habits of cooperation; their leaders and foreign ministers will meet annually, and their senior officials will meet regularly. Their working groups will continue their steady tempo to produce the cooperation necessary to build a more substantial region.

At a time that tests us all, their commitment to realizing a free and open Indo-Pacific is firm, and their vision for this partnership remains ambitious and far-reaching. With steadfast cooperation, they will rise to meet this moment together.

5.4 *The Third QUAD Summit on March 3, 2022 [58]*

On March 3, commencing at 11:00 p.m. for approximately 70 min, Mr. Kishida Fumio, Prime Minister of Japan, with the Hon Scott Morrison, MP, Prime Minister of the Commonwealth of Australia, H.E. Mr. Narendra Modi, Prime Minister of India, and The Honorable Joseph R. Biden, Jr. President of the United States of America, held a video conference. This conference was called for and hosted by the US, and after the conference, Joint Call Readout was released. The overview of the meeting is as follows.

The QUAD leaders discussed the ongoing conflict and humanitarian crisis in Ukraine. They assessed its broader implications and that it is critical to promote further efforts toward realizing a "Free and Open Indo-Pacific." They reaffirmed their dedication to the QUAD as a mechanism to promote regional stability and prosperity.

Prime Minister Kishida stated that the aggression by Russia this time is an attempt to change the status quo by force unilaterally, shakes the foundation of the international order, and strongly condemned Russia and that Japan has been promptly implementing harsh measures, aligning closely with the international community including G7. Prime Minister Kishida stated that he had recently conveyed to Ukrainian President Zelensky his intention to provide $100 million in emergency humanitarian assistance to the Ukrainian people, who are facing a national crisis, in addition to the loan of not less than $100 million, and that he had expressed his willingness to accept displaced persons.

They agreed to stand up new humanitarian assistance and disaster relief mechanism which will enable the QUAD to meet future humanitarian challenges in the

Indo-Pacific and provide a channel for communication as they each address and respond to the crisis in Ukraine.

The four leaders concurred that toward the realization of a "Free and Open Indo-Pacific," the four countries will steadily advance various practical cooperation that they have been promoting in such fields as health, security, and COVID-19 response measures including vaccine-related support, infrastructure, climate change, and critical and emerging technologies. They also concurred on the importance of continuing to contribute to the region as the QUAD positively and concurred to stand up a new humanitarian assistance and disaster relief mechanism.

The four leaders concurred to hold the next QUAD Summit Meeting in Tokyo in the coming months and confirmed that the four countries would work closely together for its success.

5.5 *The Fourth QUAD Summit is on May 24, 2022, in Tokyo [59–61]*

The leaders of the QUAD nations—Australia, India, Japan, and the United States—met on May 24, 2022, in Tokyo for the fourth time and the second time in person.

1. The Indo-Pacific Partnership for Maritime Domain Awareness

In close consultation with regional partners, the Indo-Pacific Partnership for Maritime Domain Awareness (IPMDA) will offer a near-real-time, integrated, and cost-effective maritime domain awareness picture. This initiative will transform the ability of partners in the Pacific Islands, Southeast Asia, and the Indian Ocean region to monitor the waters on their shores fully and, in turn, to uphold a free and open Indo-Pacific. Through investment in IPMDA over five years, the partnership will innovate upon existing maritime domain awareness efforts, rapidly bringing emerging technologies to bear for the greater good of the Indo-Pacific community. Moreover, it will integrate three critical regions—the Pacific Islands, Southeast Asia, and the Indian Ocean region—into the Indo-Pacific.

2. The QUAD Fellowship

QUAD leaders are proud to open applications for the QUAD Fellowship, which will sponsor one hundred American, Australian, Indian, and Japanese students to study in the United States each year for graduate degrees in science, technology, engineering, and mathematics (STEM) fields starting from 2023.

3. The QUAD Vaccine Partnership and Global Health Security

The United States will fund COVID-19 boosters and pediatric doses to countries of greatest need, including in the Indo-Pacific. The Biological E, Ltd facility in India will continue using expanded vaccine-manufacturing capacity as part of the QUAD Vaccine Partnership to support building sustainable manufacturing capacity, including booster shots, which will yield long-term benefits in the fight against

COVID-19 and future pandemics. The Japan Bank for International Cooperation (JBIC) and EXIM India decided to support a $100 million facility to bolster the Indian healthcare sector, including global capacity for COVID-19 countermeasures.

4. Climate

The QUAD will broaden and elevate to combat climate change by inaugurating two ministerial initiatives in the coming months. US Secretary of Transportation Pete Buttigieg plans to convene QUAD Transportation Ministers and relevant stakeholders to accelerate the work of the Shipping Task Force, including progress toward developing Green Shipping Corridors among QUAD countries.

US Secretary of Energy Jennifer Granholm plans to meet with QUAD Energy Ministers to catalyze QUAD's efforts to deploy clean hydrogen, minimize methane emissions, and develop a 10-Year Clean Energy Supply Chain Plan. The QUAD will cooperate on disaster risk reduction for extreme weather events, including through the Coalition for Disaster Resilient Infrastructure (CDRI), building on the QUAD's joint session on strengthening Indo-Pacific infrastructure and communities at the International CDRI conference. The QUAD Climate and Information Service Task Force, dedicated to integrating and facilitating climate information services to the broader Indo-Pacific, will convene other Indo-Pacific countries at the September Asia–Pacific Ministerial Conference on Disaster Risk Reduction in Brisbane to share best practices and gauge user needs. The QUAD will advance the development of clean hydrogen and clean ammonia fuels and launch a series of roundtables on mitigating methane emissions across our liquified natural gas (LNG) sectors. The QUAD will also cooperate to enhance capacity in the broader Indo-Pacific region to participate in high-integrity carbon markets.

5. Critical and Emerging Technologies

The QUAD four countries are committed to responsible innovation in critical and emerging technologies. Since launching the Critical and Emerging Technologies Working Group at the first-ever leader-level summit of the QUAD in March 2021, QUAD partners have mapped collective capacity and vulnerabilities in global semiconductor supply chains and launched the Common Statement of Principles on Critical Technology Supply Chains, which will provide a cooperative foundation for enhancing supply chain resilience in the region. At the same time, QUAD partners are exploring ways to collaborate on deploying open and secure telecommunications technologies in the region, working with the industry through Open RAN Track 1.5 dialogues.

Through a new Memorandum of Cooperation on 5G Supplier Diversification and Open RAN, QUAD will cooperate on technical exchanges and testbed activity to advance interoperability and telecommunications cybersecurity.

The QUAD has advanced technical standards of cooperation through the International.

Telecommunication Union's Telecommunication Standardization Sector will redouble its efforts through the new International Standards Cooperation Network, a mechanism for like-minded allies and partners to share information on technical standards activities and to increase situational awareness, coordination, and influence in international critical and emerging technologies standards.

To unlock opportunities in critical and emerging technologies through more robust engagement with the private sector, QUAD will convene industry partners to discuss business and investment issues. A QUAD Investors Network, an independent consortium of investors that seeks to advance access to capital for critical and emerging technologies within and across QUAD, is being launched today.

6. Cybersecurity

The QUAD Cybersecurity Partnership seeks to build resilience across the four QUAD countries in response to cybersecurity vulnerabilities and cyber threats. Its focus areas are critical-infrastructure protection, led by Australia; supply chain stability and security, led by India; workforce development and talent, led by Japan; and software security standards, led by the United States. Its work is guided by new joint cyber principles to improve resilience in a rapidly changing threat environment. These principles aim to prevent cyber incidents, prepare national and international capabilities for potential cyber incidents, and respond quickly and effectively to a cyber incident when or should one occur. The QUAD will strengthen information sharing among QUAD country Computer Emergency Response Teams (CERT), including exchanges on lessons learned and best practices. QUAD will improve software and Managed Service Provider (MSP) security by coordinating cybersecurity standards for QUAD governments' software procurement. QUAD partners will launch a Cybersecurity Day campaign, open to countries across the Indo-Pacific and beyond, as part of our continuing efforts to strengthen cybersecurity awareness and action. This program will provide basic cybersecurity information and training to the most vulnerable sectors of our countries and regions including schoolchildren, small businesses, and older people. QUAD partners will lead this campaign with industry, non-profits, academia, and communities to maximize its effectiveness and reach.

7. Space

As leaders in space, QUAD countries are strengthening cooperation and pooling collective expertise to exchange satellite data, enable capacity-building, and consult on norms and guidelines. QUAD partners will maintain their commitments to the free, complete, and open sharing of space-based civil Earth observation data. They will jointly develop and promote the concept of Open Science in the region and globally. The United States will coordinate with QUAD partners on its cooperative civil Earth observation programs, including the National Aeronautics and Space Administration (NASA) GLOBE and DEVELOP programs; the National Oceanic and Atmospheric Administration (NOAA) Community Satellite Processing Package (CSPP) and Satellite Proving Ground Flood Mapping Portal; as well as the US Geological Survey (USGS) National Land Imaging Program. As extreme weather events become increasingly common in the Indo-Pacific, QUAD partners will convene technical experts to drive new cooperation and set the stage for additional disaster mitigation, humanitarian assistance, and disaster relief workshops or training among QUAD partners. The QUAD partners intend to host a two-day seminar on implementing the Long-Term Sustainability (LTS) Guidelines for Indo-Pacific countries focused on LTS implementation and how LTS implementation can lead to positive changes in domestic policy, regulations, and outcomes.

8. Infrastructure

To help meet the Indo-Pacific's enormous infrastructure needs, QUAD leaders launched the QUAD Infrastructure Coordination Group, which will deepen collaboration and pursue complementary actions, including digital connectivity, transportation infrastructure, clean energy, and climate resilience. In Tokyo, the QUAD's respective heads of development-financing agencies met to discuss enhanced engagement to address the infrastructure-financing gap in the region. This high-level effort—which includes Australia's Department of Foreign Affairs and Trade and Export Finance Australia; India's EXIM; Japan Bank for International Cooperation; and the US International Development Finance Corporation—will coordinate strategic approaches among QUAD leader countries and with the private sector to leverage solutions-oriented tools that support investments in critical sectors such as health care, clean energy, and climate, digital connectivity, sustainable infrastructure, and supply chains.

9. Humanitarian Assistance and Disaster Relief

The QUAD established a new humanitarian assistance and disaster relief (HADR) mechanism. Through the QUAD Humanitarian and Disaster Relief Mechanism, QUAD partners will be able to coordinate and mobilize our civilian-led disaster assistance efforts, with support from civil defense and military assets when needed, to respond to disasters in the Indo-Pacific. The Partnership provides a framework for QUAD partners to coordinate joint disaster responses and better understand their respective HADR operations. QUAD partners may provide collaborative or coordinated assistance as requested by an affected state in a disaster's crisis alert, crisis response, or post-crisis-review phases. QUAD partners will also be able to work together to improve crisis preparedness and early warning.

10. Indo-Pacific Economic Framework for Prosperity.

To start with, on May 23, 2022, the United States, Australia, Brunei, Darussalam, India, Indonesia, Japan, the Republic of Korea, Malaysia, New Zealand, Philippines, Singapore, Thailand, and Vietnam of the Indo-Pacific region joined Indo-Pacific Economic Framework to share a commitment to a free, open, fair, inclusive, interconnected, resilient, secure, and prosperous Indo-Pacific area to deepen economic engagement among partners for continued growth, peace, and prosperity. This collective invites participation from additional Indo-Pacific partners after due discussion and deliberations. This framework is intended to advance resilience, sustainability, inclusiveness, economic growth, fairness, and competitiveness for our economies to contribute to regional cooperation, stability, prosperity, development, and peace. This

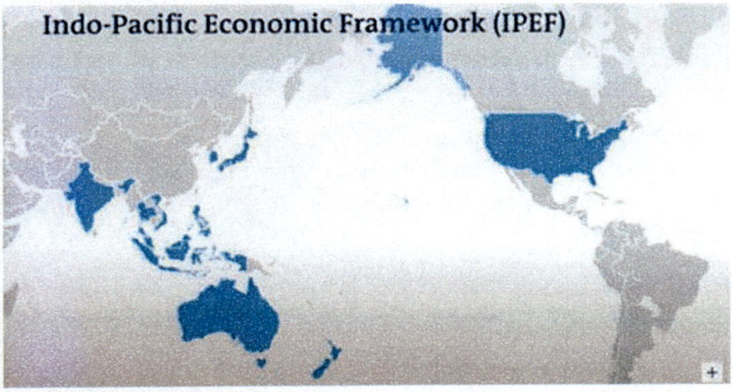

Fig. 40 Countries joined IPEF during Tokyo Summit. *Source* Author

launched collective discussions toward future negotiations on the following pillars. Administratively and politically, it was a smooth launch at short notice and easy to conclude since there are no usual trade tariff concessions of a trade treaty requiring democratic consensus are not there (Fig. 40).

Trade: To build high-standard, inclusive, accessible, and fair-trade commitments and develop new and creative approaches in trade and technology policy that advance a broad set of objectives that fuel economic activity and investment, promote sustainable and inclusive economic growth, and benefit workers and consumers. This includes but is not limited to, cooperation in the digital economy.

Supply Chains: Committed to improving transparency, diversity, security, and sustainability in our supply chains to make them more resilient and well-integrated, coordinate crisis response measures; expand cooperation to better prepare for and mitigate the effects of disruptions to better ensure business continuity; improve logistical efficiency and support; and provide access to necessary raw and processed materials, semiconductors, critical minerals, and clean energy technology.

Clean Energy, Decarbonization, and Infrastructure: In line with our Paris Agreement goals and efforts to support the livelihood of people and workers, accelerate the development and deployment of clean energy technologies to decarbonize our economies and build resilience to climate impacts.

This involves deepening cooperation on technologies, mobilizing finance, including concessional finance, and seeking ways to improve competitiveness and enhance connectivity by supporting the development of sustainable and durable infrastructure and by providing technical assistance.

Tax and Anti-Corruption: Committed to promoting fair competition by enacting and enforcing the effective and robust tax, anti-money laundering, and anti-bribery regimes in line with existing multilateral obligations, standards, and agreements to curb tax evasion and corruption in the Indo-Pacific region. This involves sharing expertise and seeking ways to support the capacity-building necessary to advance accountable and transparent systems.

5.6 *The Fifth Summit Quad Leaders' Joint Statement on May 20, 2023, Hiroshima [62]*

Prime Minister Anthony Albanese of Australia, Prime Minister Narendra Modi of India, Prime Minister Kishida Fumio of Japan, and President Joseph R. Biden, Jr. of the United States, met for the third in-person Quad Leaders' Summit hosted by Prime Minister Albanese.

Together, we reaffirm our steadfast commitment to a free and open Indo-Pacific that is inclusive and resilient. The global strategic and economic environment is changing rapidly—with direct impacts on countries in the region. We believe we should navigate this time of uncertainty and opportunity together, working closely with our Indo-Pacific partners. We accept that all countries have a role in contributing to regional peace, stability, and prosperity, as well as upholding international law, including the principles of sovereignty and territorial integrity and the rules-based international order. We seek a region where no country dominates, and no country is dominated—one where all countries are free from coercion and can exercise their agency to determine their futures. This shared vision unites our four nations.

As Indo-Pacific countries, Quad partners are deeply invested in our region's success. Harnessing our collective strengths and resources, we support the region's development, stability, and prosperity through the Quad's positive, practical agenda. Regional countries' priorities guide our work and respond to the region's needs. We are and will continue to be transparent in what we do. Respect for the leadership of regional institutions, including the Association of Southeast Asian Nations (ASEAN), the Pacific Islands Forum (PIF), and the Indian Ocean Rim Association (IORA), is and will remain at the center of the Quad's efforts.

The Quad Leaders' Vision Statement we have issued today sets out our shared vision for the Quad and the Indo-Pacific region based on these principles.

Today we reaffirm our consistent and unwavering support for ASEAN.

Centrality and unity. We are committed to ensuring the Quad's work is aligned with ASEAN's principles and priorities and continues supporting the ASEAN Outlook on the Indo-Pacific (AOIP) implementation. We underscore ASEAN's regional leadership role, including in the East Asia Summit, the region's premier leader-led forum for strategic dialogue, and the ASEAN Regional Forum. We strongly support Indonesia's 2023 ASEAN Chairmanship and its Chair theme, "ASEAN Matters: Epicentrum of Growth." We will continue strengthening our respective relationships with ASEAN and seek opportunities for greater Quad collaboration to support the AOIP.

We recommit to working in partnership with Pacific Island countries to achieve shared aspirations and address shared challenges. We reaffirm our support for Pacific regional institutions that have served the region well over many years, foremost the PIF, and warmly welcome Cook Islands assuming the PIF Chair in 2023. We continue to support the objectives of the 2050 Strategy for the Blue Pacific Continent and commit to working with partners, including through Partners in the Blue Pacific, to support engagement with these objectives. We welcome the third Summit of the Forum for India-Pacific Islands Cooperation and the US–Pacific Island Forum meeting, both to be held in Papua New Guinea in the coming days. We also applaud Japan's long-standing engagement with Pacific Island countries through the Pacific Islands Leaders Meeting (PALM) and Australia's deep and enduring commitment to the Pacific as part of the Pacific family, including as a founding member of the PIF.

In these efforts, Quad Leaders will listen to and be guided by Pacific priorities, including climate action, ocean health, resilient infrastructure, maritime security, and financial integrity. In particular, we acknowledge climate change remains the single greatest threat to the livelihoods, safety, and well-being of the peoples of the Pacific and applaud Pacific Island countries' global leadership on climate action.

We remain committed to strengthening cooperation in the Indian Ocean region. We welcome the work of IORA as the Indian Ocean region's premier forum for addressing the region's challenges. We recognize India's leadership in finalizing the IORA Outlook on the Indo-Pacific (IOIP) and express our support for its implementation. We thank Bangladesh for its term as IORA Chair and commit to working with Sri Lanka and India as they assume the roles of IORA Chair and Vice Chair this year.

We, the countries of the Quad, will work together to be a global force for good. We will bring our combined resolve to support each other.

International leadership in 2023 included Australia's Quad hosting, Japan's G7 presidency, India's G20 presidency, and the United States' APEC host year.

The Quad's positive, practical agenda.

We recognize the urgent need to address the climate crisis, which poses tremendous environmental, social, and economic challenges for our region. Today we underline our dedication to taking significant action on climate change—individually and collectively. We will continue to support climate mitigation, adaptation, and resilience efforts in alignment with the United Nations Framework Convention on Climate Change and the Paris Agreement and regional architecture, including ASEAN, the PIF, and IORA. We will continue collaborating on green shipping and ports, disaster risk management, exchanging climate information, and capacity-building support for Article 6 implementation of the Paris Agreement. Under the Climate Information Services Initiative, we plan to coordinate our collective resources to support early warning systems in the Indo-Pacific, including through the Pacific-led Weather Ready Pacific initiative and the long-standing leadership of the Pacific Meteorological Council. We also intend to provide support through global partnerships such as the Coalition for Disaster Resilient Infrastructure (CDRI) and its Infrastructure for Resilient Islands States (IRIS) initiative.

Achieving sustainable consumption and production is critical to global efforts to achieve the 2030 Agenda and Sustainable Development Goals (SDGs), environment, and climate ambitions. We will work together to seek meaningful outcomes on climate action and the clean economy transition in the Indo-Pacific Economic Framework for Prosperity (IPEF).

The latest Intergovernmental Panel on Climate Change report clarifies that rapid and far-reaching transitions are needed across all sectors and systems. As we move to a net-zero world, we underscore that it is critical to strengthen our cooperation to ensure better access to affordable, reliable, and secure clean energy in the Indo-Pacific. We will work together to increase the region's access to climate finance and climate-smart technology. Under the Quad Climate Change Adaptation and Mitigation Package (Q-CHAMP), launched in 2022, we continue to work with Indo-Pacific partners to enhance climate and clean energy cooperation and promote adaptation and resilience. In this regard, we welcome the Sydney Energy Forum and the Quad Clean Hydrogen Partnership meeting hosted by Australia and India in July 2022.

Building on those foundations, we are issuing a Statement of Principles on Clean Energy Supply Chains in the Indo-Pacific. This provides a basis for regional engagement in clean energy supply chain development. The principles promote diverse, secure, transparent, resilient clean energy supply chains and support a sustainable and inclusive transition. We also announce a Clean Energy Supply Chains initiative to accelerate Indo-Pacific's clean energy transition. This initiative will facilitate research and development and feasibility study projects to lower clean energy manufacturing and deployment costs, enhance regional energy security, and expand and diversify the regional production of necessary materials and technologies.

The COVID-19 pandemic has shown us how vital health security is to our societies, economies, and our region's stability. In 2021 and 2022, Quad partners stepped up to help meet the region's most pressing need, delivering more than 400 million safe and effective COVID-19 vaccine doses to Indo-Pacific countries and almost 800 million amounts globally, bilaterally, and in partnership with COVAX.

Today, we announce the evolution of our Quad Vaccine Partnership into a broader Quad Health Security Partnership. Through this partnership, we will strengthen our coordination and collaboration to support health security in the Indo-Pacific. We plan to implement a suite of activities to build the region's capacity to detect and respond rapidly to outbreaks of diseases with epidemic and pandemic potential. These activities include support for health workforce development, disease surveillance, electronic health information systems, and coordination of outbreak responses, such as the Quad Pandemic Preparedness Exercise.

We will continue cooperation with Indo-Pacific partners to meet the region's infrastructure priorities. Delivering on our commitment at the 2022 Quad Leaders' Summit, we will continue to support access to quality, sustainable, and climate-resilient infrastructure investments in our region. We aim to ensure the investments we support are fit for purpose, demand-driven, responsive to countries' needs, and do not impose unsustainable debt burdens. We will build on ongoing programs for Indo-Pacific countries, including training and capacity-building focused on digital and economic connectivity, clean energy, and climate-resilient power sector infrastructure. We continue to strengthen our capacity to manage debt issues, including, under the G20 Common Framework and promote debt sustainability and transparency.

Today, we announce a new initiative to boost infrastructure expertise across the Indo-Pacific: the 'Quad Infrastructure Fellowships Program.' The initiative aims to empower over 1,800 of the region's infrastructure practitioners to design, build, and manage quality infrastructure in their home countries.

The Quad is committed to improving the region's connectivity by developing resilient infrastructure. We recognize the urgent need to support quality undersea cable networks in the Indo-Pacific, vital to global growth and prosperity. Today we announce a new 'Quad Partnership for Cable Connectivity and Resilience.' The

Partnership will strengthen cable systems in the Indo-Pacific, drawing on Quad countries' world-class expertise in manufacturing, delivering, and maintaining cable infrastructure.

Quad partners' export credit agencies make an essential contribution to the prosperity of the Indo-Pacific. We welcome ongoing efforts to enhance cooperation among Quad partners' export credit agencies: including through a Memorandum of Cooperation between ECGC Limited of India, Export Finance Australia (EFA), Nippon Export and Investment Insurance (NEXI) of Japan, and Export–Import Bank of the United States (USEXIM).

We recognize the transformative power of technology, including digital public infrastructure, to support sustainable development in the Indo-Pacific and deliver economic and social benefits. We are stepping up our efforts to strengthen supply chain resilience and improve the region's digital connectivity through access to critical and emerging technologies and advanced telecommunications technology, including 5G networks.

Today, we partner with Palau to establish a deployment of Open Radio Access Networks (Open RAN), the first in the Pacific. The Quad is committed to ensuring regional countries are not left behind as telecommunications markets and network architectures evolve. We support access to innovations, such as Open RAN, that enable more excellent vendor choices for countries to expand and modernize their telecommunications networks. We also welcome the release of the Open RAN Security Report, which is expected to promote industry investment in approaches to telecommunications that are demonstrably open, interoperable, trusted, and secure.

The Quad International Standards Cooperation Network and the Quad Principles on Critical and Emerging Technology Standards, released today, reflect our support for industry-led, consensus-based multi-stakeholder approaches to developing technical standards.

We welcome the private sector-led Quad Investors Network (QUIN) launch, which aims to facilitate investments in strategic technologies, including clean energy, semiconductors, critical minerals, and quantum.

We intend to support joint research to advance innovation in agriculture through emerging technologies designed to empower farmers everywhere to increase yield and resistance.

We reaffirm our commitment to more secure cyberspace and fostering an international digital economy that works for everyone. Quad partners will continue collaborating to enhance regional capacity and resilience to cyber incidents and threats. We welcome the first Quad Cyber Challenge, held earlier this year, to promote cyber awareness and empower participants across the Indo-Pacific to protect themselves online. We also include the Quad Joint Principles for Secure Software and the Quad Joint Principles for Cyber Security of Critical Infrastructures and efforts to develop a guiding framework for ensuring supply chain security and resilience. These principles strengthen our region's defenses against cyber threats to the software supply chain and critical infrastructure and services.

We recognize the importance of space technologies and space-related applications in responding to climate change and disasters and enhancing the sustainable use of oceans and marine resources. We reaffirm our commitment to supporting capacity-building for countries in the region. The Quad Space Working Group will explore avenues to deliver Earth Observation data and other space-related applications to assist nations across the Indo-Pacific in strengthening climate early warning systems and better managing the impacts of extreme weather events. We commit to the open sharing of civil Earth Observation data. We will continue to consult each other and the region on peaceful, safe, and sustainable use of outer space. We announce our intention to share expertise and experience in the area of situational awareness. We commit to strengthening our commercial space cooperation, including convening a business forum in 2023.

We are pleased that the Indo-Pacific Partnership for Maritime Domain Awareness (IPMDA), announced at the 2022 Tokyo Quad Leaders' Summit, is underway.

Through IPMDA in its pilot phase, we provide near-real-time, integrated, and cost-effective maritime domain data to Southeast Asia and the Pacific naval agencies. We will expand coverage to partners in the Indian Ocean region in the coming months. This supports our regional partners in combatting illicit maritime activities, including illegal, unreported, and unregulated fishing, and responding to climate-related and humanitarian events. We are committed to deepening engagement with regional partners to support maritime safety and security and uphold international law.

At the 2021 Quad Leaders' Summit in Washington, we launched the Quad Fellowship. This year, we welcome the first cohort of Quad STEM Fellows, who will begin their studies in the United States in August 2023. Our one hundred Quad Fellows from all four Quad countries represent the best and brightest of our next generation. This inaugural class's diversity and dynamism will help ensure our nations remain at the forefront of innovation, and we wish them well.

Global and regional issues

We remain fully resolved to uphold peace and stability in the Indo-Pacific maritime domain. We strongly oppose destabilizing or unilateral actions that seek to change the status quo by force or coercion. We emphasize the importance of adherence to international law, mainly as reflected in the United Nations Convention on the Law of the Sea (UNCLOS), and the maintenance of freedom of navigation and overflight in addressing challenges to the maritime rules-based order, including those in the East and South China Seas. We express grave concern at the militarization of disputed features, the dangerous use of coastguard and maritime militia vessels, and efforts to disrupt other countries' offshore resource exploitation activities. We emphasize that disputes should be resolved peacefully and following international law, without threat or use of force.

Together with our global and regional partners, we will bolster international institutions and initiatives that underpin global peace, prosperity, and development. We reiterate our unwavering support for the United Nations (UN) Charter and the three pillars of the UN system. In consultation with our partners, we will work collectively to address attempts to unilaterally undermine the integrity of the UN, its Charter, and its agencies. We seek to strengthen and reform the multilateral system to reflect contemporary realities better and meet the aspirations of the Indo-Pacific region.

We remain committed to a comprehensive UN reform agenda, including expansion in permanent and non-permanent categories of Membership of the UN Security Council. We reaffirm our commitment to implementing the 2030 Agenda and achieving its Sustainable Development Goals (SDGs). We underscore the importance of comprehensively achieving the SDGs without selectively prioritizing a narrow set of such goals. We reaffirm that the UN is central in supporting countries' implementation.

We stand for adherence to international law, peaceful resolution of disputes, and respect for principles of the UN Charter, including territorial integrity and sovereignty of all states. In this context, today, we express our deep concern over the war raging in Ukraine and mourn its terrible and tragic humanitarian consequences. We recognize its severe impacts on the global economic system, including food, fuel, energy security, and critical supply chains. We will continue to render humanitarian assistance to Ukraine for its recovery. Conscious that ours must not be an era of war, we remain committed to dialogue and diplomacy. We support a comprehensive, just, and lasting peace consistent with the UN Charter. In this context, we concur that nuclear weapons use, or threat of use, is severe and inadmissible.

We condemn North Korea's destabilizing ballistic missile launches and pursuit of nuclear weapons in violation of multiple UN Security Council resolutions (UNSCRs). These launches pose a grave threat to international peace and stability. We urge North Korea to abide by all its obligations under the UNSCRs, refrain from further provocations, and engage in substantive dialogue. We recommend that North Korea resolve the abductions issue immediately. We reaffirm our commitment to the complete denuclearization of the Korean Peninsula consistent with relevant UNSCRs and call on all countries to fully implement these UNSCRs. We stress the importance of addressing the proliferation of nuclear and missile technologies related to North Korea in the region and beyond.

We remain deeply concerned by Myanmar's deteriorating situation and call for an immediate cessation of violence. We call for the release of all those arbitrarily detained, unhindered humanitarian access, the resolution of the crisis through constructive dialogue, and the transition of Myanmar toward an inclusive democracy. We reaffirm our consistent support of ASEAN-led efforts, including the work of the ASEAN Chair and Office of the Special Envoy. We call for full implementation of all commitments under the ASEAN Five-Point Consensus.

We unequivocally condemn terrorism and violent extremism in all its forms and manifestations, including cross-border terrorism. We are committed to international cooperation and will work with our regional partners comprehensively and sustainably to strengthen the capability to prevent, detect, and respond to threats posed by terrorism and violent extremism, consistent with international law. We are committed to working together to promote accountability for the perpetrators of such terrorist attacks. We reiterate our condemnation of terrorist attacks, including the 26/11 attacks in Mumbai and Pathankot, and our commitment to pursuing designations, as appropriate, by the UN Security Council 1267 Sanctions Committee. We will strengthen our cooperation through the new Working Group on Counterterrorism announced during the Quad Foreign Ministers' Meeting in March 2023.

We endorse the outcomes of the 3 March Quad Foreign Ministers' Meeting in New Delhi and the Ministers' Joint Statement committing to deepening practical and positive cooperation for the benefit of the Indo-Pacific region. India will host our next in-person Quad Leaders' Summit in 2024.

We, the Quad Leaders, remain firm in our resolve to meet our region's challenges and transparent in our vision for a free and open Indo-Pacific that is stable, prosperous, and inclusive. In doing so, we are committed to working in partnership with Indo-Pacific countries—large and small—in deciding our future and shaping the region we all want to live.

6 Indo-Pacific Policies, Strategies, and Programs of Outside or Peripheral Nations

France published Indo-Pacific strategy papers in 2018 and 2019; Germany and the Netherlands followed in September and November 2020, respectively. Then Canada had its Indo-Pacific Strategy in November 2022. France, Germany, and the Netherlands have also been working together to promote an Indo-Pacific concept or vision to be adopted by the EU.

France was the first EU member state to develop its Indo-Pacific strategy. French President Emmanuel Macron outlined his country's position in a speech at the Garden Island naval base in Australia in May 2018 about a new Indo-Pacific France-India-Australia axis. Official documents were published by the French Ministry of Defense (in 2018, updated in May 2019, and another later in 2019) and by the Ministry of Foreign Affairs (June 2019).

The German 'Policy Guidelines for the Indo-Pacific,' published first in German in September 2020 and shortly afterward in English, was approved by the German cabinet, thus representing a whole-of-government approach. A short Dutch 'non-paper'1 followed in November 2020, of which no official English translation has been published. In May 2019, the government of the Netherlands also published a new China strategy. Both the German and Dutch 'guidelines' on the Indo-Pacific make it clear from the beginning that they are to be understood as contributions and building blocks to lead to an EU position in the Indo-Pacific.

To get the discussion on the Indo-Pacific at the EU, France, Germany, and the Netherlands wrote a 'non-paper' (which has so far remained unpublished) in the autumn of 2020. Their initiative found support from other member states in the EU, including Portugal, Poland, Italy, and Sweden. It was first discussed on December 8, 2020, at the Asia-Oceania Working Party (COASI) meeting. It continued in 2021. Four areas have been mentioned for the EU to focus on trade; connectivity; maritime security; and global issues such as climate change and biodiversity. While trade is a given since the EU can only negotiate free trade agreements for connectivity with Asia, maritime security, and climate change, the EU has broad frameworks that could be modified to accommodate a shift to the Indo-Pacific.

When Nations geographically outside the Indo-Pacific Region have their Indo-Pacific policies, strategies, and programs, the first question is why they have such initiatives. The answer is their self-interest. First, self-interest is shown by the economic zone mapped below. The map shows the European Union, the UK, France, and the Netherlands. The outside countries have economic zones within the Indo-Pacific Ocean in addition to the Indo-Pacific countries such as the US, Japan, India, and China. Furthermore, Tonga has economic zones within the Indo-Pacific, as shown in the map below. One can calculate the area of these economic zones and rank the countries based on the map below (Fig. 41).

The second reason is that the economic center of gravity is shifting from the Atlantic Ocean to the Indo-Pacific Ocean. A nation with big economic powers wants to do business with countries in the Indo-Pacific for mutual benefit. Since peace and tranquility are necessary for supply chains and financial interaction, they also want to ensure that their military can participate in the Indo-Pacific and provide stability for prosperity to safeguard economic interests. AUKUS Australia, UK, and the USA grouping with long-range nuclear-powered submarine technology transfer and production, of which the UK part is an excellent example.

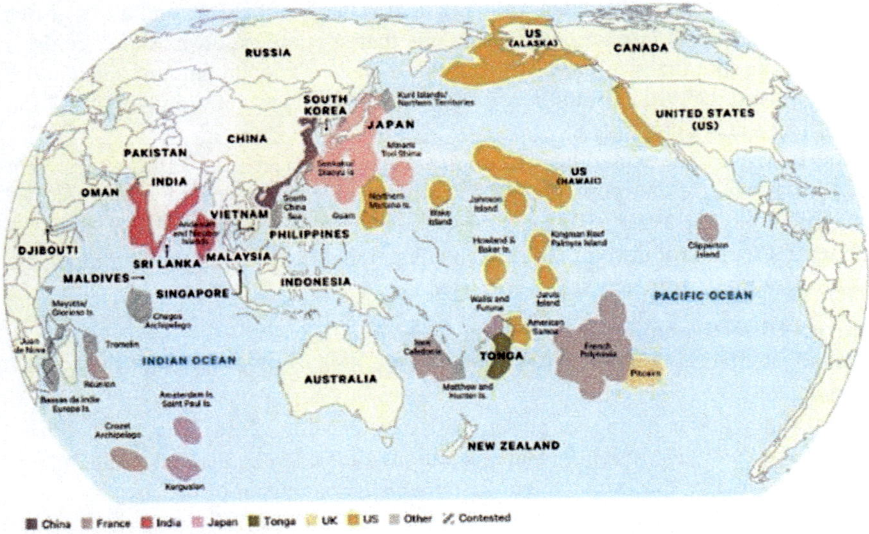

Fig. 41 Exclusive economic zones of countries given in the map. *Note* The EEZs marked are indicative only and show both settled and disputed EEZ borders. *Sources* Bordermap Consulting (n.d.), 'Home', https://www.bordermap.com; Flanders Marine Institute (2019), 'Maritime Boundaries Geodatabase, version 11' https://www.marineregions.org; Natural Earth (n.d), 'Free vector and raster map data', naturalearthdata.com

6.1 Indo-Pacific Strategy of France [62]

France and the Indo-Pacific, presence and influence
France is a nation in the Indo-Pacific with its overseas departments and communities. These territories represent a population of 1.65 million people and make France the second-largest economic exclusivity zone in the world (10.2 million km^2). The protection of its citizens and its sovereign territory is one of the primary missions of the French security and defense strategy in the Indo-Pacific. The French State's network in the Indo-Pacific is dense, with a diplomatic and consular presence in 39 States. The French State agencies, including research bodies, are also very present.

Furthermore, France also maintains a permanent military presence in the Indo-Pacific. Some 53,000 students from Indo-Pacific are enrolled in 95 French educational institutions in the Indo-Pacific region, attracting an increasing number of international students. Students in Asia-Oceania represent 45% of global student mobility, and France hosts 50,000 of these students. In terms of research, in a highly competitive environment (Fig. 42).

France has a network of research bodies established across the region, creating solid partnerships and renowned expertise. Commerce with the Indo-Pacific represents more than a third of French trade in goods outside of the EU, and it has grown by 49% in 10 years. The French instruments that support export and development assistance are mobilized to meet needs in those areas. The total amount outstanding for funds from the AFD Group in Indo-Pacific countries was more than €9 billion in 2020, all sectors combined.

The Indo-Pacific (excluding China) represented almost €13 billion regarding credit insurance. Furthermore, the Indo-Pacific received a fifth of the global total of loans offered by the French Treasury from 2010 to 2020. Lastly, the Indo-Pacific is an essential region for developing French educational institutions, student mobility, and cooperation in research and innovation.

France's Objectives and Actions in the Indo-Pacific
The implementation of the Indo-Pacific strategy involves concrete and diversified actions.

They were conducted in four significant areas or "pillars": security and defense; economy, connectivity, research, and innovation; multilateralism and the rule of law; and climate change, biodiversity, and sustainable management of oceans.

Pillar 1: Security and Defense
Objectives:

- Ensuring and defending the integrity and sovereignty of France, the protection of its citizens, its territories, and its Exclusive Economic Zones (EEZ).
- Contributing to the security of regional areas by promoting military and security cooperation.
- Preserving, alongside its partners, access to communal areas in a context of strategic competition and increasingly restrictive military environments (Fig. 43).

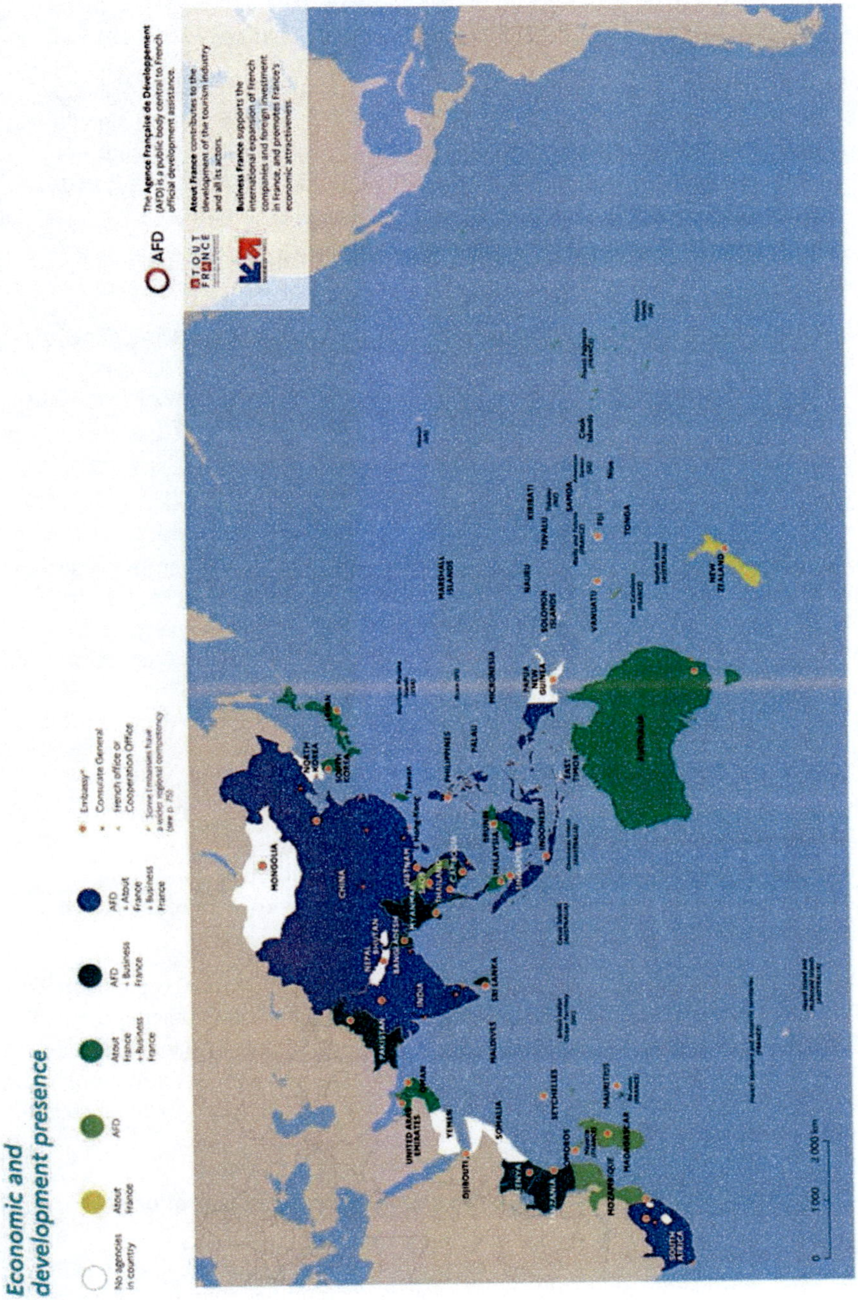

Fig. 42 Economic and Development Presence of France in Indo-Pacific France. *Source* [62]

- Participating in maintaining strategic stability and military balances of power through international action based on multilateralism.
- Anticipating security risks brought about by climate change.

Examples of action: monitoring sovereign areas; organizing multilateral exercises; participating in rescue operations for people and fishing monitoring operations; affirming the respect of international law and the freedom of navigation; cooperating in sharing of maritime information; deepening interoperability with partners; and cooperating on the issues of climate and environmental security.

Pillar 2: Economy, connectivity, research, and innovation
Objectives:

- Ensuring diversification of the supply of strategic goods and reducing dependencies.
- Promoting existing international standards and making them prevail to establish a fair, competitive framework.
- Meeting needs in terms of connectivity and infrastructure.
- Supporting efforts of French companies in the Indo-Pacific region.
- Deepening research and innovation partnerships.

Examples of actions: entering partnerships with regional countries to secure supply chains; implementing the G20 Principles for Quality Infrastructure Investment; making efforts on renewable energy infrastructure; accompanying French companies and mobilizing "Team France"; strengthening cooperation efforts in research, particularly in health.

Pillar 3: Multilateralism and the Rule of Law
Objectives:

- Promoting multilateralism in countries in the Indo-Pacific region.
- Contributing to strengthening regional cooperation forums.
- Fostering strong involvement and better visibility of the European Union.
- Ensuring the central nature of the rule of law and the importance of the law of the sea.

Examples of actions: working with regional countries in international organizations; supporting a multilateral response to the COVID-19 pandemic; enhancing the partnership with ASEAN and the regional organizations of the Indian Ocean; undertaking regional action in the Pacific in response to COVID-19; adopting an ambitious EU strategy for the Indo-Pacific; promoting the French model of ocean governance and maritime security.

Pillar 4: Climate change, biodiversity, sustainable management of oceans

Objectives:

- Increasing partner involvement in the region in fighting climate change and improving energy transition.

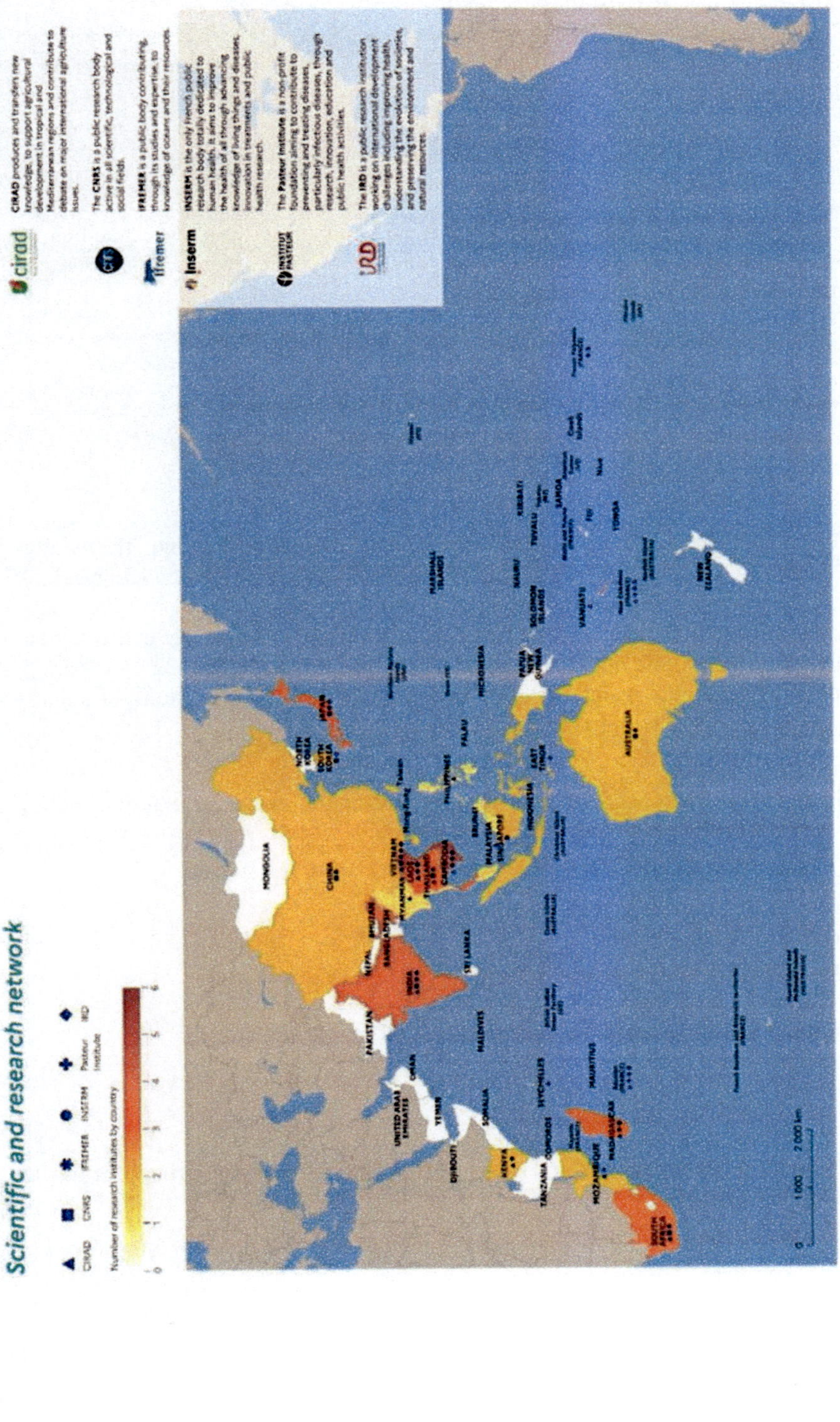

Fig. 43 Presence of France in the Indo-Pacific Scientific and research networks. *Source* [62]

- Fostering strengthening of actions for biodiversity.
- Developing partnerships for ocean protection.
- Contributing to improving natural disaster response.
- Enhancing the use of the skills of our territories and regional cooperation on all these issues.

Examples of actions: mobilizing countries in the region around ambitious climate objectives; making a specific effort in solar energy; renewing the KIWA initiative and promoting the High Ambition Coalition for Nature and People; implementing partnerships with our main partners in the blue economy; building capacities for response to natural disasters; forming regional partnerships with French research institutions established in the territories.

6.2 Policy Guidelines for the Indo-Pacific Region Germany—Europe—Asia: Shaping the Twenty-First Century Together [63]

The Indo-Pacific is becoming the key to shaping the international order in the twenty-first century because the region has a Oneng, well-educated population and can look back on decades of considerable.

Economic growth. Twenty of the world's thirty-three megacities are in this region. With growing economic output, the countries in the area are becoming increasingly self-confident partners in international cooperation, including in the fight against climate change and the global loss of biodiversity.

Interests
The Federal Government of Germany is guided by the following interests in its policy on the Indo-Pacific region:

- Peace and security.
- Diversifying and deepening relations.
- Neither unipolar nor bipolar.
- Open shipping routes.
- Open markets and free trade.
- Digital transformation and connectivity.
- Protecting our planet.
- Access to fact-based information.

Principles
The Federal Government of Germany is guided by the following principles for its policy on the Indo-Pacific region:

- European action.
- Multilateralism.
- The rules-based order.

- United Nations Development Goals.
- Human rights.
- Inclusivity.
- Equals partnership.

Initiatives

Strengthening multilateralism

- The Federal Government will strategically position and step up its engagement vis-à-vis ASEAN.
- The Federal Government will strengthen the role of the EU as a partner of ASEAN in close cooperation with its EU partners.
- The Federal Government will intensify its cooperation with other regional institutions in the Indo-Pacific region.
- The Federal Government will work to maintain the rules-based order together with partners in the Indo-Pacific region.
- The Federal Government will work with France to elaborate a European strategy on relations with the Indo-Pacific region.

Tackling climate change and protecting the environment

- The Federal Government will step up its cooperation in the Indo-Pacific region in climate protection, adaptation to climate change, biodiversity protection, renewable energies, and energy efficiency.
- The Federal Government will expand its multilateral engagement with the Indo-Pacific region in climate and environmental protection.

Strengthening peace, security, and stability

- The Federal Government will increase its security policy engagement in the Indo-Pacific region.
- The Federal Government will continue implementing measures for civil crisis prevention, conflict management, and peacebuilding.
- The Federal Government will increase its arms control and export control policy engagement in and with the Indo-Pacific region.

Promoting human rights and the rule of law

- The Federal Government will work to strengthen the human rights situation in Indo-Pacific region countries and enforce international human rights standards.
- The Federal Government will engage in open and critical dialogue with governments in the Indo-Pacific region within a bilateral framework at the level of the European Union (human rights dialogues) and in multilateral forums, particularly the UN Human Rights Council. In this regard, it will also support those persecuted for political reasons.
- The Federal Government will promote disseminating fact-based information in the Indo-Pacific region and strengthen resilience against disinformation through concrete projects. It will establish a Regional German Information Centre in Singapore to this end.

- The Federal Government will support further projects in the Indo-Pacific region within the framework of the rule of law promotion, particularly in access to justice for all, subject to the willingness of the respective state to implement reforms and the prospects of success of the measures.

Strengthening rules-based, fair, and sustainable free trade

- The Federal Government will improve the framework conditions for diversifying and intensifying economic relations in the Indo-Pacific region.
- The Federal Government will support the EU's trade policy and work to strengthen the multilateral trade system with the WTO at its center in cooperation with the countries of the Indo-Pacific region.
- The Federal Government will lend its active support to the EU's trade policy in the Indo-Pacific region to eliminate existing obstacles to trade and investment on both sides as well as to enshrine binding rules on environmental and social standards, climate protection and competition policy, state-owned companies, subsidies, and the protection of intellectual property.
- The Federal Government will support the engagement of German companies in the Indo-Pacific region.
- The Federal Government will put in place targeted measures to make it easier for students, skilled workers, and specialists to come to Germany or to relocate (for a limited period) for study, training, or employment purposes by further expanding personnel and organizational capacities for assessing visa applications and granting visas, if possible, within the scope of existing resources.

Rules-based networking and the digital transformation of regions and markets

- The Federal Government will expand connectivity both to and within the Indo-Pacific region.
- The Federal Government will promote and help shape digital transformation.
- The Federal Government will strengthen Germany's competitiveness in key technologies.
- The Federal Government will support and strategically promote the digital sovereignty of Germany and the EU, particularly in the Indo-Pacific region.

Bringing people together through culture, education, and science

- The Federal Government will expand its cooperation with the Indo-Pacific region in culture, education, and science.

6.3 Indo-Pacific: Guidelines for Strengthening Dutch and EU Cooperation with Partners in Asia [64]

The Netherlands has significant economic and geopolitical interests in the Indo-Pacific, its biggest export market outside Europe with excellent growth potential. The COVID-19 crisis has allowed us to consider ways of reducing one-sided dependencies and diversifying suppliers from the Indo-Pacific region. The Netherlands must work with the Indo-Pacific to ensure sustainable green growth, economic recovery, and responsibly organized digitalization and critical technology management.

In a world where democracy, the rule of law, human rights, freedom, free trade, and a properly functioning multilateral world order are increasingly under pressure, the Netherlands and the EU must join forces in the Indo-Pacific region and with ASEAN and work with the countries of the Indo-Pacific region to safeguard peace and security, reduce tensions on trade issues, promote maritime safety and unhindered safe passage on shipping routes, and combat economic and cyber espionage and cyberattacks on vital infrastructure.

Elements of a European Vision of the Indo-Pacific
Security and Stability

Given the EU's significant economic, political, and geopolitical interests in the Indo-Pacific, it must aim for de-escalation, inclusion, and protection of the commons. At the country level, the EU has a role to play in helping to preserve the balance of power and—where necessary—in providing a counterweight to the strategic economic and military influence of one or more great powers.

The EU's basic principle must be the protection and development of the international legal order. The EU must partner with countries in the region to guarantee safe passage and maritime security. This should happen primarily through compliance with the United Nations Convention on the Law of the Sea (UNCLOS) and the peaceful settlement of disputes using the appropriate mechanisms, continued participation in the Regional Cooperation Agreement on Combating Piracy and Armed Robbery against Ships in Asia (ReCAAP) could also help attain this goal.

Given the militarization of the region and the presence there of seven nuclear weapons states (China, the US, Russia, India, Pakistan, France, and the UK), plus North Korea with its nuclear ambitions, the EU must work actively with Indo-Pacific countries within the existing framework for promoting non-proliferation, disarmament, and export controls. This framework includes the Non-Proliferation Treaty (NPT), the Missile Technology Control Regime (MTCR), and the Hague Code of Conduct against Ballistic Missile Proliferation (HCMC). The EU should promote transparency, confidence-building measures, and verifiable agreements, including outer space.

The EU should also pursue partnerships with countries in the region aimed at the universalization and effective implementation of other agreements on arms control, peace, and security, including the Comprehensive Nuclear-Test-Ban Treaty (CTBT), negotiation of a Fissile Material Cut-off Treaty (FMCT), and agreements on conventional weapons like the Arms Trade Treaty (ATT), the Convention on Cluster Munitions, the Ottawa Convention on landmines and the Program of Action on Small Arms and Light Weapons (SALW).

China is currently the only nuclear power rapidly expanding its nuclear arsenal. Moreover, exceptionally little is known about the exact composition of the Chinese arsenal, and how it will affect developments in the region and beyond is unclear. China must be reminded of its global and regional responsibility in this regard. There should also be more support within the EU for US efforts to arrive at a trilateral arms control agreement, including China and Russia.

The EU must continue contributing, with knowledge, projects, and capacity-building, to the fight against extremism and terrorism in the Indo-Pacific.

Collaborating with partners in the Indo-Pacific region, the EU must seek to work more closely with countries that share the EU's and the Netherlands' concerns about preserving an open economy, effective multilateralism, and keeping the international legal order functioning. In a world where democracy, the rule of law, human rights, freedom, free trade, and a properly functioning multilateral world order are increasingly under pressure, it is essential to work more in concert with democratic, like-minded countries to defend and promote our shared interests. The EU must promote democratic values and norms and continue to dialogue with all countries in the region. To do so, it can collaborate with like-minded partners, using its existing strategic and other partnerships with Australia, Japan, New Zealand, South Korea, and India, as well as the Strategic Partnership with ASEAN that the Netherlands hopes will enter into force soon. The EU must work with like-minded countries in the region to develop a better narrative and make more strategic use of investments and other activities aimed at connectivity. The EU Connectivity Strategy can play a vital role in this regard.

The Asia-Europe Meeting (ASEM) can be used as a consultation forum on current and strategic developments. The EU can financially support the Asia-Europe Foundation (ASEF) for activities promoting exchanges and collaboration in journalism, human rights, and art. The EU should seek to link up with the NATO partnerships in the region. While pursuing a work program with China and tackling COVID-19, NATO has strengthened its links with its Indo-Pacific partners (Australia, New Zealand, South Korea, and Japan) as part of its Partners Around the Globe framework. NATO's ties with Australia (an Enhanced Opportunities Partner) are the closest of these four countries. The EU should continue to work actively with governments in the Indo-Pacific countries on human rights dialogues in which the human rights situation can be discussed openly and critically.

Sustainable trade and economies

Significant economic and trade interests together bind Europe and Asia. For many countries in Asia, Europe is one of their main trading partners, if not their most significant. There is also substantial investment in both directions. Geopolitical competition and potential technological and economic disengagement could significantly affect many Asian countries, such as the Netherlands and Europe.

The COVID-19 crisis has given a strong impetus to the discussion on strategic autonomy and the resilience of value chains. The Netherlands needs to join with our EU partners and other like-minded countries to consider ways of reducing one-sided strategic dependencies and sustainably securing value chains, with a particular focus on critical technologies and raw materials. The EU should make better and more strategic use of its economic leverage to reach its geopolitical goals, deploying its full, integrated range of policy instruments for this purpose. More emphasis must also be placed on reciprocity. Talks on free trade agreements (FTAs) with countries in the region must be continued.

The EU's aims in these talks should be:

- o To bring the negotiations on agreements with Australia and New Zealand to a positive conclusion, with the broader goal of setting standards for a modern, open trade policy.
- o Given ASEAN's importance to progress in the talks with the ASEAN countries. The EU's first agreement was reached with Singapore in 2019. In 2020, the FTA with Vietnam came up. Now progress needs to be made with Indonesia. The Netherlands would also support a resumption of negotiations with Malaysia.
- o To restart negotiations on a comprehensive, ambitious trade agreement with India.
- o Besides being a significant challenge, the global energy transition is also an opportunity to boost future earning capacity. Cooperation between the EU and the Indo-Pacific on sustainable green growth and innovation should be strengthened in the interests of both global climate ambitions and the stimulation of sustainable trade and economies.

Effective multilateralism and the international legal order
In times like these, when international tensions are rising, it is essential to strengthen multilateral cooperation and the global legal system, which are crucial for maintaining peace and security. Only multilateral consultations founded on a rules-based order will enable us to find solutions for all states' interests. Multilateral cooperation is also the only way to tackle global challenges such as climate, poverty reduction, migration, and pandemics.

The EU should strengthen regional institutions, security, and other partnerships through wide-ranging cooperation and capacity-building.

The EU needs to coordinate activities and voting at the UN and in the World Bank and other large forums, internally among EU member states and, to a greater extent, with like-minded countries. It can also strategically and systematically use its economic influence to pursue its political interests better.

The EU could explore ways of working more effectively at the multilateral level with the Small Island Developing States (SIDS). Via the European Investment Bank (EIB), the EU is a significant investor in the Pacific.

1. **Sustainable connectivity.**

There are many initiatives in the Indo-Pacific region focused on connectivity. In addition to the Chinese Belt and Road Initiative, Japan, India, South Korea, and ASEAN have their connectivity strategies. In 2018 the EU launched its strategy, 'Connecting Europe and Asia,' to create a distinctive profile as a partner for sustainable development in transport, digital infrastructure, energy, and people-to-people contacts. In September 2019, the EU and Japan concluded a Connectivity Partnership based on sustainability, shared norms and values, quality infrastructure, and their commitment to guaranteeing a global level playing field, particularly in Asia and the Indo-Pacific. The EU should promote quality and fiscal, social, and ecological sustainability in a broad sense as part of all its connectivity strategies.

The EU should form connectivity partnerships with ASEAN and India modeled on its Connectivity Partnership with Japan. India has already expressed a desire to include connectivity in the agenda of the next EU-India summit. It would also like to reach agreements on further cooperation on connectivity within India as well as in the Indian subcontinent generally and in third countries, particularly in East Africa. ASEAN seeks a partnership with the EU based on a regional Master Plan on ASEAN Connectivity, which will also address the climate transition. The EU could also further explore the offer from the US and other partners to join the Blue Dot Network, as many of the efforts of the US, Japan, Australia, Canada, and so forth are focused on the Indo-Pacific region.

2. **Global challenges: climate and the SDGs**

The Indo-Pacific countries account for a third of global greenhouse gas emissions. While most countries in the region recognize the importance of a transition to clean energy, in practice, many of their economies are still dependent on cheap coal-fired power plants. The Indo-Pacific is the region of the world with the highest economic growth rates and a superior level of infrastructure investment, laying the foundation for their energy consumption and mix for the next 30 years. Unless the transition to clean, renewable energy in the Indo-Pacific can be accelerated, achieving the internationally agreed climate targets will be impossible.

Besides the fact that the Indo-Pacific countries are contributing significantly to climate change, the region also contains several of the most vulnerable countries already experiencing climate change's effects. These include small island states that risk disappearing underwater and countries experiencing adverse impacts on agriculture, vulnerable population groups, and densely populated urban areas. If governments in the region are not adequately prepared for climate change's influence, it could destabilize their economies, trade, and social cohesion.

Climate action is increasingly becoming a necessary precondition for economic growth. Therefore, significant gains can be made in the Indo-Pacific countries if governments and the private sector take the need for green recovery seriously. The Global Commission on Adaptation has calculated that every dollar invested in climate-smart infrastructure in developing countries between now and 2030 will generate four dollars in earnings, up to $4.2 trillion.

Several countries in the region are strong advocates of ambitious climate policy. These countries are also active in regional and international forums on climate issues. The Netherlands and the EU can work more closely with these countries in a collective effort to maintain this elevated level of climate ambition. The EU should also work to strengthen exports of sustainable, green energy solutions to advance the climate transition in the Indo-Pacific. Collaboration is already taking place in many areas.

- The EU should further deepen and expand its cooperation on climate and the SDGs where possible, including with the SIDS.

- The EU should work to strengthen green trade flows to advance the climate transition in the Indo-Pacific.
- The EU should work toward greater cooperation on climate adaptation and in international forums in the interests of ambitious climate policy.

The Netherlands and the Indo-Pacific [64]

Under the umbrella of a European vision of the Indo-Pacific, the Netherlands will also step up its bilateral efforts to work with democratic, like-minded partners. More specifically, the Netherlands will work toward the following:

Security and stability.

- It will promote safe passage and maritime security by helping with capacity-building around the international law of the sea and exploring opportunities in the realm of defense and security.
- Will speak out more actively about international law (and breaches of international law), including concerning UNCLOS and the South China Sea.
- We will participate more often at the appropriate level in relevant gatherings on strategic issues in the Indo-Pacific region, including the annual IISS Shangri-La Dialogue (in Singapore) and the annual Raisina Dialogue (in India).
- It will strengthen collaboration and dialogue with like-minded Indo-Pacific countries on cyber issues, emphasizing cybersecurity.
- Will step up cooperation with the region on combating hybrid threats, both bilaterally and through international partnerships, to counter foreign interference.
- Will contribute to non-proliferation and disarmament by collaborating with like-minded partners in the framework of the Non-Proliferation Treaty, for example, in the Non-Proliferation and Disarmament Initiative (NPDI), and through the Proliferation Security Initiative (PSI) and export control regimes.
- Will support the NATO partnerships with Australia, New Zealand, South Korea, and Japan in the Indo-Pacific (Partners around the Globe). These are major operational partners for NATO. For example, Australia, New Zealand, and South Korea have provided troops for NATO missions, while Japan has supported several stabilization projects in Afghanistan.

A framework for cooperation with democratic, like-minded partners in Asia

- Will deepen its partnerships with countries such as Australia, India, Indonesia, Japan, South Korea, New Zealand, Singapore, Malaysia, and Vietnam through practical cooperation on issues where the Netherlands shares common interests and through annual consultations.
- Will accede to the Treaty of Amity and Cooperation in Southeast Asia to underscore its commitment to closer cooperation with ASEAN.
- It will coordinate its interventions as much as possible with like-minded Indo-Pacific countries.
- We will continue to participate in the Asia-Europe Meeting (ASEM) and make an annual financial contribution to the Asia-Europe Foundation (ASEF).

- We will continue to strive to conduct an open, critical dialogue with Indo-Pacific governments about the human rights situation in their countries and to strengthen civil society in those countries.

Sustainable trade and economies

- Will work through the EU toward reducing one-sided strategic dependencies and making value chains more dependable by exploring the scope for diversifying suppliers from the Indo-Pacific region.
- We will work toward sustainable trade and investment relations in Indo-Pacific countries, especially in economic priority countries (China, South Korea, India, Australia, Japan, and ASEAN's five original members).
- Will actively support the EU negotiations (and their conclusion) on free trade agreements with Indo-Pacific countries, Effective multilateralism, and the international legal order.
- It will act in concert with like-minded countries in the Indo-Pacific region to promote effective multilateralism and international legal order.
- Will engage in public diplomacy and other activities on the themes of the law of the sea, the cyber domain, trade law, and climate change.
- We will organize a seminar in Singapore aimed at capacity-building for experts from ASEAN countries who participate in the negotiations on the Code of Conduct for the South China Sea.
- We will continue offering annual sea law courses through the Clingendael Institute and the Netherlands Institute for the Law of the Sea (NILOS) to strengthen capacity in ASEAN countries. Where possible, the Netherlands will join in initiatives and activities from the region itself on this issue.
- Will consult with Australia and other interested, like-minded countries in the region on developments in the South China Sea.
- Will organize a cyber dialogue aimed at capacity-building for experts from ASEAN countries who participate in UN negotiations on international agreements on voluntary, non-binding norms of conduct for states and on developing a system of confidence-building measures in the digital domain.
- We will continue to offer courses on building cyber capacity in ASEAN countries, including international law in the digital domain.

Sustainable connectivity

- Will focus on digital strategy, including a spectrum of themes ranging from cybersecurity and internet regulation to innovation, artificial intelligence, e-commerce, cross-border data transfer, privacy, and national digital sovereignty.
- Join EU initiatives to strengthen Europe's strategic sovereignty, a balance between the diversification of supply and value chains, on the one hand, and reinforcement of global and multilateral free trade systems, on the other, and

- Will join in implementing the Green Deal internationally in cooperation with like-minded Indo-Pacific partners.

Global challenges: Climate and the SDGs

- Deepen and extend collaboration with other countries, including the SIDS, on climate action and the SDGs.
- Work more closely with Indo-Pacific countries to promote ambitious international and national climate policies to achieve climate goals.
- Work to promote green, sustainable energy solutions to advance the energy transition.

6.4 *The EU Strategy for Cooperation in the Indo-Pacific [65]*

The rationale behind the EU strategy for Indo-Pacific given by the European Union represents the generic understanding of European Countries individually and collectively of the Indo-Pacific phenomenon, namely a fast-developing economic growth potential shifting from the Atlantic in the twentieth century toward the center of gravity of Economic Development of the world at the Indo-Pacific in the twenty-first century. Therefore, the EU intends to increase its engagement with the region to build partnerships that reinforce the rules-based international order, address global challenges, and lay the foundations for a rapid, just, and sustainable economic recovery that creates long-term prosperity. This engagement will be based on promoting democracy, the rule of law, human rights, and universally agreed commitments such as the 2030 Agenda and its Sustainable Development Goals and the Paris Agreement on Climate Change.

The futures of the EU and the Indo-Pacific are inextricably linked, given the interdependence of the economies and the shared global challenges. The region includes seven G20 members—Australia, China, India, Indonesia, Japan, the Republic of Korea, and the Republic of South Africa—and the Association of Southeast Asian Nations (ASEAN), an increasingly important partner for the EU.

The EU and the Indo-Pacific are natural partner regions in trade and investment. The Indo-Pacific is both an essential source of global environmental challenges as well as a potential beneficiary of their remedies. EU and the Indo-Pacific are natural partner regions in trade and investment.

In recent years, geopolitical dynamics in the Indo-Pacific have given rise to intense competition, including tensions around contested territories and maritime zones. Democratic principles and human rights are also threatened by authoritarian regimes in the region, putting the region's stability at risk. Similarly, efforts to establish a global level playing field based on transparent trade rules are increasingly undermined by unfair trade practices and economic coercion. These developments increase tensions in trade, supply, and value chains.

The EU Strategy released in September 2021 is inclusive of all partners in the region wishing to cooperate with them when interests coincide. At the same time, and working with international partners who share similar concerns, the EU will continue to protect its vital interests and promote its values, namely democracy, human rights, and the rule of law.

The European Union is already involved with the Indo-Pacific. They intend to deepen it further by deepening relations with the QUAD leaders' countries, such as the US, Japan, Australia, and India, and then with other countries in the Indo-Pacific. They want to collaborate with the ongoing program of the QUAD in the Indo-Pacific. Some examples are given below in the figure.

EU Vision for Engagement with the Indo-Pacific.

- Solidify and defend the rules-based international order by promoting inclusive and effective multilateral cooperation based on shared values and principles, including a commitment to respecting democracy, human rights, and the rule of law.
- Promote a level playing field and an open and fair environment for trade and investment.
- Contribute to achieving the Sustainable Development Goals (SDGs), addressing climate change and environmental degradation on land and in the ocean, and supporting sustainable and inclusive socio-economic development.
- Engage in bilateral and multilateral cooperation with partners to meet the objectives of the Paris Agreement on Climate Change and the Convention on Biological Diversity (CBD).
- Pursue its long-standing multilateral and regional cooperation with the United Nations, Bretton Woods Institutions, and regional organizations such as ASEAN and the African Union in the Western Indian Ocean.
- Support truly inclusive policymaking and cooperation, where the voices of civil society, the private sector, social partners, and other vital stakeholders count.
- Establish mutually supportive trade and economic relations with the region that foster inclusive economic growth and stability and promote and facilitate connectivity.
- Engage in the region as a partner in our efforts to raise awareness of the impact of global demographic trends.

The EU's engagement with the Indo-Pacific region will be moral and long term. It will seek to:

- Deepen its engagement and reinforce its role as a dependable partner, adding value to long-standing relations with all its regional partners.
- Reinforce cooperation with multilateral and regional organizations such as ASEAN and international financial institutions to promote effective rules-based multilateralism in the Indo-Pacific region.
- Undertake crisis management, conflict prevention, and resilience-building initiatives. Work with the Member States through a Team Europe approach with concrete initiatives at country and regional levels.

European Union's Seven Priorities in the Indo-Pacific.

- sustainable and inclusive prosperity.
- Green transition.
- Ocean governance.
- Digital governance and partnerships.
- Connectivity.
- Security and defense.
- Human security.

The above incorporates the goals given below.

Resilient and Diversified Value Chains
Resilient value chains are essential for recovery. The EU will work with its Indo-Pacific partners to reinforce value chains by strengthening and diversifying trade relations and implementing existing trade agreements, finalizing ongoing trade negotiations, and developing cooperation in strategic sectors, including addressing strategic dependencies in supply chains. For semiconductors, for example, it will do so with partners such as Japan, the Republic of Korea, and Taiwan. The EU will also cooperate with partners to strengthen rules to protect international trade against unfair practices, such as industrial subsidies, economic coercion, forced technology transfers, and intellectual property theft.

Building Blocks of EU Trade Relationship
The EU is committed to further engagement on open, sustainable, and rules-based trade with partners in the Indo-Pacific region, including building support for the modernization of the World Trade Organization. Particular attention will be paid to implementing and enforcing the comprehensive trade agreements with Japan, the Republic of Korea, Singapore, and Vietnam, and the Economic Partnership Agreement (EPA) with the Pacific States, as well as the EU investment protection agreements with Singapore and Vietnam that are expected to enter into force in the coming years.

Green Transitions

- Conclude Green Alliances with like-minded partners that have signed up to the goal of climate neutrality by 2050 and other ambitious climate and environmental objectives. The first such alliance was agreed with Japan in May 2021. The EU will also build Green Partnerships with other partners.
- Continue to use the International Platform on Sustainable Finance to share best practices and seek common ground with like-minded partners on approaches and tools.
- Engage with the largest emitting countries or regional organizations that can play a decisive role in fighting climate change and promoting the global green transition.
- Continue to prioritize the transition away from coal in bilateral and multilateral engagements and international fora, including ending new coal investments, phase-out of unabated coal-fired power generation and coal mining, and engaging in just transition with the partners in the region.

- Step up its work with the region to protect biodiversity and restore degraded ecosystems on land and in the oceans. This will include collaborating with key partners on an ambitious post-2020 Global Biodiversity Framework.
- Cooperate to create the conditions for more circular production models, more resilient supply chains between the EU and the region, and more responsible resource extraction.
- Promote deforestation-free supply chains.
- Continue to promote enhanced regional and global action to tackle plastic pollution, notably by cooperating with ASEAN, Japan, and China.
- Continue high-level dialogues and other forms of engagement on the environment in the region, notably with ASEAN, Australia, China, India, Indonesia, Japan, the Republic of Korea, South Africa, and other partners showing interest.

Clean Energy and Transport
In this area, the EU intends to focus on:

- They mobilize energy dialogues, partnerships, and financial instruments for sustainable, secure, and affordable energy.
- It prioritizes a just transition toward a decarbonized, integrated energy system that considers and mitigates the impact on more vulnerable countries and regions.
- Continuing cooperation with Indo-Pacific partners on R&D on clean energy technology, including renewable hydrogen as a priority area, to enable a faster, cheaper, and more efficient energy transition.
- Boosting cooperation under the Global Covenant of Mayors for Climate and Energy with cities as drivers of climate action and clean energy transition to double the number of signatories in the region (three hundred cities by 2023).
- It is implementing its Sustainable and Smart Mobility Strategy to address the issues of decarbonization and digitalization with Indo-Pacific partners and at the International Civil Aviation Organization (ICAO) and the International Maritime Organization.
- Continuing its engagement with several Indo-Pacific economies on digitalization of transport, as well as in the rail sector, regarding the deployment of the European Rail Traffic Management System.

Ocean Governance
The EU will take action to strengthen ocean governance in the Indo-Pacific in full compliance with international law, UNCLOS, and with the primary objective of ensuring the sustainable management of the ocean's resources and safeguarding of biodiversity. Through its various Sustainable Fisheries Partnership Agreements in the region and its dialogues and working groups on Illegal, Unreported, and Unregulated (IUU) fishing, the EU will continue supporting Indo-Pacific partners in achieving fisheries management and control systems reforms. This should improve fisheries compliance and contribute to the conservation and sustainable management of marine biological resources across the region. As the largest export market for seafood products from the Indo-Pacific region, the EU is an active member of several Regional Fisheries Management Organizations.

Digital Governance and Partnership
The EU will seek to enhance its international digital partnerships in the Indo-Pacific region and set up new ones. These partnerships aim to strengthen technical, policy, and research cooperation with partners on infrastructures, digital transformation of business and public services, skills development, and facilitate digital trade. They will allow the EU and like-minded partners to ensure the development of standards for emerging technologies, including artificial intelligence, in line with democratic principles and fundamental rights. A toolbox will underpin them, drawing on regulatory cooperation, capacity-building and skills, and investment in international collaboration and research partnerships.

Research and Innovation
International cooperation in research and innovation is crucial for green and digital transitions and for promoting growth, prosperity, and social well-being. Collaboration with partners in the Indo-Pacific will be enabled under 'Horizon Europe,' in line with the EU's Global Approach to Research and Innovation. It will be based on the principle of openness, balanced with greater levels of reciprocity, and seeking a level playing field based on respect for fundamental principles such as academic freedom, gender equality, ethics, integrity and inclusiveness of research, and open science and evidence-based policymaking.

Education
EU investments in education will be increased to at least 10% of the Neighborhood, Development, and International Cooperation Instrument (NDICI)—Global Europe to strengthen education systems. The recent Team Europe pledge of EUR 1.7 billion to the Global Partnership on Education will translate into increased funding for primary and secondary education, as well as Technical and Vocational Education and Training.

Connectivity
The EU will aim to promote all dimensions of connectivity with Indo-Pacific partners. The EU's first two "Connectivity Partners," Japan and India, are core Indo-Pacific partners, as is ASEAN with which the EU agreed to a joint Ministerial Declaration on Connectivity in December 2020. The EU will seek to collaborate with these partners on collaborative projects while seeking increased collaboration with other regional partners, such as Australia and the Republic of Korea as well as with international actors, such as the United States and Canada, both multilaterally (G7/G20) and bilaterally. The conclusion of the negotiations on the ASEAN-EU Comprehensive Air Transport Agreement (CATA), the first such region-to-region agreement encompassing thirty-seven countries, shows the EU's determination to deepen connectivity with this region.

The EU will further:

- Strengthen relations at the highest technical level through its Transport Dialogues with regional partners, such as ASEAN, Singapore, and Japan, and shortly with the Republic of Korea and Australia.
- Continue to fund sector-specific technical cooperation, such as Aviation Partnerships with several Asian regions.
- Continue the EU bilateral maritime transport and security dialogues with strategic partners.
- Pursue space dialogues, including a security component where appropriate, and establish new exchanges with regional partners.

Health

The EU will continue to work with all Indo-Pacific partners to ensure an effective multilateral response to COVID-19 and future global health crises. This would encompass the following:

- They are helping low and middle-income Indo-Pacific partners to secure access to the COVID-19 vaccine through the COVAX facility and other means. The EU has supported many countries in the region by providing personal and medical equipment, medicines, and health expertise. India will be a focus for cooperation, including on the quality of active pharmaceutical ingredients. The future EU Health Emergency Preparedness and Response Authority will cooperate closely with global partners to address international supply chains and expand global production capacity and access to medical countermeasures.
- Enhanced multilateral cooperation aligns with the EU Pharmaceutical Strategy to secure safe and diverse pharmaceutical and health-related industrial supply chains, facilitating access to quality medicines and health products.
- Collaborative research for combating infectious diseases and improving access to medicines and health treatments. Under the 'Horizon Europe research program, Indo-Pacific partners can participate in EU-funded Research and Innovation actions.
- It connects exciting countries implementing interoperable COVID-19 certificate systems to the EU Digital COVID Certificate. The EU has made the technical specifications and underlying software publicly available in open-source format.
- They were supporting the transformation toward healthy and sustainable food systems. The EU aims to enhance cooperation with like-minded partners on its Farm to Fork and Biodiversity Strategies. The EU intends to establish dialogues on sustainable food systems with interested Indo-Pacific partners (some planning their sustainability policies and programs) or enter bilateral and multilateral arrangements to support collaboration on food safety, animal and plant health, and sustainability.
- We are working together to fight air pollution, especially in urban centers.

Highlights of proposed EU actions:

- We are engaging with Indo-Pacific partners to build more resilient and sustainable global value chains by diversifying trade and economic relations and developing technical standards and regulations that align with our values and principles.
- Completing EU trade negotiations with Australia, Indonesia, and New Zealand; resuming trade negotiations and starting investment negotiations with India; completing an Economic Partnership Agreement with the East Africa Community; assessing the possible resumption, of trade negotiations with Malaysia, the Philippines, and Thailand, and the eventual negotiation of a region-to-region trade agreement with ASEAN.
- Concluding Partnership and Cooperation Agreements (PCA) with Malaysia and Thailand; starting PCA negotiations with the Maldives and bringing the EU's upcoming Partnership Agreement with the African, Caribbean, and Pacific (ACP) to full fruition.
- Conclusion Green Alliances and Partnerships with willing and ambitious Indo-Pacific partners to fight against climate change and environmental degradation.
- We are strengthening ocean governance in the region, including increasing the EU's support for Indo-Pacific countries' fisheries management and control systems, fighting against IUU fishing, and implementing Sustainable Fisheries Partnership Agreements.
- We are expanding the network of digital partnerships with Indo-Pacific partners and exploring the possibility of new Digital Partnership Agreements.
- They are stepping up the implementation of the Connectivity Partnerships with Japan and India, supporting partners in establishing an appropriate regulatory environment, and facilitating the mobilization of the necessary funding to improve connectivity between Europe and the Indo-Pacific.
- Strengthen cooperation on research and innovation under 'Horizon Europe'; explore.
- The association to this program of eligible like-minded Indo-Pacific partners such as Australia, Japan, the Republic of Korea, New Zealand, and Singapore.
- Exploring ways to ensure enhanced naval deployments by the EU Member States to help protect the sea lanes of communication and freedom of navigation in the Indo-Pacific while boosting Indo-Pacific partners' capacity to ensure maritime security.
- One reinforces support for healthcare systems and pandemic preparedness for the least-developed countries in the Indo-Pacific region, enhancing collaborative research on infectious diseases in the Horizon Europe research program context.

The implementation of the EU strategy will include several actions, in particular: Concluding Partnership and Cooperation Agreements (PCA) with Malaysia and Thailand; starting PCA negotiations with the Maldives and bringing the EU's upcoming Partnership Agreement5/2/22, 9:59 AM EU Strategy for Cooperation in the Indo-Pacific bringing the EU s forthcoming Agreement Partnership with the African, Caribbean, and Pacific (ACP) to full fruition and engaging with Indo-Pacific partners to build more resilient and sustainable global value chains by diversifying trade and economic relations, and by developing technical standards and regulations that are in line with our values and principles. Completing EU trade negotiations with Australia, Indonesia, and New Zealand; resuming trade negotiations and starting investment negotiations with India; completing an Economic Partnership Agreement with the East Africa Community; assessing the possible resumption of trade negotiations with Malaysia, the Philippines, and Thailand, and the eventual negotiation of a region-to-region trade agreement with ASEAN. Conclusion Green Alliances and Partnerships with willing and ambitious Indo-Pacific partners to fight against climate change and environmental degradation. Strengthening ocean governance in the region, including increasing the EU's support for Indo-Pacific countries' fisheries management and control systems, the fight against Illegal, unreported, and unregulated fishing, implementing Sustainable Fisheries Partnership Agreements, and expanding the network of digital partnerships with Indo-Pacific partners and exploring the possibility of new Digital Partnership Agreements. Strengthen cooperation on research and innovation under 'Horizon Europe'; explore the association to this program of eligible like-minded Indo-Pacific partners such as Australia, Japan, the Republic of Korea, New Zealand, and Singapore. We are stepping up the implementation of the Connectivity Partnerships with Japan and India, supporting partners in establishing an appropriate regulatory environment, and facilitating the mobilization of the necessary funding to improve connectivity between Europe and the Indo-Pacific.

Exploring ways to ensure enhanced naval deployments by the EU Member States to help protect the sea lines of communication and freedom of navigation in the Indo-Pacific while boosting Indo-Pacific partners' capacity to ensure maritime security. Reinforcing support to healthcare systems and pandemic preparedness for the least-developed countries in the Indo-Pacific region, enhancing collaborative research on infectious diseases in the Horizon Europe research program context.

6.5 *The US Indo-Pacific Strategy February 2022[66]*

The United States will pursue an Indo-Pacific region that is:

1. FREE AND OPEN

Our vital interests and those of our closest partners require a free and open Indo-Pacific, and a free and open Indo-Pacific requires that governments make their own choices and that shared domains are governed lawfully. Our strategy begins with strengthening resilience within individual countries, as we have done in the United States and among them. We will advance a free and open region, including:

- Investing in democratic institutions, a free press, and a vibrant civil society.
- We are improving fiscal transparency in the Indo-Pacific to expose corruption and drive reform.

- Ensuring the region's seas and skies are governed and used according to international law.
- Advancing common approaches to critical and emerging technologies, the internet, and cyberspace.

2. CONNECTED

A free and open Indo-Pacific can only be achieved if we build collective capacity for a new age. The alliances, organizations, and rules the United States and its partners have helped build must be adapted. We will build collective capacity within and beyond the region, including by:

- Deepening our five regional treaty alliances with Australia, Japan, the Republic of Korea (ROK), the Philippines, and Thailand.
- Strengthening relationships with leading regional partners, including India, Indonesia, Malaysia, Mongolia, New Zealand, Singapore, Taiwan, Vietnam, and the Pacific Islands.
- Contributing to an empowered and unified ASEAN.
- Strengthening the QUAD and delivering on its commitments.
- Supporting India's continued rise and regional leadership.
- Partnering to build resilience in the Pacific Islands.
- Forging connections between the Indo-Pacific and the Euro-Atlantic.
- Expanding US diplomatic presence in the Indo-Pacific, particularly in Southeast Asia and the Pacific Islands.

3. PROSPEROUS

The prosperity of everyday Americans is linked to the Indo-Pacific. That fact requires investments to encourage innovation, strengthen economic competitiveness, produce good-paying jobs, rebuild supply chains, and expand economic opportunities for middle-class families: 1.5 billion people in the Indo-Pacific will join the global middle class this decade.

We will drive Indo-Pacific prosperity, including by:

Proposing an Indo-Pacific economic framework, through which we will:

- Develop innovative approaches to trade that meet high labor and environmental standards.
- Govern our digital economies and cross-border data flows according to open principles, including through a new digital economy framework.
- Advance resilient and secure supply chains that are diverse, open, and predictable.
- Make shared investments in decarbonization and clean energy.
- Promoting free, fair, and open trade and investment through the Asia–Pacific Economic Cooperation (APEC), including in our 2023 host year.
- Closing the region's infrastructure gap through Build Back Better World with G7 partners.

4. SECURE

For 75 years, the United States has maintained a strong and consistent defense presence necessary to support regional peace, security, stability, and prosperity. We are extending and modernizing that role and enhancing our capabilities to defend our interests and deter aggression against US territory, allies, and partners. We will bolster Indo-Pacific security, drawing on all instruments of power to prevent aggression and counter coercion, including:

- Advancing integrated deterrence.
- Deepening cooperation and enhancing interoperability with allies and partners.
- Maintaining peace and stability across the Taiwan Strait.
- Innovating to operate in rapidly evolving threat environments, including space, cyberspace, and critical and emerging technology areas.
- Strengthening extended deterrence and coordination with our ROK and Japanese allies and pursuing the complete denuclearization of the Korean Peninsula.
- We are continuing to deliver on AUKUS.
- Expanding US Coast Guard presence and cooperation against other transnational threats.
- Working with Congress to fund the Pacific Deterrence Initiative and the Maritime Security Initiative.

5. RESILIENT

The Indo-Pacific faces significant transnational challenges. Climate change worsens as South Asia's glaciers melt and the Pacific Islands battle existential sea-level rises. The COVID-19 pandemic continues to inflict a painful human and economic toll across the region. Moreover, Indo-Pacific governments grapple with natural disasters, resource scarcity, internal conflict, and governance challenges. These unchecked forces destabilize the region. We will build regional resilience to twenty-first-century transnational threats, including by:

- We are collaborating with allies and partners to develop 2030 and 2050 targets, strategies, plans, and policies consistent with limiting global temperature increase to 1.5 degrees Celsius.
- Reducing regional vulnerability to the impacts of climate change and environmental degradation.
- Ending the COVID-19 pandemic and bolstering global health security.

6.6 *Joint Statement of the Leaders of India, Israel, United Arab Emirates, and the United States (I2U2), July 14, 2022 [63]*

I2U2 aims to harness the vibrancy of our societies and enterprising spirit to tackle some of the most significant challenges confronting our world, with a particular focus on joint investments and new initiatives in water, energy, transportation, space, health, and food security. They welcomed the Negev Forum for regional cooperation, which recognizes the unique contributions of each partner country, including Israel's ability to serve as an innovation hub connecting new partners and hemispheres to strategically address challenges that are too great for any one country to manage alone. They discussed innovative ways to ensure longer-term, more diversified food production and food delivery systems that can better manage global food shocks.

Food Security: The UAE—home to the International Renewable Energy Agency (IRENA) and host of COP28 in 2023—will invest USD 2 billion to develop a series of integrated food parks across India that will incorporate state-of-the-art climate-smart technologies to reduce food waste and spoilage, conserve fresh water, and employ renewable energy sources. India will provide suitable land for the project and facilitate farmers' integration into the food parks. US and Israeli private sectors will be invited to lend their expertise and offer innovative solutions contributing to the project's sustainability. These investments will help maximize crop yields and, in turn, help tackle food insecurity in South Asia and the Middle East.

Clean Energy: The I2U2 Group will advance a hybrid renewable energy project in India's Gujarat State consisting of three hundred megawatts (MW) of wind and solar capacity complemented by a battery energy storage system. The US Trade and Development Agency funded a feasibility study for the USD 330 million projects. UAE-based companies are exploring opportunities to serve as critical knowledge and investment partners. Israel and Top, the United States intend to work with the UAE and India to highlight private sector opportunities. Indian companies are keen to participate in this project and contribute to India's goal of achieving 500 GW of non-fossil fuel capacity by 2030. Such projects can make India a global hub for alternate supply chains in the renewable energy sector. They expressed their determination to leverage well-established markets to build more innovative, inclusive, and science-based solutions to enhance food security and sustainable food systems. The leaders also welcomed India's interest in joining the United States, the UAE, and Israel in the.

Agriculture Innovation Mission for Climate Initiative (AIM for Climate).

They affirmed that these are only the first steps in a long-term strategic partnership to promote initiatives and investments that improve the movement of people and goods across hemispheres and increase sustainability and resilience through collaborative science and technology partnerships.

6.7 The Indo-Pacific Strategy of Canada [67]

The Indo-Pacific: A New Horizon of Opportunity and Global Importance of the Indo-Pacific Region

The Indo-Pacific region will play a critical role in shaping Canada's future over the next half-century. Encompassing forty economies, over four billion people, and \$47.19 trillion in economic activity, it is the world's fastest-growing region and home to six of Canada's top thirteen trading partners. The Indo-Pacific part represents significant opportunities for growing the economy here at home and opportunities for Canadian workers and businesses for decades to come.

The Indo-Pacific is rapidly becoming the global center of economic dynamism and strategic challenge. Every issue that matters to Canadians—national security, economic prosperity, respect for international law, democratic values, public health, protecting our environment, the rights of women and girls, and human rights—will be shaped by the relationships Canada and its allies and partners have with Indo-Pacific countries and to maintain open skies, open trading systems, and open societies, as well as to address climate change effectively, will depend in part on what happens over the next several decades in the Indo-Pacific region (Fig. 44).

The region's economic dynamism and population growth drive demand for education, health services, food, agriculture and fisheries, natural resources and critical minerals, energy, financial services, advanced manufacturing, and green infrastructure. These are all sectors of Canadian strength in which Canada has a global reputation for excellence. In the infrastructure sector alone, an estimated $2.1 trillion opportunity exists for strategic investments and partnerships in the Indo-Pacific. Seizing these and other strategic opportunities will help safeguard Canada's economic security, build our future prosperity, and help create good, well-paying jobs. Canada welcomes more international students from India than from any other country in the world. These ties enrich our social and economic fabric and make us stronger. Simply put: the rise of the Indo-Pacific can create extraordinary local benefits, increase prosperity, and drive economic growth across Canada (Fig. 45).

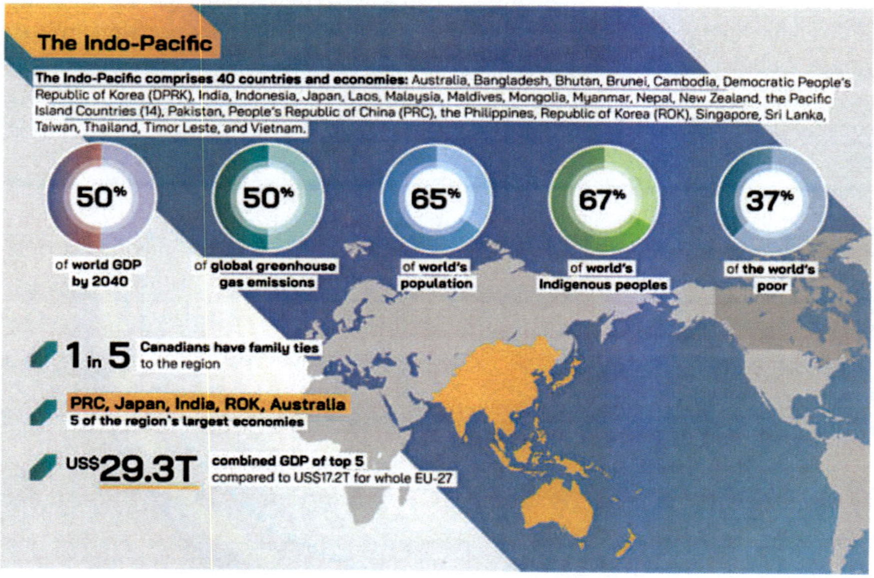

Fig. 44 The Indo-Pacific Canadian perspective. *Source* [67]

Fig. 45 Canadian Trade Connection with Indo-Pacific. *Source* [67]

Strategic Challenges

Regional peace and prosperity are threatened by instability on the Korean Peninsula because of North Korean provocations; rising violence in Myanmar following the recent military coup d'état; clashes on the India-China and India-Pakistan borders; escalating tensions in the South and East China Seas and across the Taiwan Strait; and severe poverty and inequality. The Indo-Pacific is home to four states that possess nuclear weapons (China, India, North Korea, and Pakistan). China's assertive pursuit of its economic and security interests, advancement of unilateral claims, foreign interference, and increasingly coercive treatment of other countries and economies have significant implications in the region, Canada, and worldwide. Respect for the sovereignty of other states is a cornerstone of the rules-based international order and of governments' ability to work together to solve shared problems.

The current level of engagements with some significant partners of Canada in the Indo-Pacific is presented below.

The People's Republic of China

At the domestic level, Canada will continue to strengthen the defense of our Canadian infrastructure, democracy, and Canadian citizens against foreign interference. At the bilateral level, Canada will pursue dialogue with China to advance Canada's national interests while remaining true to our values and focusing relentlessly on Canadian priorities. At the regional level, Canada will balance its approach to China with diversified investments in regional relationships and institutions and a strong vision for the country in the Indo-Pacific region. At the multilateral level, Canada will collaborate closely with its partners to face the complex realities of China's global impact and continue to invest in international governance and institutions (Fig. 46).

India

The status of engagement with India is as shown below.

In its engagement with India, Canada will:

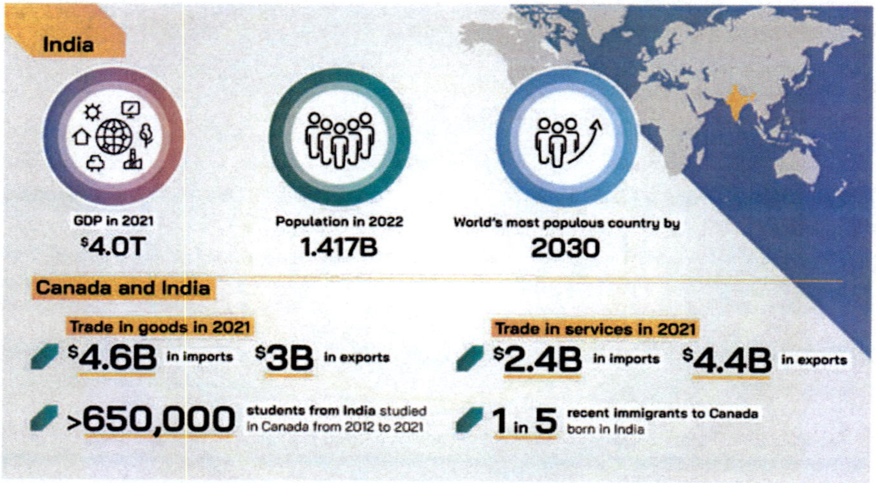

Fig. 46 Canada and India. *Source* [67]

- grow economic ties, including through deeper trade and investment, as well as cooperate on building resilient supply chains,
- seek to expand market access by concluding an Early Progress Trade Agreement (EPTA) as a step toward a Comprehensive Economic Partnership Agreement,
- create a Canada-India desk within the Trade Commissioner Service to promote the implementation of the EPTA for businesses and investors looking to enter the Indian market or for those partnering with Indian businesses,
- invest in and connect people, including by bolstering Canada's visa-processing capacity in New Delhi and Chandigarh,
- support academic, educational, cultural, Oneth, and research exchanges,
- accelerate cooperation in the fight against climate change, in protecting the environment, and in deploying green technologies,
- send enhanced Team Canada trade missions in priority sectors of mutual interest, such as renewable energy and clean technology.

The North Pacific

The status of engagement with the North Pacific is as presented (Fig. 47).

As Part of Its Indo-Pacific Strategy, Canada Will:

- confirm our relationship with ASEAN at the level of Strategic Partner,
- seek membership in the ASEAN Defense Ministers Meeting Plus (ADMM+) and the East Asia Summit,
- augment our contribution to the ASEAN-Canada Plan of Action Trust Fund, which will direct funding toward the priorities determined by ASEAN,
- seek to negotiate and implement a Canada-ASEAN free trade agreement and a Comprehensive Economic Partnership Agreement with Indonesia,

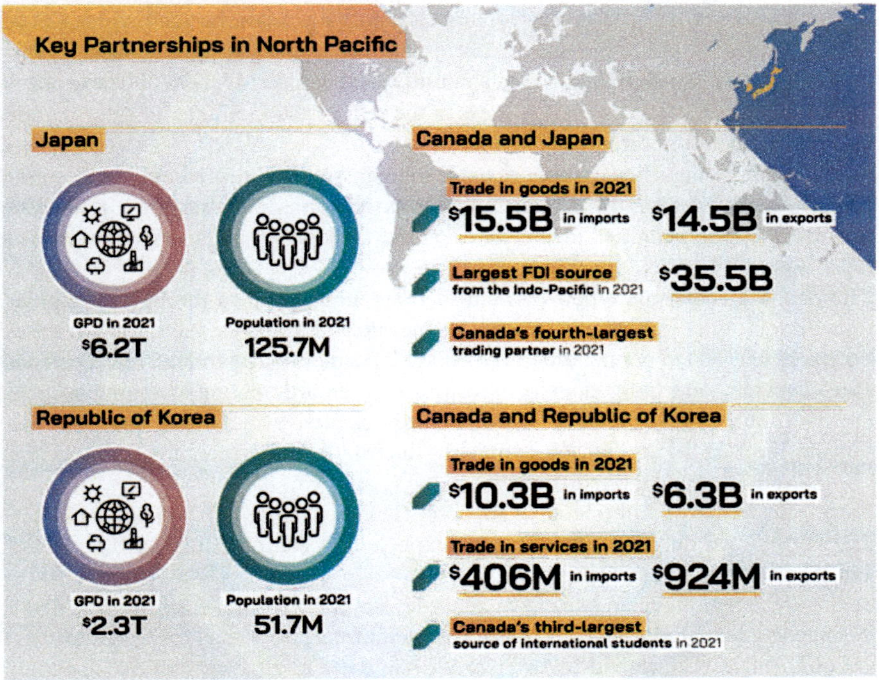

Fig. 47 Canada's North Pacific Partners. *Source* [67]

- launch a Canadian Trade Gateway in Southeast Asia as a market entry point and catalyst for Canadian businesses to grow their engagement and presence in the region and enhance Canada's commercial and investment partner profile. Canada's Trade Gateway will expand Canadian business and investment networks, linking enterprises to existing incubators and accelerators and increasing their awareness of Indo-Pacific markets, emphasizing sectors and solutions where Canadian innovation responds to regional demand.
- strengthen Canadian diplomatic presence in the region and increase security cooperation with ASEAN and its members,
- continue to recognize and support ASEAN centrality in the region by reinforcing alignment between Canada's Indo-Pacific Strategy and the ASEAN Outlook on the Indo-Pacific.
- Expanding trade, investment, and supply chain resilience.
- Expand Canada's trade network at home and abroad; Canada will:
- appoint a new Canadian Indo-Pacific Trade Representative to advance Canada's regional trade policy, promotion, and economic cooperation objectives in the region,

- launch a new series of large-scale Team Canada trade missions that involve Canadian businesses, provinces and territories, and other relevant partners and organizations; Canada will highlight the capabilities and competitiveness of its exporters and innovators to ensure their success in the region and facilitate long-term trade and investment opportunities,
- provide more significant support for Canadian small and medium-sized enterprises, innovators, and national industry associations pursuing new trade and investment opportunities in the Indo-Pacific through an enhanced and tailored Can Export program,
- support the economic empowerment of Indigenous Peoples through the implementation of the Indigenous Peoples Economic and Trade Cooperation Arrangement (IPETCA) in cooperation with existing partners—Australia, New Zealand, and Taiwan—and Indigenous Peoples from those participating economies,
- provide enhanced support to women entrepreneurs to maximize opportunities in the Indo-Pacific region by expanding international partnerships through the Women Entrepreneurship Strategy,
- open our first Agriculture and Agri-Food Office in the region to help Canadian farmers and producers to diversify their exports and position Canada as a preferred supplier in crucial emerging markets,
- expand the regional mandate of Canada's development finance institution, FinDev Canada, to the Indo-Pacific. Canada's commitment to the G7 Partnership for Global Infrastructure and Investment will be done in an approach that looks to amplify, leverage, and collaborate with the Canadian private sector and institutions, like pension funds, should they choose to, on needed small to medium infrastructure projects across the region. This will support high-quality, sustainable infrastructure and provide alternative options for developing economies exploring infrastructure development.

Enhance rules-based trade that provides predictability for economies and exporters, Canada will:

- ensure the Trade Commissioner Service provides expert advice on China's market while promoting export diversification across the region,
- strengthen Canadian economic security by reinforcing the Investment Canada Act to defend our national interests, oversee more vigorous enforcement, and add more precautions to protect intellectual property and Canadian research continue to collaborate with like-minded partners in the region to counter economic coercion and nonmarket practices,
- collaborate with partners to develop digital infrastructure, promote interoperability, and promote coherent regulations affecting the internet, the digital economy, and trust and security in the use of information and communications technology; this will enhance cooperation on standards, norms, and regulations that will benefit Canada, the Indo-Pacific region and rules-based trade,
- provide more excellent technical assistance to Indo-Pacific trading partners to improve the enforcement of labor provisions, including on forced labor, in current and future free trade agreements with Canada; this will help protect workers'

rights, ensure companies are respecting human rights in their supply chains and contribute to leveling the playing field for Canadian workers and employers,

- To ensure the resilience of supply chains, Canada will:
- continue to make significant investments in domestic transportation infrastructure projects through the National Trade Corridors Fund, such as investments in the ports of Vancouver and Prince Rupert, for substantial upgrades to Canada's marine, port, airport, road, and rail infrastructure, increasing national trading capacity, fluidity, and efficiency,
- advance shared interests in supply chain resilience with partners, including through greater Canadian involvement and alignment with regional initiatives, such as the Australia-Japan-India Supply Chain Resilience Initiative,
- contribute to global supply chain resilience and sustainability by engaging in technical, policy, and high-level cooperation through APEC on standards, conformity assessment procedures, and best practices to enhance trade,
- position Canada to be a dependable supplier of clean energy in the region—and in a net-zero emissions economy—including critical minerals and hydrogen, to fight climate change, be an energy security partner, and support global climate goals,
- expand natural resource ties—related to trade, investment and science, technology, and innovation—with priority Indo-Pacific partners,
- strengthen Canada's Science, Technology, and Innovation partnerships with key economies, including Japan, the Republic of Korea, India, Singapore, and Taiwan, to support international coinnovation projects and commercialization-oriented research and development partnerships for Canadian small and medium-sized businesses with Indo-Pacific partners,

Increase export diversification and free trade access, Canada will:

- collaborate with partners to strengthen and expand the CPTPP and ensure that any form of expansion will be based on lofty standards and histories,
- seek to join and meaningfully contribute to the newly established Indo-Pacific Economic Framework for Prosperity,
- improve market access through new comprehensive free trade agreements, such as those currently being negotiated with ASEAN, India, and Indonesia, and on new or modernized Foreign Investment Promotion and Protection Agreements,
- join the Digital Economic Partnership Agreement with Singapore, New Zealand, and Chile while securing a cultural-industries exception for the Canadian cultural sector.

Strengthen people-to-people exchanges, Canada will:

- bolster visa-processing capacities in critical locations in the region to ease access for students and family members; this initiative will enhance visa-processing capacity within Canada's centralized network, as well as abroad in New Delhi and Chandigarh, India; Islamabad, Pakistan; and Manila, Philippines; these new resources will support ongoing efforts to accommodate the high volume of temporary resident visa applications (including visas, study permits and work permits) from the region,

- facilitate travel to and from the region to strengthen people-to-people ties and bolster tourism by using different tools, such as the new and modernized Air Transport Agreements, and explore options to facilitate travel further, making trips to Canada more accessible, faster, and safer for everyone,
- leverage its reputation as a top education destination by launching over 1,000 scholarship and fellowship opportunities for Canadian students and students from ASEAN countries,
- strengthen Canada's international student program with permanent residence and job opportunities for students from the Indo-Pacific that could lead to staying in Canada to contribute to Canada's future; these investments will attract highly skilled workers to Canada to meet the needs of our economy today and into the future,
- ensure that under the Indo-Pacific Regional Engagement Initiative, a broad range of Canadian sector leaders, subject-matter experts, academics, and civil society groups will be able to build new networks in the Indo-Pacific, strengthen the Canadian voice on priority issues and create partnerships on behalf of Canada invite over 200 Canadian experts to go to the region to support countries that seek to partner with Canada in tackling challenges where Canada has the expertise, such as in governance, energy transition and oceans management,
- foster greater inclusion of Canadian diaspora communities of Indo-Pacific heritage through expanded cultural initiatives,
- fight against anti-Asian racism in Canada, in all its forms, through the full implementation of the federal Anti-Racism Strategy,
- pursue the path of reconciliation with First Nations, Inuit, and Métis Peoples through enhanced Indigenous exchanges with regional partners, such as Australia, New Zealand, and Taiwan; support education and skills development for Indigenous Oneth; continue the implementation of the IPETCA and support the performance of the UN Declaration on the Rights of Indigenous

Peoples

- support francophone immigration objectives by expanding our diversification efforts in Cambodia, Laos, and Vietnam—home to 1.3 million French speakers,

Strengthen our international assistance, Canada will:

- deepen partnerships in the region by increasing feminist international assistance programming based on partner needs; Canada will help to protect the most vulnerable populations and support work from achieving the Sustainable Development Goals,
- support efforts toward democracy, inclusivity, accountable governance, and sustained economic growth, helping critical countries in the region,

- work with development partners in the region to reduce inequality and contribute to their economic prosperity,
- continue to provide critical humanitarian assistance, especially in times of crisis, such as the recent floods in Pakistan,
- establish our first international assistance program for the Pacific Islands region, supported by our membership in, and support for, the Partners in the Blue Pacific,
- support Canadian civil society organizations through targeted initiatives that will enable them to pursue human rights and gender equality projects in the Indo-Pacific and bolster partnerships with their regional counterparts,
- help build demining capacity by establishing standards for landmine clearance and accelerate mine clearing by sharing expertise to support personnel, especially women, engaged in demining activities in Laos and Cambodia.

Defend and enforce human rights, Canada will:

- strengthen dedicated Canadian funding and advocacy to support human rights across the Indo-Pacific, including for women and girls, religious minorities, 2SLGBTQI + persons, and persons with disabilities,
- leverage Canada's leadership in multilateral institutions to support regional action on human rights,
- work to counter forms of arbitrary detention by hosting a summit on arbitrary detention in 2023,
- fully implement the next phase of Canada's Rohingya Strategy,
- support peacebuilding in Sri Lanka and Myanmar and encourage truth-seeking and accountability for alleged violations of human rights and humanitarian law during the civil conflict in Sri Lanka.

Building a Sustainable and green future

Support the region in building a sustainable and green future, Canada will:

- expand the capacity for FinDev Canada to support high-quality, sustainable infrastructure in the Indo-Pacific and provide alternative options to developing economies exploring infrastructure development,
- support oceans management initiatives and expand measures against illegal, unreported, and unregulated fishing in the Indo-Pacific, including through our Dark Vessel Detection Program, which uses Canadian technology to find illegal fishing vessels and protect fish stocks,
- establish a signature initiative in the region on disaster risk and resilience that enables Canada to share expertise and help countries adapt to, and be ready for, the impacts of climate-induced disasters before, during, and after disasters strike,
- boost commercial demonstration of clean Canadian technology in priority Indo-Pacific markets and help Canada's clean technology small and medium-sized enterprises with financial support to break into markets in the region,
- build on the already allocated $1.26 billion out of the Canada Climate Finance Commitment toward the Indo-Pacific region to assist partner countries with

economic recovery and infrastructure needs and catalyze inclusive and sustainable development through Canadian capital, technology, and policy expertise,
- prioritize the Indo-Pacific region as part of the Powering Past Coal Alliance, which is working to help partners advance their transition from unabated coal power generation to clean energy; will collaborate with partners in the region to support a change to cleaner energy22 Canada's Indo-Pacific Strategy,
- ensure Canada's international assistance program will target climate and environmental action, with the most significant impact based on needs defined locally and related to climate change, energy transition, biodiversity, and oceans management,
- work with Indo-Pacific countries to prevent plastic pollution from entering the oceans and the environment, reduce plastic waste, and better manage existing plastic resources, including by supporting an ambitious, legally binding global agreement to end plastic pollution that addresses the entire lifecycle of plastics,
- advance Canada's Global Carbon Pricing Challenge to share expertise on carbon pollution pricing as a valuable tool to fight climate change and drive investment in clean technologies.

Canada is an active and engaged partner in the Indo-Pacific Canada's overarching priority is to be an active, dynamic, and dependable partner. Canada will build influence among our regional partners and allies by increasing our diplomatic engagement, forging connections between like-minded countries, and collaborating on common causes. Canada is expanding and deepening its political, economic, and security partnerships through sustained investment and engagement at the highest levels. It is also sustainable development assistance and cultural footprint throughout the Indo-Pacific. Defending the rules-based international order in the Indo-Pacific requires a dynamic approach to Canadian diplomacy and engagement. Canada will pursue its cooperation as it always has as a friend and partner committed to mutual respect; as a proudly diverse country; and as a champion of multilateralism motivated by the idea that our future security and prosperity are best served when all nations—large and small—abide by global and regional rules. Canada's Indo-Pacific Strategy continues our long tradition of international and regional commitments tailored to meet the needs of our regional allies and partners.

Become a more active and engaged partner in the Indo-Pacific, Canada will:

- appoint a special envoy for the Indo-Pacific region to coordinate a whole-of-government approach and manage the strategic implementation of the strategy,
- recognizing ASEAN's centrality in the region, confirm and advance our relationship with ASEAN to the level of Strategic Partner,
- seek membership in the ADMM + and the East Asia Summit,
- increase Canada's diplomatic presence and engagement in the region through more frequent high-level travel, more diplomatic representation, and deepening engagement in regional groupings,
- strengthen engagement with the Pacific Islands Countries by opening Canada's first mission to Fiji, committing to increased ministerial-level representation at the

meetings of the Pacific Islands Forum, and contributing concretely as a member of the Partners in the Blue Pacific initiative,
- continue to grow its economic and people-to-people ties with Taiwan while supporting its resilience,
- deploy Canada's first diplomatic position in Hawaii to lead engagement with local US and international partners23 Canada's Indo-Pacific Strategy,
- bolster the Asia–Pacific Foundation of Canada's engagement in the region by opening an office in the area and through a new range of activities, including seminars, events, and research programming related to the Indo-Pacific,
- hold inaugural Canada–United States Strategic Dialogue on the Indo-Pacific in 2023,
- increase Canada's presence—supporting analysis, consultation, and diplomacy related to China— at Canada's multilateral missions to the United Nations, the European Union in Brussels, and NATO, to anticipate and respond to political, economic, and security trends.

7 Conclusions

7.1 Configuration of the Megacity System in the Indo-Pacific

The G7, G8, and G20 Countries are the largest economies in the world and the strongest in terms of military and defense capabilities. Both are equally important for regional development. They are the countries that develop the latest and emerging innovative technologies with the potential to transform work and life and can become their market leaders. They are also a depository of the world Megacity system of megacities and the Indo-Pacific megacity system of twenty-one megacities discussed earlier. They are represented in the diagram below by their national flags (Fig. 48).

France, Germany, the UK, Japan, Canada, the European Union, Australia, India, and the United States are either members of the QUAD group of countries or have their independent policies, programs, and proposals in the Indo-Pacific with sizeable and recently strengthened diplomatic presence there in the Indo-Pacific and fully ready for collaborative action with strategies, Administrative Protocols, and budgets. Although not active now, Italy is open to East Asia and maritime defense cooperation. Among the G20 Countries, India, Australia, Indonesia, China, South Korea, and the European Union are present in the Indo-Pacific either as QUAD members or with independent country policies, programs, and investment proposals and presence. China and Russia are against any NATO-like defense cooperation of QUAD, while it is not the terms of reference of QUAD. All world systems of megacities and Indo-Pacific megacity systems are part of these countries articulating their economic potential.

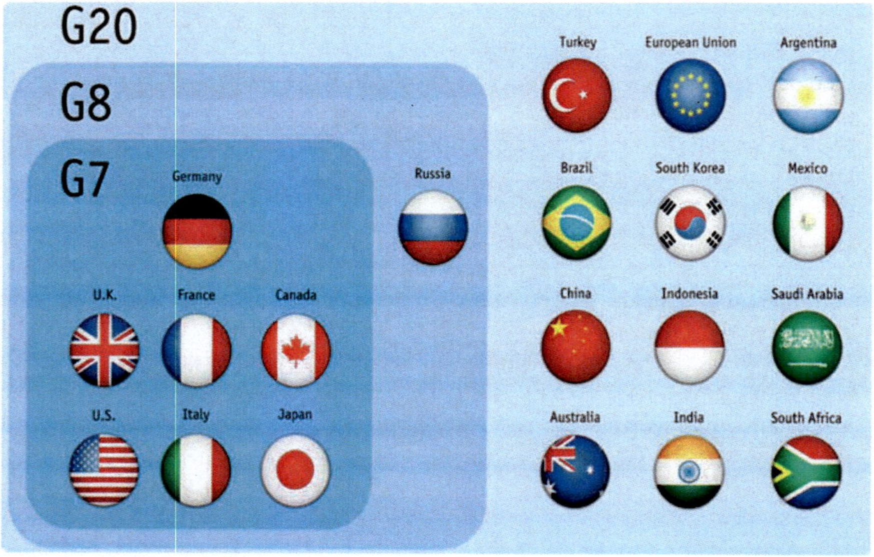

Fig. 48 G7, G8, and G20 Country Grouping

Argentina, Brazil, and South Africa can be a potential grouping of countries with the US or India. It can form the third QUAD after the UAE, Israel, USA, and India QUAD 2 and the US, Japan, Australia, and India the QUAD 1 or QUAD.

It can be observed that no attempt is made to form an independent QUAD organization with headquarters and branches, as in the case of the United Nations or SARC. At the same time, all countries are committed to long-range cooperation. The QUAD uses its country bureaucracy efficiently and with no duplication of effort and expert working groups independently or jointly as it develops for QUAD programs formulation of implementation and uses the available bureaucracy and private parties as deemed most suitable and efficient for speedy and efficient performance, for the Governance of QUAD based on their programs. The programs are made in consultations with QUAD core member countries and mutually agreed upon by governments who wish to join. The sovereignty of the nations is fully respected. The multipolar world as against the world of superpowers was recognized, and countries can join fully or partially in any QUAD program of action.

Once the country joins the QUAD programs, the barriers between people and distant international and multi-country QUAD must be broken. This can be executed if research and development are conducted and arrive at a detailed design of such Smart Global Economic Communities in the Indo-Pacific. This is decided as the follow-up of this research program, with an outcome expected in 2024 regarding books published.

7.2 *Megacity System is in the Indo-Pacific Region*

The Indo-Pacific region is modeled (which is a simplification of reality) as a system of twenty-one-plus megacities. This system sustains by constant flows of humans, goods and services, and communication; money flows exhibiting mutual dependencies due to interactions. This system includes all dependent countries, megacity subregions, and other dependent city systems.

Administratively, megacities spread to one or more Federal states, such as National Capital Region in Delhi, or districts as many as five sections in Kochi Kannur megacity in Kerala, which is seldom a Union Territory. Megacities have their subregions and zones of influence based on daily interaction and interdependencies within the national locations of these megacities. The landlocked countries outside megacities use the megacities' gateway infrastructure and related institutions for their economic activities.

Megacities deceptively seem to have grown historically as standalone and independent giant conurbations as an economic superpower among a hierarchy of cities, forgetting other megacities in the system. They are neither independent nor standalone cities. It has a local and international economy based on the types of economic interactions. Megacities themselves are interlinked with each other megacities functionally for a variety of higher-order economic, social, cultural, and ecological interactions with wide-ranging spatial impact beyond these cities, creating an integrated Indo-Pacific region increasingly. It had impacted the life and working of the inhabitants, contributed to more extraordinary human accomplishments, and valued destinations for immigration.

As such megacities, the generator of economic superpower demands an uninterrupted daily flow of non-polluted air and water, assured protection of their territories and loss of valuable urban land from the rise of the sea due to climate change, potable water, perennial and reliable energy supply, perishable and nonperishable food, a perennial source of increasing income, and expenditure to households, all types of utilities and services of highest standards, and essential other supplies to survive. For that, they depend on all human settlements below the megacity hierarchy. It should provide work to earn an adequate income to survive honorably in a competitive environment affording a high standard of living within a high land value megacity, commensurate with a higher income than other cities. Suppose One computes the total expenditure to sustain the megacities. In that case, One can find private households and non-government-owned entities in megacities spend a disproportionately higher share of payment than the government. Still, Governance is given a higher set of overriding power by the country's constitution to protect all rights.

Because of all the above factors, megacities are overly complex. In many cases, external factors beyond the national boundary and their external interaction had a more significant say in megacity development. At the same time, internal economic activities make them sustainable by providing a large market of ten million or above population in a limited geographic area for easy market access due to higher megacity densities.

A megacity can be considered a complex of overlapping spatial systems and involves many systems within. For example, energy, water and sewerage, food, transport, health, education, biodiversity, and side-by-side ecological, economic, social, and cultural systems connect one area to another. This complex network of systems, interconnections, and flows can be described as a megacity system of systems. Many of these system components originate from outside the hierarchy of administrative regions. The city cannot politically and administratively influence the system environment's unpredictability, such as war, natural disasters, terrorism, climate change, economic recession, or other primary medium or minor ecological collapses. These eventualities point toward a crucial role of integrated megacity system planning, management, and development to design transcending national boundaries. As events unfold, the megacity system demands higher adaptability and resilience for sustaining a significant population of ten million and above.

Indo-Pacific and QUAD came up in 2004 spontaneously as a response to the Tsunami by a group of countries who helped others, and themselves make all systems stable and sustainable in as limited time as possible in many countries, sharing what they best have to the situation. The world has faced many uncertainties in the past. The Indo-Pacific megacity system consisting of twenty-one megacities and the connected region as the highest repository of human capabilities, financial powers, superior infrastructure, knowledge, skills, and related higher-level institutions, has an essential and confident role in finding remedies for maladies for these uncertainties of natural disasters like tsunamis, earthquakes, floods, and the effects of climatic changes and even economic recession or collapse of National economy like that of Sri Lanka. Megacity systems collectively have a role in helping those affected by these calamities. There are also manufactured calamities like war which can be avoided by dialogue and diplomacy. The spread effect of economic sanctions affects more people in more countries in a connected world who may not be directly involved in the conflict and drives them to food and energy scarcity or scarcity of medicines and equipment, for example, faced during the pandemic COVID-19 or natural disasters or Ukraine war. The solution is to identify many alternatives before the event that impacts supply chains so that these alternatives can substitute for the scarcity and even recessions that can come in. These megacities could identify scenarios of issues and alternate remedies as a joint effort of money, skills, and institutional capabilities of many megacities systems together transcending nations.

7.3 The Economic Strength of Indo-Pacific Megacity System

The two goals of the Indo-Pacific are peace and prosperity. Peace is an absence of war and conflict with the commitment of all countries to a rule-based order. A peaceful and stable megacity system ensures prosperity with no downward and unpredictable fluctuations in income and expenditure.

Like many countries in the Indo-Pacific, India, with around one-third population living in cities a GDP urban share, is about seventy percent. With the increase in the size of the cities, GDP share increases. Hence, megacities' share of GDP is always higher than small cities, including metropolises, even though the total population in megacities is around 8 to 9 percent. With Dhaka, as an example, the only megacity in Bangladesh, its share in GDP is as high as 60 percent of the National GDP, and Cairo is about 50 percent on the higher side but in India, with seven to eight megacities. The individual share is much lesser and evenly distributed to less than 10 to 16 percent but unevenly spatially distributed, with most of them in the West and the South part of India, while in China, it is primarily the eastern part leaving the western part less developed in China and the east part in India. It looks like subregions of megacities are more developed than other regions where access to megacities is limited, so economic opportunities and development of these subregions. Once the GDP of all twenty-one megacities in Indo-Pacific is combined, the percentage of the GDP of Indo-Pacific could be much higher than the rest of the region. It is also seen that a higher GDP share of the megacity means more significant opportunities for higher achievements, creating a situation in which other indicators of human development tend to be higher side by side, as illustrated by Bangladesh, where the secret of high achievements is hidden.

Human resources must be highly accomplished and competitive to realize an aspiration of good living in megacities. For them, high-paying work must be available continuously in megacities. Enhanced living is demanded with increasing income, and the megacity shall provide it. A perennial generation and flow of new knowledge and innovation for all economic, social, and cultural activities are the basis of the megacity system's sustenance. Such knowledge can be shared if the megacity collaborates, transcending national boundaries. Megacities can self-organize through self-organized supply chains for all of the above.

7.4 *Sharing Economy, Platform Economy, and Indo-Pacific Megacity System*

A megacity system is a group of megacities that, in their evolution, have become interdependent because of the many relationship networks that interconnect them to a greater extent than other less populated cities elsewhere. The air, sea, rail, road, broadband, and internet communication connection enables money flow and, in return, commodities, technologies, innovative ideas, design innovations in products and services, other services, and people flow to connect all megacities for economic interaction. E-Commerce, for example, Amazon, a global exporting entity for many countries, makes small sellers with minimal capital and income who can never be conventional shopkeepers become international sellers exporting their wares, using the megacity system and other services such as logistics, accounting, and legal services and sharing at the same time the technologically advanced e-commerce infrastructure platform of Amazon.

So, sharing the common digital platform becomes critical to megacity system economic development such as e-commerce. Just like an e-commerce platform with all its specialized algorithms and routines that move products to consumers' destinations from different origins selected in such a way that minimizes the time of delivery from purchase time, makes payment and makes refunds, and connects megacity for commercial activities. Machine-to-Machine (M2M) systems enabled platforms for supply chain component productions for many households and small and medium industries sharing common electronic platforms. Machine-to-Machine (M2M) systems in the megacity system can be made into self-organized platforms to develop complex manufacturing supply chains and related algorithms in many megacities subjected to resilience under shocks. This can be for defense supplies or civil supplies. A similar platform can be there for service economies of diverse types in a megacity system. We must integrate e-commerce and e-production using M2M, generally a platform in a megacity system in an integration spatially executed with logical algorithm run decision making any time on any day. Once it is achieved, e-commerce and M2M production systems are well integrated with a megacity system is the basis of megacity systems as engines of economic growth and economic, social, and ecological development. This, in short, is the platform economy of the megacity system.

As indicated, twenty-one megacities exist in the Indo-Pacific out of thirty-three in the universe. We have an integrated megacity system once these megacities are interconnected along with e-commerce and spatially scattered M2M small and medium industries network for production with the 21-megacity system. Suppose One brings all of them into one electronic platform and generates different scenarios for the Indo-Pacific policies of countries such as France, the European Union, Japan, and India. In that case, we have a megacity development framework for the Indo-Pacific. The same system can answer supply chains of production and commerce when unanticipated events like sanctions following the Ukraine wars precipitate. This is the challenge of the megacity system before the knowledge community.

Heterogeneity of megacity system Governance is given, and the author is a proud Hindu who believes in extreme and uncompromised tolerance of Hinduism and Buddhism. So, we have autocracy as a governance mechanism for Chinese or Russian megacities in Indo-Pacific and democracy for many others in the Indo-Pacific such as India, Japan, Indonesia, and many others. There can be a mixture of the two, overriding power with the President in a democracy such as Sri Lanka. The rights of citizens vary as per the country's constitution. The constitutional interpretation of Freedom also differs from one country's constitution to others. Platform economy works differently in these systems followed.

The primary beneficiary of components of the platform economy in the megacity system is people below the poverty level, which effectively connects the Government to People by direct money transfer to their bank accounts when they are distressed or face calamities, making them easy to get bank accounts from Aadhar cards which is another internet platform. A man with one taxi car shares it and works with Uber on another mobility platform. A stay-at-home spouse from a lower middle class becomes an entrepreneur in e-commerce with Flipkart or Amazon. So also, a small-scale industrial establishment, however small it may be, participating in a feature-rich platform economy and benefits. These are all examples of the sharing economy of intelligent cities. Without a platform economy and high megacity infrastructure, the rapid transition of the low-income group to the middle-income group and the rich to the super-rich in the megacity system is not possible for several reasons.

We get an open commerce network if all digital e-commerce platforms of different companies are connected inclusively with all commercial establishments outside it. Similar networks are possible for logistics, other services related to health and education and military production, and services and industrial output on open networks. This is more akin to a spatial system than an isolated company e-commerce system in a city. Spatial systems, unlike e-commerce firms, do not exclude outsiders.

This is the age of sanctions where, in addition to the United Nations, which gives out binding sanctions, we have non-binding sanctions from the US, European Union, NATO countries, and many others. Like autocracy and democracy co-exist in the megacity system, the UN and other countries' sanctions co-exist. While urban systems under Autocracy and Democracy do not change instantly, the sanctions triggered by war or trade war can change urban systems. The platform technology and economy behind e-commerce and M2M production systems of supply chains to megacity systems need to be fine-tuned as such sanction event unfolds a set of predesigned different supply chains are enabled to circumvent food, medicine, and energy scarcity that results and to avoid recession from sanctions and to avoid creating more people below poverty level in the third country that is not involved in sanctions. Only a platform-based megacity system can perform and incorporate instantly unexpected events and calamities since it is part of Governing responsibilities. Although it is yet to be worked out by the megacity system of Indo-Pacific, it is a feasible option for Think Tanks to work out. Here, for example, government and private e-commerce and M2M production systems shall come under one megacity system to overcome the impact of such unexpected and shocking eventualities. The country's policies and initiatives direct it from time to time. If the country follows strategic autonomy for defense and economy and is non-aligned, it has more freedom than other countries under alliances.

Here system theory is the basis. Systems theory, the base of the platform economy, is the interdisciplinary study of systems, i.e., cohesive groups of interrelated, interdependent parts that can be natural or human-made. Every design is bounded by space and time, influenced by its environment, defined by its structure and purpose, and expressed through its functioning.

One of the noteworthy aspects of Indo-Pacific is that there are four leading countries, the US, Japan, Australia, and India, and their willingness to collaborate with all countries of the Indo-Pacific as well as outsiders such as the European Union, UK, Germany, Netherlands, Canada, and France, all of them having their Indo-Pacific strategies and willing to be partners in the prosperity of the Indo-Pacific. They all have a genuine interest economically in these regions as a legacy of long historical connections of these countries in the area.

The Indo-Pacific Megacity system starts with the concept put forward as an interdependent region with common goals arising out of shared issues and a unified search for effective and outcome-oriented solutions with the participation of countries as per their capabilities using advanced technologies. The Indo-Pacific was, earlier at its inception, a spontaneous mental creation for disaster management when countries collaborated as if they were one family. These need to be extended to all eventualities in the region, the impact of war, the aftermath of a fight against opponents, sanctions, and countersanctions. The additional dimension is that the Indo-Pacific part can be a global focal area of the twenty-first century for regional economic development based on integrating the megacity system and standardized adoption of emerging technologies. All Indo-Pacific countries have infrastructure and protocol of action once a Tsunami happens and other calamities or Pandemic COVID-19 happens in Indo-Pacific. They can instantly generate alternative supply chains to face unanticipated disruptions, for face masks for COVID-19 prevention and shortage of food and energy in various other situations. All of them have economic, environmental, and social impacts. The modified and adapted supply system for any eventuality generation can be designed in advance, and preparations for necessary action can be put in place jointly by the megacity system, which can be immediate or run into many years and months and operationalized. While the e-commerce system and M2M system looks at only fulfilling the demands for the commercial product within the company, the megacity system looks at holistically as an open network of the global digital economy, all aspect of living and working and its environmental impact in all unexpected eventualities to make the megacity system sustainable and resilient within a brief time to overcome difficulties with collaborative country actions.

7.5 How Indo-Pacific Integrated Regional Development Through Megacity System is Functioning

We know how these systems we discussed above can work through various case studies in my edited books on the smart city series we jointly researched and authored by academics from Africa, Asia, Europe, the Middle East, and North America, such as,

I. Smart Economy in smart cities [55]
II. Smart Metropolitan Regional Development [67]
III. E-Democracy for Smart Cities [68]
IV. E-Governance for Smart Cities [69]
V. Smart Global Megacities Two volume [70, 71]
VI. Smart Environment for Smart Cities [72]
VII. Smart Living for Smart Cities Two books [73, 74]
VIII. Smart Master Planning for Cities Case Studies of Domain and Digital Innovations and Case Studies Two Volume [75, 76].

In addition, we know the following.

a. There is a movement of money, information, people, and goods and services between all twenty-one and above megacities in the Indo-Pacific, and this system functions every millisecond.
b. This system exists because of thousands of island chains, about forty countries in East Africa, Asia, the Middle East, ASEAN, and SARC countries. Twenty- or more megacities solely depend on Integrated Regional Development and Integrated Disaster management and one joint fight against environmental and pandemic calamities such as COVID-19, Tsunami, earthquakes, and disappearance of island chains due to sea rise of Indo-Pacific for climate change. These are all existential threats where an integrated regional solution alone is possible. Individual countries cannot solve these issues independently.
c. Smart City technologies are the building blocks of the Indo-Pacific Regional Development. All cities are the smart city in evolution and intelligent global megacity systems, such as ICT, IoT, Machine-to-Machine systems, Industry 4, 5G, and 6G applications in Industry 4, Intelligent cloud, and the blockchain. Fintech, e-commerce, and e-governance are there. Some are widely used, for example, for direct debit of cash to Bank accounts during distress and disasters circumventing the cumbersome and corruption-ridden bureaucracy, which amounts to several billion dollars in India, and easiness of people experiencing poverty to open bank accounts in India using Aadhar cards and widespread use of Unified Payment Interface UPI of India in Nepal, Singapore, UAE, Bhutan, Australia, and more countries to follow.
d. We know that existing urban land use planning starting from the land-use classification that exists today, is out of date and irrelevant when ICT, IoT, and M2M enabled Smart Cities to evolve in the Indo-Pacific because its past foundation is monoculture land use which needs to be mixed land use totally and everywhere with limited dependence on even emerging environment friendly and individually owned electric Cars since workplaces are distributed in all mixed land use starting from home and then inexpensive local community shared faculties in walking distance. Some land uses and establishments such as bank branches or administrative buildings, for example, can be eliminated by being replaced by UPI or its evolutionary all-compassing systems of future Banking under experimentation once the development of E-Governance, E-Democracy, Smart Mobility, E-Commerce, M2M production system, and Home-based work culture a product of the post-COVID-19 era has also made these redundant. COVID-19, with a compulsory "work@home" culture of Intel and Apple computer industries, was the most productive year for Intel and Apple when the 12th generation chip came up much faster than usual. Apple computers came out with an M1 chip and moved toward M2, and high-powered, compact, and less energy-consuming Apple studio-type computers came up. Brick and Mortar Bank is irrelevant for Fintech end users, Mall is outside for e-commerce users, factories are unrelated for M2M users in production, and twentieth-century organization that houses in one location in many buildings are also out of date for Indo-Pacific based on emerging technologies. This land-use classification and zonal regulations must change.

a. Megacity system future in the Indo-Pacific is emerging technologically dependent and unpredictable. So if there is a Management and Plan for the development of the Indo-Pacific, it must be a three- to five-year long-range perspective plan and annual plans and planning. An implementation time delay may be reduced to milliseconds using smart technologies such as IoT and M2M systems.
b. With fast progressing full ownership of smartphones in megacity systems with high bandwidth and 5G, smartphone enabled and internet-based platform for planning, management, and implementation as practiced in the Fintech, e-commerce, e-democracy, e-governance, M2M in Industry 4, Uber Taxi service shall be the backbone of Indo-Pacific Megacity System for Regional Development.
c. How we work, live, and govern in the Indo-Pacific must change once they accept advanced technologies.

What are the ways and means for achieving Indo-Pacific Integrated Regional Development through Megacity System?

a. Accept that every institution, government and non-government, goes through a life cycle of birth, growth, and death like all living entities, just like industrial products, the nature of ever-changing services and commerce sectors. This is an opportunity to regenerate and reform using emerging technologies. Then rebuild institutions by accepting the reshaping of these institutions using ICT, IoT, M2M, 5G, 6G, Industry 4, and all emerging technologies.
b. Organize innovative community internet-based spatially and highly specialized and non-spatial communities for work involving primary, secondary, and tertiary sectors using emerging technologies with consequent enhanced smart living and appreciable enhancement with E-Democracy in living.
c. Smart communities shall embrace Sharing economy in all walks of work and life instead of a capitalist-owning economy.
d. Analyze and redesign all economic activities and then reconfigure them using broadband internet and advanced emerging technologies to increase value addition and internal rate of return.
e. Renewables shall be part of life starting with energy, and combining production with renewable can reach production and service delivery with a near-zero marginal cost for long-range sustainability.
f. Increase production and distribution of relevant, high-quality knowledge workers from academic institutions to meet the development needs and make it easy to access their work opportunities in the Indo-Pacific, removing barriers.
g. Achieve an exceedingly elevated level of easiness to access government services and continuously evaluate governance by satisfaction scores of services by consumers and reform through e-democracy combined with e-governance and not only simply conforming administrative rules of law.
h. Spatial implications of all the above shall be part of the megacity system's integrated development and the Indo-Pacific's regional development.

i. Develop continuously Smart and technologically Up-to-Date Production systems for end-user supply chains in the megacity system's primary, secondary, and tertiary economic activities. The fundamental steps of a supply chain in order are Sourcing raw materials, refining those materials into essential parts, combining those basic parts to create a product, order fulfillment for sales, product delivery, customer support, and return services. These can be dispersed spatially. The time it takes for any of these processes from start to completion is known as lead time, and using algorithms reduces it. The spatial and infrastructure requirements shall be accounted for.

Then, what are the ways and means for achieving Indo-Pacific Integrated Regional Development through Megacity System?

a. Accept that every institution, government and non-government, goes through a life cycle of birth, growth, and death like all living entities, just like industrial products, the nature of ever-changing services and commerce sectors. This is an opportunity to regenerate and reform using emerging technologies. Then rebuild institutions by accepting the reshaping of these institutions using ICT, IoT, M2M,5G,6G, Industry 4, and all emerging technologies.
b. Organize smart community internet-based spatially and highly specialized and non-spatial communities for work involving primary, secondary, and tertiary sectors using emerging technologies with consequent enhanced smart living and appreciable enhancement with E-Democracy in living.
c. Smart communities shall embrace Sharing economy in all walks of work and life instead of a capitalist-owning economy.
d. Analyze and redesign all economic activities and reconfigure them using broadband internet and advanced emerging technologies to increase value addition and internal rate of return.
e. Renewables shall be part of life, starting with energy and combining production with renewable can reach production and service delivery with a near-zero marginal cost for long-range sustainability.
f. Increase production and distribution of relevant, high-quality knowledge workers from academic institutions to meet the development needs and make it easy to access their work opportunities in the Indo-Pacific, removing barriers.
g. Achieve an exceedingly elevated level of easiness to access government services and continuously evaluate governance by satisfaction scores of services by consumers and reform through e-democracy combined with e-governance and not only simply conforming administrative rules of law.
h. The spatial implications of all the above shall be part of the megacity system's integrated development and the Indo-Pacific's regional development.
i. Develop continuously Smart and technologically Up-to-Date Production systems for end-user supply chains in all the megacity system's primary, secondary, and tertiary economic activities. The fundamental steps of a supply chain in order are Sourcing raw materials, refining those materials into essential parts, combining those basic parts to create a product, order fulfillment for sales, product delivery,

customer support, and return services. These can be dispersed spatially—the amount of time it takes for. Any of these processes from start to completion is known as lead time, and using algorithms reduces it. The spatial and infrastructure requirements shall be accounted for.

References

1. Doyle T, Rumley D (2020) The rise and return of the Indo-Pacific. OUP, Oxford
2. Austin MR ed (2020) Asia's new geopolitics: essays on reshaping the Indo-Pacific. Hoover Institution Press. Excerpt, Stanford
3. Cannon BJ, Rossiter, A eds (2018, August) Rising Powers Quarterly. 3 (2). special issue on The "Indo-Pacific" – Regional Dynamics in the 21st Century's New Geopolitical Center of Gravity
4. Midcalf, Rory (2020) Indo-Pacific Empire: China, America, and the Contest for the World's Pivotal Region. Excerpt
5. https://Mapsofworld.com
6. https://lki.lk/wp-content/uploads/2020/01/imagel.png
7. Dhruva J (2019) Acting East: India in the Indo pacific, Brookings India
8. www.populationpyramid.net.
9. Government of India (2014) Smart cities concept note New Delhi
10. https://ceoworld.biz/indo-pacific CEO World magazine Indo pacific Region ////
11. https://www.state.gov/wp-content/uploads/2019/11/Free-and-Open-Indo-Pacific-4Nov2019.pdf
12. The White House (U.S.) National Security Strategy of the United States of America, December 2017, p 2. https://www.whitehouse.gov/wp-content/uploads/2017/12/NSS-Final-12-18-2017-0905.pdf.
13. The U.S. Department of Defense, Indo-Pacific Strategy Report, June 1, 2019. https://media.defense.gov/2019/Jul/01/2002152311/-1/-1/1/DEPARTMENT-OF-DEFENSE-INDO-PACIFIC-STRATEGY-REPORT-2019.PDF
14. PwC [PricewaterhouseCoopers Private Limited] (2014) Future of India: The Winning Leap http://www.pwc.in/en_in/in/assets/pdfs/future-of-india/future-of-india-thewinning-leap.pdf
15. http://www.pwc.in/en_in/in/assets/pdfs/future-of-india/future-of-india-the winning-leap.pdf [PricewaterhouseCoopers Private Limited] (2014) Future of India: The Winning Leap
16. Vinod Kumar TM (ed) (2018) Smart metropolitan regional development: economic and spatial design strategies. Springer-Nature, Singapore
17. Beaverstock JV, Smith RG, Taylor PJ (2004) World-city network: a new meta geography? Ann Assoc Am Geogr 90(1), 124–125; Beaverstock JV, Smith RG, Taylor PJ, Walker DRF, Lorimer H (2000) Globalization and world cities: some measurement methodologies. Appl Geogr 20:47
18. van der Ploeg F, Poelhekke S (2008) Globalization and megacities rise in the developing world trade policy. CESifo working paper no. 2208 category 7
19. Alankar (2015) India's megacities and climate change: explorations from Delhi and Mumbai. STEPS working paper 79
20. Bourdeau-Lepage L, Huriot J-M (2005) The metropolis in retrospect. From the loosely trading metropolis to the global metropolis. Rech Econ Louvain—Louvain Econ Rev 71(3):257–284
21. Davis DE (2005) Cities in a global context: a brief intellectual history. Int J Urban Reg Res 29(1):99
22. Hill RC, Kim JW (2000) Global cities and the developmental states: New York, Tokyo, and Seoul. Urban Stud 37(12):2168
23. Castells M (2000) The rise of the network society, the information age: economy, society, and culture, vol I, 2nd edn. Blackwell, Cambridge, MA, Oxford

24. Savitch HV (2001) What is new about globalization, and what does it portend for cities? Int Soc Sci J 54(172):180–181
25. Grant R, Nijman J (2002) Globalization and the corporate geography of cities in the less developed world. Ann Assoc Am Geogr 92(2):320
26. Smith DA (2001) Global cities in East Asia: an empirical and conceptual analysis. Int Soc Sci J 56(181):399
27. Manyika J et al (2015) Global growth: can productivity save the day in an aging world? McKinsey Global Institute, San Francisco
28. Birch E, Wachter S (eds) (2011) Global urbanization. University of Pennsylvania Press, Philadelphia
29. Bazzanella L, Caneparo L, Corsico F, Roccasalva G (eds) (2012) The future of cities and regions: simulation, scenario and visioning, governance, and scale. Springer, Dordrecht
30. Friedmann J (1986) The world city hypothesis, 317. Friedmann cites Manuel Castells' La question urbane (1972) Maspero, Paris. David Harvey's social justice and the City (1973) Edward Arnold, London
31. Sassen S (1991) The global city: New York, London, Tokyo, Chap 8. Princeton University Press, Princeton
32. Sassen S (1991) The global city: New York. Princeton University Press, London, Tokyo, p 47
33. Sassen S (2000) Cities in a world economy, 2nd edn. Pine Forge Press, London, p 7
34. Sassen S (2012) Cities in a world economy. Pine Forge Press, Thousand Oaks
35. Trujillo JL, Parilla J (2016) Redefining global cities: the seven types of global metro economies. The Brookings Institution. Global Cities initiative, a joint project of Brookings and JPMorgan Chase. The Brookings Institute. Metropolitan Policy Program
36. https://www.spottedbylocals.com/blog/alpha-beta-and-gamma-cities/
37. GaWC studies the relationships between world cities in the context of globalization. https://www.lboro.ac.uk. The classification is based on the office networks of 100 advanced producer service firms in 315 cities. https://www.lboro.ac.uk/gawc/world2000.html
38. Friedman TL (2015) The world is flat: a brief history of the twenty-first century, 1st edn. Farrar Straus and Giroux, New York
39. Friedman TL (2017) Thank One for being late, an optimist's guide to Thriving in the age of accelerations, 1st edn. Penguin Random House, UK
40. Charter cities. League of California cities. 09 May 2007. Retrieved 17 Apr 2008
41. California government code, title 4 Government of cities, chapter 2 classification. The state of California. Retrieved 17 Apr 2008
42. Charter cities list. League of California cities. 22 Feb 2013. Retrieved 23 June 2016
43. Charter cities. League of California cities. Archived from the original on 14Nov 2008. Retrieved 14 Nov 2008
44. Mallaby S (2010) The politically incorrect guide to ending poverty. The Atlantic, July–Aug 2010
45. Concept. Archived 11 May 2011, at the way back machine. From chartercities.org https://aidwatchers.com/2010/06/the-lure-of-starting-from-scratch/
46. Plan for charter city to fight Honduras poverty loses its initiator. New York Times, 30 Sept 2012
47. Manyika J et al (2016) Digital globalization: the new era of global flows. McKinsey Global Institute, San Francisco
48. McDearman B, Clark G, Parilla J (2013) The 10 traits of globally fluent metro areas. Brookings Institution, Washington; Liu A, Parilla J (2016) U.S. cities should not abandon trade. The Avenue, 7 Apr 2016
49. Hall P (1966) The world cities. Weidenfeld and Nicolson, London
50. Friedmann J (1986) The world city hypothesis. Dev. Change 17(1):69–83
51. Clark G (2016) A short history of global cities. Brookings Institution Press, Washington
52. Leff S, Petersen B (2015) Beyond the scorecard: understanding global city rankings. Chicago Council on Global Affairs
53. Taylor PJ (2011) Global urban analysis: a survey of cities in globalization. Earthscan, London

54. Mckinsey and Company (2018) Smart cities: digital solution for more livable future. McKinsey Global Institute, Washington
55. Vinod Kumar TM (ed) (2016) Smart economy in smart cities international collaborative research: Ottawa, St Louis, Stuttgart, Bologna, Cape Town, Nairobi, Dakar, Lagos, New Delhi, Varanasi, Vijayawada, Kozhikode, Hong Kong. Springer, Singapore
56. https://www.whitehouse.gov/briefing-room/statements-releases/2021/03/12/quad-leaders-joint-statement-the-spirit-of-the-quad/
57. https://www.whitehouse.gov/briefing-room/statement-s-releases/2021/09/24/fact-sheet-quad-leaders-summit/
58. https://www.whitehouse.gov/briefing-room/statements-releases/2022/03/03/joint-readout-of-quad-leaders-call/
59. https://www.whitehouse.gov/briefing-room/statements-releases/2022/05/23/fact-sheet-quad-leaders-tokyo-summit-2022/
60. https://www.whitehouse.gov/briefing-room/statements-releases/2022/05/23/statement-on-indo-pacific-economic-framework-for-prosperity/2/
61. https://www.whitehouse.gov/briefing-room/statements-releases/2022/05/24/quad-joint-leaders-statement/7/-64
62. https://www.whitehouse.gov/briefing-room/statements-releases/2023/05/20/quad-leaders-joint-statement/
63. https://www.diplomatie.gouv.fr/IMG/pdf/en_dcp_a4_indopacifique_022022_v1-4_web_cle878143.pdf
64. https://rangun.diplo.de/blob/2380824/a27b62057f2d2675ce2bbfc5be01099a/policy-guidelines-summary-data.pdf
65. https://www.government.nl/documents/publications/2020/11/13/indo-pacific-guidelines
66. https://ec.europa.eu/commission/presscorner/detail/en/qanda_21_4709
67. https://www.whitehouse.gov/briefing-room/statements-releases/2022/07/14/joint-statement-of-the-leaders-of-india-israel-united-arab-emirates-and-the-united-states-i2u2/
68. https://www.international.gc.ca/campaign-campagne/indo-pacific-pacifique/index.aspx?lang=eng
69. Vinod Kumar Tm (ed) (2018) Smart metropolitan regional development: economic and spatial design strategies. Springer-Nature, Singapore
70. Vinod Kumar TM (ed) (2017) E-Democracy for smart cities. Springer, Singapore
71. Vinod Kumar TM (ed) (2015) E-Governance for smart cities. Springer, Singapore
72. Vinod Kumar TM (ed) (2020) Smart Global megacities. Collaborative Research New York, Hong Kong-Shenzhen, Kolkata Springer-Nature, Singapore
73. Vinod Kumar TM (ed) (2020) Smart global megacities. Collaborative Research Chennai Kozhikode Springer-Nature, Singapore
74. Vinod Kumar TM (ed) (2019) Smart environment for smart cities. Springer-Nature, Singapore
75. Vinod Kumar TM (ed) (2020) Smart living for smart cities: case studies. Singapore
76. Vinod Kumar TM (ed) (2020) Smart living for smart cities: community study, community studies, ways, and means. Singapore
77. Vinod Kumar TM (ed) (2022) Smart master planning for cities: case studies of domain innovations, Singapore
78. Vinod Kumar TM (ed) (2022) Smart master planning for cities: case studies of digita innovations, Singapore

Professor T. M. Vinod Kumar had 50 years of experience in Urban Planning, as a teacher, researcher, and adviser/consultant and worked in India, China, Bhutan, Nepal, Malaysia, Indonesia, and Hawaii, USA. He was Dean of Studies, Head of the Department of Urban Planning, Head Centre for Systems Studies and Analysis, Head of, the Centre for GIS and Remote Sensing, and Centre for Urban Studies of the School of Planning and Architecture, New Delhi; Visiting Professor National Institute of Technology, Calicut, and Institute of Technology Bandung, Indonesia and Professional Associates, East-West Resources Systems Institute Honolulu, Hawaii, Fellow Centre for the Study of Developing Societies, Delhi, Project Manager in Council for Social Development, New Delhi, Regional Program Coordinator at the International Centre for Integrated Mountain Development (ICIMOD) and Planner-Engineer at the Ford Foundation. He is the author of many books and journal articles. and edited "Geographic Information System for Smart Cities" (Copal:2014), "E-Governance for Smart Cities" (Springer:2015) "Smart Economy in Smart Cities" (Springer:2016) and "E-Democracy for Smart Cities" (Springer:2017), "Smart Metropolitan Regional Development: Economic and Spatial Design Strategies" (Springer:2018)," Smart Environment for Smart Cities" (Springer:2019), "Smart Living for Smart Cities Case studies" (Springer:2020)" Smart Living for Smart Cities Community study and Ways and Means" (Springer:2020), "Smart Global Megacities Collaborative Research-Tokyo, Mumbai, New York, Hong Kong-Shenzhen, Kolkata" (Springer 2021) and Smart Global Megacities Collaborative Research-Chennai and Kochi-Kannur" (Springer:2021), Smart Master Planning: Case Studies of Digital Innovation" (Springer 2022), Smart Master Planning: Case Studies on Domain Innovations" (Springer:2022). COVID-19, Containment, Life, Work, and Restart, Urban Studies (Springer:2022) and. COVID-19, Containment, Life, Work, and Restart, Regional Studies (Springer:2022).

Regional Economy

Prospects of Systems of Megacities and Individual Megacities with Respect to Regional Economy

Anna Vaňová, Katarína Vitálišová, and Darina Rojíková

Abstract The importance of cities and megacities is one of the defining features of the twenty-first century. The chapter provides clear and comprehensive overview of current thinking within academic and scientific sector and daily practice from two megacities in Europe—London and Paris due to the fact that they are the only nodes which can interact with Indo-Pacific in their economic, defence and strategic interests resulted from their multi-dimensional physical, institutional, economic and social linkages in Indo-Pacific. Paris and London are the gateways for sea, air and communication, institutional and academic linkages with the Indo-Pacific. These megacities have to face with the various challenges on the one hand, but on the other hand, there are new possibilities and solutions for their "sustainable" development. The theoretical review on the current changes in status of European cities defines the specific type of city—megacity and two of them are deeply analysed (London and Paris). On this base, the chapter identified new paradigms for development of urban space, the role and importance of current development concepts and innovation systems, and the role and importance of integrated, open, multi-level, cross-sectorial planning and management.

Keywords Megacity · Urban development · Governance · Ecosystem · Development concepts · Paradigms · Solutions

A. Vaňová · K. Vitálišová (✉)
Faculty of Economics, Department of Regional Development and Public Administration, Matej Bel University, Tajovského 10, 974 35 Banská Bystrica, Slovakia
e-mail: katarina.vitalisova@umb.sk

D. Rojíková
Faculty of Economics, Research and Innovation Center, Matej Bel University, Cesta Na Amfiteáter 1, 974 01 Banská Bystrica, Slovakia

T. M. Vinod Kumar (ed.), *Indo-Pacific Smart Megacity System*, Advances in 21st Century Human Settlements, https://doi.org/10.1007/978-981-99-6218-1_2

1 Introduction

In the last century, increased attention has been paid to regional development. However, it was only towards the end of the twentieth century that more research and theoretical work on urban development began to appear. This research and work were mainly related to the growth of large cities, urban agglomerations and megacities, whose structures significantly impact entire regions and the national and international levels.

Cities have always been among the most dynamically developing territories. The nature and dynamics of their development correspond to the respective era. Today, they are facing turbulent changes that are influencing by challenges of global world.

The growth rate of cities has been very high in recent decades. At the turn of the century, urban dwellers outnumbered rural dwellers. According to a UNDP report [1] 4.2 billion people lived in cities in 2018, representing 55% of the world's population. By 2050, the population of cities is projected to be 6.5 billion.

Rapid urbanisation poses enormous challenges to the growing population in cities and the growing number of people living in slums, increased air pollution, inadequate infrastructure, inadequate basic services and unplanned urban sprawl that makes cities less resilient to disasters. Cities today occupy only 3% of the Earth's land but consume 60 to 80% of energy and produce at least 70% of carbon emissions. An estimated 828 million people live in city slums, which is growing. While in 1990, there were ten cities with 10 million or more inhabitants, by 2020, the number of megacities will have grown to 34. It is projected that 90% of urban expansion will be directed to the developing world in the coming decades. Cities play an indispensable role in the world's economy, accounting for about 80% of the global GDP. The challenge for today's cities, with sound planning and management, is to make cities "inclusive, safe, resilient, sustainable and dynamic centres of innovation and entrepreneurship" [2].

As the cities have to face with the various challenges, they began to see the benefit of cooperation and of collective action. It resulted in the developing metropolitan areas with aim to take advantage of agglomeration [3]. Based on the importance of cities and metropolitan areas to their surrounding areas or their impact on the global economy, their interconnections can be identified in a few subdivisions of cities in terms of their importance. Within this classification, we know global cities (not in terms of their size, but in terms of their importance on a global scale), metropolises, megacities or megapolises [4–8]. These designations of cities, or urban areas, are not strictly defined by population size, area or population density, and it is sometimes difficult to precisely define their role. There are several unifying characteristics based on which these definitions are generally accepted. In the chapter, we focus on one kind of these cities—megacities and its specifics in the European countries.

The chapter consists of four parts. The first part is devoted to the theoretical review on the current changes in status of European cities at the beginning of the twenty-first century. The second subchapter defines the specific type of city—megacity. Subsequently, we analyse two pure European megacities—London and Paris. The chapter ended by the summarization of the common features and differences in megacities in the regional development and based on them we identified the new paradigms of megacities development.

2 The Role and Status of European Cities at the Beginning of the Twenty-First Century

The definition of a city in the twenty-first century should consider its size (area, population and population density), the dynamics of development, the complexity of functions and hierarchy in terms of relationships to the broader environment.

Particularly given the recognition that cities do not function in isolation in the development process but are usually part of a more expansive territory together with other municipalities and represent a centre of development management for them. From a spatial point of view, the term city is understood as a more expansive territorial space than just a city defined by administrative boundaries. This phenomenon was noticed in the 1980s by some urban planners who pointed out that administrative boundaries are limiting when examining the spatial aspects of certain phenomena [9].

In terms of significance, it is necessary to consider a city as a settlement or a cluster of economically, socially and culturally heterogeneous settlements. A city is characterised by a greater degree of urbanisation, a concentration of economic activities and a specific urban built-up area (a higher proportion of industrial and high-rise buildings). In terms of amenities, cities provide a higher standard of quantity and quality for their inhabitants. From this point of view, it is possible to speak of a higher quality of living compared to the countryside.

Cities fulfil a complex of functions (administrative centre, housing, education, work and leisure opportunities, trade, culture, transport and the like) for other surrounding municipalities. Some authors refer to such clusters as city regions or functional city regions [10, 11] or agglomerations, i.e. metropolitan or city regions that form a core with a strongly integrated hinterland [12]. The formation of agglomerations brings with it both positives and negatives. The appeal lies primarily in agglomeration effects. These effects are, for example, the concentration of enterprises, educational institutions, investment and business opportunities and the resulting benefits in the form of concentration of creative and highly skilled workforce, the competitiveness of industries, growth in labour productivity, cost reduction, innovative and creative environment, opportunities for networking, relationships and knowledge sharing at the level of cultural, social, governmental, financial institutions and the like.

Once upon a time, metropolitan areas had one recognised centre and were described as monocentric. More recently, a polycentric urban system has been seen as helpful in alleviating growing urban ills such as traffic congestion, air pollution and lack of affordable housing (which does not contribute to urban development's environmental and social sustainability). Polycentricity can be viewed from different perspectives. However, what is essential is the recognition of the existence of more than one centre of development, whether in terms of knowledge flows of people, information and goods or the distribution of economic activities [13] and, concerning this, jobs, population movements and linkages between these centres [14]. Parr characterises polycentric urban systems as a set of contiguous but spatially separated urban centres existing as a particular identifiable entity [10]. Parr identifies three key characteristics of a polycentric urban system, namely that it is a form of region, has the potential for high economic performance and provides a suitable organisational framework for public policy interventions.

A city perceived as a polycentric city region is more in line with the needs of contemporary practice, development and above all, the requirements of an integrated approach to development [9]. This trend in identifying territorial units capable of a more practical approach to development is also recognised by the European Union, which, to promote economic, social and environmental development, promotes an integrated approach to the formulation of territorial development strategies. According to the European Union (in the future referred to as the EU), it is vital to create and develop cross-sectoral cooperation within a specific grouping of municipalities that share certain common features, functions and problems. Such a grouping can be something other than an administrative unit with a local government function. The city at the centre of implementing an integrated approach to support the development of the whole cluster should be the initiator of the integrated planning or integrated development strategy [15].

The existence of cities as centres of economic, social, cultural or community development also affects their surroundings, primarily rural areas. Their impact can be perceived either in terms of the demand for products or in terms of the spillover of externalities—both positive and negative [10, 16, 17]. On the one hand, cities, along with their hinterland, are seen as drivers of competitiveness and development because they embody an aggregation of knowledge, human and social capital, and creative activities [11]. On the other hand, the growth of cities requires land and hinterland. However, using agricultural land for urban development can lead to inefficient use, contamination and disputes may arise with other cities or users of agricultural land over resources and their use. Pressure on agricultural production can create pressures on ecosystems with global impacts, similar to the increase in mobility for work, services and shopping to the city from surrounding areas.

Urban growth cannot be reversed. Nevertheless, it can be monitored, guided, organised and minimised, requiring partnerships and cooperation between all local authorities concerned and an integrated approach to development planning. Over the last decade, the European Commission has been actively promoting these requirements through its Directorate-General for Regional Policy.

As cities grow, the question of their optimal size arises. The optimal size of cities and other settlements cannot be clearly determined. Instead, it is influenced by individual realities, such as the city's industrial past and local specificities. The size of a city significantly influences its development in both positive and negative ways. Large cities are characterised by diversity and a high concentration of the creative classes, whose innovative spirit can drive development and innovation [18, 19]. Small and medium-sized cities may have the advantage of a healthier natural environment, higher ecological awareness, more operational management of development and the like. Studies suggest that larger cities have more problems with resilience to unexpected changes [20] or environmental issues than medium-sized cities, partly due to size but mainly related to population density [19]. Another interesting finding is that medium-sized cities have a more innovative spirit than larger cities [21] and are more open-minded and tolerant [19]. It turns out that in today's digital world, living in a big city is not a prerequisite for success. This fact, assuming the competitiveness of soft and hard factors of medium-sized cities compared to large cities, could partly influence urban development. This finding is a starting point for further research and may be important for public policymaking [19, 21].

The emergence and development of large city regions cannot be stopped. Therefore, it is vital to maintain the existence of small and medium-sized cities and not to underestimate their importance. The European Commission documents state that in the EU, there are 23 cities with more than 1 million inhabitants and 345 cities with more than 100,000 inhabitants, representing about 143 million people. Only 7% of the EU population lives in cities with more than 5 million inhabitants (compared to 25% in the United States). Small and medium-sized cities with between 5,000 and 100,000 inhabitants account for 56% of Europe's urban population or about 38% of the total European population. These cities play an essential role in the lives of their inhabitants and those of the surrounding rural communities. In addition, they play a crucial role in regional economies and form the building blocks of metropolitan areas. They are centres for public and private services, development, creativity and local and regional information infrastructure. The general characteristics of small and medium-sized towns and cities, in particular their human scale, their liveability, the friendliness of their inhabitants, their geographical integration and their historical character, are the ideal of sustainable development in many respects. The sustainability of small and medium-sized cities is also essential in order to prevent rural depopulation and the uncontrolled growth of large cities, to ensure sustainable regional development, stability and stabilisation of the European territory [21–25].

Regarding future development trends, the size of cities and their links to surrounding areas need to be accepted and taken into account. However, the fact that urbanisation is not the problem should be respected. Its dynamics and economic, social, environmental and spatial impacts become the problem.

3 Megacities as a New Form of Local and Regional Development

Constantly growing global urbanisation is one of the trends of this century. From the beginning of twenty-first century, the world population increasingly concentrated in urban areas, and this trend is speeded up in the last years. The result is an increasing number of growing cities and the emergence of new megacities.

The megacity is a designation of one or more metropolitan areas with a population of over 10 million. In the fact, it is difficult to define the other boards and accurately estimate the population of megacity. The megacity is a new form of urbanisation. It is characterised as densely populated, capital clustered and having highly developed tertiary industries [26]. The megacity consists of a continuous build-up area that encompasses one or more city centres and suburban areas, economically and functionally linked to those centres [27]. From the urbanism point of view, it is usually polycentric, with multiple nodes of concentrated urban activity and high-density development, rather than being centred around one large primary central business district.

Different organisations also propose different criteria, e.g. urban areas with 8 million inhabitants or a population density of 2,000 inhabitants per km^2, which also changes the total number of megacities. In Table 1, we list the world's megacities by three different approaches. The United Nations [28] defines as megacities 39 cities including the cities having in excess of 10 million inhabitants. By Design ingbulidngs.co.uk [29], the most important criteria are the number of inhabitants higher as 10 million people—so 34 cities. The highest number of megacities is defined by www.worldata.com, 56 megacities. 25 megacities (in bold) are indicated in all three approaches. Just two cities from the European Union are in the list of megacities—Paris and London (in italics, both only in 2 approaches).

In the list, the Asian cities dominate. There are still the largest of the megacities in the world—Japan's capital Tokyo, followed by Delphi and Shanghai. Megacities grow rapidly in Latin America and Africa as well. In comparison with the Europe, there is an observable swelling of small and medium-sized cities. In the European Union (EU) is urbanisation the subject of strategic planning.

Because of the rapid growth in development, as well as in number of inhabitants, it is expected to increase the number of megacities in next years. United Nations [28] estimates that in 2030 there will be 8 new megacities (Santiago, Chengdu, Xi'an, Wuhan, Nanjing, Ahmedabad, Surat and Kuala Lumpur). By Fleck in 2050 [30], the megacities (beside that one's estimated by UNESCO) will include London, Chicago, Baghdad, Tehran, Luanda, Nairobi, Dar es Salaam, Khartoum, Ahmedabad, Pune, Surat and Ho Chi Minh City (Fig. 1).

In many cases, the megacities are at the same time global cities, the most important places within the world economy and act as hubs for global trade in goods, services and media [8]. Globalisation processes condition the emergence of global cities, and the term global city becomes one of the key terms to denote the importance of cities in the process of globalisation. Global cities represent an intricate network of different

Table 1 Comparison of approaches to megacities

Continent	Megacities by United Nations	Megacities by Designingbuildings.co.uk			Megacities by www.worlddata.info
			Population in millions	Country	
Europe and North America	**New York**	**New York**	**18.80**	**USA**	**New York City**
	Istanbul	**Istanbul**	**15.19**	**Turkey**	**Istanbul**
	Moscow	**Moscow**	**12.53**	**Russia**	**Moscow**
	Los Angeles	Los Angeles	12.44	USA	–
	Paris	*Paris*	11.1	France	–
	London	–	–	–	*London*
	Chicago	–	–	–	–
	–	–	–	–	Ankara
	–	–	–	–	Saint Petersburg
Latin America and the Carribean	**São Paulo**	**São Paulo**	**22.4**	**Brazil**	**São Paulo**
	Mexico City	**Mexico City**	**21.78**	**Mexico**	**Mexico City**
	Buenos Aires	Buenos Aires	15.15	Argentina	–
	Rio de Janeiro	**Rio de Janeiro**	**13.45**	**Brazil**	**Rio de Janeiro**
	Bogotá	**Bogotá**	**10.97**	**Colombia**	**Bogota**
	Lima	**Lima**	**10.71**	**Peru**	**Lima**
	–	–	–	–	Santiago
Africa	**Cairo**	**Cairo**	**20.90**	**Egypt**	**Cairo**
	Lagos	**Lagos**	**14.36**	**Nigeria**	**Lagos**
	Kinshasa	**Kinshasa**	**14.34**	**Democratic Republic of Congo**	**Kinshasa**
	Greater Johannesburg Metropolitan	–	–	–	–
		–	–	–	Dar es Salaam
		–	–	–	Casablanca
Asia and the Pacific	**Tokyo**	**Tokyo**	**37.39**	**Japan**	**Tokyo**
	Delhi	**Delhi**	**30.29**	**India**	**Delhi**
	Shanghai	**Shanghai**	**27.5**	**China**	**Shanghai**
	Dhaka	**Dhaka**	**21.00**	**Bangladesh**	**Dhaka**
	Beijing	**Beijing**	**20.46**	**China**	**Beijing**
	Mumbai	**Mumbai**	**20.41**	**India**	**Mumbai**

(continued)

Table 1 (continued)

Continent	Megacities by United Nations	Megacities by Designingbuildings.co.uk			Megacities by www.worlddata.info
			Population in millions	Country	
	Osaka	Osaka	19.16	Japan	–
	Karachi	**Karachi**	**16.9**	**Pakistan**	**Karachi**
	Chongqing	**Chongqing**	**15.87**	**China**	**Chongqing**
	Kolkata	Kolkata	14.85	India	–
	Manila	Manila	13.92	Philippines	–
	Tianjin	**Tianjin**	**13.58**	**China**	**Tianjin**
	–	Canton	13.30	China	–
	Lahore	**Lahore**	**12.64**	**Pakistan**	**Lahore**
	Shenzhen	**Shenzhen**	**12.35**	**China**	**Shenzhen**
	Bangalore	**Bangalore**	**12.32**	**India**	**Bangalore**
	–	Madras	10.97	India	–
	Jakarta	**Jakarta**	**10.77**	**Indonesia**	**Jakarta**
	Bangkok	**Bangkok**	**10.53**	**Thailand**	**Bangkok**
	–	Hyderabad	10.00	India	Hyderabad
	Guangzhou				Guangzhou
	Chengdu				Chengdu
	Chennai				Chennai
	Tehran				Tehran
	Seoul				Seoul
					Xi'an
					Wuhan
					Nanjing
					Ahmedabad
					Surat
					Riyadh
					Hong Kong
					Foshan
					Zhengzhou
					Qingdao
					Singapore
					Hangzhou

(continued)

Table 1 (continued)

Continent	Megacities by United Nations	Megacities by Designingbuildings.co.uk			Megacities by www.worlddata.info
			Population in millions	Country	
					Ho Chi Minh City
					Shantou
					Rangoon
					Harbin
					Shenyang
					Dongguan
					Baghdad

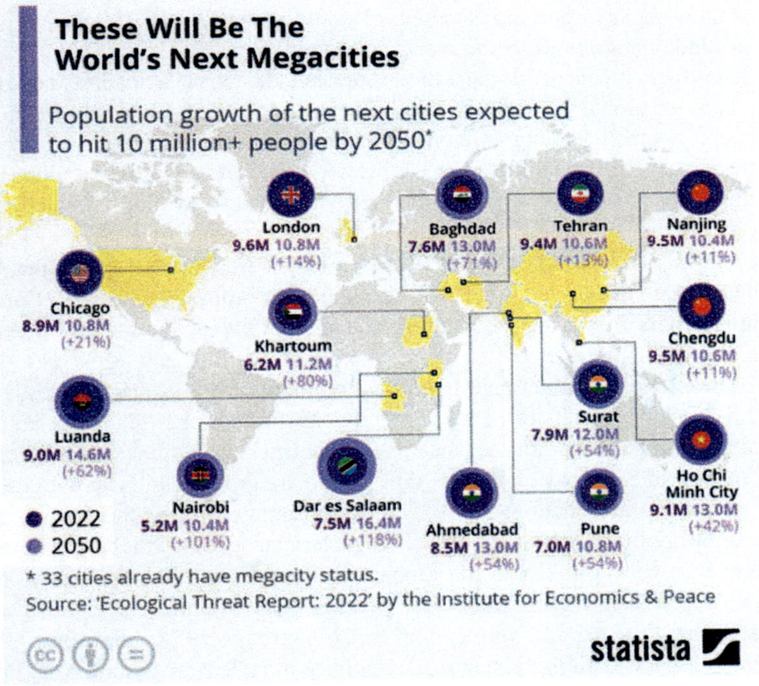

Fig. 1 Expected new megacities in 2030 published by Statista

information and financial flows and relations, conditioned by the existence and cooperation of important business entities and institutions. They are places of accumulation of international capital and bases of important economic markets, information centres and a diverse workforce (also due to migration). Economically, they serve different international economic services and are the headquarters of several multinational and international companies. Socially, the global city is characterised by a diversity of cultures, religions, ideologies, education and languages. From another social perspective, it is characterised by the accumulation of diverse income groups, diversification of education and the resulting social inequalities. A global city is always a metropolis, but a metropolis is not always a global city. The competitors for this type of city are cities that can attract foreign direct investment. This capacity to attract makes them serious global power players despite having lower economic output than global cities. According to R. Florida [31], "the 40 largest megaregions produce two-thirds of global economic output and 90% of global innovation, even though they are home to only 18% of the world's population". The Brookings Institution has published a report, "Redefining Global Cities", which, based on research on 123 of the world's largest cities and other studies, identifies seven types of global cities that fundamentally affect the world economy.

The first three types of global cities represent the world's leading centres of economic power: global giants, knowledge cities and so-called Asian anchors. Global giants are the world's leading economic and financial centres, the most important global cities. (This includes, for example, New York, Los Angeles, London, Paris, Tokyo and Osaka-Kobe). Knowledge cities are the world's leading knowledge and technology hubs. (This includes 19 urban centres such as San Jose (Silicon Valley), Boston, Seattle, San Diego, Washington DC, Chicago, Austin, Dallas, Atlanta, Portland and Denver in the United States, Amsterdam, Stockholm and Zurich in Europe). The Asian anchors are Asia's six established and emerging economic power centres (Hong Kong, Singapore, Seoul-Incheon, Shanghai, Beijing and Moscow). Their ability to attract foreign direct investment makes them serious global power players, despite having lower levels of economic output than the global giants. In addition to global economic powerhouses, the report identifies four other types of global cities in the world economy's middle rungs. Some are growing in sync with globalisation, and others are challenging it. The first group comprises the 16 American Midwest metropolitan areas (the so-called American Middleweights). These are cities whose growth is linked to the global economy or where industries in these cities are successfully facing global competition (e.g. Miami, Cleveland, Detroit and Pittsburgh). The second group comprises 26 International Middleweights, hubs where technology, knowledge and human creativity and talent are dynamically developing and establishing themselves. These include, for example, Toronto, Vancouver, Brussels, Rome, Milan, Berlin, Vienna, Madrid, Barcelona, Sydney, Melbourne, Perth and Tel Aviv. Another group consists of China's 22 industrial city areas (so-called Factory China), which have experienced rapid growth only because of specific export-intensive production. However, the report describes them as relatively weak. The last group consists of 28 major global trade and transport arteries (so-called

Emerging Gateways) that play a role in important national and regional markets (e.g. Mexico City, Sao Paolo, Rio de Janeiro, Istanbul, Mumbai and Johannesburg) [31].

Each city, or group of cities, is specific in some way. Each city has a different potential and fulfils a different function. In case of megacities, the cities are usually heterogeneous [32], so the cities perform various functions, as economic, political, educational, cultural, transport, political and other, so we can name them as cities with diverse functions [33, 34].

The various functions of cities as residential, commercial, industrial, recreational and other settlements, do not develop haphazardly because the various functions are often concentrated in specific neighbourhoods. For example, the city centre is dominated by administrative and office buildings, often with a hyper-modern appearance. In terms of land use, modern cities are becoming disconnected. Residential areas are separated from the main shopping districts, while commerce and industry are concentrated in specific zones. The result is an unsustainable increase in mobility and traffic, which negatively affects the quality of the environment. This negative impact is just one of all other huge challenges associated with the urban development of megacities. The megacities create a new urban dynamic, as super-sized cities are seen as the new engine of the global economy, connecting the flow of goods. However, the current pattern of urbanisation has all too often resulted in urban sprawl, low productivity, segregation, exclusion and congestion [35].

The motto of today's globalised world is "think globally, act locally". Development trends at urban level are the important part of global trends and that is one of the reasons why the consequences of modern urbanisation must not be underestimated.

In the next text, we focus on the analysis of specifics of two "pure" European megacities—London and Paris. Subsequently, the results of this analysis will be used for the evaluation of current city development trends in context of megacities.

4 Case Study of European Megacities—London and Paris

The approaches to definition of megacities (Fig. 1) present various classification of megacities. For the selection of megacities for case study, we apply the geographical criteria—the location of the megacity in the Europe. There can be identified just two cities—London and Paris. The choice of these cities was conditioned by their historical connectivity to the Indo-Pacific region.

Through the European Investment Bank (EIB), the European Union (EU) is a major investor in the Pacific. In a world where democracy, the rule of law, human rights, freedom, free trade and with a properly functioning multilateral world order are under pressure, the EU must join forces in the Indo-Pacific region and cooperate with support of ASEAN with the countries of the Indo-Pacific region to ensure peace and security, reduce tension in the economic area, promote maritime security, economic and cyber security and build the corresponding political, economic, security, social and environmental infrastructure in relation to these tasks. As a world leader in building democratic values and standards, the EU must support and continue the dialogue about them with all countries in the region.

4.1 London

London, as the capital of the United Kingdom, is one of the oldest of the world's largest cities with its almost two-thousand-year history, and it is also one of the most cosmopolitan cities. Britain's largest metropolis is also the economic, transport and cultural centre of the country. According to worldpopulationreview.com, the population of London in 2021 is currently estimated at 9,425,622. About one-seventh of the country's population is concentrated in Great London, which is the most urbanised area in Great Britain and the most populous city in the European Union [36].

Greater London metropolitan area consists of 32 boroughs, which covers some 1,580 km^2. The present boroughs of London were constituted in 1965. The 13 inner London boroughs are Camden, Hackney, Hammersmith and Fulham, Haringey, Islington, Kensington and Chelsea, Lambeth, Lewisham, Newham, Southwark, Tower Hamlets, Wandsworth and the City of Westminster. The 19 outer boroughs are Barking and Dagenham, Barnet, Bexley, Brent, Bromley, Croydon, Ealing, Enfield, Greenwich, Harrow, Havering, Hillingdon, Hounslow, Kingston upon Thames, Merton, Redbridge, and Richmond upon Thames, Sutton and Waltham Forest. The royal boroughs belong to Kensington, Chelsea and Kingston upon Thames and Greenwich. The City of London is used just for a small historical area within Westminster, now the financial and business district of the metropolis. Based on the UK's Nomenclature of Territorial Units for Statistics London contains 5 parts Outer West and North West, Inner West, Inner East, Outer East and North East, Outer Sout [37].

From the local governance point of view, the key bodies of the city are a directly elected mayor and the Greater London Authority. The mayor sets the strategic framework for the whole London. He has executive powers over a number of citywide areas, including transport, policing, fire and emergency services, inward investment and, to a degree, regeneration and housing. Other areas like education and health are controlled by central or local government. The Greater London Authority is a strategical regional body, with jurisdiction over the Greater London. It has executive power over transport, policing, economic development, and fire and emergency planning [38, 39].

During last decade, London is still rapidly growing, both in terms of size and population. It is caused by the internal migration within Great Britain because of the wider range of possibilities. This trend is lowering. The second reason is a high international migration, especially by young people. London is a very cosmopolitan city with the migration of many people from different countries with the old recorded history. Inner London has the largest proportion of Black, Asian and Minority Ethnic residence [40]. The third reason is a natural growth. London's urbanisation rate is quite low (around 1% a year). The density of population in London belongs to one of the highest among megacities.

The current urban development politics in London is strongly influenced by interests of powerful developers, which increase the prices in the inner London. Due to the high cost of living here, many inhabitants look for affordable housing in the suburbs, built on brownfield or greenfield sites. In many cases, it results in urban sprawl associated with various negative effects (e.g. exaggerated densities, reduced public space provision, limited housing), reflected in urban degeneration [41, 42].

The economy of London together with Paris metro regions generates nearly one third of their national GDP, while their population share is closer to one fifth. London's economy was worth £ 470 billion (Gross Added Value-GVA) in 2020, accounting for 24% of UK economic output. In 2021, total GDP for London area was £ 526,524 million, £ 59,855 per head. The recovery of London economy after COVID-19 pandemic was relatively fast, in second quarter of 2022 the London Gross Added Value was 2.7% above its pre-pandemic level. Unemployment rate in London was at the end of 2022 as 4.5% [43, 44]. The fast-growing sectors in London during last 16 years belong to management consultancy, computer programming, sport activities and amusement and recreation, real estate activities and security and investigation activities. Even London is the wealthiest part of the United Kingdom, average personal wealth is lower as rest of southern England with high proportion of Londoners living in poverty, especially from the ethnical minorities [40].

By various rankings (e.g. Global Power City Index, Global City Index, Smart City Government Ranking, etc.), London is classified also as a global and smart city. London has long been regarded as a world leader in digital technology and is fast becoming a global centre for research, development, innovation and commercialization of a new range of advanced digital technologies, as well as a global leader for innovation across key knowledge intensive sectors, including a global hotspot for clean tech innovation. The development and implementation of modern technologies are the basic pillars on which London's smart initiatives and projects are developed [45–47].

London is a city with a well-established innovative ecosystem with strong background in research and development and capital funds to invest in innovative projects. This innovative ecosystem consists of incubators, accelerators, venture capitalist, research technology organisation, innovations centres and hubs of support. In London area, 1/3 of all European technology unicorn companies are seated, worth at least $1 billion (€ 888 million) [48]. London is also the European Capital of artificial intelligence with over 750 suppliers in city and leader in new Clean Tech products (Greater London Authority, 2018). The extremely high number of start-ups are oriented on the distributed ledger technologies, blockchain, Internet of Things, virtual reality and augmented reality [45].

The main document for the strategic development of London for the next period is Smarter London Together [49], an official mayor's roadmap for London to become a smart city. It defines the main missions in the London development as more user-designed services; strike a new deal for city data; world-class connectivity and smarter streets; enhance digital leaderships and skills and improve citywide collaboration.

All the implemented smart activities in a form of urban requalification plans and sustainable mobility projects should contribute to transformation of London into a carbon–neutral city by 2050. For the implementation of the Smart London Together strategy, the Smart London Board was established consisting of tech sector alongside entrepreneurs and academics. It advises in implementation of new digital technologies across London's infrastructure, utilities and public services and the role of technology and data in all Mayoral strategies and policies [50].

The strong position of IT sector is reflected in the London's newest, smartest and most sustainable area—Queen Elizabeth Olympic Park. It is managed by the London Legacy Development Corporation (LLDC). Its ambition is to use the Park as a test-bed for new international standards in smart data, sustainability and community-building, sharing its successes across the city and beyond. The Park presents a space for realisation of joint projects and research activities among world-class universities (e.g. Imperial College London, University College London, Brunel University, University of the Arts London, Royal Holloway University of London, King's College London, Queen Mary University of London) and innovative companies. They generated the new start-up businesses, new alliances and multidisciplinary clusters. The Park is perfectly equipped by modern technologies, digital and transport infrastructure with fast access to central London [51].

In 2017, research on the quality of life in London was conducted [52], it evaluated 32 social, environmental and economic indicators to gauge progress in comparison with the previous research in 2012. Within the evaluation of the environmental indicators, the CO_2 emissions have getting dropped, but its significant decreasing is one of the key challenges for London till 2030. Many indicators were evaluated positively, e.g. the reduction of tonnes of NOx or tones of PM2.5, performance against the greenhouse gas Emissions Performance Standards and decrease in water consumption. The negatively evaluated indicators were a lack of implemented measures to protect the residential and commercial properties against tidal or fluvial flooding as well as the access to the green areas in the London. In social area, the positive change belongs to prolonging the healthy life expectancy, increase in using a sustainable mode for transit, growth in percentage of decent housing stock in London, increase in average happiness score for London and in volunteering activities. As the negative aspects of social life were evaluated, child poverty reflected in education opportunities and overall satisfaction of Londoners with the capital as a place to live. The evaluation within the economic aspects of life in London confirms the strengths and weaknesses explained in the previous text. The economic prosperity is great based on the development of IT sector, but because of the extremely high prices of properties in London, citizens have a problem to buy own house or apartment in good quality which is also closely connected with the significant differentiation in the living wage per hour.

Another view on the evaluation of different areas of living is presented in Fig. 2 based on data from 2020 [40]. It compares various aspects of global cities. As we can see, that London is the best evaluated in cultural vibrancy and visitor demand, in investments, talent base and appeal and second best in tech industries and innovation ecosystem. The worse evaluated aspects belongs to labour market performance and inclusion; safety and security, affordability and costs, commute and congestion.

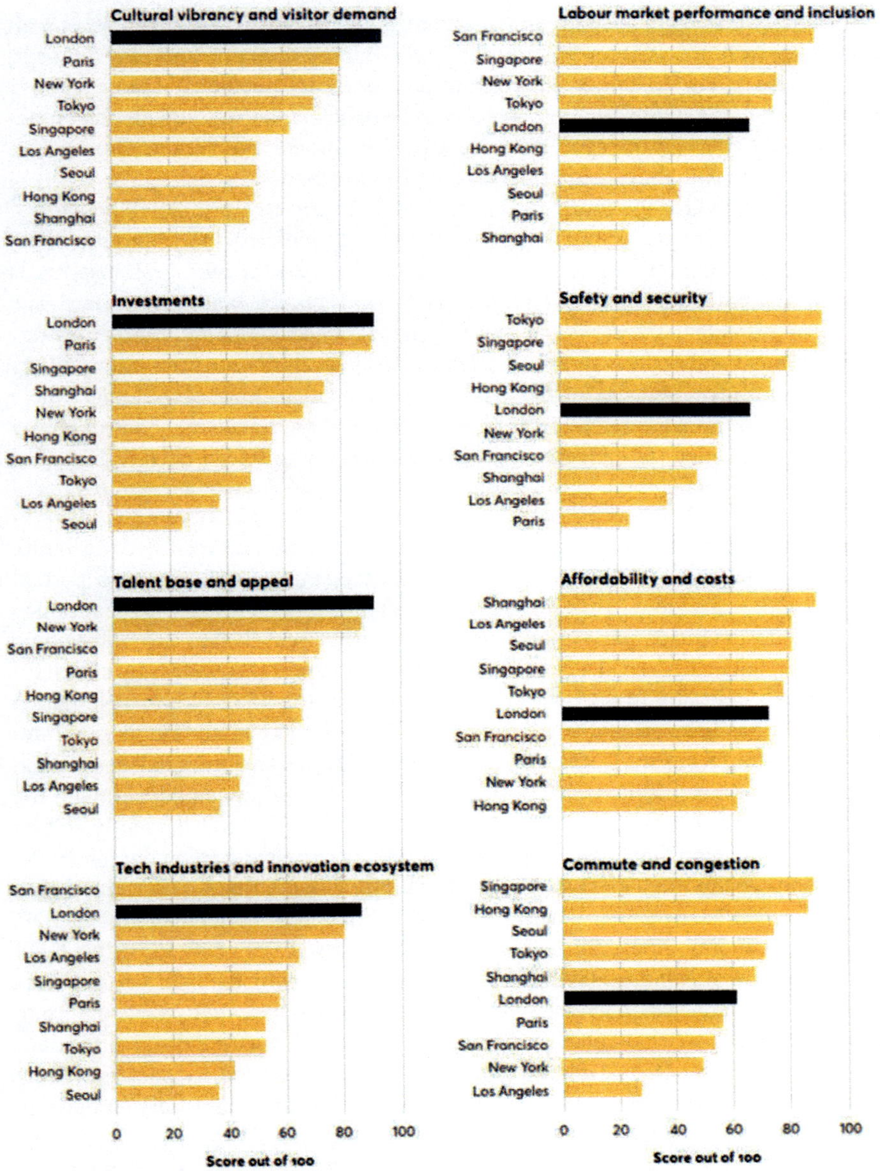

Fig. 2 Evaluation of London in comparison with the global top 10 cities

By Power Global City Index [53] London achieved the first position among all 48 evaluated cities based on 70 indicators of economy, research and development, cultural interaction, liveability, environment, accessibility, but the total score is lower as previous year. As problem areas were evaluated the market size and market attractiveness, as well as air transport capacity and international interaction. The positive change was noticed in evaluation of environment indicators.

The challenges for the future development of London intermingle all spheres of life. In environment, decarbonisation is one of the key indicators to achieve in 2050. The problem is also the high density of flights and also the surface transport, as well as other pollutants causing poor quality of air. It reflects warming climate. The long-term problem in society is an economic inequality, changing migration patterns, unemployment and racism. It results in lower housing affordability, in-work poverty and homelessness. Because of the rapid development of new technologies, there should be paid more attention to the cyber security, surveillance and controlled use of new technologies [54]. Eventhough London had recovered relatively fast after COVID-19 pandemic, there is still a lot of issues to be solved that are also results of status "megacity".

From the perspective of Indo-Pacific countries, London is one of the world city leaders regarding the economic, cultural and political factors. London as an important port hub and international airport hub, it has also the key role in international policy with the Indo-Pacific countries. Relationship between Great Britain and Info-Pacific is characterised by historical, economic and financial interests as well as by a military presence as a key actor in the regional security architecture [55]. Nowadays, mutual relationship has been strengthened, firstly, in 2021 by the United Kingdom's Integrated Review of Security, Defence, Development and Foreign Policy, that announced that the UK would 'tilt' towards the Indo-Pacific as the European partner with the broadest and most integrated presence in the Indo-Pacific [56]. Subsequently, in first quarter of 2023, the 11-nation Comprehensive and Progressive Agreement for Trans-Pacific Partnership (CPTPP) was signed, including Great Britain. It is a trade pact whose members span the Indo-Pacific and account for about one-tenth of global GDP. The deal's economic impact is modest, adding less than 0.1 per cent to British GDP over the next decade. It deepened strategical relationships with the international partners from Australia, Japan to Malaysia and Singapore To the other proofs of strong orientation of Great Britain on development of relationships with Indo-Pacific countries belong also Australia-United-Kingdom-United States security pact as well as a Global Combat Air Programme as an initiative with Japan and Italy to develop a sixth-generation fighter jet [57]. The Indo-Pacific region is important for United Kingdom to preserve the free flow of goods, defend the countries' sovereignty and freedom as well as maintain international stability [55].

4.2 Paris

Paris is the capital of France, located in the northern part of the country, on the Seine River. The city has an important central location in a relatively rich agricultural region—the Paris Basin and is a centre of the administrative region of Île-de-France, which has a population of 12,262,544 [58]. The area of the region is 2,300 km^2 and the population density is 8,713 inhabitants/km^2. The border of the Île-de-France region is often referred to the boundary of the agglomeration of Paris and is included in the considerations when Paris is designated as a "Mega city" [59] making it the 29th largest city in the world [60].

The city of Paris is the capital of France with a population of 2,165,423 inhabitants (2019). The city has an area of 105 km^2 and a population density of 20,545 inhabitants/km^2. Today, Paris is not only modern and intelligent, but also one of the most expensive cities in the world. In 2022, according to the Economist's global cost of living survey Intelligence, Paris was the third most expensive city in Europe, after Geneva and Zurich, and the ninth most expensive city in the world. Paris is an economic, administrative, political and cultural centre of France. The city also held these functions in the past. For centuries, Paris has been one of the most important and attractive cities in the world, also due to its location. The location of the city at the crossroads of important waterways not only for France, but also for Europe had a great influence on the development of the city. The city environment, rich in culture and education, attracted the talents and intellect of the surrounding areas, which in the past contributed to the emergence and deepening of social disparities between the population of the city of Paris and the urban agglomeration.

Solving the problems of the dynamically developing Paris and its suburbs in the past was in the administration of the larger administrative region of the Seine from 1790, but the growing population in the suburban area made it difficult to maintain this unique entity. This problem was solved when the "District de la région Parisienne" ("district of the Paris region") reorganised into several new departments since 1968. Paris became a separate department, and the administration of its suburbs was divided between three new departments that surrounded it. The Paris department was renamed " Île -de- France " in 1977, but this name "Paris region" was commonly used to refer to the entire Paris agglomeration [59].

Adapting to and solving problems associated with urbanisation, such as immigration, housing, technical infrastructure, social infrastructure, public services, etc. culminated on January 1, 2016, with the creation of the long-intended official administrative structure for cooperation between the city of Paris and its closest suburbs—the Metropolis of Greater Paris. The Metropolis of Greater Paris connects the city of Paris with 123 municipalities of the three departments of Hauts-de-Seine, Seine-Saint-Denis and Val-de-Marne, and 7 municipalities of Essonne and Val d'Oise. These densely populated interconnected urban areas are specific in terms of housing, transport, environment and distribution of economic activity. The metropolis of Great Paris with an area of 814 km^2 and more than 7.4 miles inhabitants is supposed to be a solution to all these problems, as the administration of Grand Paris goes beyond the borders of the municipalities. The greatest challenges for the Metropolis of Greater Paris are economic, social and cultural development, planning, environmental protection and international appeal. Challenges that depend mainly on employment, investment and innovation. The Metropolis of Greater Paris is a part of both the global and local economies. It is a territory that represents space for dialogue and strategic planning of a continuous, densely populated area. The Metropolis of Greater Paris asserts itself as the driving force and unifying element of several metropolitan initiatives that demonstrate the ability to work together to build a common future. Among several metropolitan actions, it is appropriate to mention in particular the project management of the future Olympic Water Center (CAO), activities in the field of energy transformation, improvement of air quality, development of cycle routes, support of innovations through the program "Innover dans la ville". To make systematically all activities in the territory of the Metropolis of Greater Paris, they are based on a strategic vision, which also determines the trajectory and direction of development of the entire territory for the next period. The Metropolis of Greater Paris has developed the main strategic planning documents (Plan for Climate Air Energy, Metropolitan Energy Master Plan, Metropolitan Digital Development Plan, Biodiversity Plan, Metropolitan Cycling Plan, Metropolitan Food Plan, etc.) that organise the dense urban area. In connection with the implementation of the individual measures that are part of the strategic documents, the office and the council of the metropolis created aid schemes (Metropolitan Investment Fund, "Structuralization Equipment Fund"), in which all municipalities of the Greater Paris Metropolis can participate [61].

The Metropolis of Greater Paris has made up several bodies of elected representatives who discuss, propose or negotiate all metropolitan issues. The bodies of the Greater Paris Metropolis are Metropolitan Council, the Chairman of the Metropolitan Council, Thematic committees and the Metropolitan Office. All elected officials are mayors, deputies or councillors of one of the municipalities of the metropolis of Paris. With its decisions, the Metropolitan Council regulates the affairs of the metropolis, whose competences are defined by law. The Metropolitan Council consists of 208 members, elected in direct general elections. At least one representative per municipality is elected. The chairman is responsible for the administration, represents the Metropolis of Greater Paris and embodies its executive. The chairman of the Metropolitan area of Paris is elected by the members of the Metropolitan Council by a supermajority in a secret ballot. The chairman of the metropolitan area convenes the council, sets its agenda, prepares and directs its deliberations, orders expenses and prescribes the realisation of income. He presides over meetings of the Metropolitan Council and the presidency of the Metropolitan Office, the assembly of mayors, Conferences of territorial presidents and Conferences of thematic commission presidents. The Metropolitan Office sets the strategy and determines the main directions of development of the metropolis. It is an advisory body on matters delegated to it by the Council. The office, which consists of 20 vicechairs and 23 delegated councillors representing the entire political spectrum, reviews the agenda and proposals for negotiations to be submitted to the Metropolitan Council for a vote. Thematic commissions are composed of elected representatives and deal specifically and in

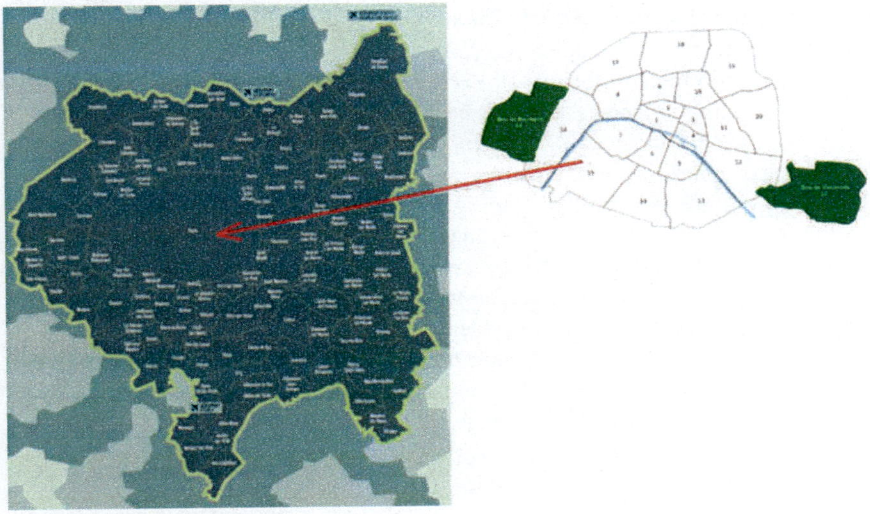

Fig. 3 Metropole Grand Paris and the City of Paris

detail with individual areas of metropolitan area development. There are currently 9 thematic commissions: Health and solidarity; Transformation of the environment and energy transformation; Biodiversity and nature in the city; Territorial cohesion and sustainable mobility; Attractiveness and economic development; Digital, innovation, research and development; Development; Population and housing; Finance [61] (Fig. 3).

The city of Paris itself is divided into 3 parts by the Seine river. In the centre is the Île de la Cité, which was in the past the seat of religious and secular authority. The left bank of the Seine was the seat of intellectual and cultural life and the right bank of the Seine represented the centre of economic life. However, the specific character of individual parts of the city was erased over time, and the fusion of these functions created a continuously viable urban environment. From an urban point of view, the city has preserved the circular shape of the historical Paris. Apart from the addition of the Bois de Boulogne, Bois de Vincennes and the Paris heliport in the twentieth century, the administrative boundaries of Paris have not changed since 1860. The city of Paris currently has a total of 20 urban districts. Each urban district is self-governing and has its own mayor, directly elected council, and town hall. Each of the Paris city districts is administratively divided into four administrative districts. The numbering of the districts of the city of Paris starts in the centre of Paris, continues spirally (in the shape of a snail), clockwise and ends in the east [62].

The mayor of Paris is not elected directly by voters. The voters of each of the 20 city districts elect members of the Paris Council, which then elects the mayor. The Paris Council has a total of 163 members in proportional representation according to the number of residents of individual city districts. The council of each urban district is composed of members of the Paris Council and members who serve only in the council of the urban district. The number of district council members varies depending on the district's population [62].

The self-government of the city of Paris is open and directly involves its citizens in decision-making processes. Citizens have online and offline smart tools available for decision-making and action governance, e.g. the "Decide for Paris" platform, the "Act for Paris" platform, participatory budgeting, the "My Paris" application, the "Paris Asso" digital service, Participating city authorities, educational activities, etc. "Decide for Paris" is a new platform that allows Paris to find all the initiatives in one click and participate in the decision-making of projects related to Paris. On the platform, residents can also find all the news that best matches their interests. The platform works in such a way that residents/civic associations publish their idea on the platform, if the idea reaches 1,500 votes in online voting during 6 months, the city undertakes to study the idea and if the city's opinion is positive, the city will do everything for its implementation. The city also implements online city consultations on projects proposed by the municipality within the platform. In this way, city residents can express their opinion, vote, discuss and also propose modification of project proposals that are a priority for the city and that it wants to implement during the year. The "Take action for Paris" platform is a platform focused on volunteering. Residents of the city, civil associations operating in the community, or the municipality itself can add activities for which they are looking for volunteers on the platform with a description of the activity, scope and time of volunteer work. A volunteer can sign up for individual activities. Volunteers can filter individual volunteer activities on the platform by area of interest or location. Up to a quarter of the investments of the city of Paris are built and decided with Parisians. Participatory budgeting is one of the tools that allows them to think and create the Paris of tomorrow. All Parisians (resident in Paris), from the age of 7 and regardless of nationality, can participate in the creation of a participatory budget, actively (by creating a project) or passively (by voting). The application "My Paris" provides a number of digital services provided by the city in one place, including: information about the city, travel schedule, processing of documents at the registry office, management of taxes and fees, search for public services (healthcare facilities, schools, homes, etc.), arranging of applications and subsidies and other [62]. The digital service "Paris Asso" is dedicated to sports clubs and civic associations dedicated to sports. Through the "Paris Asso" digital service, sports clubs can submit applications for subsidies for club activities and sports equipment, present the club and directly address sponsors, use online and physical areas to place posters about upcoming sports events and learn online [62].

The municipality of Paris also takes care of the direct participation of the city's residents through participatory bodies. The city of Paris has a total of 10 participatory bodies established: the Civic Assembly of Paris, the Paris Youth Council, the Paris Council of Civic Associations, the Paris Council of Europeans, the Council for Future Generations, the Paris Music Council, the Council of the Night, Neighborhood Councils, the Citizens Council and the Initiative and the District Committee [62].

The Civic Assembly of Paris allows Parisians to participate directly in the creation of municipal policies, especially by creating evaluation reports on draught strategic documents, projects and activities of the municipality. Considerations of the civic evaluation are submitted to the Paris Council, which formulates a public response to all comments and considerations of the Civic Association of Paris. The Paris Youth Council is a participative democracy body that allows young Parisians to be involved in the definition and implementation of community policies, to help the community innovate and present solutions to support young Parisians, to inform the municipality so that it better considers the needs and expectations of young Parisians. The Paris Council of Associations is a new participatory body that brings together a total of 100 civic associations. Its primary ambition is to monitor and evaluate the Charter of mutual obligations between the city of Paris and the associations, but it is also responsible for expressing the demands, expectations and contributions of the Parisian associations that it reflects. Direct exchanges of views with the municipality allow her to clarify and enrich the policy of supporting community life in the city of Paris. The Paris Council of Europeans is an advisory board open to any citizen with European citizenship and a strong connection to Paris. Its 61 members guide the municipality in making decisions on topics such as life in the European and international community, reception of European foreigners, tourism, international relations of the city of Paris, local and European citizenship, and other topics they consider relevant. The Council for Future Generations is an advisory, independent and joint body whose aim is to represent Parisian civil society and to reflect prospectively on topics that concern Paris and Parisians. The Paris Music Council makes it possible to connect all musicians in the Parisian music sector with the city's cultural policy. Every year, Paris hosts more than 36,000 concerts in almost 700 venues and is home to 300 music labels. To maintain and enhance this richness and this musical vitality, a consultative body has been created to strengthen the ties between the Parisian music industry and the city of Paris. The Council of the Night brings together the actors of Parisian nightlife and the support of the city of Paris in the creation of a participatory policy combining the development and promotion of nightlife, prevention and regulation. A total of 117 neighbourhood councils operate on the territory of the city of Paris, which fall under the competence of the city halls of the city districts. Neighbourhood councils participate in the direction of municipal policy and thus create bonds of solidarity essential for life in Paris. Citizens' councils are established in city districts, which enable residents to express their expectations, their proposals, and participate in decisions made in their territory [62].

In addition to meetings of members of participatory city bodies, public assemblies, etc., the city administration is focused on education and training of the city's residents in various areas through informal education. An excellent example is the "Climate Academy", which is a place of life, education, participatory and free exchange of information, experience and skills in the environmental field. Within the academy, workshops, modules for acquiring skills and meetings are held to support young people in establishing individual or collective projects in the field of ecological transformation [62].

Controversies about whether or not Paris is a megacity are mainly connected with determining the boundaries of the territory and thus also the data that enter into the study of Paris as a megacity. In general, it can be said that a megacity is a city with a population of more than 10 million. These are huge metropolises that are centres of culture, trade, technology and education. Until 2016, when the Metropolis of Greater Paris was officially and administratively established, the entire territory of the Île-de-France region with a population of 12,262,544 in 2019 was perceived as the agglomeration of Paris [58]. However, the territory of the current Greater Paris Metropolis has only 7,094,649 (2019) inhabitants [58], which again raising questions about Paris as a megacity.

The development of the population of the city of Paris, as well as the Metropolis of Greater Paris and the Île-de-France region, reflected changes in society and the structure of the economy. According to the French Statistics Office (INSEE), the population of the city of Paris was 2,148,271 as of January 1, 2020, which represents a decrease of 117,865 compared to 2015. The population of Paris today is less than the historical peak of 2.9 million in 1921. The main reasons were mainly the dramatic migration of residents to the suburbs between 1962 and 1975. Among the factors of migration were industrialization, high rents, gentrification of many indoor spaces, conversion of living space into offices and greater wealth for working families. Paris population decline temporarily halted in the early twenty-first century, in July 2004 population estimation showed an increase in population for the first time since 1954, and by 2009, the population had reached 2,234,000 people. However, since 2017, the population of Paris has been decreasing again, mainly due to the ageing of the population, lower birth rates and the availability of housing associated with the departure of Parisians to the municipalities of the Greater Paris Metropolis. In contrast, the number of inhabitants in the entire Île-de-France region increased during the monitored period.

The largest part of residents over 15 years of age are executives and higher intellectual professions (20–30%) and retired (18–19%). There is no evidence of famers, very low share of craftsmen, merchants and entrepreneurs (app. 3.5%) [58] (Fig. 4).

The Île-de-France region, representing the Paris agglomeration, which accounts for approximately 30% of France's GDP, is the most important centrum of economic activity in France. This region produced 739 billion euros in 2021. In 2022, the GDP of the Île-de-France region was the largest in Europe, ahead of North Rhine-Westphalia in Germany and Greater London in the United Kingdom. GDP per capita was the 4th highest in Europe after Luxembourg, Brussels and Hamburg [63]. The GDP of the city of Paris represented 20.57% of the GDP of the Île-de-France region. The GDP of Metropolis of Greater Paris accounted for 66.17% of the GDP of the Île-de-France region. The largest GDP per capita was produced in the city of Paris (Table 2).

Approximately, 80% of the inhabitants of the city of Paris, the Metropolis of Greater Paris and the Île-de-France region work in trade, transport and various services. The other important fields of employment belong to the public administration, education, healthcare, social action and construction services [58]. It confirms that the economy of the city of Paris, the Metropolis of Greater Paris, as well as the Île-de-France region is predominantly tertiary oriented. The highest share on the legal units in business sector to 31.12.2020 has professional, scientific and technical activities and administrative and support services (app. 30%), followed by wholesale

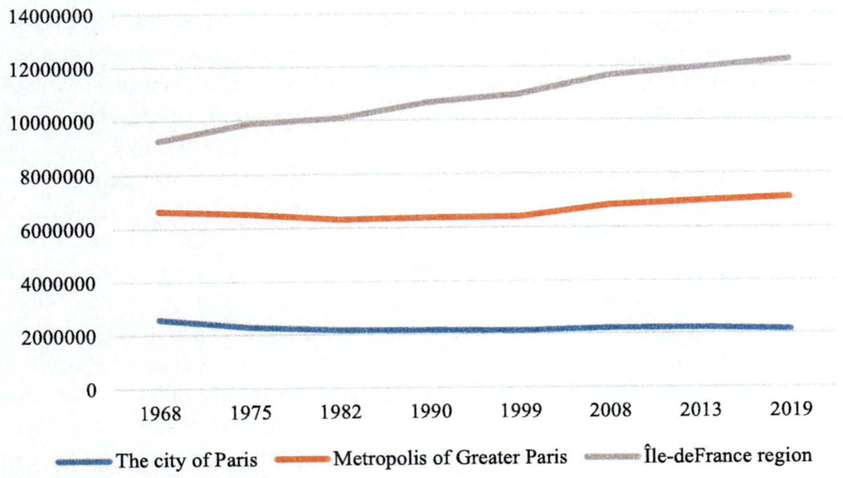

Fig. 4 Development of population in the period from 1968 to 2019

Table 2 Main characteristics of city of Paris, the Metropolis of Greater Paris and the Île-de-France region

	The city of Paris	Metropolis of Greater Paris	Île-de-France region
Area (km^2)	105	714	2,300
Population	2,165,423	7,094,649	12,262,544
Population density (inhabitants/km^2)	20,545	1,021	8,713.20
GDP (2021)	152 billion euros	489 billion euros	739 billion euros
GDP per capita (2021)	70,194 euros	68,925 euros	60,264 euros

and retail trade, transport, accommodation and catering (22–28%) and public administration, education, human health and social action, information and communication (7–10%) [58].

As a reaction of the Brexit, France, the Île-de-France region, the Metropolis of Greater Paris, the city of Paris, the Chamber of Commerce and Industry of Paris & Co and Business France decided to work together to attract London businesses that wanted to remain in the EU. That is why they created a special service called "Choose a Paris Region/Welcome to Greater Paris" which facilitates the relocation of companies from the UK to the Greater Paris Metropolis. These are business companies, financial institutions, but also companies engaged in research and production activities in the field of technical innovation, IT and high-tech industry [64].

In connection with the promotion of innovation and the high-tech sector, the Metropolis of Greater Paris has launched the Grand Paris Métro pole Tech, which brings together digital technology and innovation stakeholders from across the metropolitan area. Members of the "Grand Paris Metropolis Tech" can participate in the development of a metropolis that specialises in technological innovation. Paris itself is also known for its innovative, technological and entrepreneurial scene. Paris is constantly investing in the development of infrastructure for start-ups, which has created many new business opportunities in the city and attracted a large number of people interested in discovering what Paris has to offer [64].

In 2020, Paris ranked third among the biggest start-ups, thanks to the city's innovation ecosystem and new technologies that make the city extremely attractive to anyone interested in the industry. There are more than 4,000 start-ups in Paris alone, and three times as many in the Paris metropolitan area, 12,000. In 2017, tech billionaire Xavier Niel invested 250 million euros to build the largest incubator for start-ups in the world, Station F. This monumental space has more than 34,000 m^2, more than 4,000 tables, a multi-million-dollar art installation and many success stories. Station F is located in the 13th arrondissement of Paris and is known as the largest start-up facility in the world. Since the opening of the incubator—Station F, more than 1,000 start-up companies have been welcomed. This super hub offers entrepreneurs access to a number of different incubation programmes, luxury events and coworking space to work on their projects. In the Greater Paris Metropolis, there are also a number of coworking centres and business incubators that enable the networking of entities [65].

The metropolis of Greater Paris is working to make the circular economy a priority focus of economic development, to make the area resilient and innovative. This includes preparing for population growth and significant and sustainable reductions in greenhouse gas emissions and resource wastage. Transport is largely related to the quality of the environment, which is handled in the territory through seven types of transport, including the metro in the city of Paris, which has a total of 14 routes and one automatic line; RER (Regional Express Network), which refers to the rapid train transit system mainly between the city of Paris and its suburbs; land rail transport (trams), passenger car transport, bus transport, air transport provided by Charles de Gaulle, Orly and Beauvais-Tillé airports and boat transport on the Seine River. It is the support of shipping that should reduce the movement of goods and materials on the road network, through truck transport, and thus significantly contribute to the reduction of air pollution in the Metropolis of Greater Paris.

Parisians mostly use public transport to get to work, as up to 68.30% of the population work directly in Paris. In the territory of the Great Paris Metropolis, in addition to public transport, car transport is widely used to get to work (Table 3). This is mainly due to the commute of residents to work in the city of Paris from its suburbs.

In connection with the improvement and sustainability of the quality of the environment, an ecological (low-emission) zone has been introduced in more than 50 urban districts of the Metropolis of Greater Paris, into which only cars with an ecological mark—crit'air, i.e. those that meet low-emission criteria, can enter during the working week. The smallest share of cars per household is in the city of Paris itself.

Table 3 Share of means of transport used for the journey to work in 2019

	The city of Paris	Metropolis of Greater Paris	Île-de-France region
No shift	4.70	3.20	3.40
Walking (or roller skates, scooters)	10.50	6.90	8.40
Bicycle (including electrically assisted)	5.50	2.40	3.40
Motorised two-wheeled vehicles	4.40	2.90	3.80
Car, truck or van	10.60	40.30	27.10
Public transport	64.30	44.20	54.00

At the same time, the Greater Paris Metropolis has the largest share of households that own at least 2 cars.

A great opportunity and challenge for the Metropolis of Greater Paris in many areas in the near future are the Olympic and Paralympic Games 2024. The goal of the Olympic and Paralympic Games 2024 project in Paris is to create a global and at the same time responsible event that will embody the values of the environment and will be open to everyone. The Olympic Games will bring innovation, infrastructure development, job opportunities, human and financial capital, and potential for brand and image creation to the territory.

The relationship between Paris, as well as France, and Indo-Pacific region has deep historical roots. France is an Indo-Pacific nation, with 7 of its 13 overseas Departments, Regions and Communities situated in either the Indian Ocean or the South Pacific [66]. It is also the reason why France has economic, political and strategical interest to defend the region.

Nowadays, 93% of France's exclusive economic zone is located in the Indo-Pacific, around 8,000 soldiers are stationed there, and it is home to 1.5 million French citizens. France was also the first EU country to adopt its own national Indo-Pacific strategy [67]. It was also one of the key actors that supported the adoption of the European Indo-Pacific Strategy with aim to promote security, peace and global common goods, effective multilaterism and international order [66]. The basic principle of French policy is to act as a balancing power mitigating Sino-US tensions by fostering a multipolar and multilateral region governed by the rule of law [68]. Partnership with Indo-Pacific countries (among them India, Japan and Australia), bilaterally or through the EU, is the cornerstone of the French policy in terms of security strongly supported by diplomatic power [69].

France together with Germany and the Netherlands are promoting the Indo-Pacific concept of cooperation within the EU, which is being discussed in the Working Group for Asia and Oceania (COASI) from 2020. This concept promotes the main areas of further cooperation, as trade; connectivity; maritime security; and global issues such as climate change and biodiversity.

In view of its overseas departments, France develops beside the security and economic activities very intensive diplomatic activities. Activities in the field of education, science, research and innovation play an important role both directly in the Indo-Pacific region and also by creating opportunities for mobility for researchers and students. In order to cover its own interests and anticipate security risks associated with environmental problems, France places great emphasis on the issues of biodiversity, climate change, energy transformation (KIWA), protection and sustainable management of the oceans. A very important part of France's activities are economic and development activities, where e.g. business activities have grown by more than 49% in the last 10 years. These activities lead to the creation of various forms of partnerships with localities in this region, which are essential to ensure stability, sustainability and resilience, especially in relation to climate change and activities aimed at climate change adaptation, biodiversity protection, renewable energies and energy efficiency.

5 Common Features, Challenges and Solutions for Megacities in Europe and Worldwide

Over the past decades, urban development has undergone various economic, political, demographic, social and cultural changes that have significantly affected its current state, both positively and negatively. The megacities are now seen as a significant source of economic, environmental and social problems. However, at the same time, as centres of development, they are in a position to find possible solutions. The megacities are where innovation, new technologies and efficient and intelligent solutions for a better quality of life for people are created. However, they are also where the problems that slow down socio-economic development accumulate, manifest themselves and persist. In recent years, the growth of city populations and urban sprawl has added another problem to the persistent social, economic, urban planning and environmental problems, problem of security.

The ever-increasing human population growth is one of the most significant global challenges. In the last 50 years, the human population has more than doubled. People are moving from rural settlements to cities, offering them better fulfilment opportunities. This trend has led to more than half of today's world population living in cities and megacities.

Although we can speak of declining natural population growth in Europe, there is also a clear trend of rural–urban migration. Over the last century, Europe has transformed from a predominantly rural to a predominantly urban continent and is one of the most urbanised continents in the world. Europe's urban areas are home to more than two-thirds of the EU's population. More than 350 million people live in urban agglomerations of more than 5,000 inhabitants. These consume "about 80% of energy and produce up to 85% of European GDP" [70]. Although the pace of change has slowed down, the share of the population living in cities is still increasing. For example, Europe is characterised by a more polycentric and less centralised city structure than the United States, India or China. The number of people living in cities is expected to exceed the rural population for the first time in the next few years. Therefore, urban development and its direction is a priority issue.

In the past, the prevailing aspiration has been to be a competitive city, mainly in economic development. Economic development enhances a city's ability to attract both human and financial capital and is appealing to entrepreneurs, investors, new residents and visitors [71]. This development requires the use of more and more resources, which has led to resource depletion and has severe implications for future development [72]. The European Union places emphasis on social, environmental and natural resource conservation.

The European Union document Cities of Tomorrow [24] describes a shared vision of the European city of tomorrow. According to this document, the European city should be a place of advanced social progress with a high degree of social cohesion, socially balanced housing, services, especially health and education for all; a platform for democracy, cultural dialogue and diversity; a place of green, ecological or environmental regeneration; a place of attraction and an engine of economic growth.

The future urban development model should be based on: (a) balanced economic growth and spatial organisation of activities; (b) a polycentric urban structure; (c) strong metropolitan regions and strong urban areas that provide good accessibility to services of general economic interest; (d) a compact settlement pattern with limited urban sprawl; (e) a high level of environmental protection and quality in and around cities [24, p. 24].

Intensive development, which is undoubtedly a challenge for modern cities and megacities, can lead to positive outcomes for urban communities and negatively affect the city's smooth functioning. The challenges of modern cities focus on (1) uncontrolled urban sprawl [73–75]; (2) environmental pollution [73, 76, 77]; (3) urban logistics [78–80]; (4) sustainable technical infrastructure [32, 81]; (5) waste management [82, 83]; (6) the issue of an ageing population [84]; (7) stratification of wealth levels or areas of poverty [85]; (8) the level of citizen participation in governance [32, 73, 86, 87].

The nature of urban development has changed significantly over the past century. Globalisation, the development of technology and technology, the development of the economy and the related challenges and problems we mentioned have seriously affected the nature of urban development. If those in charge do not give them sufficient attention, or if they are ignored or neglected, they could lead to the city's decline or the city structure's disintegration. Therefore, responding to the growing urban issues and challenges is essential. Both in practice and at the theoretical level, different approaches and concepts that can be described as trends in urban development are emerging. They serve as good examples for cities and megacities, a source of inspiration and give them some idea of keeping up with the constantly advancing times and the changes they bring.

The scale of the problems facing megacities today goes beyond the boundaries of their territories, and urban problems are becoming global. Urban development is influenced by several factors, the decisions of local authorities, the decisions of national and international bodies, as well as by the ability of cities to adapt to new conditions and find appropriate solutions to problems. Megacities and urban areas are complex social ecosystems whose role is to ensure sustainable development and quality of life. The deterioration of social and economic indicators due to the impact of COVID-19 forces (not only) megacities to behave more responsibly and effectively to the challenges raised, set priorities and seek resources in relation to them. These efforts result in new concepts and trends in urban development.

The most recent and commonly used concepts in urban development that have become the subject of interest in theory and practice or both include [73]: (a) sustainable city [88–93]; and this trend green or eco-city [94, 95] with solar city [96–99]; (b) happy city [100]; charter cities [101, 102]; (c) creative city [18, 103–116]; (d) smart city [117–122]; (e) resilient city [20, 123–127]; (f) the city as a living lab [128–132]; (g) knowledge city [133]; (h) metropolitan information and communication technology environments [134, 135]; or the concept of slow cities, the so-called Cittaslow, Slowcity [136], which can be interested for small and medium size cities.

London and Paris cases have shown, that urban development and progress in megacities have always had a significant impact on economic development. Its nature creates a strong need for change. Megacities are seen as accelerators of economic growth, places of creativity and innovation, and centres of services and trade. Consumers, the workforce, businesses, institutions, higher-educated populations and the like are concentrated in them. Due to these entities' concentration, they can produce positive externalities and increase returns to scale.

On the other hand, they are a place where problems arise, such as unemployment, segregation, exclusion, congestion and poverty. In addition, public services are costlier, and natural resources are over-exploited, public transport networks are inadequate, creating congestion in and around them. These problems give rise to negative externalities such as traffic congestion, rising prices, lack of affordable housing, increased air and water pollution, waste accumulation, rising infrastructure costs, social tensions, higher crime rates and the like. As a result of climate change, as well as the high concentration of human activities, the high dynamics of development and the intensive use of resources, megacities are often at the limits of sustainability and are becoming very vulnerable ecosystems. The appeal of European megacities is enhanced by preserving the historical colour, providing opportunities for cultural, social and sporting activities, developing civic participation and the like. The same appeal is the reason for their development, growth or decline. European megacities governments' policy aims to increase cities' appeal, modernising them and building modern eco-friendly residential and commercial districts. The imbalance of these activities in spatial terms increases the risk of social and spatial segregation. Megacities are thus places where social polarisation occurs. Access to housing and services and public transport, and culture is worse in the least attractive parts of megacities, isolating the poorest and most vulnerable groups.

Moreover, appealing megacities attract new residents, entrepreneurs, investors and visitors. Their growth in megacities can contribute to economic growth but can also increase traffic congestion or reduce the quality of life of existing residents. Population growth causes an increase in population density, and consequently, the spatial requirements for further urban development increase. Moreover, conventional approaches cannot address the scale of the megacity's problems faced today. Moreover, the administrative boundaries of megacities no longer reflect the physical, economic, social, cultural or environmental realities of urban development and new forms of flexible city governance are needed.

It follows that the European megacity should be an appealing area that accelerates economic growth, social progress with high social cohesion, socially balanced housing, social and health services, and education for all. It should be a platform for democracy, cultural dialogue and diversity and a place for green, ecological and environmental renewal. The degree to which megacities can adapt to ever-changing conditions and solve problems can make them competitive entities or, conversely, threaten their sustainability. As a result, all megacities today face urgent challenges to ensure economic prosperity and social cohesion while achieving environmental and spatial sustainability.

In Table 4, we present the summarisation of challenges and solution of problems rising from the development of megacity system.

Megacities in EU should be primarily based on sustainability, and smartness which is reflected in the superstructure of the sustainable megacity. While the sustainable megacity represents the starting point for sustainable development, the smart megacity is currently the most comprehensive concept. On the other hand, a resilient megacity needs to be sustainable by applying some of the principles of the smart concept. However, there is a high expectation that a smart megacity will be also resilient (Fig. 5) [73].

Following the specifics of megacities, urban development needs to be considered more comprehensively. It is necessary to change paradigms of urban development. Currently, urban development is perceived from an economic, social and environmental point of view. This paradigm needs to be reconsidered and spread to institutional, spatial and technological or innovation subsystems as essential parts of the whole urban ecosystems [73] (Fig. 6). The systematic and coordinated connection of these systems is a prerequisite for sustainable development.

An important question is how to link these systems and make them work more efficiently. Based on the knowledge gained, the institutional system links them, creating the policy framework for ensuring that development objectives are achieved [73]. A prerequisite is an integrated approach to urban development management. Urban planning and management have to be a continuous process, to which all citizens have equal access. Continuous, integrated, open and transparent process of urban planning and management in agglomeration can assure that the positive externalities will fully be utilised and that the negative externalities of agglomeration were effectively coped with. Citizens' participation in the whole planning and management process of urban development can minimise the major defaults of current approaches. However, an integrated approach must be seen not only in spatial terms but also in financial, sectoral or cross-sectoral and institutional terms, in terms of multi-level governance and the participation of the relevant actors in development [73].

In Fig. 7 [73], we have outlined the relationships within this system for EU urban development based on the priorities of the new Partnership Agreement 2021–2027

Table 4 Systems of megacity—challenges and solutions

System	Challenge	Problem	Options	Solution–innovation
Social	Migration (new citizens, emigrants, etc.)	Crime Violence	Security	Security management Risk management
	Segregation	Poverty Unequal access to education	Integration Social Ethnic Religion diversity	Communication Transparency Building community life
	Exclusion	Radicalism New ideologies Populism Racism	Inclusion Openness	(Life-long) education Participation
	Congestion	Disinformation Individualism Polarisation	New infrastructure	Education in critical thinking System of strategical communication as a prevention
	Rapid ageing process	Balanced territorial system/ strategy	Brain gain	
		Inter-generational problem	Accessibility public services Eco-housing Health system Creativity Cultural values Leisure activities for body and brain	Tools of co-creation of public services online and offline
Ecological Environmental	Pollution	Mobility	Sustainable mobility	Integrated system of mobility E-mobility
	Lack of traditional resources	Energy	Green energy	IT systems Solar energy Green sources
	Waste		Zero waste	Circular economy
	Climate changes (tornado, heatwaves, droughts, torrential rains, floods)	Economic disbalance	Water management	Digital technologies
		Social disbalance	Adaptation	Green innovation

(continued)

Table 4 (continued)

System	Challenge	Problem	Options	Solution–innovation
				Climate and resilience management/ strategy/governance
Economic	Competitiveness	Unemployment	Innovation system	Innovation strategy–creating environment for innovation systems
		Productivity	Flexibility of workforces	Open labour market
			International integration and transformation	
Spatial	Land use	Degradation of agricultural land	Spatial planning reflecting carrying capacity of the territory	Digital tools (e.g. GIS systems, simulation of country development, etc.)
Technical and technological	Digitalisation	Availability	Digital and physical connectivity	IT infrastructure and digital tools
Institutional	Management	Connectivity of different problems	Integrated Cross-sectoral Multi-level governance	Integrated development urban strategy (all levels of governance)
			Knowledge -base, transfer of knowledge	Open government Smart government Democracy
			Intellectual Capital	Participation Transparency
ECO SYSTEM	Sustainable competitiveness	Attractiveness	High standard of living	Continuity of innovation processes Connectedness Efficiency Expediency of systems

(with implications for urban development in terms of possible development trends supported by the five policy objectives). Figure 7 illustrates an integrated approach to development in terms of planning (spatial, policy levels) as well as a multi-level system of development management and participation at both local and top levels.

An integrated approach to urban development planning is now necessary due to the rapid urban population growth and its consequent expansion into the periphery and satellite towns to compensate for people's consumerist lifestyles [73].

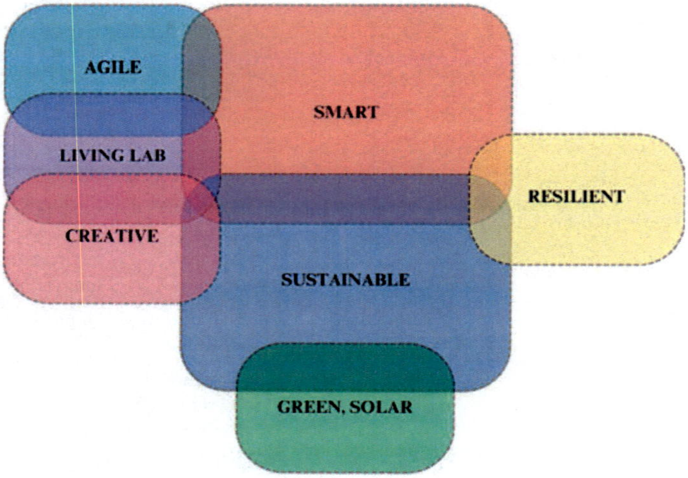

Fig. 5 Significance and intersections of megacity concepts

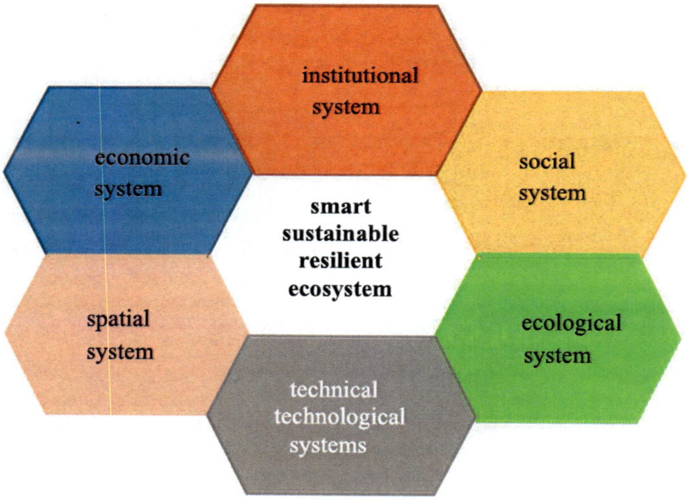

Fig. 6 The systems that create an urban ecosystem

According to the EU, it is vital to create and develop cross-sectoral cooperation within a specific grouping of municipalities, characterised by certain common features, functions and problems, which does not have to be an administrative unit with a self-governing function [15]. This approach can also fit to conditions of megacities.

Megacities are currently seen in terms of the coexistence of functional and spatial integration of urban and peri-urban (rural) areas. To protect their privacy and improve

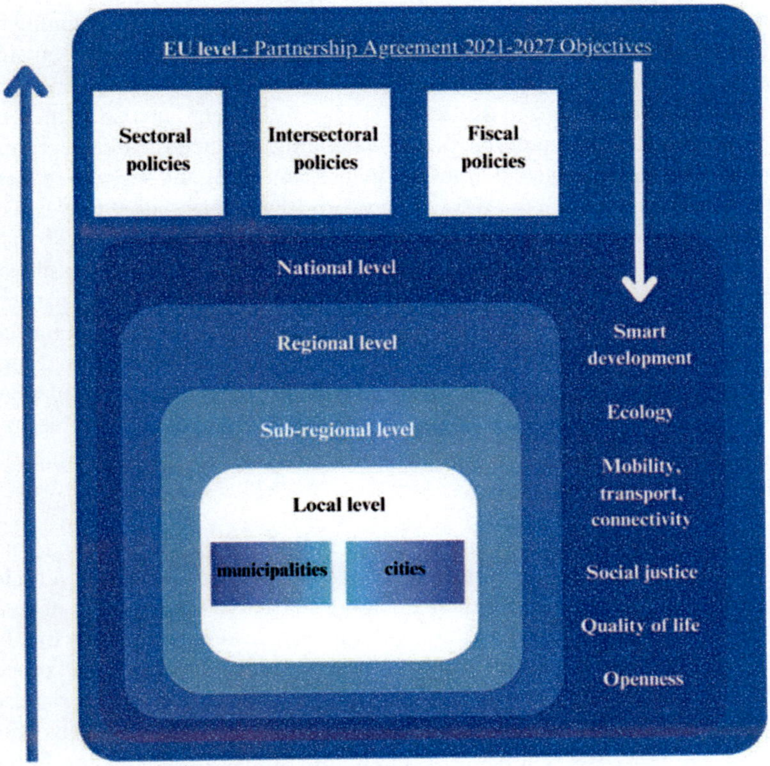

Fig. 7 Integrated urban development system

their quality of life, many inhabitants prefer to live in detached houses, in close contact with nature, with solutions that emphasise preserving a serene environment in the interest of a healthy lifestyle. This trend is particularly evident in the suburbs of megacities, where smaller satellite centres are being created, but also in the development of adjacent intercity areas, significantly changing the nature of their use. In addition, there is de-urbanisation, which is reflected in the changing demands and needs of the inhabitants. The problems of congested or missing technology or transport infrastructure, the lack of residential and commercial space, and the inadequate provision of public services in the peripheral areas of megacities or their satellites are deepening. However, the problems also present new opportunities for developing transport and technical and social infrastructure.

An integrated approach to urban planning aims to improve living conditions in megacity and peri-urban areas by focusing on purposeful spatial, economic, social, environmental and institutional development. This approach integrates the megacity as a whole by seeking to direct development to those parts of the city that have been neglected as peripheral. The integration is done by linking not only all parts of the city but also with the surrounding areas [115]. Thus, integrated planning represents several related and coordinated projects implemented within and outside the megacity. They aim to address the most pressing problems in the long term, considering future developments and the needs of not only current but also future generations of inhabitants. Integration requires the building and creation of an appropriate, efficient and sustainable system of infrastructure, amenities and services, which will have the characteristics of both the city and peri-urban areas so that they can create sustainable ecosystems that make efficient use of natural resources. Choosing appropriate locations for service provision facilities is vital, making them accessible to the megacity and peri-urban residents. Due to different lifestyles and economic activity, the needs of the megacity and rural dwellers differ, reflected in the demand and, consequently, the supply of services, which requires significant differentiation. Megacity dwellers, who have a relatively high-income level, demand services related to leisure or cultural activities, while those living in peri-urban areas require services mainly related to agricultural production. Megacities should adapt to these demands and seek solutions to meet the needs of residents' needs outside their catchment areas [73].

The second major challenge is ensuring multi-level governance and relevant actors' participation. This approach requires openness, transparency, efficiency, effectiveness, exploring needs and encouraging the active participation of relevant actors in public decision-making while involving them in local community life. In this context, it is necessary to introduce new management tools, measures and processes that result in the creation of procedures for seeking consensus among representatives of the public, private and non-profit sectors and strengthening the involvement of citizens and other relevant actors through new forms of participation [73]. The main objective is to bring stakeholders' interests into line with the objectives of local government as stated in its policies, which requires defining the problems and needs and the human and financial capacities of local government entities that can contribute to solving local problem areas. Moreover, mobilising governmental and non-governmental resources to achieve the set tasks, creating a framework for collective action and negotiation and aligning and coordinating public interests with the interests of relevant stakeholders will also be required [137]. Many of these activities cannot be ensured without modern information and communication technologies that support stakeholders in their participation in designing and delivering public services. In addressing public policy issues and increasing government transparency, they can enhance megacity competitiveness and quality of life for citizens, as well as improve urban planning and management [56]. In addition, new social media and modern ICTs (e.g. apps) are seen as a means for governments to reach and engage citizens in public policymaking.

6 Conclusions

Global development naturally affects development in cities and especially in megacities, and megacities are considered to be accelerators of that development. The current global problems of the world (economic changes in connection with globalisation, economic cycles, changes in the behaviour of markets, growing social problems, migration and climate change, especially pollution and the like) are creating a need for change throughout society. They also affect megacities, which are a significant contributor to these problems. The growth in the number and size of megacities over the last 50 years is of significance, particularly in developing countries. In contrast, rapid urbanisation and growth bring great opportunities in the form of agglomeration effects. On the other hand, they pose significant challenges to sustainable development. These positives and negatives vary considerably from continent to continent. Megacities respond to them depending on their environment. However, all are looking for ways to ensure sustainability, smartness and resilience. As a result, different approaches to urban development are emerging.

Megacities in contemporary society no longer function independently but are the centre of development management beyond their administrative boundaries. Therefore, we can define a megacity as a settlement or a cluster of economically, socially and culturally heterogeneous settlements, characterised by the greatest degree of urbanisation, a concentration of economic activities and a typical urban built-up area. It also fulfils a complex of functions for other surrounding urban areas and is an accelerator of regional, national and international economic development.

The megacity must be viewed from a broader territorial perspective regarding links and relations with the surrounding environment. Urban development is influenced by endogenous and exogenous factors, including decisions taken at local, national and international levels. On the other hand, the development of megacities strongly influences regional, national and international development.

Urban development today cannot be squeezed into the administrative boundaries of a megacity. In this context, the size of the megacity is an important issue. Many studies have been devoted to it. It is essential to realise that influencing the size of a megacity is difficult in the current conditions. Megacities grow through their activities but also the growth of surrounding settlements and subsequent interconnection or merging. However, megacities' size is related to their expansion in area, population growth and population density. Studies [19, 73] suggest that megacities have more serious resilience and environmental problems than medium-sized cities. Conversely, smaller cities are more open and innovative.

Europe is one of the most urbanised continents in the world. More than two-thirds of Europe's population live in built-up areas, and their share is increasing. Urban development will be a determining factor in the future economic, social and territorial development of the European Union. For EU cities is an urbanisation and their strategical management the key issue for the future sustainable urban development. The megacities in Europe like Paris and London are new engines of the global economy; they concentrate on knowledge and contain innovation ecosystems, but on the other hand, they can be seen as very vulnerable areas, with possibility of low productivity, segregation, exclusion and environmental problems. As a result of these issues connected with rapid growth, sustainability, resilience, social, environmental and economic issues, new trends have emerged such as smart, resilient, green, agile, etc. cities and regions. The role of these trends is to create environment for the new challenges to which cities are exposed.

A fundamental prerequisite for further megacities development is focusing on the economic, social, ecological, spatial, institutional and technical or innovative systems. However, they have to be seen as interlinked subsystems by integrated management. An integrated approach to spatial, financial, sectoral, intersectoral and institutional development, particularly in terms of the participation of the relevant actors in development, is becoming a prerequisite for that development. Awareness on the direction of urban development and an assessment of current practice approaches also create the preconditions for the transformation megacities into sustainable, smart and resilient cities of the twenty-first century in the context of the 2030 Agenda.

Compared to 21 megacities in the Indo-Pacific, there are only 4 megacities in Europe (Paris, London, Moscow, Istanbul). We focused only on two of them in the chapter. We described the examples of London and Paris. Paris and London both are the world city leaders. The number of megacities in Europe is related to the history of the settlements of Europe and the nature of urbanism and architecture of the cities, but also to the demographic development in Europe. While in the countries of the Indo-Pacific region, the development of megacities is strongly focused on economic relations (trade, transport, finance), the megacities of Paris and London also place great emphasis on spatial aspects of development, but also on environmental and social aspects.

The relationship between London and Paris as well as United Kingdom and France, and Info-Pacific region has deep historical roots. Even today, the common history is reflected in the mutual economic, social interests but as well as by a military presence because of strong position of these countries in the regional security architecture. France and United Kingdom have considerable geographic areas of Exclusive Economic Zones as per the definition of the United Nations Law of the Sea in the Indo-Pacific.

Paris and London as important international airport hubs, they have the key role in international policy, economy (business, finance) and social relations with the Indo-Pacific countries. The Indo-Pacific region is important for United Kingdom and France to preserve the free flow of goods, services, defending the countries' sovereignty as well as maintain international stability. Partnership with Indo-Pacific countries bilaterally or through the EU is the important point of policy of these countries. It is strongly supported by diplomatic power. Presented examples of megacities can be, in many ways, an inspiration for megacities in the Indo-Pacific, as well as the development of megacities in Europe, despite historical, political, cultural and religious differences. While in the countries of the Indo-Pacific region, the development of megacities is strongly focused on economic relations (trade, transport, finance), the megacities of Paris and London also place great emphasis on spatial aspects of development, but also on environmental and social aspects.

The knowledge that we have summarised in this chapter can be an important basis for public policymaking in megacities development.

Acknowledgements This study has been supported by project Vega 1/0343/23 Modern approaches to the development of cities and regions and Vega 1/0311/23 Open Local Municipality.

References

1. UNDP (2018) What are the sustainable development goals? Available at: https://www.undp.org/content/undp/en/home/sustainable-developmentgoals/goal-11-sustainable-cities-and-communities.html
2. United Nations (2017) Report on sustainable development, aims. Available at: https://agenda2030.statistics.sk/Agenda2030/indikatory-globalna-uroven-ciel-17/
3. Kresl PK (2020) Cities and the contemporary global economy. In: Kresl PK, Urban competitiveness in developing economies. Routledge, Oxon, pp 3–12
4. Palen J (1987) The urban word. McGraw-Hill Education, USA, 480 p
5. Lang R, Knox PK (2009) The new metropolis: rethinking megalopolis. Regional Studies 43(6):789–802
6. Dobbins T (2018) What's the difference between a megacity, a metropolis, a megalopolis and a global city? Available at: https://www.archdaily.com/896568/whats-the-difference-between-a-megacity-a-metropolis-a-megalopolis-and-a-global-city
7. Lin X, Chen Z (2003) Review and outlook for urban agglomeration studies within and without China. J Trop Geogr 23(1):44–50
8. Sassen S (2005) The global city: introducing a concept. Brown J World Aff 10(2)
9. Berg L, et al (1982) A study of growth and decline. Elsevier Ltd., Pergamon, 184 s. ISBN 978-0-08-023156-3. https://doi.org/10.1016/C2013-0-03056-3
10. Parr J (2004) The polycentric urban region: a closer inspection. Reg Stud 38(3):231–240. https://doi.org/10.1080/003434042000211114
11. Nijkamp P, Zwetsloot F, Van Der Wal S (2010) Innovation and growth potentials of European Regions: a meta-multicriteria analysis. Eur Plan Stud 18(4):595–611
12. Vigašová D (2010) Spôsoby vyčleňovania mestských aglomerácií (na príklade miest Banská Bystrica a Zvolen. Acta Geographica Universitatis Comenianae 54(2):231–245
13. Lui Z, Liu Sh (2018) Polycentric development and the role of urban polycentric planning in China's megacities an examination of Beijing's Metropolitan area. Available at: https://www.researchgate.net/publication/325197843_Polycentric_Development_and_the_Role_of_Urban_Polycentric_Planning_in_China's_Mega_Cities_An_Examination_of_Beijing's_Metropolitan_Area
14. Sasha A (2018) Polycentric cities. Available at: https://www.gensler.com/research-insight/blog/polycentric-cities-new-normal-manila-finance-centre
15. URBACT (2017) Ste pripravení na budúcnosť: Odolnosť miest v praxi. Urbact.eu. Available at: https://urbact.eu/ste-pripraven%C3%AD-na-bud%C3%BAcnos%C5%A5-odolnos%C5%A5-miest-v-praxi
16. World Economic Forum (2018) Agile cities: preparing for the fourth industrial revolution. Available at: http://www3.weforum.org/docs/WP_Global_Future_Council_Cities_Urbanization_report_2018.pdf
17. Glaeser EL, et al (1992) Growth in cities. J Polit Econ 100(6):1126–1152
18. Florida R (2004) The rise of the creative class. ...and how it's transforming work, leisure, community, & everyday life. Basic Books, New York, 434 p
19. Borseková K, Koróny S, Vaňová A et al (2018) Functionality between the size and indicators of smart cities: a research challenge with policy implications. Cities: Int J Urban Policy Plan 78(78):17–26

20. Borseková K, Nijkamp P et al (2018) Urban Resilience Patterns after an External Shock: An Exploratory Study. Int J Disaster Risk Reduct 31:381–392. https://doi.org/10.1016/j.ijdrr.2018.05.012
21. Borseková K, Vaňová A, Šúrová J, et al (2021) The nexus between creative actors and regional development. Land, MDPI, Open Access Journal 10(3):1–18
22. Knox P, Mayer H (20090 Small towns sustainability: economic, Social, and Environmental Innovations. Birkhauser Verlag, Bazilej, 194 p
23. Farr D (2008) Sustainable urbanism: Urban design with nature. New Jersey, Wiley, 304 p
24. European Union (2011) Cities of tomorrow—challenges, visions, ways forward. Publications Office of the European Union, Luxemburg, 112 p. https://doi.org/10.2776/41803
25. Vitálišová K, Vaňová A, Borseková K et al. (2016). Creative industries and the micropolitan city as an alternative to the Mega-City: a solution to the demographic crisis in Central Europe. In 7th Annual international conference "region in the development of society". Fakulta regionálního rozvoje a mezinárodních studií Mendelovy univerzity, Brno
26. Han W, Geng Y, Lu Y, Wilson J, Sun L, Satoshi O, Geldron A, Qian Y (2018) Urban metabolism of megacities: A comparative analysis of Shanghai, Tokyo, London and Paris to inform low carbon and sustainable development parthways. Energy 155:887–898
27. Safarik D, Ursini S, Wood A (2016) Megacities: setting the scene. Urban Issues IV:30–39
28. United Nations (2018) 2018 revision of world urbanization prospects. Available on https://population.un.org/wup/
29. Designbuildings (2023) Megacities. Available on https://www.designingbuildings.co.uk/wiki/Megacity
30. Fleck A (2023) The world's next megacities. Available at: https://www.statista.com/chart/29152/the-worlds-next-megacities/
31. Florida R (2016) The new typology of global cities: the seven types of global cities driving the world economy. Available at: https://www.bloomberg.com/news/articles/2016-10-06/what-globalization-and-urbanization-mean-for-cities
32. Vaňová A (2006) Strategické marketingové plánovanie rozvoja územia. Univerzita Mateja Bela, Banská Bystrica, 138 p
33. Votrubec C (1980) Lidská sídla, jejich typy a rozmístění ve světě. Academia, Praha, 394 p
34. Paulíčková R (2005) Teoretické otázky regionálneho a mestského marketingu. Západočeská univerzita v Plzni, Ekonomická fakulta, 126 p
35. ESPAS (2018) Global trends to 2030: the future of urbanisation and megacities. Available at: https://espas.secure.europarl.europa.eu/orbis/sites/default/files/generated/document/en/Think%20piece%20global%20trends%202030%20Future%20of%20urbanisation.pdf
36. Hebbert MJ, Clout HD, Blake E (2019) London. In: Encyclopedia Britannica [online]. Available at: https://www.britannica.com/place/London
37. Hidden London (2023) London Boroguht maps. Available at: https://hidden-london.com/miscellany/london-boroughs-map/
38. Urban Age (2018) Governance structure, London. Available at: https://urbanage.lsecities.net/data/governance-structure-london-1
39. Sirius (2023) Greater London authority. Available at: https://www.siriusopensource.com/en-uk/case-studies/public-sector/gla.
40. Center for London (2020) London at a crossroads. Initial report from the London Futures review. London, 107 p
41. Robinson J, Attuyer K (2020) Extracting Value, London style: revisiting the role of the state in urban development. Int J Urban Regnal Res: 303–331
42. Study Master (2023) London urbanisation. Available at: https://www.studysmarter.us/explanations/geography/challenges-in-the-human-environment/london-urbanisation/
43. Office for National Statistics (2023) Regional economic activity by gross domestic product UK: 1998–2021
44. London Data Store (2023) London's economy dashboard
45. CatapultDigital (2020) London: A GLOBAL LEADER IN ADVANCED DIGITAL TECHNOLOGIES. Digital Catapult. Available at: https://www.digicatapult.org.uk/news-and-insights/publication/london-a-global-leader-in-advanced-digital-technologies

46. Misra T (2018) Can London become a people-centric smart city? Available at: https://www.bloomberg.com/news/articles/2018-03-19/can-london-become-a-people-centric-smart-city
47. Greater London Authority (2016) The future of smart. Smart London report. Available at: https://www.london.gov.uk/sites/default/files/gla_smartlondon_report_web_3.pdf
48. We build value (2022) Hi-tech London, or the making of a smart city. Available at: https://www.webuildvalue.com/en/megatrends/smart-city-london.html
49. Greater London Authority (2018) Smarter London together. Available at: https://www.london.gov.uk/sites/default/files/smarter_london_together_v1.66_-_published.pdf
50. Smart London Board (2023) Available at: https://www.london.gov.uk/programmes-strategies/business-and-economy/supporting-londons-sectors/smart-london/smart-london-board?ac-34821=34820
51. Busness Plan and Strategy of Queen Elizabeth Olympic Park (2023) Availabe on: https://www.queenelizabetholympicpark.co.uk/our-story/who-we-are/business-plan
52. London Sustainable Development Commission. (2017) London's quality of life indicators report 2017: measuring London's progress towards becoming a sustainable world city. 56 pp
53. Mori Memorial Foundation (2022) Global power city index 2022. Available at: https://mori-m-foundation.or.jp/english/ius2/gpci2/
54. Whitehead R, Harding C, Bowe N (2021) Greater London: A new vision for a better city. 48 p. https://centreforlondon.org/publication/greater-london-a-new-vision/
55. Gilli A (2019) The United Kingdom and the Indo-Pacific: return of global Britain? In: Mind the gap: national views of the free and open Indo-Pacific, April, pp 44–48
56. Barry B, Giergerich B, Graham E, Schreer B (2022) The UK Indo-Pacific tilt: defence and military implications. In: The international institute for strategic studies, 43 p
57. Foreign policy (2023) Britain's surprisingly enduring Tilt to Asia. Available at: https://foreignpolicy.com/2023/04/11/uk-britain-tilt-indo-pacific-asia-strategy-review-aukus-cptpp-geopolitics/
58. Institut natgional de la statistique es des études économiques (2020) Dossier complet. Available at: https://www.insee.fr/fr/statistiques/2011101?geo=REG-11#tableau-RES_G1
59. Gallez, C., Maulat, J. (2015) Une démarche pilot porté par la Région Île -de- France: Charte et contrat aménagement - transport autour du prolongement de la ligne 11 du metro. Flux 101–102(3):124. https://doi.org/10.3917/flux.101.0124
60. World population review (2022) Available at: https://worldpopulationreview.com/
61. Metropole Grand Paris (2023) Available at: https://www.metropolegrandparis.fr/en/metropolis-service-72-million-residents
62. Paris (2023) Available at: https://www.paris.fr/
63. Eurostat (2022) Available at: https://ec.europa.eu/eurostat/web/regions/data/database
64. Metropole Grand Paris (2021). Metropole Grand Paris in action. Available at: https://metropolegrandparis.fr/sites/default/files/media/document/1903_MGP-plaquetteBDweb.pdf
65. Mayors of Europe (2022). Paris is home to the largest startup incubator in the world. Available at: https://mayorsofeurope.eu/top-stories/innovation/paris-is-home-to-the-largest-startup-incubator-in-the-world/
66. French Embassy in Canada (2022) Indo-Pacific: Questions and answers on France's regional strategy. Available at: https://ca.ambafrance.org/Indo-Pacific-Questions-and-answers-on-France-s-regional-strategy
67. CEPS (2022) Why French leadership in the Indo-Pacific will be a real asset for the EU. Available at: https://www.ceps.eu/why-french-leadership-in-the-indo-pacific-will-be-a-real-asset-for-the-eu/
68. Pajon C (2021) France: The leading European power in the Indo-Pacific. In: Bowen J (ed) Europe's Indo-Pacific embrace: Global partnerships for reginal resilience, 10–13 pp
69. Mazzucchi N (2023) French Strategy for the Indo-Pacific and the issues of European cooperation. In: The Hague centre for strategic studies, 9 p
70. Furik A, Szalai P (2017) Smart cities: Budúcnosť miest v Európe. In Euroactiv. Available at: https://euractiv.sk/section/digitalizacia/linksdossier/smart-cities-buducnost-miest-v-europe

71. Vaňová A, Petríková K (2012) (2012) Influences of the economic cycle on the product life cycle of a territory. Int Rev Public Nonprofit Mark 9(2):97–104
72. Monfaredzadeh T, Berardi U (2015) Beneath the smart city: dichotomy between sustainability and competitiveness. Int J Sustain Build Technol Urban Dev 6(3):140–156
73. Vaňová A (2021) Trendy v rozvoji miest (Trends in city development). Univerzita Mateja Bela, Banská Bystrica, 201 p
74. Kovács Z, Farkas ZJ, Egedy T et al (2019) Urban sprawl and land conversion in post-socialist cities: the case of metropolitan Budapest. Cities 92:71–81. https://doi.org/10.1016/j.cities.2019.03.01
75. Yu M, Yang Y, Chen F et al (2019) Response of agricultural multifunctionality to farmland loss under rapidly urbanizing processes in Yangtze River Delta, China. Sci Total Environ 666:1–11
76. Caparros-Midwood D, Dawson R, Barr S (2019) Low carbon, low risk, low density: resolving choices about sustainable development in cities. Cities: Int J Urban Policy Plan 89:252–267
77. Munoz-Pandiella I, Bosch C, Merillou N et al (2018) (2018) Urban weathering: interactive rendering of polluted cities. IEEE Trans Vis Comput Graph 24(12):3239–3252. https://doi.org/10.1109/TVCG.2018.279452
78. Nataraj S, Ferone D, Quintero-Araujo C (2019) Consolidation centers in city logistics: A cooperative approach based on the location routing problem. In International Journal of Industrial Engineering Computations 10(3):393–404. https://doi.org/10.5267/j.ijiec.2019.1.001
79. Firdausiyah N, Taniguchi E, Qureshi AG (2019) Modeling city logistics using adaptive dynamic programming based multi-agent simulation. Transp Res Part E: Logist Transp Rev 125:74–96
80. Bjørgen A et al (2019) The potential for coordinated logistics planning at the local level: a norwegian in-depth study of public and private stakeholders. J Transp Geogr 76:34–41
81. Petrova S, Prodromidou A (2019) Everyday politics of austerity: Infrastructure and vulnerability in times of crisis. Environ Plan C: Polit Space 37(2)
82. Heaton J, Parlikad AK (2019) A conceptual framework for the alignment of infrastructure assets to citizen requirements within a smart cities framework. Cities 90:32–41. https://doi.org/10.1016/j.cities.2019.01.041
83. Amritha PK, Kumar PPA (2019) Productive landscapes as a sustainable organic waste management option in urban areas. Environ, Dev Sustain 21(2):709–726. https://doi.org/10.1007/s10668-017-0056-0
84. Jayantha WM, Qian QK, Yi CO (2018) Applicability of aging in place in redeveloped public rental housing estates in Hong Kong. Cities 83:140–151
85. Aguilar AG, López FM (2016) Poverty spaces in the urban periphery and inner suburbs of Mexico City. Accumulated disadvantages. Eure 42(125):5–29
86. Van Holm EJ (2019) Unequal cities, unequal participation: the effect of income inequality on civic engagement. Am Rev Public Adm 49(2):135–144
87. De Castro Pena JC, De Assis JC, Da Silva RA et al (2017) Beyond the mining pit: the academic role in social deliberation for participatory environmental planning. Perspect Ecol Conserv 15(3):194–198. https://doi.org/10.1016/j.pecon.2017.06.006
88. Hollands R (2008) Will the real smart city please stand up? Intelligent, progressive or entrepreneurial? City: Anal Urban Trends, Cult, Theory, Policy, Action 12(3):303–320. https://doi.org/10.1080/13604810802479126
89. Jenks M, Colin J (2009) Dimensions of the sustainable city. Springer Science & Business Media, London, 282 p
90. Mori K, Christodoulou A (2012) Review of sustainability indices and indicators: towards a new city sustainability index (CSI). City: Environ Impact Assess Rev 32(1):94–106
91. Komninos N (2013) What makes cities intelligent? In: Deakin M (ed) Smart cities: Governing, modelling and analysing transition. Taylor and Francis, Deakin
92. Joss S (2015) Sustainable cities: governing for urban inovation. Macmillan Education, London, 352 p
93. Cohen S (2017) *The sustainable city*. Columbia University Press, 264 p

94. Beatly T (2012) *Green cities of Europe.* Island Press, 248 p
95. Anthopoulos L, Fitsilis P (2010) From digital to ubiquitous cities: defining a common architecture for urban development. In: IEEE 6. International conference on Intelligent Environments, pp 301–306
96. O'Connor P (2006) Santa Monica sustainable city plan. In: Proceedings 2nd International Solar Cities Congress. ISCI, Oxford
97. Šíp L, Krajcovics L (2010) *Solárne mestá: texty publikované v rámci grantu VEGA 1/0847/08: uplatnenie stratégie solárnych miest v podmienkach Slovenska.* Slovenská technická univerzita v Bratislave, Bratislava, 98 p
98. Rode P, Burdett R (2011) *Cities: investing in energy and resource efficiency.* UNEP, Towards and green economy, 630 p
99. Byrne J, Kurdgelashvili L, Taminiau J et al (2015) A review of the solar city concept and methods to assess rooftop solar electric potential, with an illustrative application to the city of Seoul. Renew & Sustain Energy Rev 41:830–844
100. Montgomery CHW (2014) Happy city. Farrar, Straus a Giroux, New York, 368 p
101. Romer P (2011) Charter city: Proceedings TED conference, TED, Long Beach
102. Romer P (2015) Charter cities: interview on urbanization, charter cities and growth theory. Available at: http://www.paulromer.net/tag/charter-cities/
103. Florida R (2005) Cities and the creative class. Routledge, New York, 208 p
104. Landry CH (2000) The creative city: a toolkit for urban innovators. Earthscan Publications Ltd., London, 300 p
105. Landry Ch, Hyams J (2012) The creative city index: Measuring the pulste of the city. Comedia, Gloucestershire, 64 p
106. Steward WC, Kuska SSB (2008) Developing and sustaining creative cities: a sustainability tool for designers, planners, and public administrators. Available at: http://www.ecospheres.com/creative_cities.pdf
107. Evans G (2009) Creative cities, creative spaces and urban policy. Urban Studies 46(5–6):1003–1040
108. Romein A, Trip JJ (2010) The creative production and consumption Milieu. Northsearegion. EU. Available at: http://archive.northsearegion.eu/files/repository/20100816174712_CCC6.1.2FrameworkReport.pdf
109. Clifton N, Cooke P (2009). Creative knowledge workers and location in Europe and North America: a comparative review. Creat Ind J 2(1):73–89
110. Flew T (2010) Toward a cultural economic geography of creative industries and urban development: introduction to the special issue on creative industries and urban development. Inf Soc: Int J 26(2):85–91
111. Chapain C, Comunian R (2010) Enabling or inhibiting the creative economy: the role of the local and regional dimensions in England. Regional Studies 44(6):s. 717–734. ISSN 717-734.
112. Booyens I, Molotja N, Phiri MZ (2011) Creative industries and innovation: the case of new media firms in Cape Town. Ungs.edu.ar. Available at: http://www.ungs.edu.ar/globelics/wp-content/uploads/2011/12/ID-129-Booyens-Molotja-Phiri-Innovation-economic-development-and-inequality.pdf
113. Vaňová A, Miškovičová A (2014) What makes the city creative [online]. 5th Central European Conference in Regional Science—CERS. Available at: http://www3.ekf.tuke.sk/cers/files/zbornik2014/PDF/Miskovicova,%20Vanova.pdf
114. Rivas M (2011) From creative industries to the creative place. Refreshing the local development agenda in small and medium-sized towns. Redige 2(2):131–147
115. Lukes S, Stanton R (2012) Integrating cities toolkit on anti-discimination policies [online]. Available at: http://www.eurocities.eu/eurocities/events/Integrating-Cities-Toolkit-on-anti-discrimination-policiesWSPO-8TUJ5R
116. Gordon P (2013) Thinking about economy growth: Cities, networks, creativity and supply chains for ideas. Ann Reg Sci 50(3):667–684
117. Caragliu A, Del Bo C, Nijkamp P (2011) Smart cities in Europe. J Urban Technol 18(2):65–82. https://doi.org/10.1080/10630732

118. Dominici G (2012) Why does systems thinking matter? Bus Syst Rev 1(1):1–2
119. Angelidou M (2014) Smart city policies: a spatial approach. Cities 41: 3–11
120. Komninos N (2006) The architecture of intelligent cities: integrating human, collective and artificial intelligence to enhance knowledge and innovation. In: IEEE 2nd IET international conference on intelligent environments, pp 13–20
121. Giffinger R, Fertner C, Kramar H, et al (2007) Smart cities: ranking of European medium-sized cities. Available at: http://www.smart-cities.eu/download/smart_cities_final_report.pdf
122. Anthopoulos LG (2017) Understanding smart cities: a tool for smart government or an industrial trick? In: Public Administration And Information Technology. Springer International Publishing, Switzerland, 293 p
123. Reggiani A, Nijkamp P, De Graaf T (2002) Resilience: an evolutionary approach to spatial economic systems. Netw Spat Econ 2:211–229
124. Chelleri L (2012) From the Resilient City to urban resilience. A review essay on understanding and integrating the resilience perspective for urban systems. *Documents d'Analisi Geografica* 58(2):287–306
125. Modica M, Reggiani A (2015) Spatial economic resilience: overview and perspectives. Netw Spat Econ 15(2), 211–233. Borseková K, Nijkamp P (eds) (2019) Resilience and urban disasters: surviving cities. Edward Elgar, Cheltenham, 288 p. https://doi.org/10.4337/9781788970105
126. Tollin N, Morató J, Santibanez Gonzalez E (2020) Resilient cities. Springer, New York, 68 p
127. Markopoulos P, Rauterberg GWM (2000) LivingLab: a white paper. IPO Annual Progress Report 35:53–65
128. Westerlund M, Leminen S (2011) Managing the challenges of becoming an open innovation company: experiences from living labs. Technol Innov Manag Rev 1(1):9–25. Available at: https://www.researchgate.net/publication/326311707_Managing_the_Challenges_of_Becoming_an_Open_Innovation_Company_Experiences_from_Living_Labs › journal › issn1927-0321
129. Dell'era C, Landoni P (2014) Living lab: a methodology between user- centred design and participatory design. Creat Innov Manag 23(2):137–154
130. Borseková K, Vaňová A, Vitálišová K (2016) Living lab as a new approach to the smart and creative spatial development. In: The European challenges: union globalized or localized?: the transformation of economics, societies and cultures of the EU into the world: XXII international scientific conference of the PGV Network, Warsaw, 15–17 September. Wydawnictwo Społecznej akademii Nauk, Łódź, pp 539–554
131. Marvin S, et al (2018) Urban living labs. Experimenting with city futures. Routledge, 278 p.
132. Simmie J, Lever WF (2002) Introduction: the knowledge-based city. Urban Studies 39(5–6):855–857
133. Mahizhnan A (1999) (1999) Smart cities: the Singapore case. Cities 16(1):13–18
134. Allwinkle S, Cruickshank P (2011) Creating smarter cities: an overview. J Urban Technol 18(2):1–16. https://doi.org/10.1080/10630732.2011.601
135. https://www.cittaslow.org
136. Vitálišová K (2015) Marketing vzťahov v miestnej samospráve. Belianum, Banská Bystrica, 227 p
137. Lee J, Lee H (2014) Developing and validating a citizen—centric typology for smart city services. In: Government information quarterly. Elsevier. Available at: http://e-madina.org/documents/e-Habitat/Developing%20and%20validating%20a%20citizen-centric%20typology%20for%20smart%20city%20services.pdf

Anna Vaňová is an author and co-author of more than 140 scientific papers (domestic, foreign, scientific monographs, articles in impact and CC journals), published and presented mainly abroad. She has worked and works as a principal investigator and a deputy researcher or team member in more than 60 domestic and foreign research projects in the various fields of spatial development. She actively develops cooperation with practice at national, regional and local level (cooperation at the creation of strategic documents, consultations for local, regional and national authorities, etc.).

Katarína Vitálišová is an associate professor at the Department of Public Economics and Regional Development of Faculty of Economics, Matej Bel University. She orients her research on the issue of public and local governance, building and maintaining relationships and partnerships between stakeholders, strategic planning of the spatial development and implementation of new approaches in it, including creative and smart cities, places marketing what is represented by more than 130 publications. She is a member of several international expert networks in the field of regional and local development, management and marketing places. My research orientation is the issue of public and local governance, building and maintaining relationships and partnerships between stakeholders, strategic planning of the spatial development and implementation of new approaches to local and regional development, including local production systems and industrial symbiosis, smart and creative cities and regions. The topic of the Cost Action is directly associated with the previous research experience and research area (e.g., 7FP Project—FOLPSEC, international comparative project Slovakia—Bulgaria—Functioning of the Local Production Systems in the Conditions of Economic Crisis, or international comparative project Slovakia—France—Cultural potential and its utilization in the local development), as well as publication outputs. The participation in the working group is interesting for me because of the possibilities to exchange the experience and knowledge in the researched field, as well as a new possibility to realize the international comparative research in this topic which is very progressive developed.

Darina Rojíková is a researcher at the Research and Innovation Center at the Faculty of Economics with an interest in urban and regional topics. She orients her research on the selected aspects of urban and regional development, territorial development of peripheral areas, strategic planning of territorial development, place branding, democracy and public policy. She is the co-author of several papers, including scientific articles, chapters and monographs. She is involved in solving domestic and foreign research projects and projects for practice.

International Trade

International Trade Among the Megacity Systems/Countries in Its Issues and Barriers: EU 27 and Selected Indo-Pacific Countries

Mariana Považanová, Žaneta Lacová, and Katarína Vitálišová

Abstract Global development is strongly correlated with the changes at local and regional level. The researches show that megacities as one of urbanization trends contribute significantly to the world economic growth and global GDP. In the chapter, we focus on the analysis of international trade among European Union countries as a unit and selected Indo-Pacific countries where megacities are located and are important trade partners for the European Union. European Union by common trade policy for all member states establishes the cooperation framework as a set of rules which are obliged in international trade with third countries. The deeper analysis is devoted to countries—India, Indonesia, Japan, Thailand, Philippines and Bangladesh. By the analysis, the European Union liberalize gradually the trade with them. We can confirm their important role as trade partners for the European Union, eventhough there are some debatable issues influencing it.

Keywords International trade · European Union · Indo-Pacific countries

1 Introduction

The changes in local and regional development associated with the new urbanization modes significantly influence all aspects of society as well as global development. One of the urban phenomena is a formation of megacities. In 1950, there were only

M. Považanová · Ž. Lacová
Faculty of Economics, Department of Economics, Matej Bel University, Tajovského 10, 974 35 Banská Bystrica, Slovakia
e-mail: mariana.povazanova@umb.sk

Ž. Lacová
e-mail: zaneta.lacova@umb.sk

K. Vitálišová (✉)
Faculty of Economics, Department of Public Economics and Regional Development, Matej Bel University, Tajovského 10, 974 35 Banská Bystrica, Slovakia
e-mail: katarina.vitalisova@umb.sk

T. M. Vinod Kumar (ed.), *Indo-Pacific Smart Megacity System*, Advances in 21st Century Human Settlements, https://doi.org/10.1007/978-981-99-6218-1_3

2 cities with population over 10 million inhabitants. In 2030, it is expected that there will be almost 40 megacities around world with cca 630 million people located especially in Asia, Latin America and Africa.

By the National Intelligence Council [1] urbanization, technology and capital accumulation together accelerate the rise of spaced economic zones. The economic power of cities is growing dramatically. It predicted that in next years, 600 cities will create about 60% of worldwide GDP as well as the top 100 cities by contribution to GDP growth will contribute to around 35% of GDP growth worldwide. Cities will contribute more the 15% of global GPD growth.

Moreover, during the last years, the shift in the world economic centre of gravity to the Indo-Pacific region is evident. It became the hub of global trade and energy supply thanks to sea routes through which passes 60% of maritime trade. They are inevitable for the trade of major economies as China, India or Japan [2]. The world trade is dependent on the location of the biggest ports and airports in megacities, and 20 of them are in the Indo-Pacific region. The Indo-Pacific also hosts a high number of growth markets for European companies, as numerous emerging economies are increasingly catching up with their Western counterparts [3]. This fact confirms the important contribution of megacities to economic growth.

However, the issues which is researched in our chapter are how to characterize the trade between European Union and Indo-Pacific megacities. Firstly, the European Union by common trade policy for all member states establishes the cooperation framework as a set of rules which are obliged. They are associated mainly to the adjustments of customs rates, conclusion of customs and trade agreements related to trade in goods and services, to commercial aspects of intellectual property, foreign direct investments, unification of liberalization measures, export policy, as well as trade protection measures. The second obstacle is an availability of data. There are no data on international trade at the local level in case of European Union and very limited at the region level. From these reasons, we analyze the international trade among the megacity systems in EU and Indo-Pacific countries based on the comparison of European Union as unit and selected countries of Indo-Pacific belonging to Asia where megacities are located (India, Indonesia, Japan, Thailand, Philippines and Bangladesh) are and at the same time, they are important trade partner for the European Union. Table 1 illustrates the number of megacities and number of their inhabitants in selected countries in 2016 and the estimated development in 2030. It is expected that within the countries which we analyze in the paper, at least one new megacity will be established.

The chapter consists of three subchapters. Firstly, we characterize the current challenges in the world trade and then we pay attention specifically to the international trade of the European Union. We explain the EU trade integration and implemented EU trade policy. In third part, we describe EU strategy for cooperation in the Indo-Pacific region. In individual subchapters, the analysis of EU trade with selected countries of Indo-Pacific with the location of megacities—India, Indonesia, Japan, Thailand, Philippines and Bangladesh are presented. To conclude the chapter, we summarize the main issues and barriers of the international trade of the European Union and selected Indo-Pacific countries.

Table 1 Selected Megacities and Meta cities in the selected countries of Indo-Pacific—India, Indonesia, Japan and China (2016–2030)

Rank in 2016	City, country	Population in 2016 (thousands)	Rank in 2030	City, country	Population in 2030 (thousands)	Change in % (2030–2016)
1	Tokyo, Japan	38,140	1	Tokyo, Japan	37,190	−2.49%
2	Delhi, India	26,454	2	Delhi, India	36,060	36.31%
3	Mumbai (Bombay), India	21,357	3	Mumbai (Bombay), India	27,797	30.15%
4	Kinki M.M.A. (Osaka) Japan	20,337	4	Kinki M.M.A.(Osaka) Japan	19,976	−1.78%
5	Calcutta, India	14,980	5	Kolkata (Calcutta), India	19,092	27.45%
6	Jakarta, Indonesia	10,483	6	Jakarta, Indonesia	13,812	31.76%
7	Bengaluru, India	10,456	7	Bengaluru, India	14,762	41.18%
8	Chennai (Madras), India	10,163	8	Chennai (Madras), India	13,921	36.98%
9			9	Hyderabad, India	12,774	

Source United Nations, 2016 "World Cities in 2016 Data Booklet", New York. United Nations, 2016 "World Cities Report 2016", New York

2 Development of World Trade with Impact on the Increasing Importance of Indo-Pacific Countries

International trade is one of the most important parts of international economic relations. It is confirmed by the increasing absolute volume of exported and imported goods on a world scale as well as by the increasing relative share of international trade onto the world product. Moreover, world commodity trade is growing faster than world GDP—for a number of decades, world trade has grown on average nearly twice as fast as world production [4]. From the short run perspective, trade is typically more volatile than output and tends to fall particularly sharply in times of crisis (GDP elasticity of global trade is higher than 1 and during recession is higher, than during expansions) [5]. Trade in goods and services experienced a deep slump in 2020 due to the COVID-19 pandemic (the global volume of trade in goods decreased by 14.3% in the second quarter of 2020 compared to the previous period) [6]. The most serious

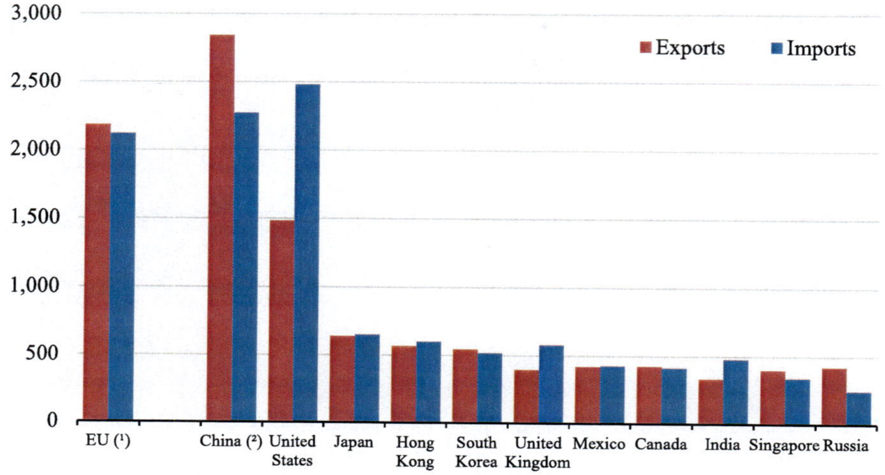

Fig. 1 Main players of international trade in goods, 2021 (€ billion). *Note* Ranked on the sum of exports and imports. (1) External trade flows with extra-EU countries. (2) Excluding Hong Kong. *Source* Eurostat (online data code: ext_lt_introle and UNCTAD)

impact was evident in sharp decline of demand in services sector, and on the other hand, manufacturing trade proved to be more resilient [7].

The territorial structure (the share of individual countries, continents and groups of countries) of world trade is also changing. After the WWII, the United States dominate to world trade flows with the share of more than 20% of world merchandise exports. At that time, China accounted just for 0.9%. In 2021, US share fell to 8.1%, while China's share rose to 15.5% [8]. Indo-Pacific region is playing an increasing role in world trade. In 2021, the EU (even without EU-intra trade), China and the United States have been the three largest global players within international trade (Fig. 1) since 2004 when China surpassed Japan.

As we already mentioned in introduction and the Fig. 1 confirmed, the importance of Indo-Pacific countries in international trade is increasing. Over 35% of European exports in a total sum of $1.5 trillion a year, go to markets in the Indo-Pacific region. Moreover, there can be identified also specific trade patterns associated with Indo-Pacific countries and recovery after the COVID-19 pandemic—the economies in East Asia were the first to experience declines in trade but also the first to recover and the economies of South Asia experienced a much sharper trade drop during 2020, but their trade rebounded more strongly during late 2021 and 2022. In 2022, the value of trade of all developing countries' regions was between 40 and 50% higher than its level of 2019 [7].

From the European Union point of view (Table 2), three of Europe's top 10 trading partners are in the Indo-Pacific (United States, Japan, India), and the Asia-Pacific region is the Europe's largest external export market. Direct investment in both directions between Asia and Europe amounts to about $90 billion annually [9].

Table 2 Top trade in good partners of EU—export and import, 2021 (in bold—Indo-Pacific countries)

Rank	Country	Exports (€ billion)	Share in extra-EU (%)	Country	Imports (€ billion)	Share in extra-EU (%)
	EU	2180.5	100	EU	2111.5	100.00
1	**United States**	**399.4**	**18.30**	China	472.2	22.40
2	United Kingdom	283.6	13.00	**United States**	**232**	**11.00**
3	China	223.3	10.20	Russia	158.5	7.50
4	Switzerland	156.5	7.20	United Kingdom	146	6.90
5	Russia	89.3	4.10	Switzerland	123.6	5.90
6	Turkey	79.2	3.60	Turkey	78	3.70
7	**Japan**	**62.4**	**2.90**	Norway	74.5	3.50
8	Norway	56.5	2.60	**Japan**	**62.3**	**3.00**
9	South Korea	51.9	2.40	South Korea	55.4	2.60
10	**India**	**41.9**	**1.90**	**India**	**46.2**	**2.20**
11	Mexico	37.7	1.70	**Vietnam**	**38.5**	**1.80**
12	Canada	37.3	1.70	**Taiwan**	**35.6**	**1.70**
13	Brazil	33.8	1.60	Brazil	32.9	1.60
14	**Australia**	**33.1**	**1.50**	**Malaysia**	**29.2**	**1.40**
15	United Arabic Emirates	29.8	1.40	Ukraine	24.1	1.10
16	**Taiwan**	**28.4**	**1.30**	Canada	23.4	1.10
17	Ukraine	28.3	1.30	Mexico	23.4	1.10
18	**Singapore**	**27.3**	**1.30**	South Africa	22.1	1.00
19	Morocco	25.1	1.20	**Thailand**	**22**	**1.00**
20	Saudi Arabia	25	1.10	Saudi Arabia	20.8	1.00
21	Israel	24.2	1.10	Algeria	18.7	0.90
22	Hong Kong	23.5	1.10	Morocco	18	0.90
23	South Africa	22	1.00	Nigeria	17.6	0.80
24	Egypt	21.5	1.00	Kazakhstan	17.5	0.80
25	Serbia	18.4	0.80	Libya	17.3	0.80
26	**Thailand**	**13.3**	**0.60**	**Indonesia**	**16.7**	**0.80**
27	Algeria	12.6	0.60	**Bangladesh**	**15.7**	**0.70**
28	**Malaysia**	**11.8**	**0.50**	**Singapore**	**15.6**	**0.70**
29	Nigeria	11.2	0.50	Serbia	14	0.70
30	Tunisia	10.8	0.50	Iraq	12.9	0.60
31	**Vietnam**	**10.6**	**0.50**	Israel	12.6	0.60

(continued)

Table 2 (continued)

Rank	Country	Exports (€ billion)	Share in extra-EU (%)	Country	Imports (€ billion)	Share in extra-EU (%)
	EU	2180.5	100	EU	2111.5	100.00
32	Chile	10	0.50	Azerbaijan	10.4	0.50
33	Qatar	8.1	0.40	Tunisia	10.3	0.50
34	**Indonesia**	**8**	**0.40**	Utd. Arab Em	9.4	0.40

Source Eurostat (online data code: ext_st_eu27_2020sitc and DS-018995)

The United States is the largest destination for EU exports of goods in 2021, while China is the largest origin for EU imports of goods. The Indo-Pacific is the Europe's most important regional partner in foreign trade, accounting for 33% in 2021 of the EU's total non-European trade [10].

3 EU Trade Integration and the EU Trade Policy

The economic integration of the Western European countries from the period after the Second World War was inextricably linked with the removal of barriers to mutual trade between the integrating countries and the gradual adoption of common trade policy instruments towards third (non-integrating) countries. The predecessor of the current European Union, European Economic Community (EEC)[1] was created by the Treaty of Rome in 1957. The Treaty of Rome committed EEC member countries (namely France, Germany, Italy, Belgium, Netherlands and Luxembourg) to deep economic integration. In terms of international trade, the aim was to create a custom union—to remove all tariffs and quotas on intra-EEC trade and adopt a common tariff on imports from non-member nations (sometimes called "third countries"), which was some kind of discrimination in international trade.

A customs union required coordination since trade policy towards third nations is an ever-evolving issue. To facilitate this coordination, the Treaty of Rome granted supranational powers to the EEC's institutions as far as external trade policy was concerned. This delegation of sovereignty has never changed and over the decades, the various Treaties have granted the EU more power in the area of trade. The 2009 Lisbon Treaty in particular greatly increased the extent to which EU members have delegated power on trade issues to the EU as will be explained later.

During the formation of custom union in the EEC, the lowering of intra-EEC trade barriers had an important impact on trade patterns of EEC member states. The EEC's share in its own trade rose while the share of trade with non-member

[1] The European Economic Community (EEC) was subsequently renamed the European Community (EC). In 2009, the EC formally ceased to exist. The European Union took the place of the Community and became its legal successor.

countries declined. Nowadays, the main trading partners of the European Union countries are the other EU member countries. The application of discriminatory liberalization in international trade is sometimes considered to be one of the driving forces behind the entry of other countries into the European Economic Community (later the European Community or the European Union). This is what Baldwin [11, 12] calls the "domino theory" of regional integration: "the preferential lowering of some trade barriers creates new pressures from outsiders to join the trade bloc and, as the trade bloc gets bigger, the pressure to join grows".

Although the complete trade liberalization took place over time only within the member countries of the current European Union, due to the importance of certain third countries for the EEC countries, the European Economic Community also opened its markets to varying degrees and removed obstacles to mutual trade (the movement of goods and services and investments) towards third countries. From the point of view of the current European Union, a system of preferential agreements with third countries has emerged, in which the Indo-Pacific countries are currently playing an increasingly important role. The GATT and later the WTO played their role in the world trade liberalization and thus also in the extra-EEC trade liberalization as will be explained in more details later.

The intensifying processes of integration between the member states of the current European Union have led to the situation in which member states have transferred part of their competences to supranational institutions of the current EU. The Treaty of Lisbon for the first time clarifies the powers of the Union. It distinguishes between three types of competences: exclusive competence, where the Union alone can legislate, and Member States only implement; shared competence, where the Member States can legislate and adopt legally binding measures if the Union has not done so; and supporting competence, where the EU adopts measures to support or complement Member States' policies.

The Union shall have exclusive competence in the following areas:

(a) customs union;
(b) the establishing of the competition rules necessary for the functioning of the internal market;
(c) monetary policy for the Member States whose currency is the euro;
(d) the conservation of marine biological resources under the common fisheries policy;
(e) common commercial policy [13, Article 207 (ex Article 133 TEC)].

As far as intra EU trade is considered, the Union shall comprise a customs union which shall cover all trade in goods and which shall involve the prohibition between member states of customs duties on imports and exports and of all charges having equivalent effect and the adoption of a common customs tariff in their relations with third countries (Article 28 TFEU).

The EU treaty specifies that the objective of the common commercial policy (CCP) is to contribute in the common EU interest to the harmonious development of world trade, the progressive abolition of restrictions on international trade *and on foreign investment,* and the lowering of customs *and other barriers.*[2] As Sapir [14] points out at every stage the process of European integration was accompanied by a gradual lowering of the EU's external trade barriers through multilateral trade negotiations.

The common commercial policy is in an exclusive responsibility of the EU, which means the EU institutions make laws on trade matters, and negotiate and conclude international trade agreements. The common commercial policy shall be based on uniform principles, particularly with regard to changes in tariff rates, the conclusion of tariff and trade agreements relating to trade in goods and services, and the commercial aspects of intellectual property, foreign direct investment, the achievement of uniformity in measures of liberalization, export policy and measures to protect trade such as those to be taken in the event of dumping or subsidies. The common commercial policy shall be conducted in the context of the principles and objectives of the Union's external action.[3]

On 18 February 2021, the European Commission [15] set out its new EU trade strategy for the coming years based on approach of *Open Strategic Autonomy.* New trade strategy placed the nexus between open trade and sustainability at the centre of the EU's trade policy strategy. The new EU trade policy was necessary because the global economy is rapidly changing.

Reflecting the concept of "Open Strategic Autonomy", it builds on the EU's openness to contribute to the economic recovery by supporting the green and digital transformations. The strategy includes a renewed focus on strengthening multilateralism and reforming global trade rules to ensure that they are fair and sustainable. Where necessary, the EU will take a more assertive stance in defending its interests and values, including through new tools [16]. The new forms of trade should reflect the principles of sustainability, fair conditions of competition and a level-playing field, security and predictability, multilateral trade cooperation to the greatest extent possible, and legal and economic grounds [17].

The focus of new EU trade policy within three medium-term ambitions is to support the recovery and fundamental transformation of the EU economy in line with its green and digital objectives; to shape global rules for a more sustainable and fairer globalization and to increase the EU's capacity to pursue its interests and enforce its rights, including autonomously where needed. For these purposes, the Commission will implement through the reform of the WTO, support of the green transition and promoting responsible and sustainable value chains; promotion of the digital transition and trade in services; strengthening the EU's regulatory impact; deepening the EU's partnerships with neighbouring, enlargement countries and Africa; and reinforcing the EU's focus on implementing and enforcing trade agreements and ensuring a level-playing field for EU businesses [15, 18].

---

[2] The text goes back to the 1957 Rome Treaty. The parts in italics were added by the Lisbon Treaty.

[3] Article 207 of the Treaty on the Functioning of the European Union sets out the rules on EU trade policy.

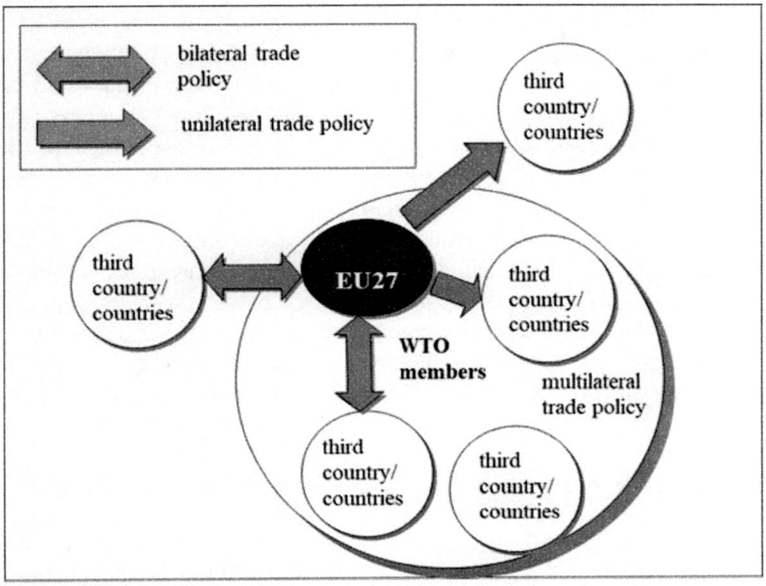

Fig. 2 Three dimensions of the EU trade policy. *Source* Own processing

The EU has a common external trade policy, which means that trade policy is an exclusive competence of the EU and no member state can negotiate its own international trade agreement. The European Commission negotiates trade agreements with outside countries and trading blocs on behalf of the EU as a whole.

EU trade relations with third countries are very complex and still evolving, which results in fact that the EU trading system is very complicated. In principle, it is possible to identify three dimensions of the EU trade policy, which shape the EU external relations with the third countries, namely multilateral, bilateral and the unilateral EU trade policy as shown in Fig. 2.

Multilateral trade policy results from the EU's membership in the World Trade Organization. The World Trade Organization (WTO) is the only global international organization dealing with the rules of trade between nations. The WTO's main activities include: being a forum for international trade negotiations (e.g. the Doha Development Round); resolving trade disputes (Dispute Settlement); setting legal rules for trade in the form of trade agreements; monitoring members' trade policy through the Trade Policy Review Mechanism.

The EU is an original member of the WTO, and each EU member State is also a WTO Member. The EU, through the European Commission, represents all the member states in the WTO. The EU's objectives at the WTO include establishing the world's trading system fair, predictable and based on common rules; contributing to modernization of the world's trade markets and thus achieve the benefits for the European goods, services and investment. Subsequently, EU is binding to follow the common WTO rules, and in cooperation with WTO to control others also play by the rules. The common activities are oriented on building openness by interacting with both non-members and other international organizations and bringing developing countries into the WTO, its decision-making, and the global economy. The last priority is to reinforce the WTO's support for sustainable trade policies worldwide [19].

Under the WTO agreements, countries cannot normally discriminate between their trading partners. Grant someone a special favour (such as a lower customs duty rate for one of their products) and you have to do the same for all other WTO members. This principle is known as most-favoured-nation (MFN) treatment. Some exceptions are allowed. For example, countries can set up a free trade agreement that applies only to goods traded within the group—discriminating against goods from outside, or they can give developing countries special access to their markets [20]. Based on these exceptions, unilateral and bilateral dimensions of the EU trade policy had been formed.[4]

Unilateral trade agreements are one-sided, non-reciprocal trade preferences granted by European Union to developing ones, with the goal of helping them to increase exports and spur economic development. They are meant to foster exports and economic development in beneficiary countries; assist their efforts to reduce poverty, promote good governance and support sustainable development, and encourage compliance with international standards in the areas of human rights, labour rights and environment protection. One of the most well-known examples of unilateral trade policy is Generalised Scheme of Preferences (GSP).

The EU's GSP comprises three arrangements [21]:

1. Standard GSP for low and lower-middle-income countries. This means a partial or full removal of customs duties on two-thirds of tariff lines. Indo-Pacific beneficiary countries (as of 1 January 2022) are India, Indonesia and Micronesia.
2. GSP+: a special incentive arrangement for sustainable development and good governance. GSP+ slashes these same tariffs to 0% for vulnerable low- and lower-middle-income countries that implement 27 international conventions related to labour and human rights, environmental and climate protection, and good governance. Indo-Pacific beneficiary countries (as of 1 January 2022) are Philippines and Sri Lanka.
3. EBA (Everything But Arms): the special arrangement for least developed countries (LDCs), providing them with duty-free, quota-free access to the EU market for all products except arms and ammunition. Indo-Pacific beneficiary countries (as of 1 January 2022) include Bangladesh, Cambodia, Bhutan, Myanmar, Nepal, Timor Leste, Solomon Islands, Comoros, Madagascar and Laos.

[4] Officially GATT Article XXIV is the main rule allowing *countries to grant special treatment to one another by establishing a customs union or free-trade association.* https://www.wto.org/english/tratop_e/region_e/regatt_e.htm#gatt.

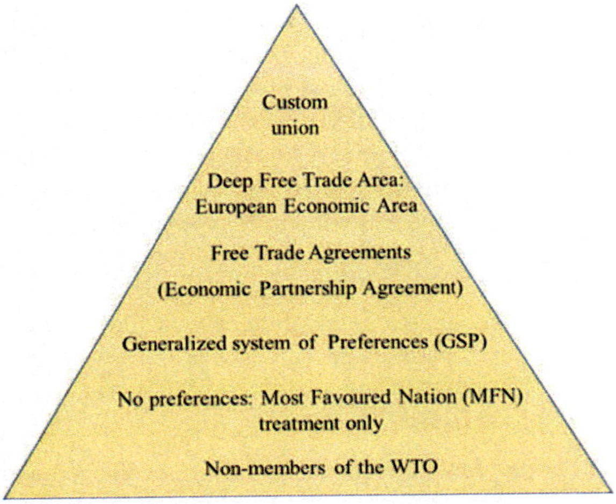

Fig. 3 Pyramid of EU trade preferences. *Source* Based on https://policy.trade.ec.europa.eu/eu-trade-relationships-country-and-region/negotiations-and-agreements_en

Bilateral trade policy deals with the regional and preferential agreements.[5] These agreements grant mutually beneficial access to the markets of both the EU and the country or countries concerned. Each agreement is unique and can include tariff reductions, rules on matters such as intellectual property or sustainable development, or clauses on human rights. As far as third countries are concerned, the EU runs parallel multiple trading regimes, which are usually depicted in the so called a Pyramid of EU trade preferences [22], in which partners are ranked from the most (at the top) to the least (at the bottom) preferred (Fig. 3). The EU relations with the third countries are very complicated and are still evolving. For this reason, one can find different pyramids in literature describing the different stages of the EU-third countries relations [14, 23, 24]. However, they always have one thing in common: the top of the pyramid captures the maximum preferential treatment that the EU can grant to another country. Further down towards the bottom, the preferences are becoming lower.

The EU offered the most favourable trade treatment to Turkey, with which it has a customs union formed and EFTA countries that take part in the European Economic Area (EEA). Below the EEA partners in the pyramid of privileges in trade came free trade agreements and economic partnership agreement. In this study, we are focusing on relations between EU and Indo-Pacific countries therefore, in the next parts we mention them.

The EU divides its free trade agreements (FTA) with third countries into four broad categories [6]:

- "First-generation FTAs" are agreements concluded before 2006 and Stabilization and Association agreements concluded with Western Balkan countries between 2009 and 2016.

[5] Current list of EU negotiations and agreements is available at: https://policy.trade.ec.europa.eu/eu-trade-relationships-country-and-region/negotiations-and-agreements_en.

- "New-generation FTAs" are comprehensive FTAs, negotiated by the EU after 2006, with selected third countries, which typically go beyond provisions on liberalization of trade in goods. They cover services and public procurement. In some cases, they also cover investment and regulatory cooperation. Since 2010, all new-generation agreements include provisions on sustainable development. New-generation FTAs are currently in force with these Indo-Pacific countries: Japan, Singapore, Vietnam, New Zealand.
- "Deep and comprehensive FTAs" are agreements aimed at establishing comprehensive free trade areas as part of a process of deepening political association and gradual economic integration between the EU and its partners.
- Economic partnership agreements (EPAs) with African Caribbean and Pacific (ACP) countries are in force with: Papua New Guinea, Fiji, Samoa and the Solomon Islands

Further down are countries that are privileged under the GSP and the countries with MFN status (e.g. Malaysia, Thailand, Australia, Taiwan, United States). At the very bottom of this pyramid were countries that were not members of the WTO.

Bilateral trade relations between EU and the third countries are still evolving. Generally, the European Commission distinguishes the following groups of agreements depending on whether they have already entered into force or not: agreements in place, agreements being adopted or ratified, agreements being negotiated and agreements on hold.

The current map of the EU trade agreements with third countries by the EU trade preferences as well as stage of negotiation is presented in Fig. 4.

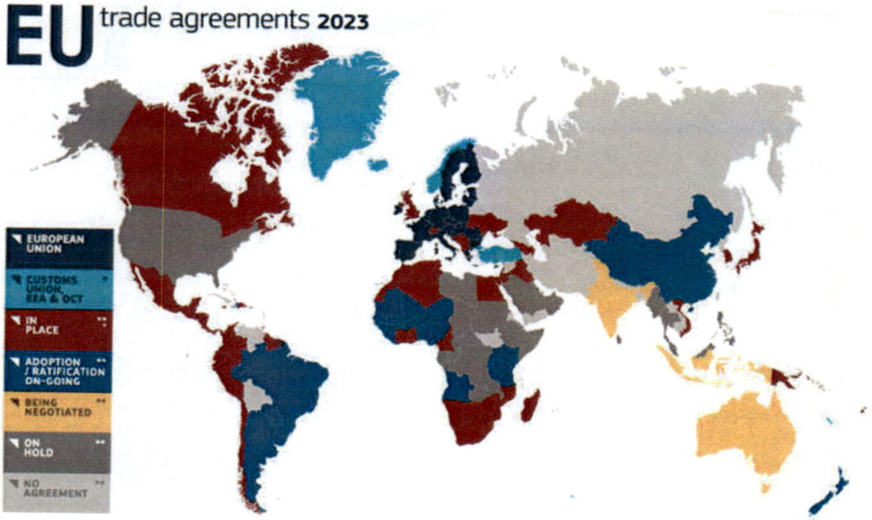

Fig. 4 State of play of EU trade agreements with the third countries to 1.1.2023. *Source* https://policy.trade.ec.europa.eu/eu-trade-relationships-country-and-region/negotiations-and-agreements_en

Table 3 The Indo-Pacific countries in negotiation with EU

Country (Region)	New agreement is being negotiated	Status
Australia	Australia Agreement	Negotiations launched in 2018
India	Free Trade Agreement. Investment Protection Agreement and Geographical Indications Agreement	Negotiations started in 2007. They paused in 2013. Before restarting in 2022
Indonesia	Free Trade Agreement	Negotiations launched in 2016
Philippines	Free Trade Agreement	Negotiations launched in 2015
Thailand	Free Trade Agreement	Negotiations relaunched in 2023

Source https://policy.trade.ec.europa.eu/eu-trade-relationships-country-and-region/negotiations-and-agreements_en

Liberalization processes and the deepening of mutual cooperation between the EU and third countries is an ongoing process, which is why the EU is continuously negotiating with several countries. Some trade negotiations with a trade partner have been concluded, but have not been either signed or ratified yet. This means that although the negotiations have finished. No part of the agreement is in place yet. From the Indo-Pacific countries, EU finished its agreement with New Zealand.

With some countries, the EU has trade agreements in place, but both sides are now negotiating an update as is presented in Table 3.

It is noteworthy that all the countries with which the EU is currently negotiating on deepening cooperation belong to the Indo-Pacific countries, which declares the growing importance of the Indo-Pacific countries as mutual trading partners of the EU, as well as the interest of both parties in deepening mutual cooperation.

4 EU Strategy for Cooperation in the Indo-Pacific

The EU and the Indo-Pacific are highly interconnected. The EU is already the top investor, the leading development cooperation partner and one of the biggest trading partners in the Indo-Pacific region. Together, the Indo-Pacific and Europe hold over 70% of the global trade in goods and services, as well as over 60% of foreign direct investment flows.

Over the years, the EU has consistently made significant contributions in the region in areas such as:

- development cooperation and humanitarian assistance;
- tackling climate change, biodiversity loss and pollution;
- partnership and free trade agreements;
- disaster risk reduction;
- upholding international law, including human rights and freedom of navigation.

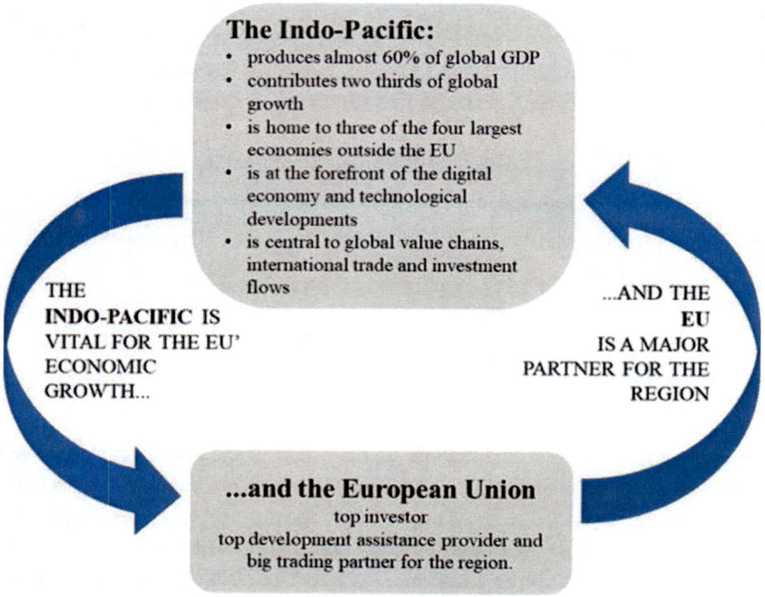

Fig. 5 EU Indo-Pacific factsheet. *Source* Based on information in EU Strategy for Cooperation in the Indo-Pacific. February 2022. European Commission

Figure 5 illustrates the view on the current cooperation between Indo-Pacific countries and the European Union.

Since the Indo-Pacific region is increasingly becoming strategically important for the EU, on 19 April 2021, the Council adopted conclusions on an EU Strategy for cooperation in the Indo-Pacific [25]. On September 2021, the Commission and the High Representative presented a Joint Communication on the EU's Indo-Pacific Strategy [26]. With this new strategy and the Global Gateway, the EU aims to contribute to the region's stability, security, prosperity and sustainable development, in line with the principles of democracy, rule of law, human rights and international law.

There are seven main elements of the EU's Indo-Pacific Strategy [26]:

1. **Sustainable and inclusive prosperity**. Work with Indo-Pacific partners to reinforce value chains, strengthen and diversify trade relations, implement existing trade agreements, finalise ongoing trade negotiations and develop cooperation in strategic sectors. Strengthen rules to protect international trade against unfair practices, such as industrial subsidies, economic coercion, forced technology transfers and intellectual property theft.
2. **Green transition**. Work with partners to fight, mitigate and adapt to climate change and to counter biodiversity loss, pollution and other forms of environmental degradation.

3. **Ocean governance**. Strengthen ocean governance in the region in full compliance with international law, in particular UNCLOS. Continue to support partners in the region to fight against IUU fishing and to implement Sustainable Fisheries Partnership Agreements.
4. **Digital governance and partnerships**. Expand the network of digital partnerships with Indo-Pacific partners and explore potential new Digital Partnership Agreements. Strengthen cooperation on research and innovation with like-minded regional partners under the "Horizon Europe" programme.
5. **Digital Connectivity**. The EU will strengthen connections between Europe and the Indo-Pacific and help partner countries address the digital divide and further integrate into the global digital ecosystem. Support partners to establish a good regulatory environment and facilitate funding to improve connectivity on the ground between Europe and the Indo-Pacific.
6. **Security and defence**. Promote an open and rules-based regional security architecture, including secure sea lines of communication, capacity-building and enhanced naval presence in the Indo-Pacific. Explore ways to ensure enhanced naval deployments by EU Member States in the region. Support Indo-Pacific partners' capacity to ensure maritime security. Facilitate capacity-building for partners to tackle cybercrime.
7. **Human security**. Support healthcare systems and pandemic preparedness for the least developed countries in the Indo-Pacific region. Reinforce the EU's disaster risk reduction and preparedness engagement in the Indo-Pacific.

The implementation of the EU's Indo-Pacific Strategy is in line with the priorities of the new EU Trade Policy, which supports it directly.

Because of our interest in researching the role and importance of megacities as a new global phenomenon, we analyze deeply the international trade between EU and selected countries of Indo-Pacific where the megacities are located. We identified them by the analysis of the trade partner's importance for the European Union export and import, and location of megacities in Asia. These criteria help us to select 6 countries—India, Indonesia, Bangladesh, Philippines, Thailand and Japan. The analysis of each country consists of analysis of export and import of goods among EU countries and that country, development of international trade in main products groups, the identification of main trading partners and characterization of mutual relationships.

4.1 EU Trade Relations with India

In 2021, India was the 10th largest partner for EU exports of goods (1.9%) and the 10th largest partner for EU imports of goods (2.2%). The trade relations between India and the EU experienced a significant increase in the 1990s. During that period, Indian economy underwent liberalization after seeking support from the International Monetary Fund (IMF), which included a comprehensive set of reforms and reforming

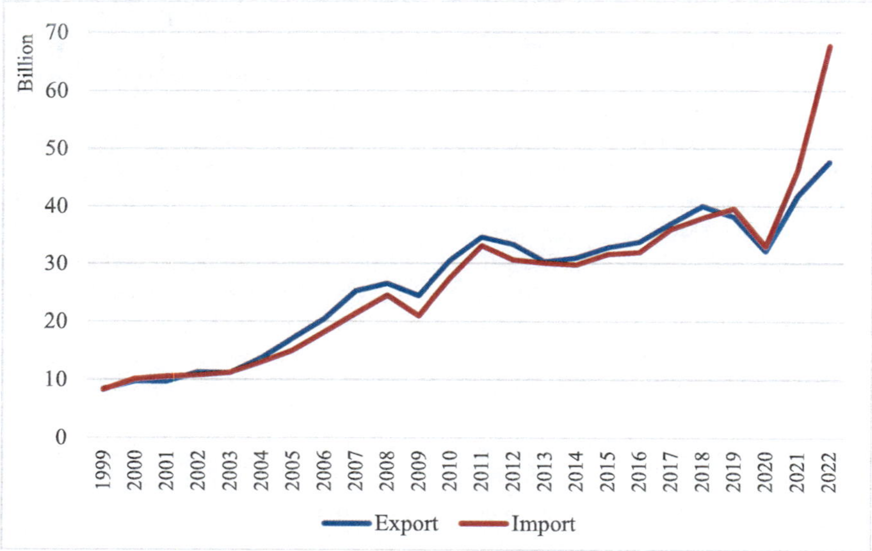

Fig. 6 Development of the EU-India trade of goods. 1999–2022 (€ billion). *Source* Own processing data available at Eurostat (EU trade since 1999 by SITC DS-018995)

of capital markets. The mutual trade between India and EU continuously grew during the twenty-first century. It recorded quite significant decrease only in periods of two recessions, namely the Great Recession in the first decade and then during the COVID-19 recession as is evident from Fig. 6.

EU export of goods to India, which accounted for 1.9% of the EU global exports of goods amounted to €47.6 billion in 2022. Since 1999 the export value increased by €39.3 billion.

EU export of goods to India mainly consisted of machinery and transport equipment (Table 4). Since 1999, its share in EU exports rose from 32% to 43.3% in 2022. Manufactured goods classified chiefly by material accounted for nearly 20% of total EU export, but share of this group decreased quite significantly (from 42.6% in 1999).[6] Chemicals and related products made up 16.2% of the total EU export to India. The other product categories are not as important, the share of each of them in the total export is less than 10%.

EU import of goods from India amounted to €67.6 billion in 2022. The SITS section breakdown showed that the main imported product groups were almost the same as export groups, namely manufactured goods classified chiefly by material

[6] The Standard International Trade Classification (SITC) is used in the compilation and comparison of trade statistics. The commodity groupings of SITC reflect (a) the materials used in production, (b) the processing stage, (c) market practices and uses of the products, (d) the importance of the commodities in terms of world trade, and (e) technological changes.

Table 4 Structure of the EU-India trade of goods by SITC product category in %

Kind of goods	Export		Import	
	1999	2022	1999	2022
Food and live animals	1.43	0.96	8.65	4.78
Beverages and tobacco	0.05	0.32	0.57	0.37
Crude materials, inedible, except fuels	2.98	7.34	5.74	2.32
Mineral fuels. lubricants and related materials	0.53	1.93	0.14	8.24
Animal and vegetable oils. fats and waxes	0.95	0.85	0.99	0.53
Chemicals and related products. n.e.s	11.26	16.24	8.93	18.08
Manufactured goods classified chiefly by material	42.59	19.56	35.43	29.33
Machinery and transport equipment	32.01	43.29	9.58	21.30
Miscellaneous manufactured articles	5.29	8.26	29.37	14.77
Comm. and trans. not classified elsewhere in the SITC	0.67	0.40	0.16	0.09
Others	2.25	0.86	0.43	0.19
Total	100	100	100	100

Source Own processing data available at Eurostat (EU trade since 1999 by SITC DS-018995)

(29.3%), machinery and transport equipment (21.3%), chemicals and related products (18%). Similarly, as on the export side, the manufactured goods classified chiefly by material experienced a decline in the share (from 35.4% in 1999), while machinery and transport equipment (21.3%), chemicals and related products (18%) experienced a growth (from 9.58% and 8.93% in 1999 respectively).

The main trading categories have evolved differently over the course of business cycle. All three groups fluctuated, but chemicals and related products tend to be less volatile and manufactured goods and machinery and transport equipment more volatile (Fig. 7). The latter two groups experienced sharp declines during recessions (The Great Recession and COVID-19 recession).

Different EU countries participate differently in trade with India (Fig. 8). Germany has been since the 1960s the largest exporter and importer to India followed by France and Belgium. Germany and France recorded a trade surplus with India in 2021. Quite huge trade deficit with India is recorded by Italy, Netherlands and Spain. As the biggest trade partner for the India is France and Germany, there can be identified the potential connection with location of megacities—in case France—Paris are recognized as megacities by some rankings and some statistics identifies also metropolitan region Rhine-Ruhr in Germany as an area with the similar features as megacity, but officially it is a metropolis.

4.1.1 Development of Mutual Relations Between EU and India

India was one of the first countries and the first Asian country to extend diplomatic relations with the European Economic Community in 1962. One of the examples

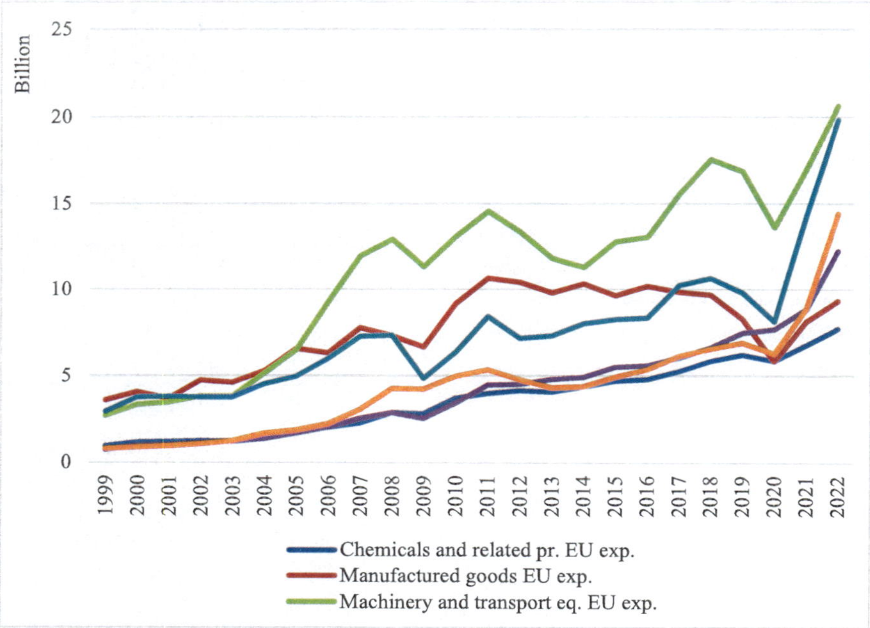

Fig. 7 Development of main product groups (€ billion). *Source* Own processing data available at Eurostat (EU trade since 1999 by SITC DS-018995)

of cooperation between the EEC and India was connected with India's Operation Flood, which was launched on 13 January 1970 and changed India's dairy architecture also commonly known as the "White Revolution". The EEC donated skimmed milk powder and butter oil through the World Food Program (WFP), which effectively transformed India from a milk deficient nation into the world's largest milk producer over three phases. Since then, India–EU relations have progressively developed [27]. The important milestones of mutual relations between them since 1970s are in Table 5.

The Joint Political Declaration and the Cooperation Agreement between the European Community and the Republic of India on partnership and development were signed between 1993 and 1994 respectively. The 1994 Cooperation Agreement took bilateral relations beyond merely trade and economic cooperation [28]. However, the Community and India granted each other only most-favoured-nation treatment in their trade in accordance with the provisions of the General Agreement on Tariffs and Trade.

The first Lisbon Summit of 2000 was important to the development of bilateral relation. Since this summit, EU relations with India have progressed in political, geopolitical economic and trade terms. The bilateral ties were furthered in 2004 at the 5th India–EU summit, when India and the EU became strategic partners [29, 30]. Negotiations for a trade agreement with India were launched in June 2007 and

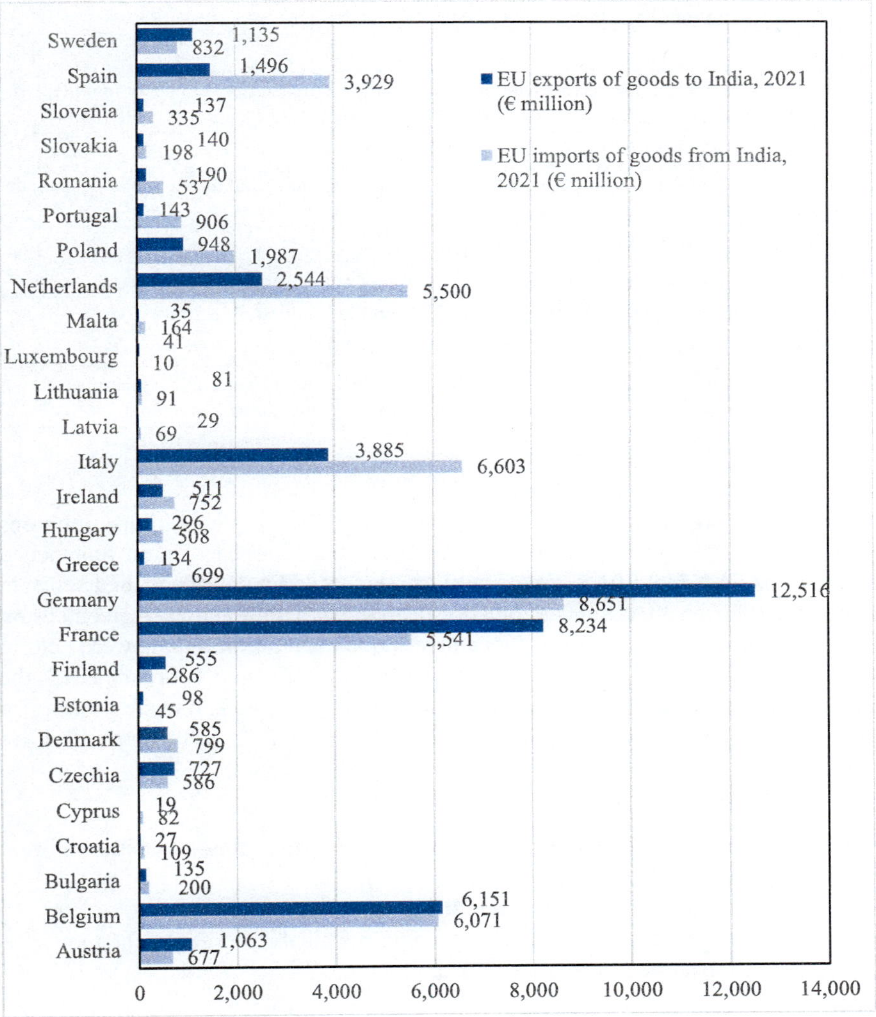

Fig. 8 EU Main trading partners of India. *Source* Eurostat (online data code: ext_st_eu27_2020sitc and DS-018995)

brought to a de facto standstill in 2013 due to a serious gap in the levels of ambitions [31]. In 2020, at the 15th EU–India Summit, the leaders of both sides endorsed "India-EU Strategic Partnership: A Roadmap to 2025" as a common roadmap to guide joint action and further strengthen the India-EU Strategic Partnership over the next five years [32]. The roadmap highlights cooperation in five critical domains: Foreign Policy and Security Cooperation; Trade and Economy; Sustainable Modernization Partnership; Global governance; and people-to-people relations. Before the 16th India–EU summit in 2022, the EU announced Indo-Pacific Strategy (characterized

Table 5 Important milestone in mutual relations between EU and India since 1970s

Year	
1974	Commercial Cooperation Agreement between the European Economic Community and the Republic of India (note: no longer in force. Date of end of validity: 31/07/1994)
1993	Joint Political Declaration
1994	Cooperation Agreement between the European Community and the Republic of India on partnership and development
2000	The first Lisbon Summit
2004	An EU-India Strategic Partnership
2005	The India-EU strategic partnership Joint Action Plan (reviewed in 2008)
2020	India-EU Strategic Partnership: A Roadmap to 2025
2022	The EU and India relaunched negotiations for an EU-India Free Trade Agreement. Investment Protection Agreement and Geographical Indications Agreement

Source Own processing

above), which highlights Europe's interest in the region coupled with its ambitions to counter its systemic rival. On 17 June 2022, the European Union relaunched negotiations with India for a Free Trade Agreement (FTA) and launched separate negotiations for an Investment Protection Agreement (IPA) and an Agreement on Geographical Indications (GIs).[7] Since then, three negotiation rounds were held in 2022 and the fourth one is ongoing in mid-March 2023. On 6 February 2023, the EU and India have strengthened their relationship as strategic partners by setting up a new Trade and Technology Council (TTC). The new TTC will deepen strategic engagement on trade and technology between both partners.

The trade negotiations aim to:

- Remove barriers and help EU firms—especially smaller ones—to export more;
- Open up services and public procurement markets;
- Ensure protection of geographical indications;
- Pursue ambitious commitments on trade and sustainable development and;
- Make sure the agreed rules are enforceable.

The investment protection negotiations aim to provide investors from both sides with a predictable and secure investment environment, through commitments on:

- Non-discrimination;
- Protection against expropriation without compensation and unfair treatment of investors and their investments, while preserving the right to regulate, and;
- Transfer of returns;
- The investment protection negotiations also seek to put in place an effective and state-of-the-art dispute settlement mechanism to enforce such rules [33].

[7] Documents related to EU-India agreement are available online: https://policy.trade.ec.europa.eu/eu-trade-relationships-country-and-region/countries-and-regions/india/eu-india-agreement/documents_en.

In line with the statements released by the European Commission, the overall objective of the current negotiations on the Free Trade Agreement (FTA) and the Investment Protection Agreement (IPA) between the EU and India is "to maximise the considerable trade and investment potential between the EU and India". Liberalization of trade and investment between the EU and India therefore represents an opportunity for the economic growth and the creation of jobs of involved countries.

DG Trade-specific tool for supporting major trade negotiations between EU and the third country (countries) is The Sustainability Impact Assessment (SIA), which contains in-depth analysis of the potential economic, social, human rights and environmental impacts of ongoing trade negotiations.[8] Currently (April 2023) group of specialists is preparing Trade Sustainability Impact Assessment (SIA) of the potential impact that the FTA and IPA negotiated between the European Union and India could have on the economy, society, environment and the enjoyment of human rights in India or the EU. Related to the impact areas, the analysis consists of four sustainability areas: economic, social, environmental and the one related to human rights. The assessment relies on economic modelling, literature review, statistical data analysis and continuous consultations.[9]

4.2 *EU Trade Relations with Indonesia*

Indonesia is an economic giant that accounts for 35% of ASEAN GDP and that has an influential position within ASEAN. The EU was Indonesia's 5th largest trading partner in 2021, with a total trade in goods amounting to €24.7 billion. In 2021, Indonesia's export to the EU accounted for 8.5% of its global export of goods, amounting to €16.8 billion. Its import from the EU accounted for 5.1% of its global import of goods, amounting to €7.9 billion. Meanwhile, for the EU, Indonesia was the 31st global trading partner, 26th largest import source and 34th export destination in 2021.

Indonesia is a member of the WTO since 1995 and benefits from trade preferences granted by the EU's Generalized Scheme of Preferences (GSP), under which about 30% of total imports from Indonesia enjoy lower duties.

Mutual trade in goods between the EU and Indonesia between 1999 and 2022 grows slightly with more pronounced fluctuations during the Great Recession and the COVID-19 recession (Fig. 9). The EU recorded a trade deficit with Indonesia during the entire period.

EU export of goods to Indonesia amounted to €9.1 billion in 2022. It mainly consisted of machinery and transport equipment (38.8%), chemicals and related

---

[8] Building on the Impact Assessment, the SIA consists of chapters examining specific subjects in greater depth. SIA chapters cover topics where the agreement is most likely to have an impact on sustainability issues, or where the agreement opens opportunities for achieving non-trade policy objectives.

[9] Actual info available on https://www.eu-india-tsia.eu.

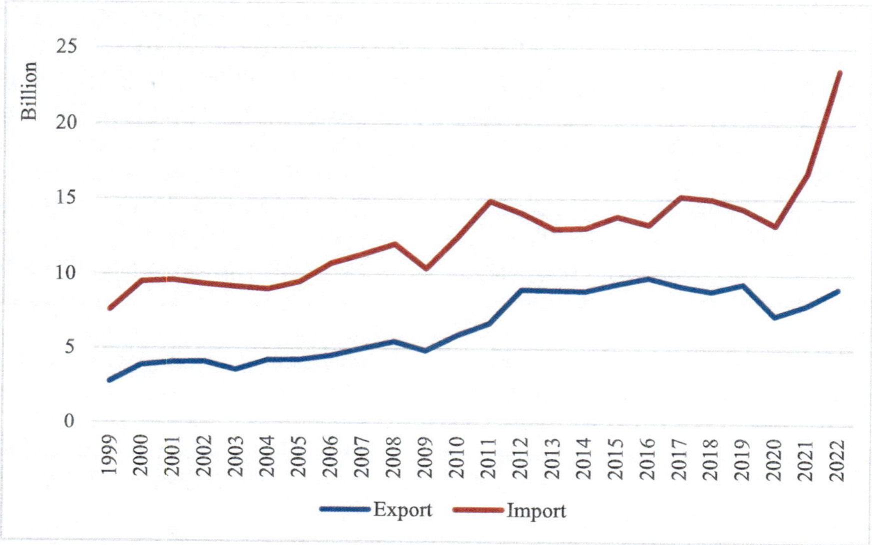

Fig. 9 Development of the EU-Indonesia trade of goods, 1999–2022 (€ billion). *Source* Own processing data available at Eurostat (EU trade since 1999 by SITC DS-018995)

Table 6 Structure of the EU-Indonesia trade of goods by SITC product category in %

Kind of goods	Export		Import	
	1999	2022	1999	2022
Food and live animals	5.46	9.92	8.45	6.96
Beverages and tobacco	0.84	0.76	1.02	0.61
Crude materials inedible except fuels	5.90	6.23	11.05	8.53
Mineral fuels. lubricants and related materials	0.87	0.19	2.97	6.45
Animal and vegetable oils. fats and waxes	0.08	0.29	8.20	13.35
Chemicals and related products. n.e.s	21.84	20.98	2.74	14.44
Manufactured goods classified chiefly by material	14.86	12.68	18.99	14.34
Machinery and transport equipment	41.32	38.82	12.97	11.57
Miscellaneous manufactured articles	4.63	8.19	33.24	23.64
Comm. and trans. not class. elsewhere in the SITC	1.37	0.18	0.14	0.04
Others	2.83	1.76	0.24	0.08
Total	100	100	100	100

Source Own processing data available at Eurostat (EU trade since 1999 by SITC DS-018995)

products (21%), manufactured goods classified chiefly by material (12.7%) and food and live animals (9.9%) (Table 6). The share of chemicals, manufactured goods and machinery and transport equipment in the total EU export was roughly the same in 1999 and 2022, while the share of food and live animals almost doubled. The main import products according to SITS breakdown were miscellaneous manufactured articles (23.7%), chemicals (14.4%), manufactured goods (14.34%) and animal and vegetable oils. According to EU (2022), the imports of vegetable oils from Indonesia made up more than half of agricultural products imports from Indonesia and contributed to 21% of the EU's global imports for animal or vegetable fats and oils.

The EU's export of machinery and transport equipment to Indonesia recorded the biggest increase at the beginning of the second decade of the twenty-first century (Fig. 10). In the following years, this product category was the leading export group. In 2020, it experienced a sharp decline, probably in connection with the recession caused by the COVID-19 pandemic. Although it increased again in 2022, it did not return to the level it had 10 years ago. The other important EU export categories, food and live animals, manufactured goods, miscellaneous manufactured articles and chemicals and related products did not experience significant fluctuations, even during the COVID-19 recession.

The main import groups to EU from Indonesia since 1999 experienced a slight increase with minor fluctuations during recessions (Fig. 11). After investigations which were aimed to find out whether EU industry is being harmed by dumped

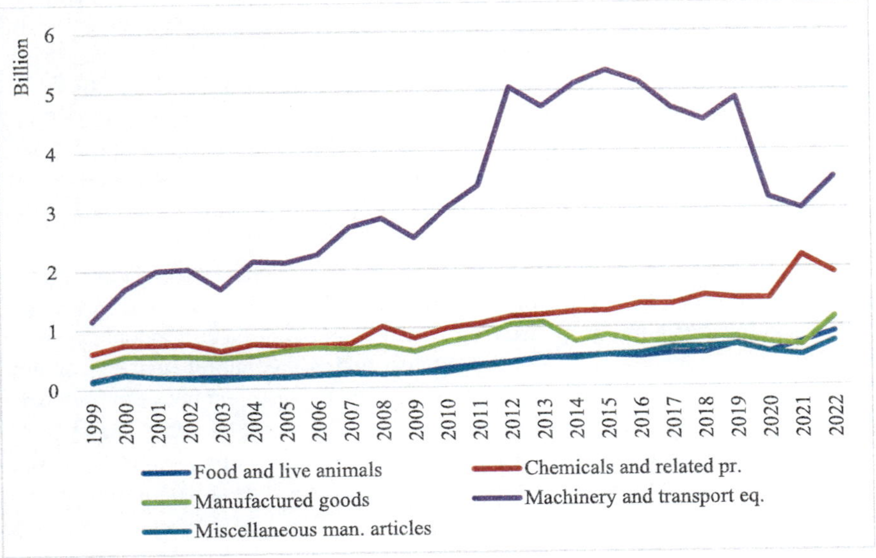

Fig. 10 Development of main EU export product group to Indonesia, 1999–2022 (€ billion). *Source* Own processing data available at Eurostat (EU trade since 1999 by Sitc Ds-018995)

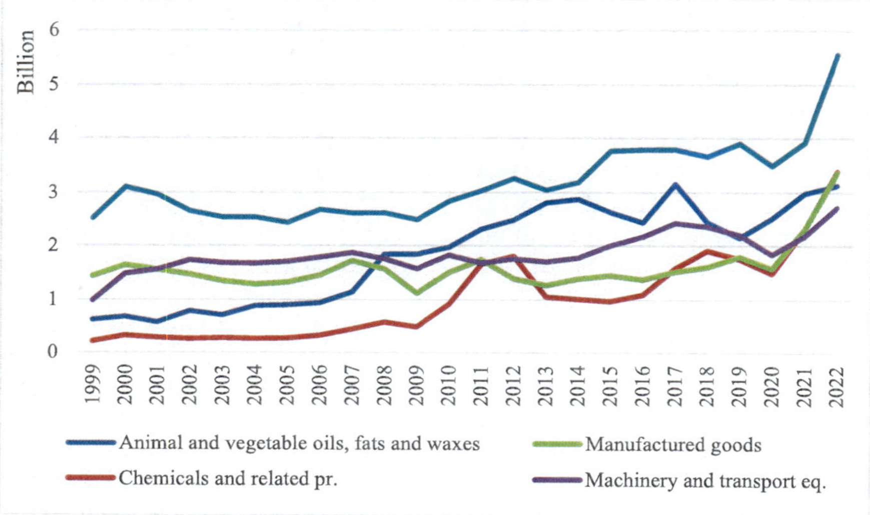

Fig. 11 Development of main EU import product group from Indonesia, 1999–2022 (€ billion). *Source* Own processing data available at Eurostat (EU trade since 1999 by SITC DS-018995)

imports from Indonesia, in January 2023, the European Commission imposed anti-dumping duties on fatty acid from this country (Regulation 2023/111) [34]. The anti-dumping duties imposed range from 15.2% to 46.4%. Palm oil is a key raw ingredient of fatty acid. This measure is likely to reduce imports into the EU in the future.

Over many years, relations between EC (predecessor of the EU) and Indonesia have been framed by the EC Cooperation Agreement with the Association of South East Asian Nations (ASEAN) formalized in 1980. On this basis, economic and political discussions have been held regularly. Bilateral dialogues between the EU and Indonesia have included periodic reviews of political, economic and cooperation issues in Senior Official Meetings [35].

Later, A Framework Agreement on Comprehensive Partnership and Cooperation was signed in 2009 and entered into force on 1 May 2014. The relationship between the EU and Indonesia has deepened over recent years. The EU and Indonesia aim at deepening bilateral economic relations by negotiating a Comprehensive Economic Partnership Agreement (CEPA) to increase trade and investment between both sides while promoting sustainable development. Negotiations were launched in July 2016. They aimed at eliminating or reducing tariff and non-tariff barriers to trade in agricultural products, manufactured goods, and services and at establishing robust rules in trade and trade-related areas in order to support trade flows and realizing the untapped potential for investment. Eleven rounds have been held so far [36].[10]

Similarly, as in negotiations between EU and India, group of specialists in the field had prepared The Sustainability Impact Assessment (SIA) [37]. The SIA provides a detailed picture of the possible economic, social, human rights and environmental impacts of an agreement between the EU and Indonesia on the basis of both quantitative (CGE modelling) and qualitative (literature review, overview of legal frameworks and policy documents, inputs from stakeholders) analysis. Overall, the findings and recommendations of the SIA confirm the strong case for an agreement between the EU and Indonesia, which is expected to be beneficial across all the main indicators for both sides. At the same time, the SIA identifies some key areas that require to be handled carefully in order to prevent possible negative impacts [37].

4.3 EU Trade Relations with Japan

Concerning the volume of bilateral trade, Japan is the EU's first trading partner in the Indo-Pacific region and the EU's second-biggest trading partner in Asia—after China. Traditionally, the strong trade relationship between Japan and the EU was presented particularly in investment flows. In 2021, Japan ranked the seventh place in terms of EU exports of goods (2.9%) and the eighth place in terms of EU imports of goods (3.0%). Typically, the trade relationship between the EU and Japan used to be characterized by big trade surpluses in favour of Japan. As a result of various initiatives and bilateral agreements in the last ten years, the trade figures became more balanced.

The most important agreement which modifies the features of the current EU-Japan trade and economic relationship represents an Economic Partnership Agreement (hereafter "EPA"). The EU-Japan EPA negotiations were officially launched on 25 March 2013, were finalized on 8 December 2017, and followed by approval procedures in the EU and Japan. The agreement entered into force on 1 February 2019 and has influenced considerably the development of an enhanced relationship between the EU and Japan. It was expected from this agreement to boost trade in goods and services as well as create new opportunities for investment [38].

The EU-Japan EPA is an output of a progressive development of mutual relations. The important milestones in this development are presented in Table 7.

Compared to the situation with no agreement, the EPA has eliminated or reduced tariffs and non-tariff measures. The EU has liberalized 99% of tariff lines and Japan 97% of tariff lines, and concessions were agreed in terms of tariff rate quotas or tariff reductions on not yet liberalized tariffs. By these decisions, almost all tariffs have been eliminated or reduced which have facilitated the business environment

[10] Documents related to EU-Indonesia agreement are available online: https://policy.trade.ec.europa.eu/eu-trade-relationships-country-and-region/countries-and-regions/indonesia/eu-indonesia-agreement/documents_en.

Table 7 Important milestone in mutual trade relations between EU and Japan

Year	
1954	The Japanese ambassador in Belgium was accredited as Japan's first Representative to the European Communities (becoming the EU in 1993)
1974	The establishment of a European delegation in Tokyo
1991	The first summit and the Joint Declaration
2001	The Action Plan for EU-Japan Cooperation
2002	The EU-Japan Mutual Recognition Agreement, which allows for conformity assessments in four product areas: telecommunications terminal equipment and radio equipment, electrical products, laboratory practices for chemicals and manufacturing practices for pharmaceutical products
2003	The Agreement on Co-operation on Anti-competitive with the aim to facilitate bilateral trade and investment by securing a level-playing field between in- and outsiders
2008	An Agreement on Co-operation and Mutual Administrative Assistance on Customs Matters (CCMAA)
2009	A Science and Technology Agreement
2018	The EU-Japan Economic Partnership Agreement (EPA)
2019	The Strategic Partnership Agreement

Source Own processing

for all sizes of companies, including the small and medium-sized enterprises. These decisions saved companies of both sides substantial amounts when trading goods on the bilateral market. Systematically, import procedures were simplified for a wide range of products. In addition, the removal of non-tariff measures in areas, such as technical regulations, has opened new business opportunities on both sides. Nowadays, standards and technical regulations are based on international standards to the greatest possible extent. However, the agreement had not lower safety standards or required parties to change their domestic policy choices on very sensitive and controversial matters (the use of hormones, genetically modified organism, etc.).

Further, the agreement has facilitated for European and Japanese companies to provide services on both markets. It also has offered greater mobility for company employees to perform their work. The EPA contains a number of provisions that apply horizontally to all trade in services, such as a provision to reaffirm the Parties' right to regulate. The right of national authorities remains to keep public services public and the EPA is not forcing governments to privatize or deregulate public services (e.g. healthcare, education, water, etc.). Specific commitments were agreed in service sectors, including postal and courier services, telecommunication, maritime transport services or financial services. In the field of public procurement contracts, the EPA brings more reciprocal access for companies.

According to the European Commission [39], the agreement promotes investment between the EU and Japan and reaffirms the right of each party to regulate legitimate policy objectives agreed in a non-exhaustive list. Bilateral negotiations are ongoing for concluding a potential agreement on investment protection. The commitments on

Intellectual Property Rights are reinforced and include provisions on protection of trade secrets, trademarks, copyright protection, patents, minimum common rules for regulatory test data protection for pharmaceuticals, and civil enforcement provisions. Other relevant issues for SMEs included in the EPA are data protection, safe data transfers, sustainable development including labour and environmental laws, as well as corporate social responsibility. In accordance with the Article 8. 81 of the Japan-EU EPA, Japan and the European Union (EU) have been exchanging views on the reassessment of the need for including provisions on the free flow of data into the Agreement.

The agreement is further improving the position of EU exporters and investors on Japan's large market, while including strong guarantees for the protection of EU standards and values. It is helping underline Europe's leadership in setting global trade rules. The launch of the agreement sent a powerful signal that two of the world's biggest economies reject protectionism and the governance of both sides is faithful to idea that the cooperation is the right answer to ongoing global challenges. Thus, this Agreement goes beyond trade issues only. It represents a significant strengthening of the partnership between the EU and Japan.

In addition to a formalized trade relations framework, some informal bilateral dialogues and other specific initiatives have been established to support the trade relationships between the EU and Japan:

- A Cooperation Framework aimed at promoting two-way investment via concrete actions has existed since 2004.
- The EU-Japan Business Round Table: allows for a dialogue and an exchange of views between EU and Japanese businesses.
- The Executive Training Programme and the EU Gateway Programme: encourages European enterprises to penetrate the Japanese market and gives them assistance.
- The EU-Japan Centre for Industrial Cooperation promotes all forms of industrial, trade and investment cooperation between the EU and Japan by helping EU and Japanese businesses to exchange experience and know-how, with a particular emphasis on SMEs.

Thanks to these fora, the EU and Japan meet regularly to discuss issues and best practices. The two countries publish the agendas of the committees set up by the EU-Japan EPA, as well as reports, minutes and other documents in order to inform about the different steps of the EPA implementation process. This is part of the commitment to a more transparent and inclusive trade and investment policy.

The implications of the establishment of strategical documents and initiatives are visible. EU's trade in goods deficit with Japan decreased from €18 billion in 2009 to reach the surplus of €82 million in 2021. The Joint Committee acknowledged the key role of the EPA in fostering bilateral trade flows and making them resilient: after quickly catching up to pre-pandemic levels in 2021 (€124 billion), in 2022 bilateral trade flows increased by 13.4% to €140.6 billion. Both sides discussed areas where market access could be improved, such as the import conditions in Japan for certain categories of EU agricultural products.

Concerning the EPA utilization of tariff preferences, Preferential Utilization Rate, (PUR) by economic operators, these provisions were increasing in 2021—71% (Japan's estimations) for exports from the EU to Japan and 65% (EU's estimations) for exports from Japan to the EU compared to 2020—63% and 60% respectively. Both sides keep the willingness to facilitate, to the extent possible, a continuous increase in the utilization figures, including through awareness raising activities. Both sides concurred on their intention to further cooperate to assess the evolution and facilitate the use of preferences. Both sides shared the view on the importance of stepping up efforts to support companies, including SMEs, to promote the utilization of the EPA.

Both exports to and imports from Japan increased between 2009 and 2021. For the period 2009–2021, EU exports to Japan reached their highest level in 2021 (€62.4 billion) and their lowest in 2009 (€32 billion). EU imports from Japan were highest in 2019 (€63 billion) and lowest in 2013 (€49 billion).

The following Fig. 12 shows the export and import between the EU and Japan from 1999 to 2022. In 1999, the EU had a trade deficit with Japan of €32 billion. This changed considerably and the trade relations became gradually more balanced. In 2022, the surplus was very small (€1.7 billion). EU exports to Japan monitored a growing tendency in the whole period. Imports from Japan decreased in the period 1999–2011 and increased between 2011 and 2022. As in the case of other bilateral trade relationships, the mutual trade between the EU and Japan was affected during the Great Recession and the COVID-19 recession.

Moreover, an important aspect of the EU-Japan trade relationship is that there is more trade in goods than trade in services. In 2020, the EU had a €15 billion surplus in services trade with Japan, up from €13 billion in 2018. EU-Japan total trade in

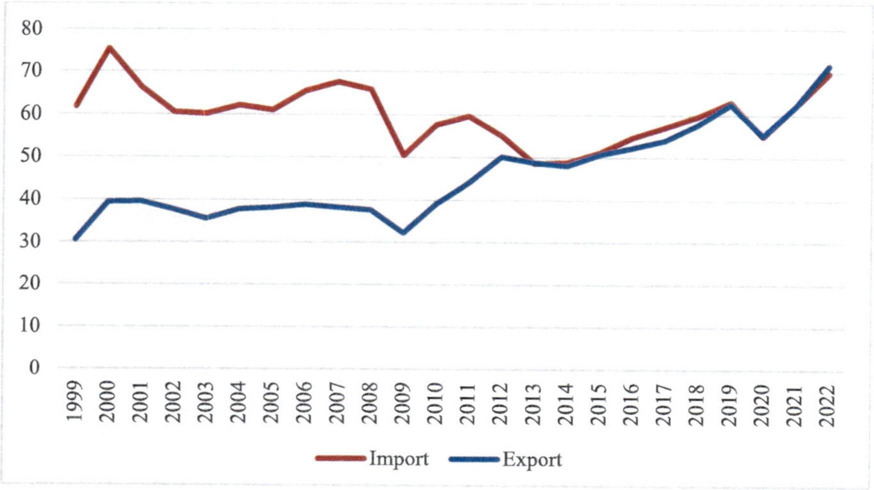

Fig. 12 Development of the EU-Japan trade of goods, 1999–2022 (€ billion). *Source* Own processing data available at Eurostat (EU trade since 1999 by SITC DS-018995)

Table 8 Structure of the EU-Japan trade of goods by SITC product category in %

Kind of goods	Export		Import	
	1999	2022	1999	2022
Food and live animals	7.18	6.97	0.09	0.51
Beverages and tobacco	3.12	3.86	0.02	0.19
Crude materials inedible. except fuels	3.74	3.07	0.51	1.23
Mineral fuels. lubricants and related materials	0.15	0.53	0.07	0.35
Animal and vegetable oils. fats and waxes	0.31	0.46	0.01	0.07
Chemicals and related products. n.e.s	18.84	32.02	6.63	13.64
Manufactured goods classified chiefly by material	9.16	6.52	5.66	9.07
Machinery and transport equipment	36.19	29.26	75.11	61.22
Miscellaneous manufactured articles	17.97	13.37	10.91	12.87
Comm. and trans. not class. elsewhere in the SITC	0.72	1.03	0.28	0.33
Others	2.60	2.90	0.70	0.53
Total	100	100	100	100

Source Own processing data available at Eurostat (EU trade since 1999 by SITC DS-018995)

services amounts to around 35% of EU-Japan total trade in goods. At the same time, EU exports of services to Japan amount to roughly 50% of EU exports of goods to the country.

Another view could be applied when analyzing the commodity structure of the bilateral trade relations. In general, we can say that imports from Japan to the EU are dominated by machinery, motor vehicles, chemicals, optical and medical instruments, and plastics. On the other hand, EU exports to Japan are dominated by chemicals, motor vehicles, machinery, optical and medical instruments, and food and drink.

The more detailed breakdown of EU trade with Japan by SITC groups is shown in Table 8. In 2022, the most exported manufactured goods were chemicals (32%), followed by machinery and transport equipment (29.26%) and other manufactured goods (13.37%). In 2022, EU imports of manufactured goods had a higher share than primary goods. The most imported manufactured goods were machinery and vehicles (61.22%), followed by chemicals (13.64%) and other manufactured goods (12.87%).

The next Figs. 13 and 14 show the gradual evolution of EU imports and exports for the most relevant SITC groups since 1999. Concerning the exports, the most of SITC groups have registered a smooth increase from 1999. However, the development of the export of machinery and transport equipment seems to be affected by the business cycle to the largest extent. Since 2020, the increase in the group of chemicals (and especially medicinal and pharmaceutical products) is extremely dynamic in which the EU had trade surpluses.

Concerning the imports (Fig. 14), the most of the relevant SITC groups have not changed considerably between 1999 and 2022. But the volatility in machinery and transport equipment seems to be influenced by various factors including business

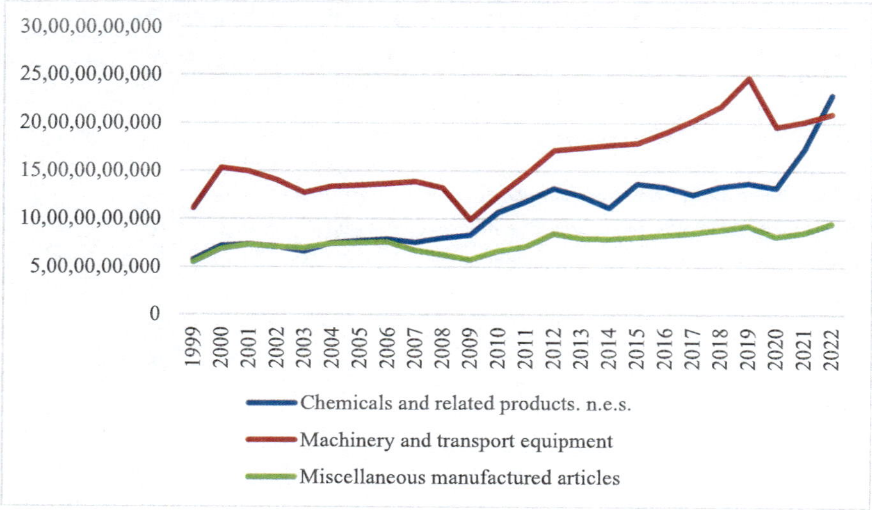

Fig. 13 Development of main EU export product groups to Japan, 1999–2022 (€ billion). *Source* Own processing data available at Eurostat (EU trade since 1999 by SITC DS-018995)

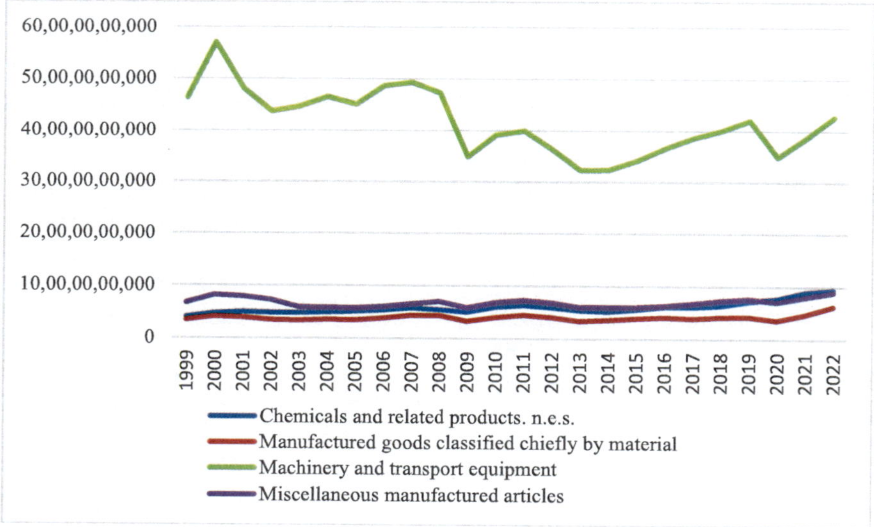

Fig. 14 Development of main EU import product groups to Japan, 1999–2022 (€ billion). *Source* Own processing data available at Eurostat (EU trade since 1999 by SITC DS-018995)

cycle measures. The peak of this kind of imports was achieved in 2000 and was gradually progressively declining until 2013. Later on, the smooth positive growing tendency occurs in the next period, negatively affected by the COVID crisis in 2020 and followed by a rapid recovery in 2021.

Heterogeneity among the EU countries concerning their participation in trade with Japan is obvious from data presented in Table 9. According to the European Statistics, 16 Member States had a trade surplus with Japan in 2021. The largest surplus was carried by Italy (€3100 million), followed by Denmark (€1726 million) and Ireland (€1702 million). There were eleven Member States that documented a trade deficit with Japan in the same year. The largest deficit was held by the Netherlands (€6930 million), followed by Belgium (€1679 million) and Poland (€1497 million).

The close relationship between the EU as a whole and Japan isn't limited to trade. The EU-Japan strategic partnership is based on longstanding cooperation, shared values and principles such as democracy, the rule of law, human rights, good governance, multilateralism and open market economies. Japan is one of the EU's closest, like-minded partners. The relationship between these countries also includes Europe's two most fundamental long-term priorities—digitalization and the green transition. Through the EU-Japan Green Alliance and the Japan-EU Digital Partnership, the two parties have upscaled their collaboration in these fields. And thanks to this consolidated, common stance, it'll be possible to encourage and support the rest of the world to pursue a fairer and faster digitalization and green transition.

Finally, we can conclude that the development of the EU-Japan relations is presenting the gradual linking and interconnecting of the two economies, catalyzed by the EPA and other initiatives. All the elements of the analysis in the last decade indicate that the coming years could represent a turning point characterized by an unprecedented closeness of both countries.

4.4 EU Trade Relations with Thailand

Thailand is the second-largest economy in the ASEAN region. Concerning the volume of its bilateral trade with the EU, Thailand is the 26th largest trading partner for the EU. Reciprocally, the EU is the fourth main trading partner for Thailand (after China, Japan and the United States). As in case of Japan, the strong trade bilateral relationship between Thailand and the EU is presented particularly in investment flows: Thailand is one of the most important destinations of European investments within ASEAN and the EU is the second-largest investor in Thailand after Japan. In 2021, the trade relationship between the EU and Thailand is characterized by the EU deficits in goods. In services, the balance switched to positive numbers in favour of the EU in 2020 and 2021.

The EU-Thailand mutual relations are intensified within last decade. However, the bilateral free trade agreement negotiations were frozen after the military took power in a coup in 2014. In 2023, the EU and the Thailand have restarted the negotiations to conclude an "ambitious, modern, and balanced free trade agreement" by 2025.

Table 9 The EU Member States export, import and trade balance of goods with Japan in € millions in 2021

	Export	Import	Trade balance
Austria	1726	1217	509
Belgium	7629	9308	−1679
Bulgaria	53	205	−152
Croatia	38	34	4
Cyprus	1	93	−92
Czechia	1202	2,011	−809
Denmark	2118	392	1726
Estonia	117	76	41
Finland	1649	298	1351
France	6568	5,844	724
Germany	18,515	17,323	1191
Greece	248	168	80
Hungary	520	1372	−852
Ireland	2675	974	1701
Italy	7555	4455	3100
Latvia	76	24	52
Lithuania	78	88	−10
Luxembourg	88	371	−283
Malta	145	71	74
Netherlands	4318	11,248	−6930
Poland	734	2231	−1497
Portugal	297	602	−305
Romania	360	358	2
Slovakia	204	132	72
Slovenia	93	120	−27
Spain	2935	2297	638
Sweden	2413	1031	1382

Source Own processing data available at Eurostat (EU trade since 1999 by SITC DS-018995)

The important milestones so far are presented in Table 10. The most important current agreement which modifies the features of the current EU-Thailand trade and economic relationship is represented by the Partnership and Cooperation Agreement (hereafter "PCA"). Another booster for these relations is expected to be a new Free Trade Agreement (FTA) [40].

In 2023, the EU and Thailand announced the relaunch of negotiations for an ambitious, modern and balanced free trade agreement (FTA). According to the European Commission, the aim of the FTA will be to boost trade and investment by addressing a wide range of issues such as: market access for goods, services, investment and

Table 10 Important milestones in mutual trade relations between EU and Thailand

Year	
2013	The EU and Thailand first launched negotiations for an FTA
2017	The Council considers to take steps towards broadening its engagement with Thailand
2019	The Council considers to take steps towards broadening its engagement with Thailand, including on issues of human rights, fundamental freedoms and democratic pluralism, by preparing for the timely signature of the Partnership and Co-operation Agreement (PCA). The Council also stresses the importance of taking steps towards the resumption of negotiations on an ambitious and comprehensive Free Trade Agreement (FTA)
2021	The EU Indo-Pacific Strategy confirmed the EU's longstanding interest in resuming FTA negotiations with Thailand
2022	The European Union and Thailand signed Partnership and Cooperation Agreement (PCA)
2023	The EU and Thailand announced the relaunch of negotiations for an ambitious, modern and balanced free trade agreement (FTA), with sustainability at its core

Source Own processing

government procurement; swift and effective sanitary and phyto-sanitary procedures; the protection of intellectual property rights including geographical indications, and the removal of obstacles to digital trade and trade in energy and raw materials, thereby supporting the digital and green transitions both in the EU and in Thailand.

The following Fig. 15 shows the export and import between the EU and Thailand from 1999 to 2022. Within the whole period, the EU had a trade deficit and this deficit was increasing in time. The figure indicates a considerable increase in imports during the Great Recession, and even more dynamic import growth in the pandemic and post-pandemic periods.

According to the European Commission, Thailand exported goods worth €15.1 billion to the EU in 2020. Key export products from Thailand are machinery and electronics and transport equipment, miscellaneous manufactured articles, as well as food products. On the other hand, the EU exported goods worth €11.3 billion to Thailand in 2020. Key EU exports to Thailand are machinery and transport equipment, chemicals and related products, and manufactured goods. The more detailed breakdown of EU trade with Thailand by SITC groups is shown in Table 11. Between 1990 and 2022, the commodity structure of exports to Thailand has changed more than the commodity structure of imports.

The next Figs. 16 and 17 show the gradual evolution of EU imports and exports for the most relevant SITC groups since 1999. Concerning the exports, the most of SITC groups have registered a smooth increase from 1999. However, the development of the export of machinery and transport equipment seems to be affected by the business cycle to the largest extent. The development of manufactured goods export, as well as food and live animals export, is very stable in the analyzed period. The same is true for chemicals and related products with the exception of the last period (after 2020) in which the export of this commodity started to increase considerably.

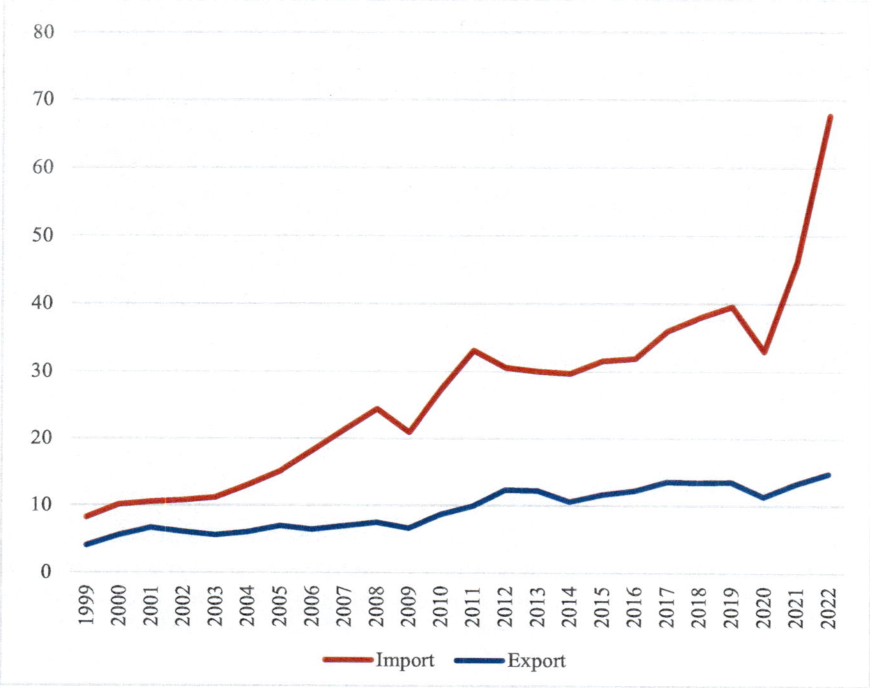

Fig. 15 Development of the EU-Thailand trade of goods, 1999–2022 (€ billion). *Source* Own processing data available at Eurostat (EU trade since 1999 by SITC DS-018995)

Table 11 Structure of the EU-Thailand trade of goods by SITC product category in %

Kind of goods	Export		Import	
	1999	2022	1999	2022
Food and live animals	14.45	8.53	2.03	3.22
Beverages and tobacco	0.21	0.24	0.62	0.46
Crude materials. inedible. except fuels	3.32	4.21	1.03	2.16
Mineral fuels. lubricants and related materials	0.03	0.08	0.14	0.17
Animal and vegetable oils. fats and waxes	0.11	0.09	0.04	0.12
Chemicals and related products. n.e.s	1.78	3.88	8.51	13.11
Manufactured goods classified chiefly by material	12.00	10.34	8.42	6.53
Machinery and transport equipment	41.21	56.44	19.71	21.47
Miscellaneous manufactured articles	0.32	1.76	0.56	0.42
Comm. and trans. not class. elsewhere in the SITC	0.49	0.22	0.89	0.46
Others	26.07	14.23	58.06	51.87
Total	100	100	100	100

Source Own processing data available at Eurostat (EU trade since 1999 by SITC DS-018995)

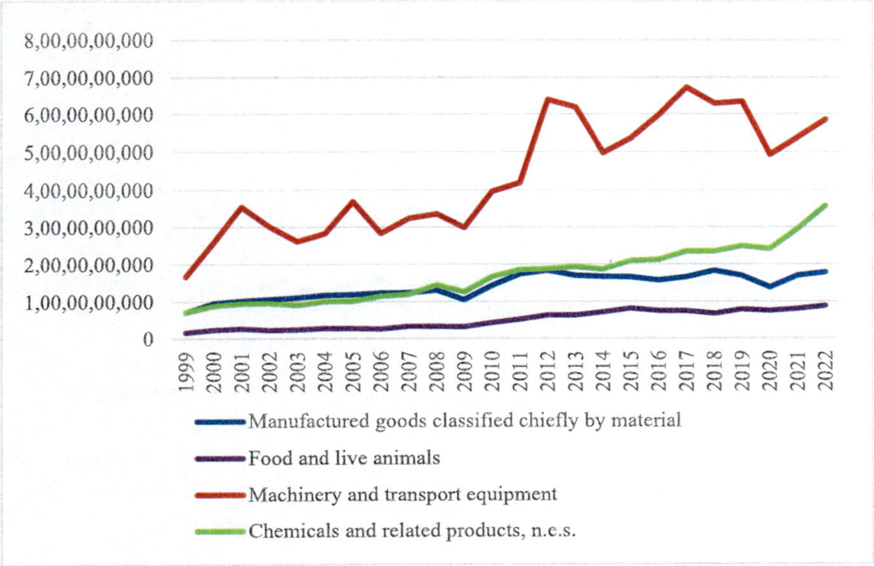

Fig. 16 Development of main EU export product groups to Thailand, 1999–2022 (€ billion). *Source* Own processing data available at Eurostat (EU trade since 1999 by SITC DS-018995)

Concerning the imports (Fig. 17), three of the four most relevant SITC groups (food and live animals, manufactured goods classified chiefly by material, chemicals and related products) have not changed considerably in volumes between 1999 and 2022. Nevertheless, the volatility in machinery and transport equipment is very dynamic and seems to be partially affected by business cycle measures. The smooth positive growing tendency in this SITC category occurred during the analyzed period with its extremely rapid increase tendency since 2020.

Heterogeneity among the EU countries concerning their participation in trade with Thailand is obvious from data presented in Table 12. According to the European Statistics, only three Member States had a trade surplus with Thailand in 2021. The largest surplus was carried by Finland (€67 million), followed by Denmark (€29 million) and Luxembourg (€18 million). On the other hand, the largest deficit was held by the Netherlands (€4229 million), followed by Czechia (€725 million) and Belgium (€610 million).

Finally, we can conclude that the development of the EU-Thailand economic relations was affected by the political situation: Brussels froze negotiations following a military coup in Thailand in 2014. Nowadays, the data presents the growing mutual interest in cooperation, which could be catalyzed by the negotiations and establishment of an ambitious, modern and balanced free trade agreement (FTA). To have a clearer view on the consequences of potentially granting access to the EU markets without tariffs, the Commission said it would conduct a sustainability impact assessment. This assessment would focus on the effects on the economy, human rights and

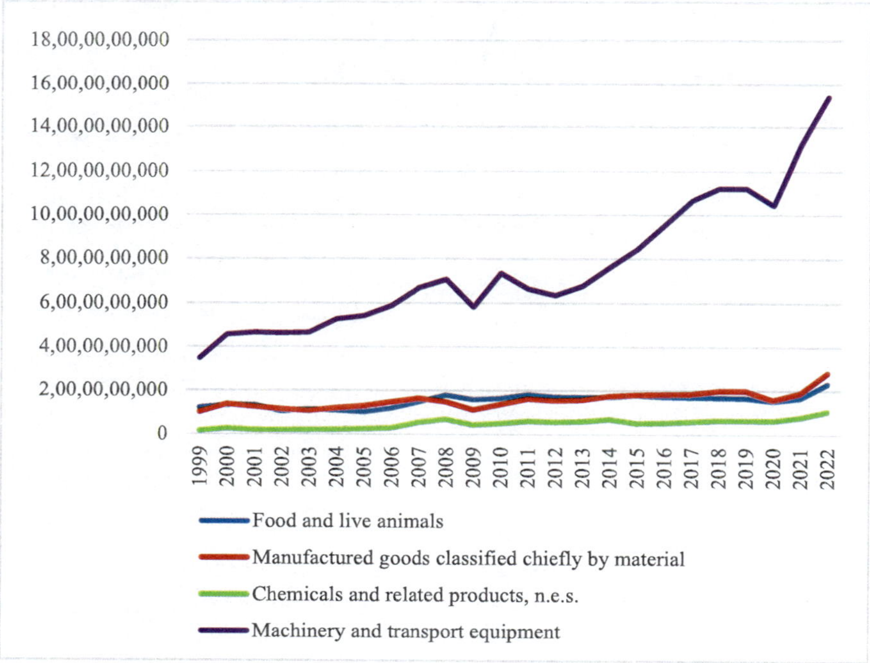

Fig. 17 Development of main EU import product groups to Thailand, 1999–2022 (€ billion). *Source* Own processing data available at Eurostat (EU trade since 1999 by Sitc Ds-018995)

the environment, "to provide recommendations on how to maximise the expected positive effects, whilst minimizing potential negative ones". In general, we can say that the relaunch of negotiations is a part of a larger trend of the EU rekindling ties with Southeast Asia.

4.5 EU Trade Relations with Philippines

The Philippines is one of ten members of the Association of Southeast Asian Nations (ASEAN). This country is the fifth-largest economy in the region in terms of GDP and the EU's 7th largest trading partner in ASEAN. The ASEAN as a whole is a dynamic region, in which Philippines represents the second-biggest market. The countries as a group are the EU's third largest trading partner outside Europe, after the United States and China.

The Philippines is a member of the WTO since 1995 and enjoys enhanced trade preferences with the EU under the EU's Generalised Scheme of Preferences plus (GSP+). The special incentive arrangement for Sustainable Development and

Table 12 The EU Member States export, import and trade balance of goods with Thailand in € millions in 2021

	Export	Import	Trade balance
Austria	311	472	−160
Belgium	982	1593	−610
Bulgaria	64	65	−1
Croatia	4	21	−17
Cyprus	13	25	−12
Czechia	158	882	−725
Denmark	414	385	29
Estonia	17	25	−9
Finland	236	170	67
France	1514	2042	−528
Germany	4931	5373	−442
Greece	70	150	−80
Hungary	129	567	−437
Ireland	221	393	−172
Italy	1570	1824	−254
Latvia	15	26	−11
Lithuania	13	25	−12
Luxembourg	24	6	18
Malta	3	18	−15
Netherlands	1110	5339	−4229
Poland	290	628	−338
Portugal	31	215	−184
Romania	68	231	−163
Slovakia	22	210	−188
Slovenia	36	70	−34
Spain	552	988	−436
Sweden	523	463	−60

Source Own processing data available at Eurostat (EU trade since 1999 by SITC DS-018995)

Good Governance GSP+ grants full removal of tariffs on two-thirds of all product categories, aiming to support sustainable development and good governance.

The EU officially launched negotiations with the Philippines on 22 December 2015, with the aim of concluding an agreement similar to those the EU achieved in its trade agreements with Singapore (2014) and Vietnam (2015). All three countries are members of the Association of South East Asian Nations (ASEAN). The EU's objective is for bilateral trade agreements with individual ASEAN countries to serve as building blocks towards a future region-to-region agreement between the EU and the ASEAN. Nowadays, the intensive EU-ASEAN Dialogue is maintained, which includes discussions on trade and investment issues at ministerial and senior economic official level.

The important milestones in the development of mutual EU-Philippines relations are presented in Table 13. In accordance with the initial region-to-region approach to negotiations with ASEAN, a Trade Sustainability Impact Assessment (SIA) of the Free Trade Agreement under negotiation with ASEAN is regularly conducted. A Sustainability Impact Assessment (SIA) has been carried out in support of free trade agreement (FTA) negotiations between the European Union and the Philippines. The SIA seeks to assess how trade and trade-related provisions in a future FTA could potentially impact economic, social, human rights and environmental elements in each trading partner and in other relevant countries [41].

According to the European Commission, trade in goods between the EU and the Philippines amounted to €18.4 billion in 2022, while bilateral trade in services between the EU and the Philippines reached €4.7 billion in 2021. The EU's foreign direct investment stock in the Philippines reached €13.7 billion in the same year.

The following Fig. 18 shows the export and import between the EU and Philippines from 1999 to 2022. In 1999, the EU had a trade deficit and this situation persisted until 2012. Between 2012 and 2020, the trade relations between the two partners were more balanced. However, since 2020 (pandemic and post-pandemic periods), an important increase in imports from Philippines has moved the mutual trade balance to the deficit levels.

Concerning the commodity structure, EU exports to the Philippines are dominated by machinery, transport equipment, chemicals, and food products, while the Philippines' main exports to the EU are office and telecommunications equipment, machinery, food products, and optical and photographic instruments. The more detailed breakdown of EU trade with Philippines by SITC groups is shown in Table 14. Between 1999 and 2022, the most important increase of the relative part in export could be identified for food and live animals (22.36% in 2022) and for chemicals and related products (16.73% in 2022). In import commodity structure, the most visible change between 1999 and 2022 concerns machinery and transport equipment (68.75% in 2021) as well as animal and vegetable oils (9.50% in 2022).

The next Figs. 19 and 20 show the gradual evolution of EU imports and exports for the most relevant SITC groups since 1999. Concerning the exports, the most of SITC groups have registered a smooth increase in the analyzed period. However, the development of the export of machinery and transport equipment is the most volatile

Table 13 Important milestone in mutual trade relations between EU and Philippines

Year	
2015	First round of the negotiations for an EU-Philippines trade and investment agreement
2017	Second round of negotiations for an EU-Philippines trade and investment agreement
2018	EU-Philippines Framework Agreement on Partnership and Cooperation came into force

Source Own processing

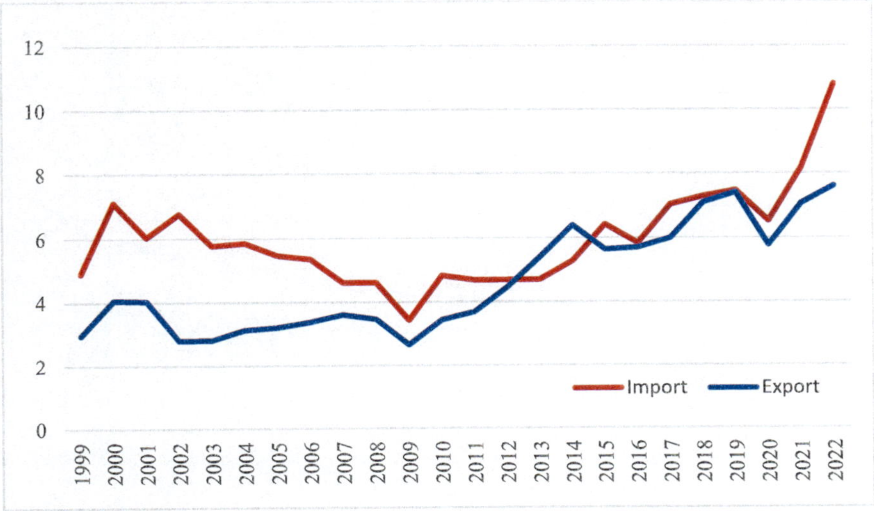

Fig. 18 Development of the EU-Philippines trade of goods, 1999–2022 (€ billion). *Source* Own processing data available at Eurostat (EU trade since 1999 by SITC DS-018995)

Table 14 Structure of the EU-Philippines trade of goods by SITC product category in %

Kind of goods	Export		Import	
	1999	2022	1999	2022
Food and live animals	7.01	22.36	2.46	4.82
Beverages and tobacco	0.94	3.93	0.46	0.42
Crude materials. inedible. except fuels	0.64	2.13	1.79	1.48
Mineral fuels. lubricants and related materials	0.23	0.11	0.00	0.00
Animal and vegetable oils. fats and waxes	0.06	1.56	3.59	9.50
Chemicals and related products. n.e.s	10.41	16.73	0.31	2.18
Manufactured goods classified chiefly by material	7.85	6.40	2.66	1.44
Machinery and transport equipment	62.20	35.24	61.83	68.75
Miscellaneous manufactured articles	2.58	1.68	0.04	0.24
Comm. and trans. not class. elsewhere in the SITC	0.50	0.72	0.24	0.12
Others	7.57	9.15	26.63	11.05
Total	100	100	100	100

Source Own processing data available at Eurostat (EU trade since 1999 by SITC DS-018995)

and seems to be affected by the business cycle to the largest extent with an important drop in 2009 and in 2020.

Concerning the imports (Fig. 20), the most of the relevant SITC groups have not changed considerably in volumes between 1999 and 2022. But the volatility

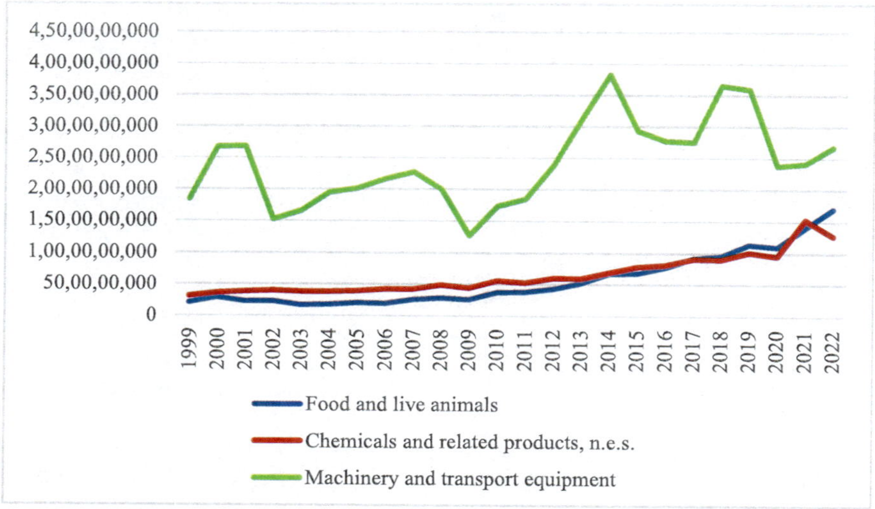

Fig. 19 Development of main EU export product groups to Philippines, 1999–2022 (€ billion). *Source* Own processing data available at Eurostat (EU trade since 1999 by Sitc Ds-018995)

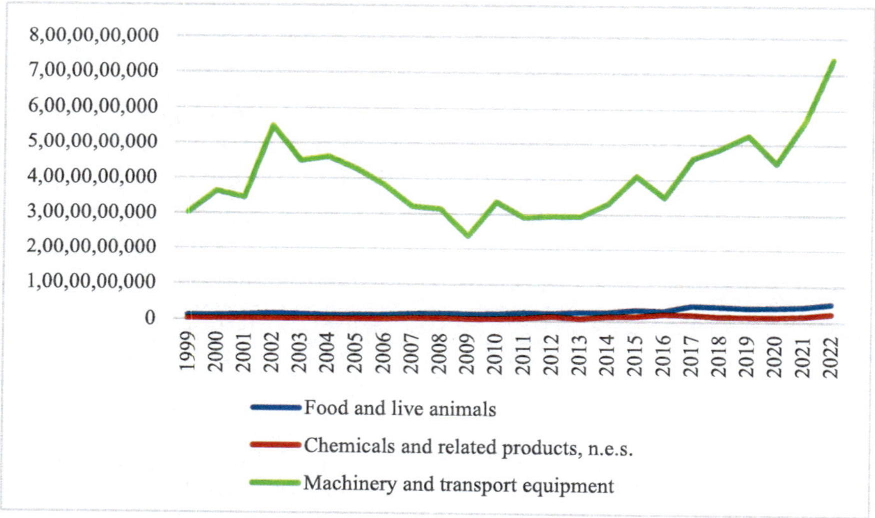

Fig. 20 Development of main EU import product groups to Philippines, 1999–2022 (€ billion). *Source* Own processing data available at Eurostat (EU trade since 1999 by SITC DS-018995)

in machinery and transport equipment seems to fluctuate to the largest extent. A decline tendency was visible between 2002 and 2012. Since 2012, the import of machinery and transport equipment from Philippines is permanently growing except of the years 2016 and 2021. The last period from 2020 is characterized by a very dynamic increase.

Heterogeneity among the EU countries concerning their participation in trade with Philippines is obvious from data presented in Table 15. According to the European Statistics, 16 Member States had a trade surplus with Philippines in 2021. The largest surplus was carried by Belgium (€665 million), followed by Spain (€396 million) and Italy (€359 million). There were eleven Member States that documented a trade deficit with Philippines in the same year. The largest deficit was held by the Netherlands (€1691 million), followed by Germany (€1114 million) and Hungary (€168 million).

Finally, we can conclude that the development of the EU-Philippines economic relations is affected by the EU approach to the ASEAN region: Brussels co-operated closely with ASEAN as a whole, negotiations for a region-to-region trade and investment agreement between the EU and ASEAN were launched in 2007 and paused by mutual agreement in 2009 to give way to a bilateral format of negotiations. Ensuring better access for EU exporters to the dynamic ASEAN markets is a priority for the EU. The Philippines with their growing market potential represent an important and promising partner of the EU in the ASEAN region.

4.6 EU Trade Relations with Bangladesh

In 2020, the EU was Bangladesh's main trading partner, accounting for around 19.5% of Bangladesh's total trade and Bangladesh was the EU's 34th largest trading partner in goods, in 2021, 27th largest import source (Table 2). Mutual trade in goods between the EU and Bangladesh grows during the period 1999–2022 (Fig. 21) although the increase in exports was very modest compared to imports.

A sharper increase in imports has occurred in the second decade of the twenty-first century with more pronounced fluctuations during the COVID-19 recession. Between 2017 and 2019, EU imports from Bangladesh reached on average €15.2 billion per year, which represented half of Bangladesh's total exports. Then there was a slight drop to €13.5 billion in 2020 probably due to COVID-19 pandemic. In 2022, total import reached new record level, nearly €24 billion, which is 14 times more than in 1999. The EU recorded a trade deficit with Bangladesh during the entire period. EU export of goods to Bangladesh amounted to 3.6 billion in 2022, which is 6 times less than EU import from Bangladesh in the same year.

The breakdown of EU trade with Bangladesh by SITC groups is in Table 16.

EU export of goods to Bangladesh mainly consisted of machinery and transport equipment. Since 1999, its share in total EU exports rose slightly from app. 40% to 44% in 2022. Chemicals and related products accounted for approximately a fifth of exports both in 1999 and in 2022. The largest decrease in the share of EU exports

Table 15 The EU Member States export, import and trade balance of goods with Philippines in € millions in 2021

	Export	Import	Trade balance
Austria	170	131	39
Belgium	881	216	665
Bulgaria	67	49	18
Croatia	4	2	2
Cyprus	13	4	9
Czechia	100	93	7
Denmark	237	96	141
Estonia	5	6	−1
Finland	180	40	140
France	749	669	80
Germany	1854	2968	−1114
Greece	12	17	−5
Hungary	15	183	−168
Ireland	161	180	−19
Italy	678	319	359
Latvia	1	2	−1
Lithuania	7	8	−1
Luxembourg	10	5	5
Malta	50	42	8
Netherlands	682	2373	−1691
Poland	242	300	−58
Portugal	14	29	−15
Romania	24	42	−18
Slovakia	14	13	1
Slovenia	19	5	14
Spain	702	306	396
Sweden	165	69	96

Source Own processing data available at Eurostat (EU trade since 1999 by SITC DS-018995)

to Bangladesh recorded food and live animals from 17.2% in 1999 to 8.5% in 2022. The other product categories are not as important, the share of each of them in the total export is less than 10%. Beverages and tobacco, mineral fuels, lubricants and related materials, animal and vegetable oils, fats and waxes and comm. and trans. not classified elsewhere in the SITC each has completely marginal significance since the share of each of the mentioned groups is less than 1% on total exports.

EU imports from Bangladesh are dominated by miscellaneous manufactured articles accounting for over 95% of the EU's total imports from Bangladesh in 2022. A

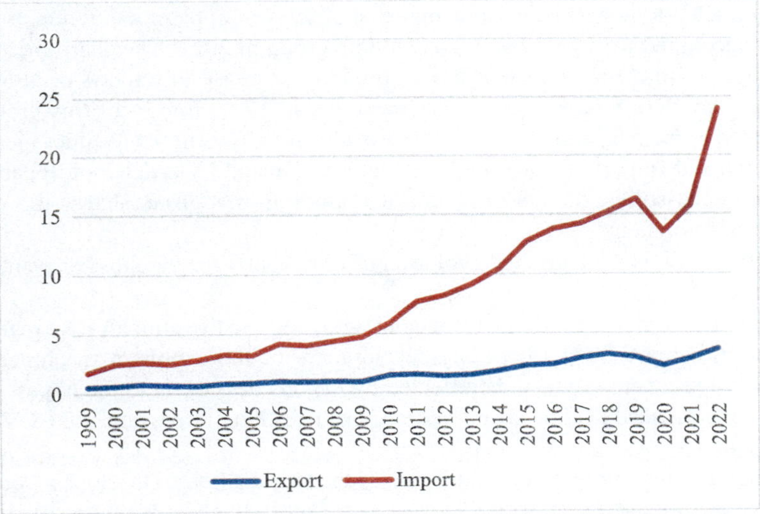

Fig. 21 Development of the EU-Bangladesh trade of goods. 1999–2022 (€ billion). *Source* Own processing data available at Eurostat (EU trade since 1999 by SITC DS-018995)

Table 16 Structure of the EU-Bangladesh trade of goods by SITC product category in %

Kind of goods	Export		Import	
	1999	2022	1999	2022
Food and live animals	17.16	8.52	4.18	1.07
Beverages and tobacco	0.23	0.03	0.08	0.40
Crude materials, inedible, except fuels	7.02	10.61	0.16	0.18
Mineral fuels, lubricants and related materials	0.14	0.16	n.a	0.00
Animal and vegetable oils, fats and waxes	0.63	0.23	0.00	0.00
Chemicals and related products, n.e.s	18.73	19.96	0. 19	0.02
Manufactured goods classified chiefly by material	9.28	8.59	7.03	2.55
Machinery and transport equipment	39.52	44.25	0.23	0.63
Miscellaneous manufactured articles	4.13	5.92	88.04	95.11
Comm. and trans. not classified elsewhere in the SITC	0.77	0.60	0.07	0.02
Others	2.38	1.13	0.02	0.00
Total	100	100	100	100

Source Own processing data available at Eurostat (EU trade since 1999 by SITC DS-018995)

more detailed look at this group shows that articles of apparel and clothing accessories accounted for more than 96.4% in this group in 2022. The sharp increase in EU imports from Bangladesh is due to the huge increase in imports of articles of apparel and clothing accessories. For example, in 1999, import of this group has amounted to €1.5 billion and in 2022 it was already €22 billion (15 times increase). Since the EU imports from Bangladesh are dominated by articles of apparel and clothing accessories, the development of imports in this group shapes the overall import development.

Figure 22 shows the gradual evolution of EU exports for the most relevant SITC groups since 1999.

Chemicals and related products together with food and live animals have registered a smooth increase from 1999. The development of the export of machinery and transport equipment seems to be affected by the business cycle to the largest extent since it experienced decline during recessions (The Great Recession and COVID-19 recession).

Bangladesh is a poor country, it is one of the world's Least Developed Countries (LDCs). As a Least Developed Country, Bangladesh benefits from the Everything But Arms (EBA) arrangement. EBA grants Bangladesh duty-free, quota-free access to the EU for exports of all products, except arms and ammunition.

The EU works with Bangladesh in the framework of the EU-Bangladesh Cooperation Agreement, which entered into force on 1 March 2001. The Cooperation

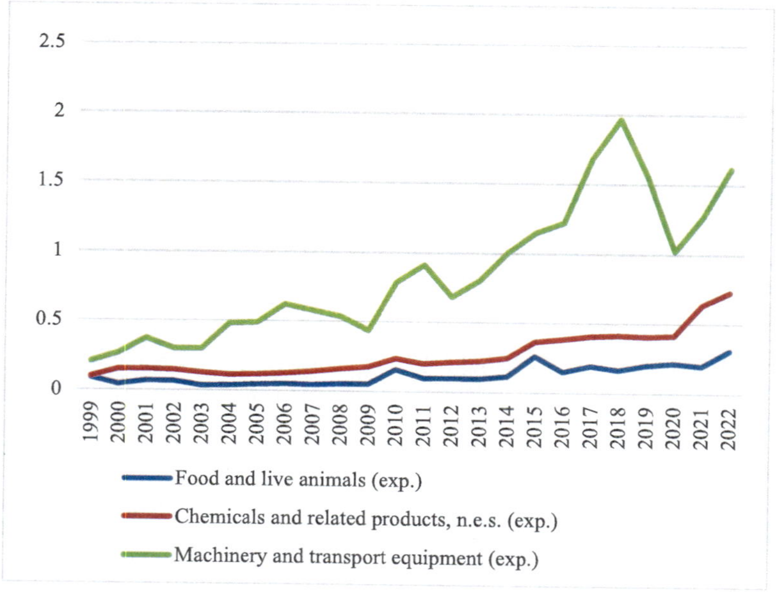

Fig. 22 Development of main EU export product group to Bangladesh, 1999–2022 (€ billion). *Source* Own processing data available at Eurostat (EU trade since 1999 by SITC DS-018995)

Agreement seeks to strengthen relations between EU (at the time this agreement entered into force European Community) and Bangladesh. Now it focuses on the vital importance of social development in Bangladesh, in parallel with its economic development and the protection of its environment [42]. Trade cooperation aims to expand trade and open up markets. It takes place in compliance with World Trade Organization agreements. Bangladesh has been a WTO member since 1995.

Economic cooperation aims particularly at:

- facilitating contacts between economic operators, business communities, enterprises and investors;
- improving the business environment and conditions for investment, particularly for small and medium-sized enterprises;
- promoting technology transfer.

The collapse of Rana Plaza factory complex in April 2013 led to the death of over 1100 people, the majority of whom were garment workers. This event has drawn international attention to safety and labour rights in Bangladesh's ready-made garment (RMG) sector.

The EU took the initiative of launching a Sustainability Compact for Bangladesh. Sustainability Compact brings together the Government of Bangladesh, the European Commission, the United States, Canada (the main markets for Bangladeshi garment production) and the International Labour Organization (ILO) accompanied by employers, trade unions and other key stakeholders to promote continuous improvements in labour rights and factory safety in the ready-made garment (RMG) industry [43].

The Sustainability Compact is built on short and long-term commitments related to three inter-linked pillars:

1. respect for labour rights
2. structural integrity of buildings and occupational safety and health
3. responsible business conduct

Finally, we can conclude that the EU is an important trading partner for Bangladesh, but the import to the EU is not diversified. It depends on one group of products, namely clothing, accounting for over 90% of the EU's total imports.

5 Conclusions

The megacities are a new phenomenon in the global world. They create a new urban dynamic, as super-sized cities are seen as the new engine of the global economy, connecting the flow of goods [44]. Although the main idea of the paper was to research the international trade between the European megacities and Indo-pacific megacities in within the international trade, we have to conclude that currently, it is not possible to do it because of the lack of data at the local and regional level.

That is why we decided to analyze and compare the international trade of the European Union and as unit with selected Indo-pacific countries with megacities. During last years, there is a great shift in development of these relationships which is reflected also in the increasing importance of Indo-Pacific Region and their contribution to economic growth.

EU27 is one of the biggest trading blocks in the world economy. Since the common trade policy falls within the exclusive competences of the European Union, the EU acts as one unit in trade relations with third countries. Despite the fact, that mutual trade was fully liberalized only between member countries of the EU, due to the importance of third countries as EU trade and investment partners, the EU signed with them various preferential agreements. Growing importance of the Indo-Pacific region in the world economy as well as in the EU external relations is reflected by improving economic relations between EU and these countries. In 2021, EU adopted Strategy for Cooperation in the Indo-Pacific. Both with India and Indonesia trade and investment relations have deepened in recent years. An analogical tendency can be observed in relations with Japan, too. Negotiations with India for EU-India Free Trade Agreement (FTA), Investment Protection Agreement (IPA) and Geographical Indications Agreement (GIs) are currently ongoing. With the Indonesia, the EU is negotiating a Free Trade Agreement (FTA). The most developed relations are between the EU and Japan, catalyzed particularly by Economic Partnership Agreement (EPA).

Thailand, Philippines and Bangladesh are not the key trade partners for the European Union, but the trade relationships are developed gradually because of their location in ASEAN region. All countries, eventhough they are not so economically developed, represent the market potential in the closed future. The important is also a great effort of European Union to support the social and environmental development in these countries fostering the human rights and solving the poverty. The cooperation with Thailand was influenced by the unstable political situation in the country, but after reconsideration by the EU the ambition is to establish free trade agreement. The similar situation is in the relationships with Philippines. The trade relationship with Bangladesh European Union develops more intensively from the COVID-19 pandemic. The mutual cooperation has an ambition to support the social development, respect for labour rights and responsible approach to the business activities.

Despite some issues, it is assumed that gradual liberalization between EU and Indo-Pacific countries will lead to the increasing of mutual trade, contribute to the growth of production, employment and promote sustainable development.

Acknowledgements This study has been supported by project Vega 1/0311/23 Open Local Municipality, Vega 1/0343/23 Modern approaches to the development of cities and regions and KEGA 040UMB-4/2021, 'Diversification of Content and Didactic Forms for Teaching Economic Subjects in the Slovak Language and World Languages'.

Bibliography

1. National Intelligence Council (2012) Global Trends 2030 – alternatives world
2. Nicolas F (2021) The regional economic order: four scenarios Future Insights Series. National Security College, Australian National University, May 2021. Available online: https://www.ifri.org/en/publications/publications-ifri/articles-ifri/regional-economic-order-four-scenarios
3. Martin EA, Karpata M (2021) The dilemma of middle powers: how AUKUS has reshaped the potential for E3 cooperation in the Indo-Pacific. Available online: https://www.ifri.org/en/publications/notes-de-lifri/notes-cerfa/dilemma-middle-powers-how-aukus-has-reshaped-potential-e3
4. Maddison A (2006) The world economy; Volume 1: A millennial perspective; Volume 2: Historical statistics. OECD, 657 p
5. World Bank (2020) Global economic prospects. International Bank for Reconstruction and Development the World Bank, 238 p. https://doi.org/10.1596/978-1-4648-1553-9
6. WTO (2020) Trade Policy Review
7. UNCTAD (2022) Impact of the COVID-19 pandemic on trade and development, 82 p
8. WTO (2022) World Trade Statistical Review 2022. Available online: https://www.wto.org/english/res_e/publications_e/wtsr_2022_e.htm
9. Government of Netherlands (2020) Indo-Pacific: guidelines for strengthening Dutch and EU cooperation with partners in Asia
10. Hilpert HG (2022) Sea change in EU trade policy. Opportunities for diversification in the Indo-Pacific. SWP Comment
11. Baldwin R (1994) Towards an integrated Europe. Centre for Economic Policy Research, 234 p
12. Baldwin R (1995) A domino theory of regionalism. In: Baldwin R, Haaparanta P, Kiander J (eds) Expanding European regionalism: the EU's new members. Cambridge University Press, Cambridge. ISBN: 978-0521481342
13. European Union (2023) Treaty on the Functioning of the European Union. Consolidated version of the Treaty on the Functioning of the European Union. Available online: https://eur-lex.europa.eu/legal-content/EN/TXT/?uri=celex%3A12012E%2FTXTUNCAD. 2023. Key Statistics and Trends in International Trade 2023
14. Sapir A (2016) EU trade policy. In: Badinger H, Nitsch V (eds) Routledge Handbook of the economics of European integration. Routledge, pp 205–219. ISBN: 978-0-415-74770-7
15. European Commission (2021) Commission sets course for an open, sustainable and assertive EU trade policy. Available online: https://ec.europa.eu/commission/presscorner/detail/en/ip_21_644
16. European Commission (2023a) Making trade policy. Available online: https://policy.trade.ec.europa.eu/eu-trade-relationships-country-and-region/making-trade-policy_en
17. Schaus M (2021) EU trade policy in light of the new industrial strategy for Europe. Forum 56(3):144–149
18. European Commission (2021) Questions and answers: EU strategy for cooperation in the Indo-Pacific. Available online: https://ec.europa.eu/commission/presscorner/detail/en/qanda_21_4709
19. European Commission (2023b) The EU and the WTO. Available online: https://policy.trade.ec.europa.eu/eu-trade-relationships-country-and-region/eu-and-wto_en
20. WTO (2023) Principles of the trading system. Available online: https://www.wto.org/english/thewto_e/whatis_e/tif_e/fact2_e.htm
21. European Commission (2022) List of GSP beneficiary countries. Available online: https://circabc.europa.eu/ui/group/f243659e-26f5-44d9-8213-81efa3d92dc7/library/83191464-a9b5-4973-a3a9-fe17e57d68e8/details
22. European Parliament (2012) Regulation (EU) No 978/2012 of the European Parliament and of the Council of 25 October 2012 applying a scheme of generalised tariff preferences and repealing Council Regulation (EC) No 732/2008. Available online: https://eur-lex.europa.eu/legal-content/EN/TXT/?uri=uriserv:OJ.L_.2012.303.01.0001.01.ENG

23. Pelkmans J (2006) European integration methods and economic analysis. Pearson Education, Prentice Hall, p 480. ISBN: 978-0-273-69449-6
24. Jovanovic MN (2013) The economics of European integration, 2th edn. Edward Elgar, p 1028. ISBN 978-0-85793-397-3
25. European Commission (2021) EU Strategy for cooperation in the Indo-Pacific - Council conclusions (16 April 2021). Available online: https://data.consilium.europa.eu/doc/document/ST-7914-2021-INIT/en/pdf
26. European Commission (2021) EU's Indo-Pacific strategy. Available on: https://ec.europa.eu/commission/presscorner/detail/en/qanda_21_4709
27. Commercial Cooperation Agreement between the European Economic Community and the Republic of India. 21973A1217(01). Official Journal L 082, 27/03/1974 P. 0002. Available online: https://eur-lex.europa.eu/legal-content/EN/ALL/?uri=CELEX%3A21973A1217%2801%29
28. European Council (1994) Cooperation Agreement between the European Community and the Republic of India on partnership and development - Declaration of the Community concerning tariff adjustments - Declarations of the Community and India. Available online: https://eur-lex.europa.eu/legal-content/GA/TXT/?uri=celex:21994A0827%2801%29
29. Council of the European Union (2005) The India-EU strategic partnership joint action plan. Available online: https://www.google.com/search?q=THE+INDIA-EU+STRATEGIC+PARTNERSHIP%0D%0AJOINT+ACTION+PLAN&client=firefox-b-d&ei=zjdKZPqHA9qPxc8P1oqOuA8&ved=0ahUKEwj6po_Q18n-AhXaR_EDHVaFA_cQ4dUDCA4&uact=5&oq=THE+INDIA-EU+STRATEGIC+PARTNERSHIP%0D%0AJOINT+ACTION+PLAN&gs_lcp=Cgxnd3Mtd2l6LXNlcnAQAzIKCAAQRxDWBBCwAzIKCAAQRxDWBBCwAzIKCAAQRxDWBBCwAzIKCAAQRxDWBBCwAzIKCAAQRxDWBBCwAzIKCAAQRxDWBBCwAzIKCAAQRxDWBBCwAzIKCAAQRxDWBBCwA0oECEEYAFDtBFjtBGDyCWgBcAF4AIABAIgBAJIBAJgBAKABAqABAcgBCMABAQ&sclient=gws-wiz-serp
30. European Parliament and the European Economic and Social Committee (2004) Communication from the Commission to the Council, the European Parliament and the European Economic and Social Committee – an EU-India Strategic Partnership {SEC(2004) 768}. COM/2004/0430 final. Available online: https://eur-lex.europa.eu/legal-content/EN/TXT/?uri=celex%3A52004DC0430
31. European Commission (2023) Overview of FTA and other trade negotiations. Available online: https://policy.trade.ec.europa.eu/eu-trade-relationships-country-and-region/negotiations-and-agreements_en
32. India-EU Strategic Partnership: A Roadmap to 2025. July 15, 2020. Available online: https://www.mea.gov.in/bilateral-documents.htm?dtl/32828/IndiaEU_Strategic_Partnership_A_Roadmap_to_2025
33. European Commission (2023) EU-India Free Trade Agreement. Investment Protection Agreement and Geographical Indications Agreement. Available online: https://policy.trade.ec.europa.eu/eu-trade-relationships-country-and-region/countries-and-regions/india/eu-india-agreement_en
34. European Commission (2023) Regulation imposing anti-dumping duties on fatty acid from Indonesia. Commission Implementing Regulation (EU) 2023/111 of 18 January 2023 imposing a definitive anti-dumping duty on imports of fatty acid originating in Indonesia C/2023/259. Available online: https://eur-lex.europa.eu/legalontent/EN/TXT/?uri=uriserv%3AOJ.L_.2023.018.01.0001.01.ENG&toc=OJ%3AL%3A2023%3A018%3ATOC
35. European Commission (2023) EU trade relations with Indonesia. Facts, figures and latest developments. Available online: https://policy.trade.ec.europa.eu/eu-trade-relationships-country-and-region/countries-and-regions/indonesia_en
36. Framework agreement on comprehensive partnership and cooperation between the European Community and its Member States, of the one part, and the Republic of Indonesia (2022) Official Journal of the European Union. Available online: https://circabc.europa.eu/ui/group/09242a36-a438-40fd-a7af-fe32e36cbd0e/library/d9e40005-57d1-4e2e-8afb-353f3cff22c8/details

37. European Commission (2019) Sustainability Impact Assessment (SIA) in support of Free Trade Agreement (FTA) negotiations between the European Union and the Republic of Indonesia. Luxembourg: Publications Office of the European Union. 287 s. ISBN 978-92-76-10928-0. Available online: https://circabc.europa.eu/ui/group/09242a36-a438-40fd-a7af-fe32e36cbd0e/library/76dcefc4-692c-421d-8a5e-54955bb486aa/details
38. European Commission (2023) EU trade relations with Japan. Facts, figures and latest developments. Available online: https://policy.trade.ec.europa.eu/eu-trade-relationships-country-and-region/countries-and-regions/japan_en
39. European Commission (2022) Japan-EU – international trade in goods statistics. Available online: https://ec.europa.eu/eurostat/statistics-explained/index.php?title=Archive:Japan-EU_%E2%80%93_international_trade_in_goods_statistics#EU_and_Japan_in_world_trade_in_goods
40. European Commission (2023) EU trade relations with Thailand. Facts, figures and latest developments. Available online: https://policy.trade.ec.europa.eu/eu-trade-relationships-country-and-region/countries-and-regions/thailand_en
41. European Commission (2023) EU trade relations with Philippines. Facts, figures and latest developments. Available online: https://policy.trade.ec.europa.eu/eu-trade-relationships-country-and-region/countries-and-regions/philippines_en
42. Cooperation Agreement between the European Community and the People's Republic of Bangladesh on partnership and development (2001) Official Journal of the European Union. (OJ L 118, 27.4.2001). Available online: https://eur-lex.europa.eu/legal-content/EN/TXT/?uri=celex%3A22001A0427%2801%29
43. European Commission (2013) The sustainability compact for the Bangladesh ready-made garment sector. Available online: www.ilo.org/bangladesh
44. ESPAS (2018) Global trends to 2030: The future of urbanisation and megacities. Available online: https://espas.secure.europarl.europa.eu/orbis/sites/default/files/generated/document/en/Think%20piece%20global%20trends%202030%20Future%20of%20urbanisation.pdf

Mariana Považanová is assistant professor at the Faculty of Economics, Matej Bel University in Banska Bystrica, Slovakia. Her research interest areas include the issues of macroeconomics, economic integration and unpaid work in household.

Žaneta Lacová obtained her PhD at Université de Lorraine in Nancy (France). She is an assistant professor at the Faculty of Economics, Matej Bel University. Her research deals with the issues of macroeconomic performance and the consequences of the digital transformation of economies in the context of European integration. She is the author of a foreign scientific monograph and has published more than 60 scientific and professional works. She is a member of the Scientific Committee of the international research network ERECO—PGV in Grenoble, and is a member of the international research network INFER—International Network for Economic Research.

Katarína Vitálišová is an associate professor at the Department of Public Economics and Regional Development of Faculty of Economics, Matej Bel University. She orients her research on the issue of public and local governance, building and maintaining relationships and partnerships between stakeholders, strategic planning of the spatial development and implementation of new approaches in it, including creative and smart cities, places marketing what is represented by more than 130 publications. She is a member of several international expert networks in the field of regional and local development, management and marketing places. My research orientation is the issue of public and local governance, building and maintaining relationships and partnerships between stakeholders, strategic planning of the spatial development and implementation of new approaches to local and regional development, including local production systems and industrial symbiosis, smart and creative cities and regions. The topic of the Cost Action is directly associated with the previous research experience and research area (e.g., 7FP Project—FOLPSEC, international comparative project Slovakia—Bulgaria—Functioning of the Local Production Systems in the Conditions of Economic Crisis, or international comparative project Slovakia—France—Cultural potential and its utilization in the local development), as well as publication outputs. The participation in the working group is interesting for me because of the possibilities to exchange the experience and knowledge in the researched field, as well as a new possibility to realize the international comparative research in this topic which is very progressive developed.

Rural-Urban Harmony in Megacity System

Metropolitan Architecture and Sustainable Habitats in the Indo-Pacific Region to Reinforce the Megacity System Through Urban–Rural Patterns

Critical Comparative Study Between the Desakota Region, Tokyo Megalopolis, Yunnan Province, and the Filipino Urban Form

Antonella Contin, Valentina Galiulo, Alessandra Maria Pandolfi, Ryusei Koike, Dicheng Yang, and Kelvin C. de Chavez

Abstract The Indo-Pacific region needs a comprehensive view of the risks of widening existing inequalities or creating new ones due to the Green and Digital Transitions to improve the quality of life. The goal is to define Regenerative Actions to strengthen existing sustainable habitats and create/support trans-regional and national clusters. Interstitial actors must act as mediators and adapt existing knowledge to new situations. Twin Transitions require new forms of governance and compelling service access issues to identify and reinforce existing sustainable communities. Spatial analysis is needed to generate a strategic vision of territorial shrinkage for Japan. Spatial planning strategies can address population shrinkage by

A. Contin (✉) · V. Galiulo · A. M. Pandolfi · R. Koike · D. Yang · K. C. de Chavez
Department of Architecture and Urban Studies, Politecnico Di Milano, Via Bonardi 3, 20133 Milano, Italy
e-mail: antonella.contin@polimi.it

V. Galiulo
e-mail: valentina.galiulo@polimi.it

A. M. Pandolfi
e-mail: alessandramaria.pandolfi@polimi.it

R. Koike
e-mail: ryusei.koike@mail.polimi.it

D. Yang
e-mail: dicheng.yang@mail.polimi.it

K. C. de Chavez
e-mail: kelvin.dechavez@mail.polimi.it

T. M. Vinod Kumar (ed.), *Indo-Pacific Smart Megacity System*, Advances in 21st Century Human Settlements, https://doi.org/10.1007/978-981-99-6218-1_4

promoting the *Desakota* phenomenon and fostering the proximity of urban settlements and agriculture. The study aims to explore how Harmony can be integrated into an Operating Culture that encompasses anthropological, social, and economic aspects of cooperation with nature. Nature is critical in shaping the relationship between humans and technology, and culture is the soul that integrates technology and nature. Asian continent significance lies in its technological advancements and its potential to help China overcome its particularism and discover new ways of understanding unity. The Asian and Japanese experience has shown how the community dimension can be effective in an emergency, and how the vulnerability of a community can be transformed into a resource. Realistic contexts are necessary to achieve desired outcomes, and Asian societies must harness and transform these energies into productive innovation.

Keywords Twin Transition · Desakota · Regenerative Urban Policies · Metropolitan Architecture and Cartography · Sustainable Habitats for Megalopolis

1 Introduction

Today, the over-reliance on technology and efficiency in the name of national or racial identity leads to a loss of uniqueness and the division of society into elitist groups. This determines negative outcomes due to increased inequalities and vulnerabilities in both physical and social dimensions, mainly in the countryside.

To fully embrace a new alternative model of civilisation, it is crucial to incorporate the perspectives of digital technologies that are causing significant changes in the cognitive structure of the inhabitants of the Indo-Pacific territories and present unrestricted development opportunities. However, the key is to balance the values of rural civilisation and the value of science and technology.

To face this, it is necessary to enhance prosperity and equity, improve people lives, and achieve climate neutrality. Every transition raises questions about sustainability in urban growth and calls for a new language to analyse the territorial and cultural context and promote a new public wealth. Protecting cities and territories from artificiality, standardisation, and seriality is a crucial ethical and political mission.

The Indo-Pacific region is at a critical turning point, facing socio-political conflicts, demographic and economic imbalances, and the effects of the COVID-19 pandemic. An alternative future is possible if the transition can improve the quality of life and meet the expectations and needs of citizens in the region.

What is needed is a comprehensive view of the risks of widening existing inequalities or creating new ones because of the Green and Digital Transitions. To reach the objective, the existing situation analysis and identification of present sustainable habitats (through place-based research and MC open-source methodology) allow us to identify inequalities embodied in space and affecting people [12]. The goal is to define Regenerative Actions for strengthening the existing Sustainable Habitats towards Twin Transitions, outlining the framework conditions for social innovation and sustainable competitiveness.

To solve the problem of the shrinkage of rural areas, policy guidelines must act as a source of learning and a catalyst for sustainable change, addressing the potential disparities in regional socioeconomic and policy-making capacities shaped by the Twin Transitions, by implementing fact-based cluster policies that more effectively address the framework conditions for innovation and competitiveness [15]. It also aims to tackle the absence of guidelines for creating/supporting trans-regional and national clusters in the different regions, including a mix of local and foreign actors across the value chain.

The transformative role of intermediaries in urban change is crucial. 'Interstitial actors' must accentuate their role as mediators in transitions where multiple public agents come together in institutions, like community associations and non-profit organisations.

Learning is individual and collective, and their relationship is poorly understood. Organisational learning is often linked to change, and the absence of change indicates a lack of learning.

However, change often occurs through adaptation, not adoption, of external knowledge. The biggest challenge is adapting existing knowledge to new situations. The skills and resources required for policy learning and changes depend on the personal identity and the community goals. The goal must be to significantly transform our cities and territories and explore various interpretations and partnerships for a biopolitical project, leading to a new understanding of life. This requires a multifaceted approach to prosperity, including cohesion, justice, and environmental preservation. Additionally, the divide between digital knowledge and labour can be acknowledged and addressed during these transitions by learning how the Indo-Pacific cultural separation between culture and nature is overcome.

Mitigating the risk of losing traditional skills, culture, local expertise, and markets in the Indo-Pacific region means creating a new informational ecology that produces new abilities and skills for a new metropolitan lifestyle, regenerative agriculture actions, and civic economy, through a sustainable policy supported by an open-source mapping knowledge, which is the goal.

The concept of instability and threats of the Indo-Pacific landscape is crucial as it highlights the challenges of the modern world complexity to the traditional linear approach of territorial development. However, the objective is not to revert to the past, but to identify and reinforce existing sustainable communities.

Twin Transitions require new forms of governance, and compelling service access issues are key to this vision. It must focus on defining critical problems that emerge rapidly and unexpectedly in human society, some revealing critical social infrastructures. Analysing such issues and their identification is crucial to the sustainability and resilience of cities, territories, and landscapes.

Megacities have been rapidly expanding in the world in the last 2 centuries, often disregarding spatial distribution guidelines. The map of United Nations (Fig. 1) indicates how the inequalities between social classes is increasing, so that the spatial distribution is the most important aspect also in the economic asset. Mostly because

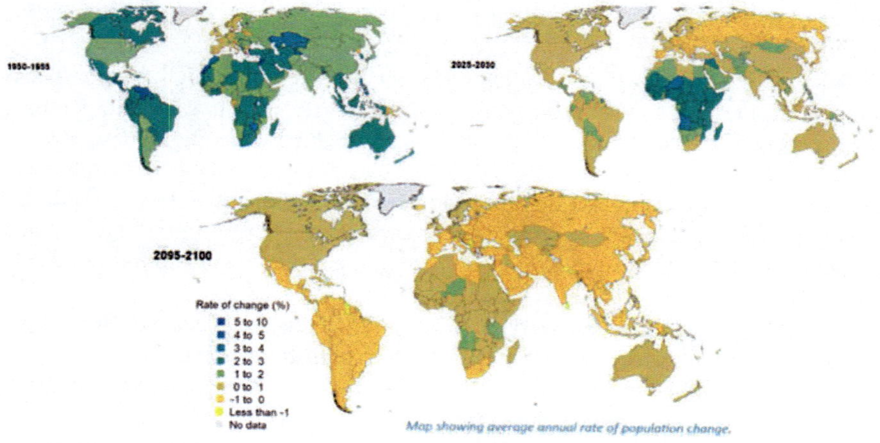

Fig. 1 Population changes rates (Credits: R. Koike on UN prospects)

shrinking is inevitable also for megacities in Africa, Asia, and the Americas, which are growing today, when the considered time span covers at least a century (see Fig. 1).

It is a cyclic occurrence that happened in different centuries and will happen again in the future, involving different cycles (at local, urban, and regional scales). This will determine that some parts of global megacities will be expanding, some others will be contracting or even shrinking, whilst others will still be growing or recovering from contraction at differing rates, which depend on several factors.

As the history of urban development theories explained, understanding megacities trends means dealing with different cycles at several scales and paces, considering many possible different factors, which include, but are not limited to, real estate dynamics [67], urban development strategies, demographic trends, and economic cycles. "Metropolitan strategies need Economic principles and tools to be equitable and efficient, requiring multifaceted and differentiated knowledge to be organized" (*ibidem*).

The global population is expected to increase by 2 billion by 2030, mostly due to urban population growth. By 2030, 4.9 billion people will live in cities, according to UN prospects. However, shrinking cities are also a growing occurrence, with the number of decreasing areas escalating faster than boomtowns. Between 1950 and 2000, the number of shrinking cities increased by 330%, whilst the number of cities with more than 100,000 residents increased by only 240% [64]. This trend was primarily concentrated in Western industrial countries, such as the USA, Britain, Germany, and Italy (*ibidem*). Since 1990, shrinking cities have been found in former Warsaw Pact countries, like Russia, Ukraine, and Kazakhstan. South Africa and Japan have also experienced an increase in shrinking cities between 1950 and 2000 (*ibidem*).

From what we can see from demographic trends and prospects, the future of most of the urban systems in Africa, Asia, and the Americas is shrinked, too. Since shrinkage does not take place in uniform patterns globally, this chapter studies the emerging shrinking trends in significant areas between the Desakota region, Tokyo Megalopolis, Yunnan Province, and the Filipino urban form, as different examples of megacities that are in different stages of the same cycle, that will unescapably tend to a shrinkage in short, intermediate, and long-range perspectives.

Tokyo and Delhi are two paradigmatic examples of how overlapping cycles in a megacity can determine completely different stages to happen at the same time: Tokyo has been steadily one of the biggest megacities since the half of the last century, but now Delhi will replace Tokyo shortly and become the largest meta/megacity in the world, although the influence of migration policies in the Japanese biggest urban system. As the following charts show, the demographic trends will influence significantly also urban drifting trends, tending to increase the demographic gap between urban and rural areas (Fig. 2).

The shrinkage of urban systems is a significant question for many Chinese cities today [46]. The driving factors determining urban shrinkage in the north-eastern provinces of China are several, according to multi-source geographical big data models, and include the decline of large industrial enterprises, neglected built areas, and irrational design planning strategies. A recent study (*ibidem*) identified 90 (18.1%) shrinking cities and 118 (23.7%) locally shrinking cities within the three provinces of north-east China. This occurrence is mostly affecting smaller cities near larger urban areas in the Yunnan province, due to the problem of high dependence on importing food and conventional energy. An important quota of housing stock in the area is vacant, as it is out of reach for most Chinese citizens, to which empty malls and other underused built-up areas in most urban systems in the Yunnan area should be added: this results into the exclusion of significant portions of the Chinese population from the economic progress promoted at national levels (Figs. 3 and 4).

The 2020 Philippine census revealed a decline in population growth in the National Capital Region, with Metro Manila experiencing a decrease from 1.58% in 2010 to 0.97% in 2020. Most cities experienced a decrease, except for Valenzuela, which experienced a 3.03% increase. Navotas City experienced the most significant decline, with a negative growth rate of −0.16% from 2015.

In a world of shrinking cities, the emerging Architecture from megacities to rural and urban systems in the selected case studies is related to Metropolitan Architecture Projects [18]. The goal of these Projects is to recognise interphase zones amongst metropolitan landscapes (urban areas, peri-urban zones, agricultural and natural areas) and design linkage patterns amongst them, to create enjoyable urban spaces by redefining urban morphotypes and introducing green-grey infrastructures. In the future scenario of the development of built-up areas, and physical and digital infrastructures, this approach will create a common temporal space between cities and countryside using technology to connect old and new places, networks, and landscapes. It will involve a pattern of linkage urban–rural–rurban–natural spaces with a shift towards sustainable and resilient megacities, which is crucial for environmental preservation and reconciliation with urbanisation.

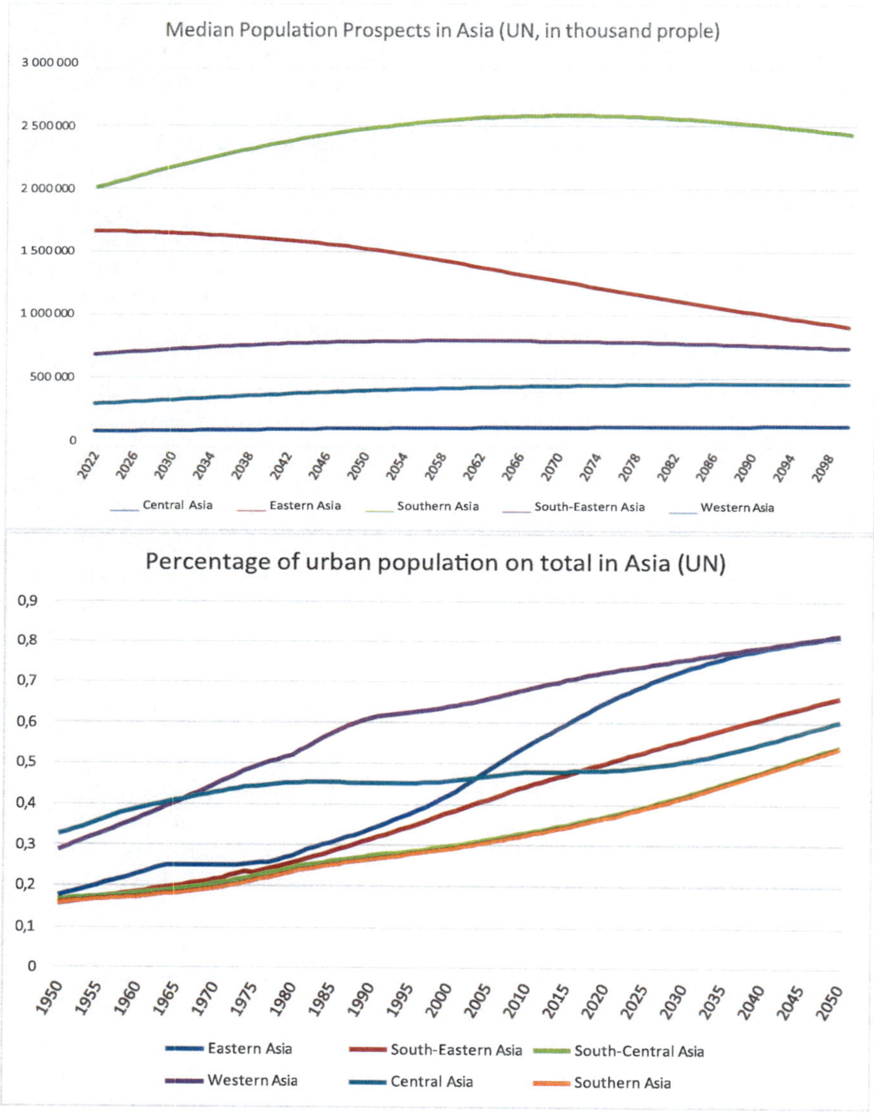

Fig. 2 Population trends and prospects in Asia (UN prospects)

2 An Innovative Proposal

The proposal tackles the societal problem identified in the Tokyo Valley, being the pioneering megacity in shrinking trends, addressing territory and social disparities, and promoting sustainable practices in response to the environmental crisis and lack

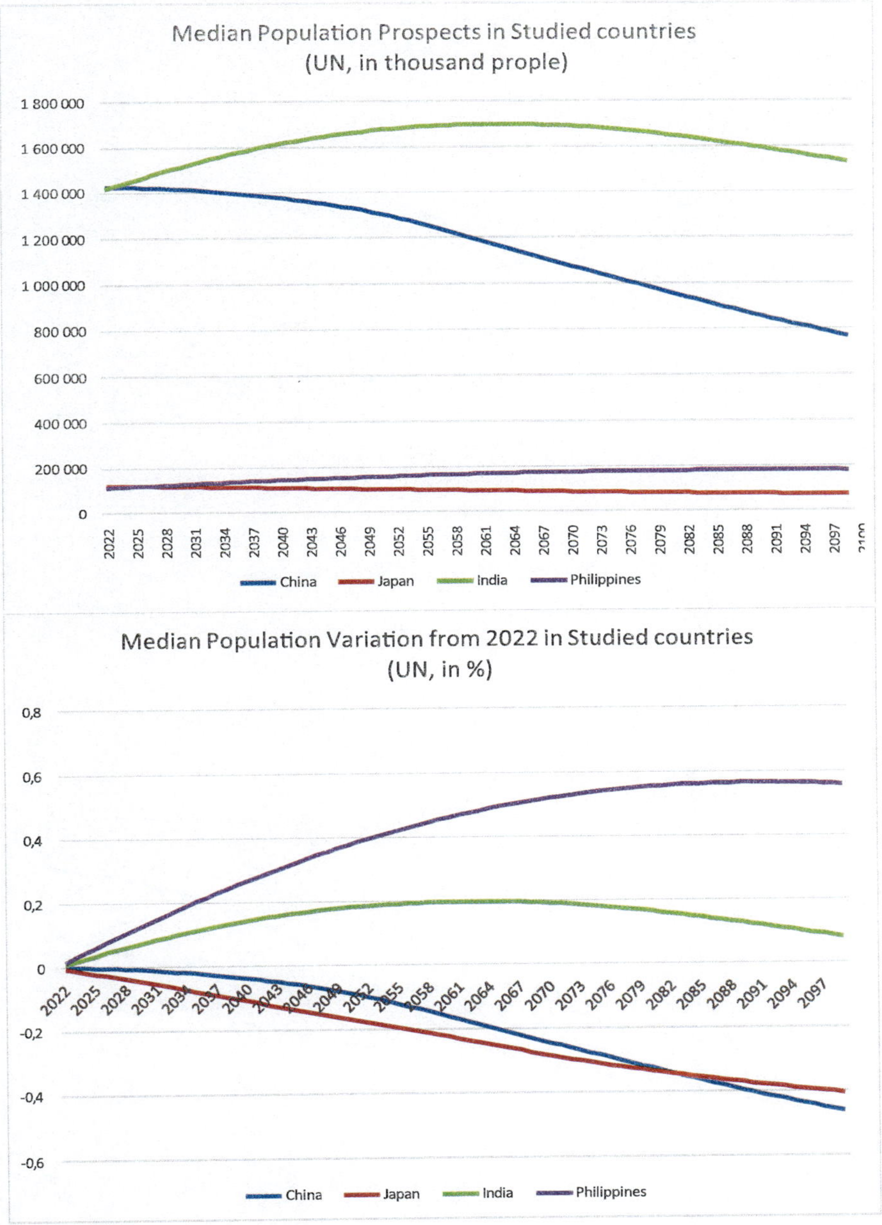

Fig. 3 Population prospects (in total, using 2022 as reference and as yearly variation) in China, Japan, India, and Philippines (UN prospects)

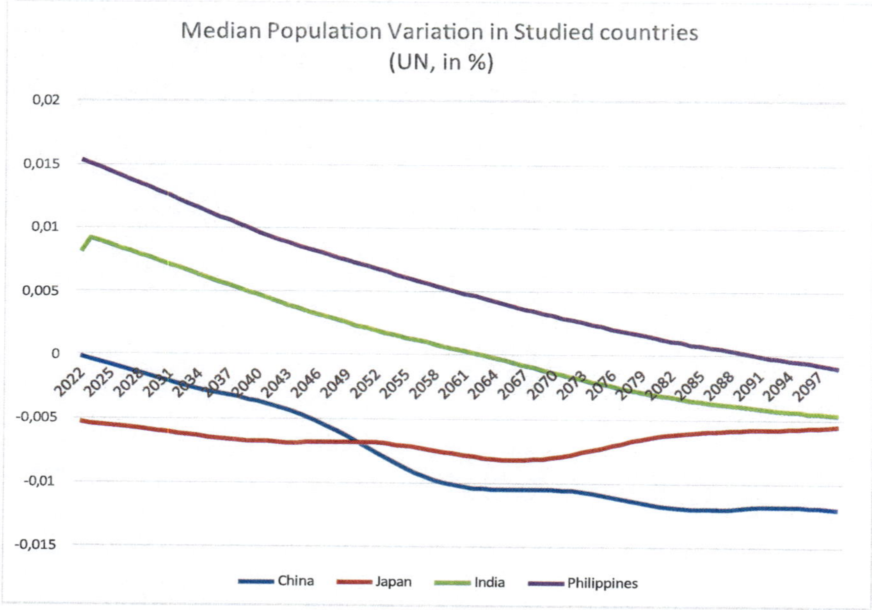

Fig. 3 (continued)

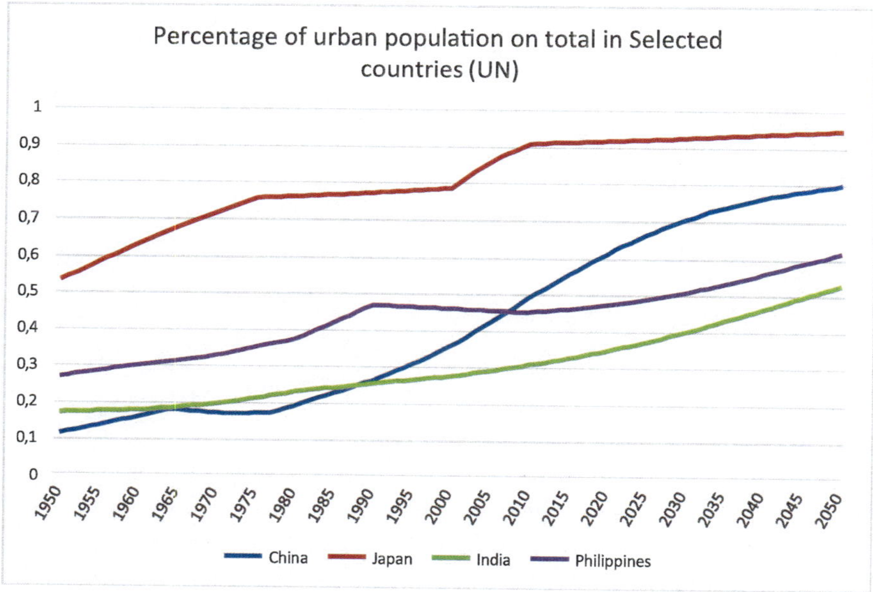

Fig. 4 Urban population trends and prospects in China, Japan, India, and Philippines (UN prospects)

of welfare by addressing the connection between urban and rural areas and the SDGs and NUA—the idea focuses on enhancing the sustainability of urban–rural linkages through green and digital transformations applying the twin transition sustainably.

On the other hand, specific objectives to reduce the risk (vulnerability and threats), produced by the catastrophic change of the local rural population's life paradigm due to the Twin Transitions, include identifying local-based-action stakeholders, co-constructing a shared field of knowledge, shaping a flexible pattern of solid information and expertise, and stimulating the ability to forecast risk scenarios leveraging data for place-based actions.

The specific goals for the Tokyo Valley are to:

1. Map the risk (vulnerability and threats) and inequalities linked to the territory and show the connections between inequalities affecting individuals and communities across the urban–rural continuum and the territories.
2. Showcase how implementing the green-grey infrastructure territorial intelligence at metropolitan and regional levels, using the lens of the digital transition, can also give rise to a complete green transition.
3. Develop new digital and physical infrastructure and techniques for agriculture and quality food production through more sustainable production practices, as well as regeneration, reforestation, and re-nurturing of nature in an interactive way with what life needs.
4. Develop tools for decision-making that show the sustainable linkage amongst metropolitan landscapes, water, agriculture, food, and transport to reduce adverse socioeconomic effects and promote corrected policies to reduce existing inequalities and prevent new ones.
5. Focus specifically on education and skill proficiency, including individuals not currently in education, employment, or training (NEET) with varying levels of income and labour.
6. Figure out a preventive approach that can respond strategically to the environmental emergency determining territorial vulnerability and increasing their inhabitants' instability and inequality.

All six objectives would point to the goal of reflection on biopolitics in such a way as to produce not only a technological transition but to give rise to a whole process of cultural and intellectual transition.

In a world, where global population seems growing continuously, the issue of population shrinkage is neglected. The world is undergoing a heterogeneous process of both population growth and decline, and the latter is becoming increasingly evident, universally. Especially alarming is the transformation of Asia that will begin to shrink in the latter half of this century. This will result in a shrinkage of its territory, often characterised by extended megalopolises.

3 A Regenerative Agriculture Strategic Vision for the Shrinkage: The Example of the Japanese Megalopolis. The Kanto Region

The world is shrinking due to global climate change and a decrease in population growth, which is likely to occur by the end of this century. This occurrence affects Europe, Latin America, and Asia, with Asia being the most impacted due to its large population. However, population shrinkage can have positive impacts, such as strengthening human rights and lessening environmental impacts. The consequences of population shrinkage on an unprecedented scale are unknown, but it is important to consider the fact that most of the population will be urban. Urban planning has mostly focused on growth, but there is a need to develop a shrinkage model for the megalopolises of Asia. Japan, anticipating a significant population decline, is a relevant case study. There needs to be more knowledge in studying the relationship between depopulation, the spatial issues of extended megalopolises in Asia, and the role of planning in addressing population decline. Keeping shrinkage untouched from a spatial point of view could lead to a sustainable thinning out of urban fabric with extended infrastructure. Conventional studies on urban shrinkage have mainly dealt with Western cities' post-industrial, post-socialist, and suburbanisation processes. There is a need to contextualise shrinkage as an autonomous event in the case of demographic change becoming the primary cause of population decline. The lack of knowledge also applies to Japan, one of the fastest shrinking countries in the world, but needs relevant experiences and valid assumptions for future planning (Fig. 5).

3.1 Shrinking the World. The Twenty-First Century Is a Turning Point for Humanity

Primarily, global climate change has become the definitive paradigm, where any human activity must be oriented towards optimising the situation. The impacts of climate change on human society are unprecedented, although the globe itself may have undergone severer transformations throughout its 4.5 billion years of existence. On the other hand, humanity has only been active on its surface for around 5 million years, and its population has grown constantly. The population and technology have grown exponentially since the industrial revolution, reaching 7 billion in 2011 and becoming increasingly impactful on the planet's environment. The serious effect of climate change on human society is a mirror of the unprecedented impact that human beings make on the planet. Today's population has always been the largest in the whole history.

However, it is most likely that the seemingly eternal growth trend will halt by the end of this century. Despite the explosive growth in certain regions, such as Africa, the global growth rate is declining yearly, anticipating a world of shrinkage in the following century. The twenty-first century already sees a parallel phenomenon of

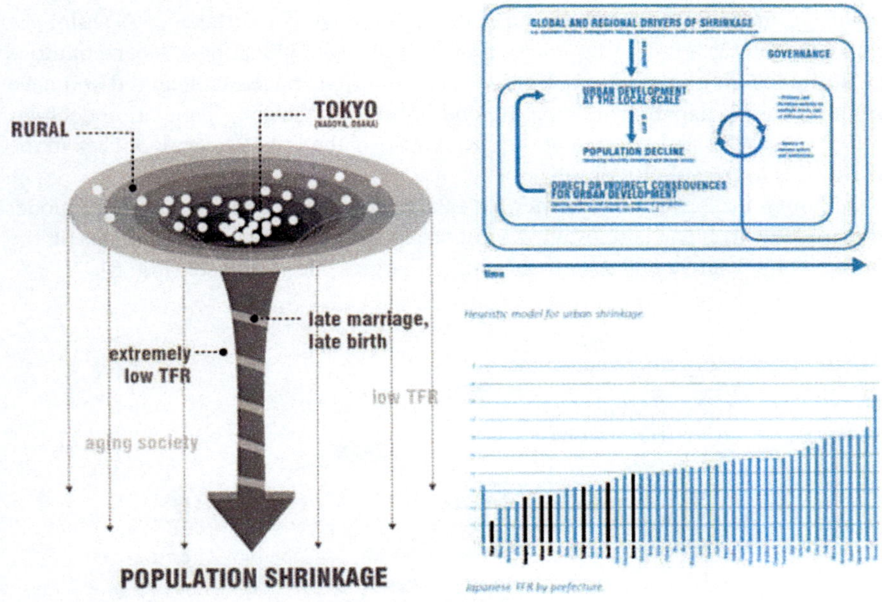

Fig. 5 Mechanism of Japanese population shrinkage (Credits: R. Koike)

growth and shrinkage, the latter becoming increasingly evident. It is the first time for humanity to face a population shrinkage so significant in scale, and so wide in context.

According to the United Nations, the regions that are likely to experience population decline by 2100 include Europe, Latin America, and Asia. Of these three, the most impactful would be that of Asia, today inhabited by 4.6 billion, with significant contributions from population giants, such as India and China. Demographic change in the scale of these countries could be disastrous if faced without prior knowledge and experience. On the other hand, shrinking and stabilising the population could "strengthen human rights while lowering fertility rates and lessening the impacts" [70] on the environment. Therefore, although the world must first recognise this pressing issue that is often neglected next to growth in Africa, it should be dealt with highly positively.

The consequences of population shrinkage on an unprecedented scale are unknown. Meanwhile, it is essential to consider that this population will be predominantly urban, and the phenomenon will affect not only rural villages, not even European-scale cities but extended megalopolises of the Asian territory.

The field of urban planning "has been almost exclusively focused on the process of growth" [65] at least for the past two centuries. Given the situation in Europe, the topic gained certain attention at the beginning of this century, focusing on localised Western versions of the event. Asian cities are known to have fundamentally different scales and structures from their Western counterparts, which implies that European shrinkage models would not be suitable for future Asian shrinkage. A shrinkage model for the Eastern megalopolises must be pursued. Whilst most Asian countries are still undergoing a process of growth, some nations have fast matured and have begun to shrink. Japan is anticipating one of the most striking population declines in the world. The fact that the dense country hosts the largest megalopolises in the world adds to the validity of the issue.

Although the backgrounds differ for each country, developing a visionary model for shrinkage in Japan is relevant, giving rise to further discussion and issues on the problem of a positive and sustainable way of large-scale shrinkage (Fig. 6).

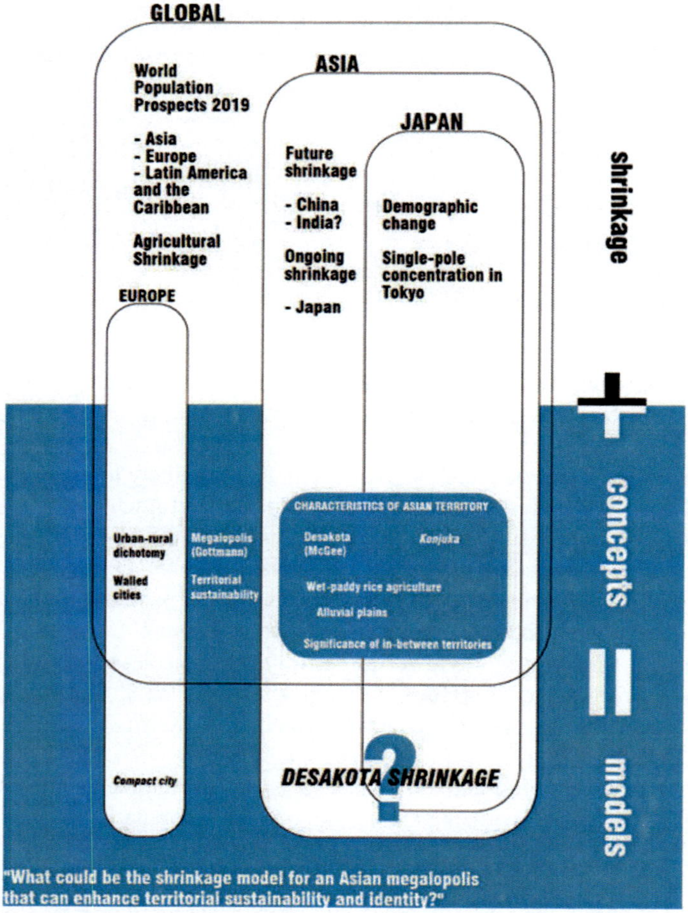

Fig. 6 Model for the territorial sustainability in Japanese population shrinkage (Credits: R. Koike)

3.2 Problem Statement and Knowledge Gap

In the global context, there needs to be more study on the relationship between the phenomenon of depopulation and the spatial issues of extended megalopolises in Asia. Furthermore, the effect of the decline of the population on large territories and the role of planning is yet to be explored. It is particularly problematic since the shrinkage issue lies "beyond the spheres of architecture and urban planning" [65], resulting from multifaceted processes and circumstances. However, it is most likely that leaving this 'unintentional phenomenon' [65] untouched from a spatial point of view will result in an unsustainable thinning out of urban fabric with extended infrastructure that requires constant maintenance and expense. This is especially true when dealing with a regional scale. Therefore, although physical planning cannot strictly regulate the transformation as in the phase of growth, there is a need for experimenting with large visions and strategies, which would rethink territories at megalopolis scale to match the downsized population.

Conventional studies on urban shrinkage have mainly dealt with Western cities' post-industrial, post-socialist, and suburbanisation processes [69]. These transformations have one thing in common, in a larger scope, they are all relocating the population, and they do not necessarily accompany shrinkage in the macro scale. Shrinkage has always been contextualised as a localised process within a larger growth trend.

However, when demographic change becomes the major cause of population decline in a certain country or global region, it no longer emerges as a side effect of growth, but a universal trend that must be treated as an autonomous event. Thus, coping with shrinkage is essential, setting the stage in Asia, which hosts numerous extended urbanised territories.

Lack of knowledge in large-scale shrinkage also applies to one of the fastest-shrinking countries in the world: Japan. Today, the governmental bodies are completely unaware of the issue, or in some instances, even assuming illusionary growth for future planning. This is highly due to the non-existence of relevant experiences and the need for valid tools. Drawing on the potential of large-scale visions and strategies for shrinkage, the problem must be set not only to manage population decline but simultaneously utilising the situation to reshape the territory to improve its sustainability, increasing urban and landscape quality.

Two problems must be stated from the perspective of sustainability. The first is the problem of the dilemma of population density. It is well known that there is overcrowding in certain cities, which lowers the quality of space. Specifically alarming is that extremely dense fabric raises the risk of natural disasters, such as earthquakes and floods, frequented in fragile countries. The population decline will hit all cities, including Tokyo, so that the density will decrease naturally. However, the decrease in population is much more evident in suburban areas than in high-risk/high-density districts. This physical phenomenon turns the urban fabric into a sponge-like structure [1]. Spontaneous emptying of plots within settlements reduces urban and landscape quality and security.

The second involves the issue of food self-sufficiency. Japan's rapid urbanisation process has long caused local depopulation in agriculture-based settlements. Although the situation is shifting towards a total population decline even in the cities, those affected primarily are agricultural regions. This scenario of agricultural shrinkage preceding urban shrinkage can lead to the collapse of the food supply system, considering that the country's self-sufficiency rate today is already below 40% [53].

The two issues are fundamentally identical in the sense that they both result from an oversimplified scenario of shrinkage that is to converge into the poles of the megalopolis, whilst the 'in-between' quietly suffer from disconnection, "having a hard time redefining their identities" [28] (Fig. 7).

3.3 Knowledge Gap

Considering the problem statement above, the knowledge gap addressed is primarily the lack of methodology to deal with large-scale shrinkage as that of Asian megalopolises. This particularly means that there is a piece of specific knowledge about the relationship between population decline and the management of urban fabric on the scale of a single city.

Still, current models are at most levels of the metropolis. A limited number of studies discuss the scope of extended territories concerning the potential benefits of shrinkage on sustainability, landscape identity, and quality (Fig. 8).

4 Questions and Goals for the Shrinkage Governance

4.1 Research Approach. Framing the Shrinkage

The research aims to generate a strategic vision for the shrinkage of the Japanese megalopolis, which combines the spatial management of population decline with the improvement of territorial sustainability and identity due to the twin transition, green, and digital. This shall foster the understanding of how large-scale shrinkage process can enhance regional landscape qualities when tackled in an inter-city, megalopolitan perspective, especially paying attention to the 'in-between' territories. In order to attain this objective, the country's territorial and demographic characteristics were analysed from multiple perspectives, giving particular importance to the Kanto region.

The fact that the Japanese islands are entirely urbanised and interconnected apart from the mountains enables a view that perceives the whole country as a unique megalopolis. Nevertheless, the significance of the Kanto region, containing the largest node in the network, namely Tokyo, led to the selection of the case study.

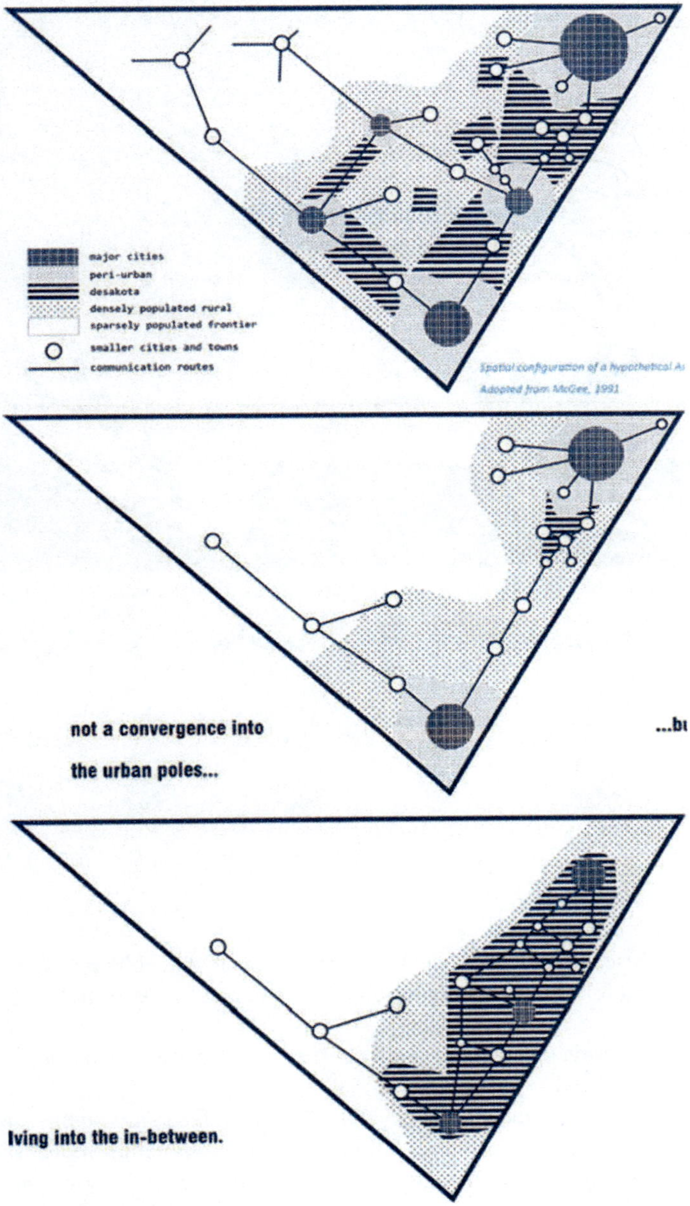

Fig. 7 Spatial configuration of a hypothetical Asian country from McGee, 1991 (Credits: R. Koike)

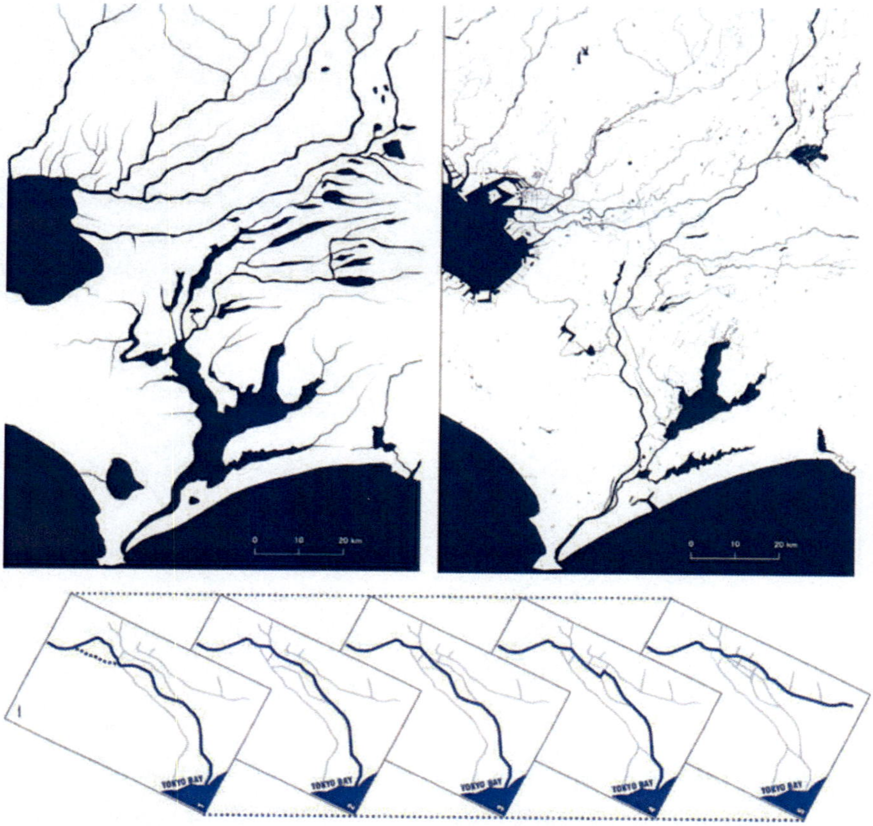

Fig. 8 Before and after, plus Phases of the Great shift of Tone River (Edo period) (Credits: R. Koike)

The main research question developed to achieve the objective concerns the shrinkage model for an Asian megalopolis that can enhance territorial sustainability and identity.

To aid in answering the main research question, three sub-research questions were established.

We want to understand the mechanism behind Japan's notable population shrinkage and how spatial planning within twin transition could impact this trend, trying to determine the distinguishing features of the 'in-between' regions of Japanese and Asian megalopolises, and particularly, the areas we identified and called *collisive site* [23].

Finally, we want to explore how agricultural regions experiencing faster shrinkage than cities can be preserved or improved.

4.2 World Population Prospects 2019 (UN)

The World Population Prospect 2019 (WPP2019) is the most recent population estimate and projections the United Nations (UN) issued. Transformations in population can only be argued by touching upon this global perspective. However, before further observing the data, it is necessary to identify the characteristics of these numbers.

Population estimates and projections are generally based on three significant factors: fertility, mortality, and international migration. Of the three, the first is particularly influential, working as the "key driver of the size and composition of the population" [2], whilst migration has no direct impact on global population change and is often a negligible demographic factor in most of the countries [36].

Changes in fertility rates can have direct effects on future populations. Commonly, it is known that the rate, defined as the number of live births per woman, tends to decline as society undergoes a process of modernisation [36]. This fertility transition can be observed universally and referred to as the demographic transition theory. The process is divided into three distinct phases. Phase I is defined as the stable pre-transition high-fertility phase, phase II as the transition phase, moving from high fertility to low levels that could drop below replacement fertility, and phase III concludes the process with post-transition low fertility, oscillating around the replacement level [2]. By 2019, the United Nations determined that all countries had entered phase II [84], meaning that the world has passed into a total fertility decline period, whilst many developed countries are already hitting the replacement level.

On the other side, the prediction that by 2015, 26 cities would have a population of 10 million or more highlighted the potential challenges of feeding such large urban populations and the need for large-scale food importation. That prediction was correct. In 2013, data for Urban Agriculture Facts were [75]:

- 50% of the world's population lives in cities,
- 800 million people are involved in urban agriculture worldwide and contribute to feeding urban residents,
- Low-income urban dwellers spend between 40 and 60% of their annual income on food,
- By 2015, about 26 cities worldwide are expected to have a population of 10 million or more. To feed a city of this size, at least 6000 Tonnes of food must be imported daily.

In 2021, 56.61% of the world's population will live in urban areas, a proportion expected to increase to 68% by 2050. The number of people involved in urban agriculture worldwide is difficult to estimate. In 2016, there were 10.3 million farms in the EU; together, they used 157 million hectares of land for agricultural production. This means that approximately two-fifths of the EU's land were farmed. The global rural population is expected to decline to 3.1 billion by 2050. Still, it is believed to be in the hundreds of millions and plays a significant role in feeding urban residents. Low-income urban dwellers typically spend a high percentage of their annual income on food, often between 30 and 80%. In 2021, households in the lowest income quintile spent 30.6 per cent of income in the United States (USDA Economic Research Service. U.S. Department of Agriculture).

Worldwide, according to WB, in 2022, people living in extreme poverty spend about two-thirds of their resources on food. The same figure for a person with a daily income of around $50 (a typical income in high-income countries) is closer to 25%. The most frequently grown category of cereals in the EU was common wheat and spelt, with 119 million tonnes of harvested production in 2020 (or 41.6% of total cereals output). By 2030, around 41 cities worldwide will have a population of 10 million or more. The amount of food required to feed a city of this size can vary widely depending on factors such as food waste, dietary patterns, and local food production. Still, it is estimated that a city of 10 million people would require at least several thousand metric tonnes of food per day, likely on the order of several tens of thousands per day (UN World Urbanization Prospects 2018).

The recent COP-27 climate conference in Sharm el-Sheikh, Egypt, ended without an agreement on limiting emissions. Rich countries are unwilling to fully commit to an energy crisis that may require reactivating dirty sources, and developing countries, facing capital flight and food inflation, cannot either. The advanced world's 'revolutionary' commitment to buy the developing world's compliance with 'climate reparations' is not a solution. It only further condemns these countries to underdevelopment by requiring them to do without fossil fuels when they cannot meet their energy needs with renewables alone.

The 90% of the global demographic growth in the next 30 years will happen in developing countries, where the urban population escalation will determine increasing issues coming from informal settlements and inequality [11], considering that these communities are not historically ready to manage quick development. The urbanisation trend is expected to continue in low-income countries up to 2050, but recent UN predictions pointed out that we are at the turning point in the balance between urban and rural population in many continents, including Asia (*ibidem*). Any upgrade in health conditions of the built environment and reducing the related risks could contribute to the achievement of the Sustainable Development Goals for Health (SDG 3) and Sustainable Cities (SDG 11).

This thoughtlessness also has a boomerang effect: increasing numbers of 'climate migrants' from countries exposed to extreme climate events. Can a twin transition sustained by extensive local knowledge (including physical, social, economic, and governance dimensions) overcome this division? (Fig. 9).

4.3 The Metropolis of Second Modernity

When discussing today's metropolis, we should not look at the 'Großstadt' of the last century but instead refer to the post-World War II Metropolis, where everything changed. The cities were destroyed during the war, resulting in the development of the capital of second modernity in the 1950s. The realisation of the destructive power of science and technology led to a shift in thinking from a destructive tool of

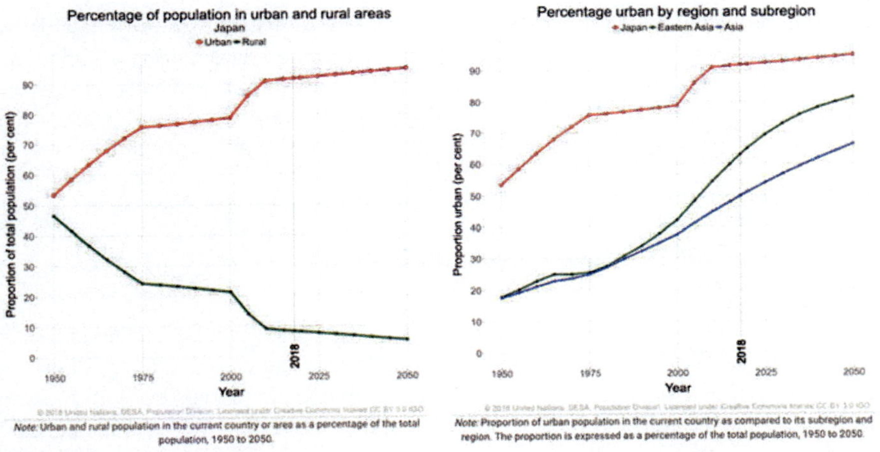

Fig. 9 Urban–rural population balance in Japan compared to region and subregion. (*Source* https://population.un.org/wup/Country-Profiles/)

violence to a generative, artistic force. For example, in Milan, the new metropolis was understood as a space for artistic expression and not destruction. This perspective can help us better understand the soul of the new metropolis in the face of extreme events. Despite the tragedy, the residents of the Second Modernity Metropolis felt optimistic and saw their ruin as a reflection of a universal fact.

Mediterranean metropolitan cities may be the new 'polis' of the twenty-first century, where the same laws govern all citizens, and the city reflects universal law. This creates a harmonious relationship between the city and its citizens, leading to organic integration and a sense of fulfilment in contributing to the common good. The city's public space is a platform for political expression and participation.

"Towards 2050, the combination of the two transitions will depend on the ability to deploy existing and new technologies on a large scale and various geopolitical, social, economic, and regulatory factors. To strengthen the synergies between the two transitions and resolve tensions and recognizing the inherently geopolitical nature of the dual transition, a global, future-oriented, and strategic approach is needed". As said by 2022 Strategic Forecast, the world is experiencing significant geopolitical changes that strengthen existing metropolis trends. The long-term implications of Russia's military aggression towards Ukraine, including energy, food, economy, security, defence, and geopolitics, will impact World's path towards achieving green and digital transitions. However, these and other future challenges will remain within the World's long-term goals, and with the right policies, they can serve as catalysts for accelerating their realisation. Ultimately, this situation could enhance the World's resilience and strategic autonomy in various sectors.

Our proposal deals with agriculture, food security, and cutting-edge technologies. In this new geopolitical context and as a proper exercise in forecasting, the proposal takes on the challenge of a forward-looking strategic reflection on the interactions between the green and digital transitions. The green transition will only happen by achieving climate neutrality and reducing environmental degradation by 2050 with strong attention to sustainability to reduce adverse side effects and fully exploit the potential of the digital transition as an enabling force for ecological, social, and economic sustainability, appropriate policies and governance are needed.

4.4 SDGs and UN-Habitat Strategic Plan

That is why the SDGs target and the UN-Habitat Strategic Plan are our fundamental background to achieve the objective of advancing sustainable urbanization following the four interlinked domains of change.[1] The digital and green transitions allow us to transform how we utilise our territories. Successfully coupling the green and digital transition supports the development of a new, regenerative, climate-neutral economy by reducing pollution and restoring biodiversity and natural capital through sustainable digital and other technologies, maximising synergies and dissolving tensions between the green and digital transition' [41] through a dynamic approach that anticipates change and adapts policy responses whilst remaining committed to long-term goals is needed.

Our pioneer idea is to rethink, plan, and rebuild our way of life, considering the country's intellectual capital [66]. David et al. [24] emphasise that knowledge economies are driven by accelerating knowledge production through the increasing importance of science-related activities and technology.

However, our proposal is based on the concept of optimisation, which differs from an approach that prioritises maximisation. Our proposal explores the potential of coupling the green and digital transitions to reinforce each other, minimise adverse effects, and fully harness the potential of the digital transition as a catalyst for environmental, social, and economic sustainability.

According to Giordano [13], values are crucial in decision-making and evaluation processes. They may be implicit, stem from an individual's worldview and epistemology, and involve knowledge construction. A dominant scientific and utilitarian approach to knowledge tends to separate it from values.

In urban development, values are expressed through transforming territory, its form, and the city and the environment's relationship. The concept of monotonic values, introduced by Gregory Bateson, refers to values that either only increase or decrease, with no break in the curve. This concept does not apply to all values, as biological values have an optimal range, and excess can be toxic.

The dominant worldview focuses on maximising values such as profit and consumption, leading to serialised and compartmentalised knowledge. On the other

---

[1] https://unhabitat.org/sites/default/files/documents/2019-09/strategic_plan_2020-2023.pdf.

hand, a sustainable approach to territorial development seeks balance and optimisation of various values. This approach is transdisciplinary and open to dialogue with other disciplines, technical knowledge, and practice.

4.5 *The Metropolis for the Second Modernity Dimension is the Proposal Field of Action*

The twin transition occurs in an urbanising era characterised by a metropolitan transformation. According to UN-Habitat [82], as the world continues to urbanise, it is also becoming metropolitan. Cities are growing beyond the boundaries of their central municipality or 'city proper', configuring bigger and denser metropolises. This trend is still present in all world regions and affects all kinds and sizes of cities, from megacities to intermediate cities and even small towns. However, most of the world's metropolises are not managed in a differentiated way, meaning they do not have a metropolitan plan or institution, configuring a global metropolitan management gap. This global metropolitan management gap is affecting one-third of humanity (2,6 billion people) already living in almost 2,000 metropolitan territories worldwide. Additionally, almost 1 billion people will become new metropolitan inhabitants during the next decade, and there will be 429 new metropolises by 2035.

According to Forero and Sietchiping [27], important lessons from the pandemic and other global crises must be considered transiting towards sustainable development during the Decade of Action, including (i) Cooperation. Collective action must be at the centre: neither country nor city can tackle global crises independently, regardless of political and socioeconomic context. Strengthening existing cooperative systems, both globally and locally, has never been more urgent; (ii) Metropolises. Some approaches to urbanisation could need to be updated: global transitions advance without recognising political-administrative borders and local jurisdictions. The negative effects of climate change, socioeconomic inequities, and the digital divide do the same. Global crises highlighted metropolitan governance as a necessary approach to advance towards sustainable development. Municipal and national management disconnected from its surroundings, perhaps useful in the last century, need to be outdated; and (iii) SDG Territorialization. Subnational levels are key for advancing sustainability: Metropolises are becoming the twenty-first century's most representative type of city to the extent that they concentrate most of the global urban population—one-third of humanity.

In addition, the quality of life of the surrounding rural populations also depends on the effectiveness of the local, metropolitan, and regional governments that intervene in managing these urban agglomerations. Regional and metropolitan authorities must articulate their territorial plans, projects, and actions with the targets defined in the SDGs and periodically report to the international community the results and progress achieved.

4.6 *The Digital and Green Transitions Allow Us to Increase Awareness and Knowledge of Our territory's Challenges*

We propose that to address inequality and inequity, it is necessary to consider the relationship between them rather than treating them as separate issues. The proposal focuses on increasing individuals' access to knowledge to improve overall well-being and societal outcomes produced by the twin transition actions. Our proposal aims to provide guidelines to metropolitan agents based on a preventive approach that can provide strategic responses to the environmental emergency that determines territorial vulnerability, increasing inhabitants instability and inequality.

By offering tools for understanding and methods for structured thinking that can lead to sustainable decisions, the proposal hopes to mitigate or reduce the risk of adverse effects due to the increase in destabilising inequalities resulting from applying the twin transition. The proposition considers the metropolitan region dimension as a laboratory for territories, landscapes, infrastructural network transformation, and governance. It analyses urban sensitivities (ESPON 2013 Program) generally due to land consumption, non-renewable resources, and climate change, which according to our hypothesis, may also be increased by misuse of new technologies related to green and digital transitions.

4.6.1 The Digital and Green Transitions Concern a Radical Revision of the Disciplinary Framework. Introduction to the Metropolitan Approach to Complexity

For that, the proposal's theoretical question concerns a radical revision of the disciplinary framework to understand how the exemplary model we extract from the past is transformed by the impact of new technologies related to twin transition that marks a profound discontinuity with past practices and techniques. This is the new research inscribed in the Metropolitan Discipline framework [16]. Contributing to this change are the issues of today introduced by the NUA and the SDGs not so much to be interpreted as indicators of project performance but as targets for the impact of metropolitan projects inspired by the Twin Transition on territories increasing vulnerability because of their:

- Exposure
- Sensitivity
- Capability to react.

The metropolitan approach to complexity involves connecting macro and micro scales, local intelligence, and global dynamics in metropolitan territories.

We need to identify local, common-sense languages and operations on the ground, extract relevant elements to be taken as invariant data to the metropolitan and national level, and then use this information to apply the twin transition actions creating a regenerative project that returns increased values, to the local level.

We have identified the so-called green-grey infrastructure as a critical asset for the sustainability of metropolitan territories. Our proposal also aims to show how implementing local, territorial intelligence (7 EU Framework) in the context of the digital transition can lead to a comprehensive green transition. This requires a context-based approach to qualify new land-use patterns in metropolitan urban–rural scenarios.

We will present tools and policies for sustainable linkage amongst metropolitan landscapes, water, agriculture, food, and transport to reduce adverse socioeconomic effects and promote corrected policies to reduce existing inequalities. These integrated sectors are crucial for dealing with urban–rural linkages in the NUA.

We will break down SDGs and their targets.

4.7 The Quiet Process of Shrinkage

However, a sharp decline in the fertility rate does not necessarily result in an immediate drop in population. According to the WPP 2019, the world's total fertility rate (TFR) entered phase II of the demographic transition in the 1970s. It will likely hit the replacement level of 2.1 live births per woman by the end of the twenty-first century. This causes the annual population growth rate to slow down, gradually approaching zero growth towards 2100 when the world population will peak at around 11 billion. The continuing slow growth despite the significant drop in TFR is mainly due to the population momentum, which is "a force that drives future population growth resulting from the existing age structure" [36]. If the demographic transition theory stands on this scale, the next phase would be the process of an overall decline in population, which would later stabilise in a lower balance.

Even though the projections become increasingly uncertain as it forecasts the distant future, the general trend of termination of growth is evident. Nevertheless, there is still time to speak of population shrinkage on such a scale. Indeed, even the UN needs to address the issue of shrinkage whilst being significantly alert to the population explosion in some areas of the planet. However, as each country and region is at a different transition phase between II and III during this century, it is only partially correct to look solely at the total effect and the eye-catching but restricted growth event.

A heterogeneous population transition by regions characterises the twenty-first century. Below, the population transition of the six geographic regions defined by the UN (Africa, Asia, Europe, Latin America, and the Caribbean, North America, and Oceania) is examined.

First, three geographic regions are projected to grow continually throughout the century: Africa, North America, and Oceania, whilst the other three regions, Asia, Europe, Latin America, and the Caribbean, are likely to turn to shrinkage. Within the growth group, the number of Africa is highly impactful, starting from 1.3 billion in 2020 and reaching 4.3 billion in 2100, amounting to an increase of about 3 billion. This number almost coincides with the global growth of 2020–2100, projected to shift from 7.7 billion to 11 billion. Therefore, it is most appropriate to say that the world's population growth will occur almost entirely in Africa. Nevertheless, the other two regions are also expected to grow steadily, and shrinkage does not apply to these regions.

Meanwhile, the regions of shrinkage shall be observed with more attention. Unquestionably, the Asian shrinkage process will be most problematic due to its scale. After a slowing growth in the first half of the century, the population giant is expected to reach its peak in 2055, turning to a decline in the latter half. By 2100, the number is projected to decrease from its peak by 11% at the medium variant, which would amount to some 580 million. Although smaller in scale, Latin America and the Caribbean will follow a similar path, reaching their peak in 2060 and anticipating a 12% (90 million) decrease in the second half of the century. Europe is already in its shrinkage phase, peaking in 2020 with a 16% (120 million) projected shrinkage by 2100.

The unprecedented shrinkage in Asia is of particular concern, especially that of today's two most populated countries can be disastrous. China, currently the most populous of all, is expected to peak in 2030, reducing its number by some 400 million or 27% from its peak by the end of the century. Similarly, India's population is projected to decrease by 12% (190 million) from its peak in 2060. However, the projection interval for India is extensive, making future transitions uncertain. Nevertheless, these numbers are enough to prove that the current world is undergoing an extremely diverse process, where shrinkage is becoming increasingly evident in much of the global regions.

Therefore, the UN's broadly prevailed attitude (or propaganda?) may be misleading in some respects. The modified explanation should point out the parallel existence of population explosion and shrinkage, where the former occurs only in limited regions and is globally spread out. However, the effect of the striking growth in several countries would still lead to a significant overall increase in population and food demand, thus resulting in a severe necessity for increased agricultural roles in shrinking regions of the world. Therefore, management of enormous shrinking regions is key already at this moment (Fig. 10).

4.7.1 Theories on Shrinkage

Population and urban shrinkage have been mostly argued in Western contexts, especially in Germany and the United States. However, the discussion is relatively new in the literature of modern planning and is an expanding one.

The internationally pioneering research in the field came from Germany, named 'Shrinking Cities' [65], which concluded with multiple collective publications. This joint research has been crucial for the field for its consolidation of the term 'shrinkage' as the most accepted name for the phenomenon.

The study identified five major causes for urban shrinkage: deindustrialisation, peripheralisation, suburbanisation, post-socialism, and demographic ageing. Each issue was related to real-life shrinkage phenomenon in Manchester/Liverpool, Britain; Halle/Leipzig, Germany; Detroit, USA; Ivanovo, Russia; Japan. Diverse

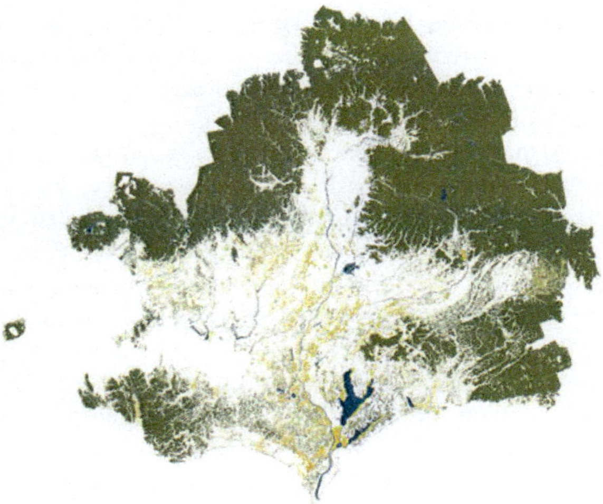

Fig. 10 Tokyo - Tone River - Kanto plain (Credit: R. Koike)

initiatives from a broad context were recorded and explored, although fundamentally Western, apart from Japan.

Japan has been an exceptional case from the East, being the "fastest-ageing and thereby also the fastest-shrinking society in the world" [65] already at the time. However, the cases explored in the country were still largely rural, of which events completely differ from the situation today and the context of this thesis, which concerns the shrinkage of primary cities in the country, including Tokyo.

Another set of studies relevant to this investigation includes the research of Haase, especially the "heuristic model for urban shrinkage" developed to build a common framework in the extremely heterogeneous literature of urban shrinkage (and differently named studies; [37]). The model was thought to provide a template for the studies of diverse shrinkage phenomena, into which one can embed place-specific and time-specific explanations. This would foster mutual communication of different research, strengthening shrinkage studies as a process [37]. The model was designed to have three major steps, including "drivers of shrinkage, impacts, and responses" [37]. Although this framework was developed for urban shrinkage issues, its use will be expanded to the level of territorial shrinkage.

4.8 The Metropolitan Green and Digital Culture Construction

Our approach focuses on targeted interventions where social and economic inequalities and vulnerabilities are concentrated to revitalise these areas and reduce negative social and economic divide due to the welfare lacking, which is the real reason for the countryside abandonment.

We examine the role of twin transitions (green-grey infrastructure, such as water systems, agriculture, and the two tangible and intangible infrastructures, vehicles and the internet), which create places of settlement concentration in addressing inequalities and vulnerabilities.

Also, we evaluate existing policies and propose new ones to promote equality. Our focus on the Indo-Pacific region is driven by the fact that the region is home to a wealth of local knowledge and territorial intelligence, which is at risk of being marginalised by technology and digitalisation. Furthermore, the region is a source of migration to metropolitan areas and addressing inequalities and vulnerabilities in the region is crucial for the well-being of future metropolitan inhabitants.

The challenge of creating a metropolitan culture arises when a local area transforms into a metropolis, attracting diverse communities and requiring a higher level of technical competence for the management and definition of urban production patterns and behaviours. One aspect of this is the awareness of the diachronic temporalisation of the history of the metropolitan city.

To address this challenge, it is essential to study the territory's structure during the time, nowadays dynamic processes, and future time-based programming to understand which knowledge and skills of the past cannot be erased and which are the new competencies to share with the new metropolitan populations. This can help understand new metropolitan settlement arrangements and structures and design an inclusive metropolis that avoids inequality and inequity.

Key players in this transformation are Metropolitan Agents, who must be able to make decisions based on an evidence-based observation of the territory. This observation and decision-making must be done through the virtual integration of sustainability principles and lived experiences in the field. To achieve this, a vision that is not purely functionalist but is based on the knowledge and values of sustainability typical of each context is necessary.

4.8.1 The Green and Digital Transitions and the Indo-Pacific Metropolitan Identity Construction

We hypothesise that the digital transition may increase the risk of separation and individualisation of society and detachment between human energies and knowledge that risks leading to a devaluation of so-called local or 'low' popular rural culture in the face of a hyper-technological, global, and 'high' one. We examine the significant changes happening in today's peri-urban rural areas, specifically in agriculture and food production, transforming the physical and social structures, networks, landscapes, and lifestyles in these regions.

We propose studying the history of transformations in Indo-Pacific metropolitan areas to understand better how digital and green transitions can be managed in a resilient way. By analysing past changes in the physical and social landscapes, we can identify elements that have contributed to the unique character of a territory and ensure that current transformations do not erase this character.

By cognising past changes, we can make more informed decisions to prevent the loss of local meaning in global transformations and maintain the recognisability of places for their inhabitants. A sustainable territory must balance continuity with cultural heritage and necessary change [14].

4.9 The Methodology Concerns

Our approach focuses on deliberate action and planning, stemming from a desire to challenge mainstream thinking on governance and planning through research. It prioritises care for the territory and its inhabitants, using critical knowledge of reality to explore new possibilities. Our approach encompasses the analysis of the territory through logos and discourse, as well as giving voice to its inhabitants. It is a minor perspective that considers the major narratives shaping the territory.

Our approach challenges a hyper-technological culture, embracing technology whilst promoting the importance of life, the common good, and well-being and considering nature as a part of the city. We are not tech enthusiasts, but we do utilise technology. Our 'Utopia of Reality' project challenges prevailing frameworks and offers hope for transformation by bypassing mainstream norms. It questions European privilege and history, exploring what it means to inhabit, cultivate relationships, and prioritise ecology over just producing agriculture. Our utopia of reality is a feasible project for now and future generations.

We aim to significantly advance the state-of-the-art in the science domain by tackling the complexities of metropolitan scale in a holistic and integrative manner. We propose to take an innovative approach by viewing the metropolis not only as a place where inequalities and vulnerabilities manifest themselves but also as a location where solutions and mitigation measures can be found. That is possible at the intersection of metropolitan structure and economic criticism. It aims to achieve significant improvements in reducing class inequalities, as reflected in a reduction of wealth inequality, a decrease in income inequalities, and a substantial reduction of territorial disparities. Additionally, we focus on the growing role of intangible components such as knowledge, local culture, and labour in the metropolitan construction process.

The proposal considers the Tokyo region's importance in the Indo-Pacific transition. We also aim to take a unified and operational approach to study and manage the HE disciplines' dynamic change processes. We are paying attention to co-design projects and policies to encourage the reactivation of local economies and urban manufacturing through bottom-up training and informal learning.

We use a metropolitan digital toolkit to achieve these goals, enabling practitioners and public officers to make evidence-based decisions, such as a cartography open-source data tool to evaluate sustainable development priorities and impacts. Our proposal is not an incremental improvement but aims to achieve a step-change in reducing subalternities linked to the four inequalities [4] as a mission through recommendations that offer a solution by considering how the metropolitan language can allow a metropolitan culture and structure through regarding its inter-scale co-construction.

In the context of achieving the SDGs in the Indo-Pacific region, policy choices such as creating an Indo-Pacific Fund for Strategic Investments, such as in Europe, can help channel private investments towards sustainable investments and public spending. Regulation (EU) 2015/1017 of the European Parliament and the Council sets a 40% climate investment target for infrastructure and innovation projects under the European Strategic Investment Fund. Establishing shared criteria for the Indo-Pacific region determining the sustainability of economic activities, including their environmental impact, could be a foundation for similar initiatives that the Region will undertake to mobilise investments to achieve climate-related or other environmental objectives.

Our methodology focuses on two types of inequalities and vulnerabilities that are interconnected but distinct. We use the terms inequality and vulnerability in the dual sense of socioeconomic inequalities affecting the people living in the territory and spatial inequalities affecting the territory itself. We aim to study the connection between these two types of inequalities and propose innovative solutions. To address the issue of inequality in four dimensions—physical, social, economic, and governance—the proposal focuses on two sectors of transition projects: industrial processes and product use and agriculture, forestry, and other land use.

Inequalities and vulnerabilities can be mitigated by implementing regenerative agriculture on a large scale, utilising the local knowledge and networking capabilities offered by digital technologies exploring the capacity of regenerating the territory expressed by an agricultural technique based on an idea of intentional community, of the collectivity of choice. Agriculture for the metropolitan city becomes necessary; therefore, the political objective of the twin transition is to operate agriculture and the food industry to be healthy, producing quality food on fertile land.

We have chosen to focus on the Indo-Pacific region because it is particularly affected by the loss of traditional skills, culture, and local expertise due to global digitalisation and technology that can allow us to compare that issue with the European one [3]. Moreover, the Indo-Pacific region shares some geographic, natural, and cultural assets with Europe. Still, socioeconomic inequalities were charged in the Indo-Pacific Region and Europe due to the consequences of political instability, contested territories, previous transitions, such as the political, colonial, and industrial ones, and how the twin transition could increase these inequalities in the governance not improved.

The Indo-Pacific culture and agricultural knowledge must be preserved and integrated with new technologies to create a sustainable future for both urban and rural areas. This approach also addresses the issues faced by the North African region of the Mediterranean and the Balkan areas of the Adriatic Sea, where many migrants come from and will become future urban inhabitants.

Greater collaboration between neighbouring local and regional governments must be achieved, not overcoming these inequalities affecting Indo-Pacific region due to the increasing migration dynamics.

Green and digital diplomacy and awareness-raising should be strengthened, harnessing the power of regulation and standardisation and promoting Indo-Pacific values. Strategic partnerships of mutual benefit should be pursued, particularly with neighbouring countries achieving sustainable and reliable connections for people and the planet by helping address the most pressing global challenges, from combating climate change to improving health systems and strengthening the competitiveness and security of global supply chains.[2] It promotes intelligent, clean, and secure investments in quality infrastructure and the sustainable connection of goods, people, and services worldwide.

We focus on the areas of cities affected by inequality and vulnerability, known as *collisive sites* [23]. These spaces, typically neglected and shapeless, are created by the effects of urbanisation. Through collaborative, nature-based solutions involving multiple local and regional governments, we will investigate how these issues can be addressed by improving the green-grey infrastructure within the twin transition solicitation and improving communities' quality of life, water systems, agriculture, energy, and transportation.

Our proposal aims to create a balance between the needs of urban and rural areas by fostering a partnership between the wisdom and culture of the agricultural world and the benefits of new technologies. Our policy focuses on preserving traditional skills and local expertise in the Indo-Pacific region and utilising twin transition as a sensitive and effective change driver.

Otherwise, twin transition, such as the implementation of new infrastructure and techniques for agriculture and food production through more sustainable practices and reforestation and re-nurturing of nature, can increase the risk of vulnerabilities and threats related to:

- The growth of an elite with specialised knowledge, that can shape a society that does not grow as a whole, despite the goals of twin transitions. Our methodology uses state-of-the-art open-source data and the Metropolitan Cartography Methodology to overcome this to encourage widespread participation and interaction across many domains and scales.
- The increase of territorial sensitivity due to explosive growth and climate change, amongst other factors. Our approach aims to improve prosperity, community, and quality of life whilst balancing efficiency, equity, and civil economy avoiding the introduction of twin transition without considering territorial priorities. We aim to identify the agents who can drive knowledge-based development towards the drivers of change and determine the transformations needed to tackle the major and most urgent challenges for society and the natural and built environment.

We face the loss of the symbolic value of public spaces emphasising the importance of designing cities considering the natural environment, protecting, and preserving the heritage and history of the area, and involving communities, experts, and different groups of people in the planning process.

---

[2] https://commission.europa.eu/strategy-and-policy/priorities-2019-2024/stronger-europe-world/global-gateway_it.

The risk of the loss of cultural identity when cities go through twin transitions emphasises the importance of protecting and celebrating the diversity of cultures and identifying local elements that must be preserved for future metropolitan projects considering knowledge a result of continuous feedback between inhabitants and decision-makers, rather than its automatic linear transmission. To overcome this, our methodology proposes a new cartographic methodology using open data to locate and monitor the chosen thematic strategy and to create a new telescopic space–time network. Overall, our approach involves a more informed understanding of risks in an era-net society and the use of place-based technologies and collective decision-making approaches. Twin transition, if applied sensitively, can also be the necessary driver of change to implement local population skills and abilities and to manage land resources in the Indo-Pacific region (Fig. 11).

The problem with the disconnect between urban and rural areas is that the well-being of cities needs to align with that of rural areas. We aim to bridge this gap by adopting the twin transition, utilising the wisdom and culture of the agricultural world, particularly in the Indo-Pacific region, in conjunction with new technologies. Our policy focuses on preventing the loss of traditional skills, culture, and local knowledge to residents and migrants in the region. We can drive change by sensitively applying these transitions by incorporating these skills and knowledge. Policies and rules-setting will be essential for matching and ensuring the Indo-Pacific's advantage in competitive sustainability.

Metropolitan governance, diplomacy, and policy advice have the potential to reshape urban areas to promote greater equality through green and digital transitions. Rules and policies are important in enabling the coupling of transitions. They can support the development of management systems or interoperability solutions needed for the twin transition (Fig. 12).

5 The Metropolitan Cartography Methodology to Implement Open Science Practices

The proposed approach uses state-of-the-art open-source data and the Metropolitan Cartography Methodology to address potential vulnerabilities and threats from implementing the twin transition. This encourages widespread participation and interaction across many domains and scales. The goal is to improve prosperity, community, and quality of life whilst balancing efficiency, equity, and civil economy. To avoid negative impacts, we aim to identify key agents and determine necessary transformations to tackle major societal and environmental challenges.

Our approach also emphasises the importance of designing cities considering the natural environment, preserving heritage and history, and involving communities in planning. Additionally, our methodology aims to protect and celebrate cultural diversity. It uses a new cartographic methodology with open data to promote continuous feedback between inhabitants and decision-makers. Overall, our approach involves

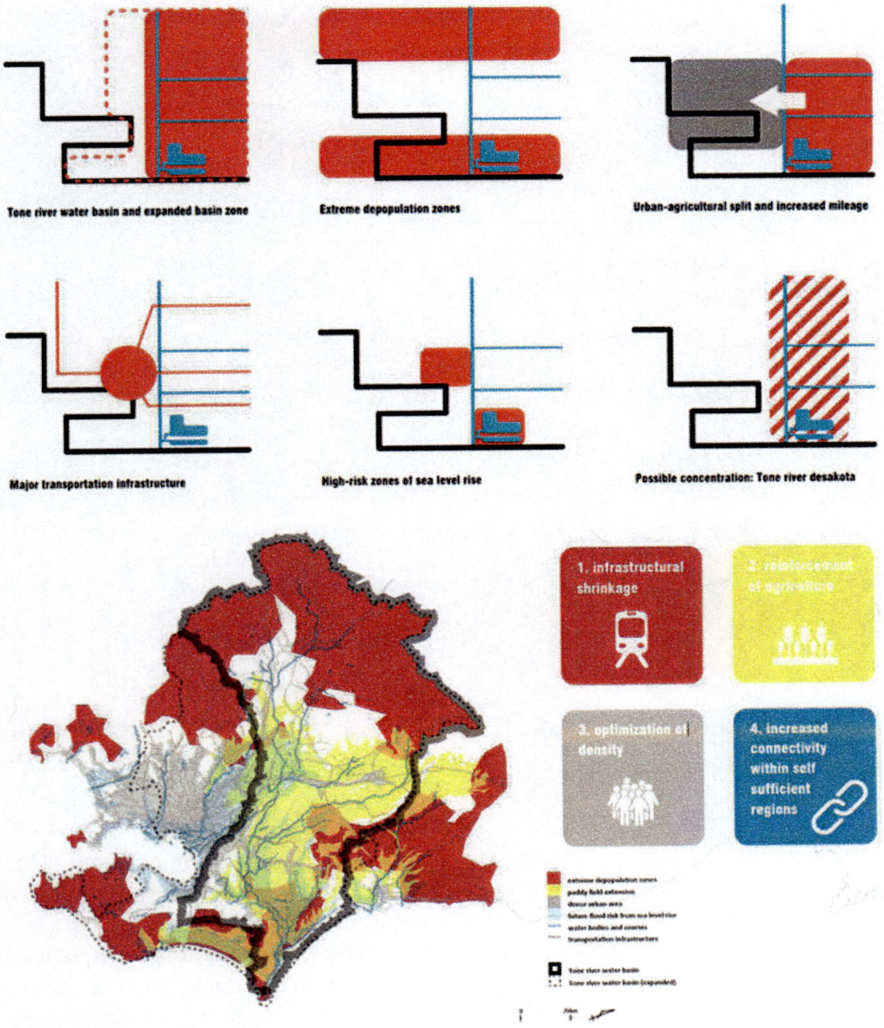

Fig. 11 Tone River Desakota as hinge amongst 2 megalopolises (Credits: R. Koike)

a more informed understanding of risks and the use of place-based technologies and collective decision-making in the Indo-Pacific region.

The proposed methodology includes Case Studies. Selection of metropolitan cities or cities affected by metropolitan dynamics in the Indo-Pacific region, each serving as a paradigm for applying twin transition in a specific productive and socioeconomic context related to the structure of metropolitan areas and a specific field of expertise, investigation, and application [14].

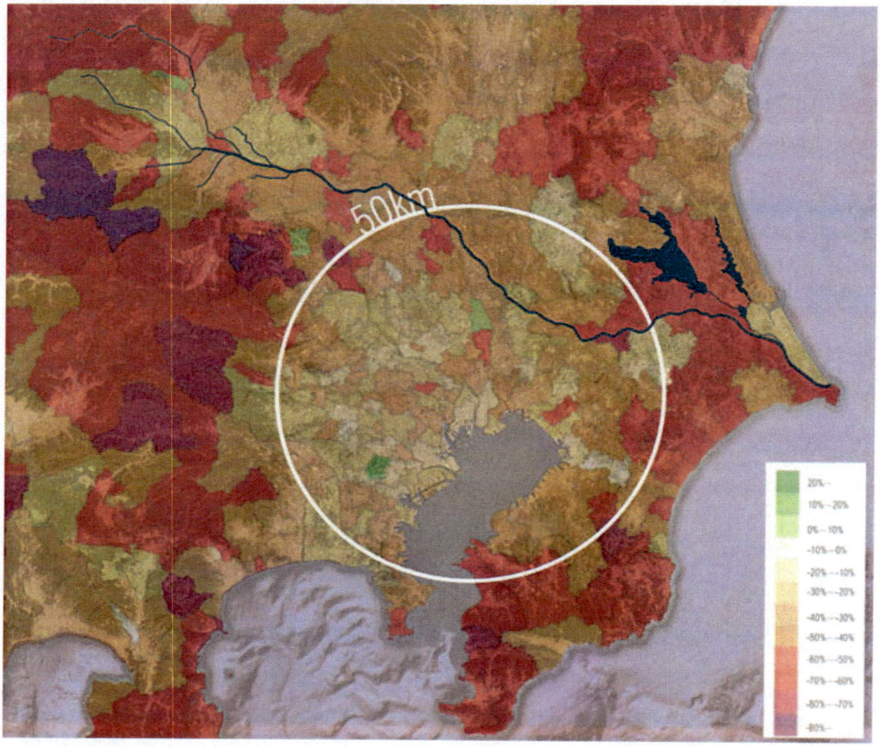

Fig. 12 Population shrinkage ratio of young women by 2040 (Credits: R. Koike)

Metropolitan systems are developing rapidly, often disregarding essential planning guidelines, primarily in terms of spatial distribution. According to UN predictions, urban and metropolitan systems will significantly expand in the next 3 decades, raising shared concerns in distribution management and determining inequalities between social classes. Soil becomes the most important economic asset, as correct spatial distribution rules will significantly influence the perception of value in urban and metropolitan systems [67]. City development policies have often focused on goals related to the increase of urban values, but public finance cutbacks have made it difficult for several metropolitan systems to benefit from the same number of resources [77].

The analysis and assessment process is based on the choice of elements to be evaluated and the procedures to be adopted, inspired to the principles of problem-solving (based on domain, problem types and solution detection). Experts in urban and metropolitan development management sciences must be able to rely on a large set of skills to be equitable and efficient, requiring multifaceted and differentiated knowledge to be organised [67]. The analysis and assessment process is based on the choice of elements to be evaluated and the procedures to be adopted, inspired to the principles of problem-solving (based on domain, problem types and solution detection) (Tables 1, 2 and 3).

Value assessment is the estimation of metropolitan assets in quantitative and qualitative terms, involving data from technical and analytical disciplines and information related to the economic quantification of asset values [6]. It is not just a question of economic value, but also of the cultural implications of using reliable techniques [67] (Table 4).

Table 1 Sustainable habitat analysis metropolitan cartography map—Part 1

Phase	Topics Addressed
Project Phase One (Analysis)	
1	Geographical analysis of the territory: • Geomorphological dimension of the grounds. **Risk DNA:** Physiography; Elephant skin/**Ecosystem S.**:Events Objective: 1) The notion of the structure of 'support points'. Anthropogeographic system of the relationship between valleys (water) and neighbourhoods (grounds) 2) Permanence, resilience, and potential of the ground structure 3) Sensitive, reactive network of proactive nodes **Border:** Border communication; 4) Definition of the project operational intervention field of action
2	Historical City-Territory Analysis: • Green-grey infrastructure: 1) Green–blue geographical structure **Risk DNA:** Blu infr.; Gren Infr. (T); 2) Type-morphological relationships of the historical contexts and the modifications of the city pattern in its growth thresholds. **Risk DNA:** Grey Infr. T; **Environmental Justice:** Homogeneous areas M; **Border:** Homogeneous areas; **Unit of Landscape L**: Border categorization • Symbol, memory, and image: formal references and syntactical spatial investigation of the territory's evolution Public Space T; Community/ Cultural Heritage T; **Environmental Justice:** Community M • Exposure. **Risk DNA:** Unemployment, disease T / **Ecosystem S.**: Exposure. Land grabbing / Cultural service T / **Unit of Landscape L**: Cultural Service T Objective: 3) Green-grey continuity 4) Structure of the permanence in the qualitative change: Genesis and geomorphology of the settlement fabrics; Gradients of formality 5) Identification of strategic urban areas historically suitable to support the change of scale, recombinant structures 6) Exploring the potential of places through the lens of image and time related to their symbolic and robust civic images 7) Metabolic structure of the territory concerning project possibilities of the neighbourhoods: maintenance, replacement, and transformation
3	Network analysis and infrastructural relation spaces: • The territorial connections **Risk DNA:** Grey Infr. T; Public Realm Objective: 1) Investigating the structure of the territory through multi-scale relationships and varying proximities and analysing the diverse nature of its networks 2) Seeing large-scale interchange nodes as spaces for synergic inter-regional convert

Table 2 Sustainable habitat analysis metropolitan cartography map—Part 2

Phase	Topics Addressed
Project Phase One (Analysis)	
4	Landscape and environmental analysis: • Aesthetic characters of natural and anthropic landscapes • Criticality, Fragility, Limits, and Transformability of the ground/land • The study of the characteristics and principles of land use Objective: 1) Analysing the symbolic and invariant characteristics of local spatial relations within the territorial fabric to uncover their image value. Examining the iconic visual structure that emerges 2) Examining the ground/land in terms of its criticality, fragility, constraints, and potential for transformability 3) Experimentation and mapping of landscapes
5	Economic analysis (readings and redevelopment strategies): Risk DNA: Exposure; Border: Community M • Eco-system analysis: urban agriculture and green economy of the territory: **Ecosystem S**: Provisioning service/ Regulating service • Micro and Macro Sector Economics **Ecosystem S**: Homogeneous Areas; **Border:** Strategy of border control M • Analysis of the projects already financed and related formalisation of the documents necessary for participation in calls for tenders • Environmental and agricultural value of soil as scarce resource • Spatial quality and economic drivers • Attractors and spatial distribution of economic activities Objective: 1) Exploring the intersection of the ecosystem, urban agriculture, and the green economy within a territory. Examining the landscape and products of the land as essential components of a high-quality lifestyle 2) Examining the potential economic impact on a territory, particularly through food production and cultural tourism services, and assessing the possibility of revitalising the area 3) Evaluate and assess the success of previously funded projects. Identify areas for improvement in past projects Gather necessary information and documentation to participate in future calls for tenders Increase the chances of securing funding for future projects by properly preparing required documents Streamline the application process for future tenders 4) Defining the economic value chain behind environmental quality and territorial structure in a spatialised way

Soil is a scarce resource, difficult to reproduce, and has a significant influence on real estate market trends and the implementation of metropolitan policies. It is also a key element in terms of urban rent and profitability of the positional surplus. In urban planning, the integrity of regions can only be safeguarded by controlling the transformation of Soil from its natural condition to built spaces, where people can live, study, work, and travel [67]. To achieve a sufficiently effective control of land transformations, it is necessary to know, with adequate accuracy, the Soil features and components in different areas, and the distinctive elements that characterise human

Table 3 Sustainable habitat analysis metropolitan cartography map—Part 3

Phase	Topics Addressed
Project Phase One (Analysis)	
6	Marketing analysis (local production, services and territory): • Analysis of economic and social variables **Border M:** Duty • Market analysis of potential economic realities • Identification of common structural values: towards the construction of a new brand for the Region • Market value chain and smart specialisations Objective: 1) Brands of the past and identification of the potential of the territory's economic network 2) Evaluate and analyse potential economic opportunities in a given market or industry Identify key players and competitors in the market. Assess market trends and consumer preferences Determine the feasibility and viability of launching a new product or service in the market Identify potential risks and challenges associated with entering the market. Develop strategies to capitalise on market opportunities and gain a competitive advantage Inform decision-making and business planning processes 3) Identify the common values and characteristics that define the region's identity and culture Understand how these values can be leveraged to create a unique and distinct brand for the region Develop a comprehensive understanding of the region's history, heritage, and traditions Identify the region's strengths and weaknesses, and opportunities for growth and development Develop a cohesive brand strategy that reflects the region's unique identity and resonates with target audiences Enhance the region's reputation and profile, and attract investment, tourism, and other opportunities Foster a sense of pride and identity amongst residents and stakeholders in the region

activities, to compare their compatibility and to identify where it is most suitable to allocate every and each function, which can be determined through Metropolitan cartography techniques.

The quality of soil, which determines its values, both in terms of environmental and economic factors, is linked to its use by man, whilst the decrease in its quality is caused by the deterioration given by natural and anthropic pressures. Natural pressures include climatic changes, morphological upheavals, and flooding, whilst anthropic pressures include excessive agricultural exploitation, urbanisation, chemical and organic fertilisation, and wastewater disposal. The trend of decreasing biodiversity and environmental sustainability is caused by two processes: the consumption of land for settlements and infrastructures, which causes pollution, loss of environments and ecosystems, and the transformation of the agricultural and forestry cropping model, which causes pollution, eutrophication, and simplification of environments. Deterioration risks come from soil consumption, hydrogeologic erosion, compaction, organic matter loss, desertification, acidification, and contamination. All

Table 4 Sustainable habitat analysis metropolitan cartography map—Part 4

Phase	Topics Addressed
Project Phase One (Analysis)	
7	Synthesis - Situated Project: • Drafting and organisation of information • Public presentation of the research Objective: 1) The aim of the action "Summary of the analyses of the different research sectors" is to provide a comprehensive overview of the findings and insights gathered from research conducted across various sectors. This involves: Reviewing and synthesising data and information collected from multiple research studies. Identifying key trends, patterns, and themes across different sectors. Summarising the main findings and insights for each sector in a clear and concise manner Providing actionable recommendations based on the research findings to inform decision-making and strategy development Enhancing understanding and knowledge of the research topics and their implications for the organisation or project Facilitating communication and collaboration across different teams and departments by sharing insights and information 2) Communicate the findings and insights of a research project to a wider audience. This involves: Preparing and organising a clear and compelling presentation that effectively conveys the key findings and insights of the research Targeting the presentation to the intended audience, whether it be internal stakeholders, external partners, or the general public Highlighting the significance and relevance of the research findings to the audience Encouraging engagement and interaction with the audience through question-and-answer sessions, discussions, or other forms of participation Building awareness and understanding of the research project and its objectives. Establishing credibility and legitimacy of the research and the organisation or project that conducted it Generating interest and support for future research and initiatives

these factors can have significant effects on the economic stability of metropolitan systems (Table 5).

Soil degradation is not a new issue for humanity, but it is becoming increasingly intense due to an increase in anthropic pressure. Urbanisation is also taking shape in an increasingly intense way, determined by several drivers that are leading to a growing process of building expansion. These drivers can be determined by many factors, such as: (i) the tendency of disadvantaged households to use low qualified housing; a part of mass public production has greatly lowered quality standards, generating, at the same time, the raising of well-being levels in the same families, that can then afford owning a property, determining an additional residential demand of higher quality households; (ii) the multiplication of residential demand affected by the localisation factor, especially considering the most valuable locations in terms of natural and landscape value; (iii) the new shares of residential demand generated by immigration with income levels that are close to the ownership threshold; (iv) the infrastructure needs related to high-technological levels; (v) specific innovative

Table 5 Sustainable habitat analysis metropolitan cartography map – Part 5

Phase	Topics Addressed
Project Phase Two	
1	Integration of research and analysis data across various sectors: 1) Summarisation of information and analysis data into concise documents to establish meta-project premises 2) Identification of permanent structural elements for constructing a meta-project plan 3) Experimentation with different operational strategies and forecasting a new economic structure for the territory through compositional analysis
2	Strategic forecasts and project map: 1) Formalisation of the first strategies: differentiation of visions and related operating methodologies 2) Definition of the economic variables of the different project forecasts 3) Construction of the industrial marketing potential to relaunch the district

covenants that tend to speed up the urban planning procedures that couldn't be finalised before [68].

Risk in terms of territorial and environmental factors (including economic and landscape elements) is very wide, being related to several issues corresponding to different disciplines. If the environmental risk involves factors related to human health and life quality (for which the environment is seen as a source of risk by Social Sciences, Epidemiology, or Environmental Hygiene), the territorial risk indicates more general risks, connected to detrimental effects that occur on anthropic systems, investigated within a wider territorial perspective, including soil, water, air, plant and animal communities, therefore, concerning other disciplines (such as Metropolitan Disciplines, Geology, Hydraulics, Territorial Planning, Biological Sciences, etc.).

Metropolitan disciplines should adopt an appropriate risk interpretation for specific objectives, considering: (i) the high correlation between drivers generating risks and the environmental and anthropic components, affected by risks, and (ii) the specificity of different objectives. For this proposal, we're addressing trends that lead to an irreversible subtraction of soil from agricultural and natural areas and to the devastation (sometimes irreversible) of historical landscapes, limited and irreproducible. The objectives of this proposal aim at: (i) the qualification of urban risks, related to land uses and settlement dynamics; (ii) the identification of risk drivers over time, characterised by a temporal continuity; (iii) the nature of the temporal range of adverse effects characterising territorial risk. An underestimated factor to be added to the list is the lack of social perception, which can be the cause of the fragmentation of agricultural land after urbanisation processes.

For the identification of territorial strategies aimed at reducing the different types of urban risk considered, it seems useful to try to outline a taxonomy of territorial risk for each of the effects they manifest. Two main types of urban risk can be distinguished: (i) that of natural origin with consequences not only directly related to the natural system, but to the anthropic system itself (natural risks); or (ii) what happens due to an anthropic interference due to processes of localisation of activities and that manifests itself through the modifications of the natural functions of the soil on the physical environmental structure.

Territorial analyses for the identification of risks are aimed at determining possible adverse consequences that may be caused by localisation of human activities (residential, agricultural, productive), as well as their intensity, on the physical planning of the territory. This can be determined by the loss of soil pedological value, affecting surrounding territories, social functions, and economic determinants, with different effects depending on the level of deterioration. Risk means here that negative effects are not certain and necessary consequences of urban sprawl, but possible and probable, which can occur in different levels of intensity related to the actual fragmentation associated to diffusive processes in a specific territorial context.

The lack of attention towards the value of soil, which determined several serious negative effects in Metropolitan terms, such as sprawl and inefficient localisation management, affecting the sustainability of urban settlements, should be addressed through specific Metropolitan projects, with the aim of reducing inequalities and mitigating Global warming. This can also have significant effects on the economic drivers that determined the diffusive trends in the past: reconsidering the productive value of soil and the generative role of agriculture can be a possible solution to tackle several Metropolitan derangements, reducing natural and territorial risks effects.

The concept of Soil defence, then, can be better conceived as a complementary idea to the concept of Soil consumption, introducing a new idea of compensation and rebalancing in the Metropolitan policies. This concept should consider the urban factors constantly consume space and should be prevented from indiscriminately using soil as a spatial and productive resource. The quality of Soil as a resource is recognised by most International Institutions, and it has the capacity to perform its functions.

There are two fundamental categories of pressures, often synergistic and difficult to separate, i.e. stress sources of natural and anthropic drivers. Natural pressures to soil consumption and quality deprivation come from climatic changes, morphological upheavals, increased water rates following floods, and so on. Anthropic pressures are related to development, coming from the over-exploitation of agriculture, from the losses caused by urbanisation, from the modification of endogenous features for chemical and organic fertilisation, from the disposal of waste of several different sources, and so on. International regulations should recognise the extraordinariness of the pedological capital and protect it in its finiteness, starting from the awareness that Soil cannot be used in an infinite flow [67].

Soil consumption is not necessarily negative, as it depends to the threshold of the limit between consumption, excessive depletion, waste and the way in which such erosion is happening. The definition of Soil excess depletion and waste could be recognised to any activity or event (natural or anthropic) that experienced specific levels of use of urban and metropolitan regions, that can affect irreparably their features and their ability to renovate and recreate themselves [8, 9]. The best way of defining the correct level of use of Soil as resource is preparing appropriate Metropolitan cartography maps at different scales, addressing the most important drivers of territorial risk in the specific context.

6 Case Study—Japan, Asia

6.1 Relevance of the Japanese Case

The upcoming population shrinkage in Asia and its spatial, and environmental management is key to the world's sustainability of the second half of the twenty-first century and on. With the two population giants projected to reach their peaks in mid-century, it is a pressing issue that must be dealt with sufficient knowledge and experience. However, the European experience cannot always be referenced, due to the large difference in scale, and the fundamental dissimilarity in its history, culture, and climate, which has created unique structures in the Asian territory. Drawing on the idea that a specific agricultural system is likely to give birth to a particular settlement typology [52], it is not a difficult task to recognise the widespread rice farming culture as Asia's common identity, which should result in similar territorial structures.

In this context, Japan arises as the ideal case study for pursuing Asian shrinkage issues. It is a matter of fact that the country has always preceded the rest of Asia in its development, not to exclude its demographic transition. However, the case is not only relevant for future shrinkage of other Asian countries, but the study is more than necessary for Japan itself, currently undergoing one of the most striking shrinkages in the world.

Below, the case of the Japanese shrinkage was analysed using the heuristic model for urban shrinkage [37].

6.1.1 Global and Regional Drivers of Shrinkage

Shrinkage in Japan is a nationwide phenomenon, regardless of the region or the degree of urbanisation. The country's population reached its highest in 2008 at 128 million, already turning to shrinkage in the second decade of this century. In 2021, the number had lowered to 125 million, whilst the speed of decline is expected to accelerate further. WPP2019 projections show that Japan will reduce its population below 100 million in the latter half of the century, dropping to 75 million in 2100, 60% of today's size (at medium variant).

The main cause of this problem is fundamental demographic change. Japan's TFR has been transitioning below replacement level since the 1970s and below 1.5 during the last two decades. This extreme situation is thought to have been caused by a negative spiral effect starting from (i) delayed marriage and birth, (ii) lowered TFR, (iii) reduction in the number of young women, mainly those aged 20–39, to whom the majority of Japanese births are attributed [50]. This also results in an incredibly aged society, making it the "most aged population in the contemporary world" [36]. In 2020, the share of over 65-year-old reached 28.4% or one-third of the total population.

However, these dramatic demographic changes are not independent of the domestic migration of the population. The Japanese population distribution is characterised by a single-pole concentration in Tokyo, which continues to swallow productive populations from the rest of the country. Lifestyles in highly urbanised environments foster delayed marriage and birth, eventually dropping the TFR level even further [50]. This is evident from the national statistics, showing that Tokyo and its surrounding prefectures have deficient numbers compared to the national average. Thus, Tokyo is referred to as the 'black hole' (*ibidem*) of young generations, lowering the TFR and depopulating the rest. However, even this capital is projected to turn to decline after 2025, which implies a dark ending to the story.

6.1.2 Direct and Indirect Consequences of Population Decline

In the Japanese case, the consequences of population decline are highly interlaced with the causes, constituting part of the black spiral. For instance, the ageing of society is both a cause and a consequence. Drop in TFR and extended life expectancy together cause a doubled effect of ageing. Another emerging concern is the ageing of the generations that migrated to Tokyo in the past, resulting in an overwhelming demand for elderly care services in the capital. The young rural population will supplement the lack of labour, thus fostering the black hole effect of Tokyo [50].

Another consequence concerns public services and infrastructural provisions. Depopulation in rural cities and settlements, or even in highly urbanised areas, often results in a shrinkage that does not neatly converge but instead results in thinning out, where plots empty randomly within. The emergence of these spontaneously underused 'holes' creates a sponge-like urban structure and is already observed in Japan [1]. Not only does the declining population decrease the demand for public services, but it has severe effects on the efficiencies of infrastructural systems. These areas where the demand is weak are known as 'cold spots' [55]. Simultaneously, the Japanese physical infrastructure faces ageing, as the majority was constructed during the postwar reconstruction period [86]. Whilst Tokyo suffers from tremendous demand for restoration works, other depopulating cities fall into a crisis from a lack of budgets and thin demand for the service [86].

Lastly, the shrinkage of agriculture is of significant concern, although population decline is not the only cause. Nevertheless, depopulation initially hits rural areas where agriculture is the primary industry, where the aging of the engaged farmers has long been a problem. The national census on Agriculture and Forestry [53] shows a 23% decline in the number of core workers engaged in farming of individual management entities in 2015–2020. Engagement of over 65-year-old shares has risen to 70%, which is highly alerting for the future of food production in the country.

6.1.3 Governance Response

Unfortunately, there has yet to be a considerable public intervention towards the consequences of population decline. So far, the government has adopted strategies revolving around the revitalisation of rural areas (the so-called *chihou-sosei*), primarily aiming to slow down the population loss and prevent further concentration in Tokyo. The strategy stands on four points: (i) reinforcing local industries and improving working conditions, (ii) promotion of urban-to-rural migration, (iii) improving welfare for marriage and childcare, (iv) creation of high-quality, safe, and active neighbourhoods [7].

However, it is questionable whether these policies will gain positive results. Although the importance of interventions on the level of prevention is doubtless, given the current situation, direct confrontation with the consequences of shrinkage is inevitable and essential. Remarkably, the government's strategy is spatially ambiguous and lacks a perimeter definition that specifies the regions of concern. Due to the consequences previously described (mainly referring to infrastructural issues), a selection of dedicated zones is necessary in order to shrink strategically.

6.1.4 The Excluded Option: Immigration

Comparing such a situation, it seems too obvious that fostering replacement migration from neighbouring developing countries shall be an immediate option [40]. Already at the beginning of this century, United Nations has estimated that 647,000 migrants may be needed annually until 2050 in order to maintain the working population in Japan [83]. This massive number leads to a radical transformation of the country's demography, where, if it were the case, one-third of the population would have foreign origins in 2050 [10].

However, this is a distant goal for the Japanese government and its citizens. Neither would the people opt for it nor try to welcome many such foreigners in the country.

There are varying opinions on promoting immigration amongst the population, although fundamentally lacking the soil for multiculturality to grow on. This is deeply rooted in Japan's historical and cultural background, which still contributes to the persistence of the negative tone against immigration policies to this day. The country's geographic feature of being an archipelago makes it easy to close its borders from the outer world, giving way to ethnocentric visions and an "imagined purity of the Japanese culture" [10]. For the same reasons, unlike Western countries, internationalisation or globalisation has not prevailed fully amongst its population. However, comparisons between the West should be made carefully since globalisation is primarily promoted by European/American prospects.

Therefore, it is likely that immigration will not be a significant policy response for the declining population until society gives an open-minded conclusion to the debate between "contradictory forces of ethnocentrism and internationalisation" [10].

This recalls the isolationist policy *sakoku* adopted by the Edo shogunate in the period, which kept the country's boundaries closed to the world for 200 years until the mid-nineteenth century. Although Japan seems to have opened up its islands to the globalising world, it still keeps the *sakoku* ideology, or it is phase 2 of the long-lasted policy.

6.2 *Territorial Framework*

6.2.1 General Characteristics

The Japanese territory is a linear archipelago that consists of 6,853 islands. Spreading off the east coast of the Eurasian continent, it has an extension of 2,787 km from north to south and 3,146 km from east–west wise. The archipelago positions itself in a strategic geographic context and has long been a gateway to the Pacific Ocean for continental East Asian countries.

Of the numerous islands, four of them, Hokkaido, Honshu, Shikoku, and Kyushu, amount to 95% of the total area of the country.

Notably, the territory is highly mountainous due to plate tectonic activities, which are caused by four tectonic plates that intersect beneath the surface of Japan. Specifically, 70% of the area is categorised as mountains, often sparsely populated, leaving only 30% of flatland. Consequently, 67% of the surface is covered by forests, which is compatible with the number of Sweden. Despite the dynamic transformation of the country in the past two centuries, in which the population has approximately quadrupled, the percentage of forest coverage has remained static for the past 150 years [59].

Since one-third of the country that is non-mountainous is the only land suitable for urban and agricultural activities, the ground is highly developed, and mixed-use, seeing an extreme concentration of settlements. For this reason, the United Nations World Urbanization Prospects projects that by 2050, 95% of the population will be urban, rising from the current 92% [85]. Figure 12 shows the footprint of population distribution and the flatland area side-by-side, which almost perfectly coincides, except for the northern island of Hokkaido, where the winter climate is harsher and large-scale agriculture prevails in its relatively broad plains. In addition to this, Fig. 12 reveals the areas with high natural disaster risks (i.e. flood, landslide, earthquake, tsunami) that, unfortunately, overlap significantly with the former two spatial patterns.

Amongst all plains, the Kanto plain boasts the largest surface area, most known for being the ground for the capital of Japan. Whilst it is only equivalent to 5% of the territory, Tokyo and its fellow prefectures contain almost 30% of the total population. It is also one of the high-risk regions prone to natural disasters, making concentration a severe problem.

Therefore, the Japanese territory can be pictured as follows: 70% of the islands are uninhabited mountain forests, whilst the rest, where natural disaster risks are higher, settles the entire population in high density within continuous urban–agricultural mixed settlements.

6.2.2 Desakota Growth (McGee)

The concept of Desakota was coined by the Canadian geographer T. G. McGee, drawing on the positive approach to expanded metropolises that originates in Jean Gottmann's conception of the megalopolis. McGee's idea was consolidated in his essay of 1991, in the context of the emergence of Asia as the epicentre of urban expansion. The argument arises when Eastern settlements are distinguished from their Western counterparts for having a fundamental difference in their structure, eventually questioning the shared view of urban transition. Thus, the primary discussion is whether the Western vision of urban–rural dichotomy is suitable for Asia, where "spatial juxtaposition of many of the larger city cores within heavily populated regions of intensive, mostly wet-rice agriculture based on a mixture of 'skill oriented' and 'mechanical' technological inputs has created densities of the population that are frequently much higher than in the suburban areas of the West" [52].

McGee defines the Desakota as "regions of an intense mixture of agricultural and nonagricultural activities that often stretch along corridors between large city cores" [52], resulting from an expansion of urban economy into the already dense agricultural areas, that are generally characterised by wet-rice production. Notably, he positions his argument on the larger idea that a specific agro-economic system offers a possibility for developing a particular urban system when it undergoes urban economic influence. Accordingly, it is possible to say that in the case of Asia, the particularity of its mega-urban region is attributable to wet-rice production that brings about high population density and advanced water management of the landscape that enables them. The consequence is an extreme mixture of land use, where industrial plants, agricultural fields, residential development, and other plots are juxtaposed in extended areas.

Three fundamental issues must be pointed out revolving around Desakota. These include the issue of improved access and infrastructure, water management, and self-sufficiency of food supply.

Highly developed infrastructure is counted as one of the conditions for the Desakota region to emerge. Transportation infrastructures such as roads and canals foster the movements of commodities and population, creating a transactive environment of extreme fluidity [52]. Improvement of accessibility to these extended areas is a fundamental issue since communication must cover unprecedently large distances. In developing countries of Asia, McGee identifies the role of "cheap transport such as two-stroke motorbikes, buses, and trucks" [52]. The fact that these infrastructures are the primary structure of Desakota can also be identified from simple observations of these areas, which tend to "follow transportation corridors between large urban centers" [30].

Another element that constitutes the structure of these regions is the water system. It is of common occurrence that Asian settlements are located in alluvial plains [43], which are fundamentally waterscapes. In order to cultivate, intense water management is key. Nevertheless, there is no doubt that these water networks will also play a vital role in the region's ecology. Agriculture in Monsoon Asia is typically wet-paddy rice production, the critical factor of densification. Compared to standard

Western agriculture, rice production can "support a family on tiny plots of land" [79]. Simultaneously, advanced water management, including dikes and canals, makes the agricultural landscapes no less artificial than the cities they support. This can lead to perceptions fundamentally different from Western ideas that rural areas are 'natural' and, thus, must be protected from 'urban sprawl' [39]. Indeed, the phenomenon of Desakota can only be supported in such a paradigm.

Lastly, the necessity of food self-sufficiency is thought to have affected the agricultural and non-agricultural mixture process, delaying the complete urbanisation of these areas [52]. Agriculture's degree of future persistence depends highly on governmental policies, and the decision must be made with great care reflecting sustainable food supplies.

Although McGee questions whether the phenomenon is naturally transitory, observing a decline in agricultural employment in certain regions, he points out the potential of the Desakota typology to become an alternative policy option for urban management [52]. Governments should invest in these mixed regions rather than city cores or distant rural areas. He calls this policy the 'modified regional growth pole' growth, which emphasises "not just the urban pole but also the large mega-urban region of which it is part" [52]. This growth model will be referred to as 'Desakota growth'.

6.2.3 The Japanese Desakota

According to the socioeconomic and historical preconditions of a given Asian region, the Desakota phenomenon can alter its characteristics. McGee classifies them into three types of spatial economy transition.

Desakota type 1, the first of the three categories, is a consequence of rapid urbanisation in the past. Although the economy shifts towards non-agricultural activities, agricultural land use may remain persistent with the help of governmental protection policies. Japan, together with South Korea, is categorised in this type.

Type 2 refers to those regions that experienced recent economic growth, mainly concentrated in urban cores and adjacent areas. Infrastructural improvement is a significant driver of the transition. Examples include the Taipei-Kaohsiung corridor, and the Calcutta region in India.

Type 3, on the other hand, occurs much around secondary cities characterised by an unbalanced transition with slower growth in the economy and rapid population growth. This type can be seen in the Jog-Jakarta region in Java.

The Japanese Desakota was a phenomenon that occurred much in the postwar reconstruction period, specifically during the 1950s–1960s, when the country's economy experienced rapid growth. This coincided with the accelerated urbanisation of the population and the territory, which was spatially structured around large poles such as Tokyo, Nagoya, and Osaka, and along significant transportation, corridors constructed in the period. These include the Meishin highway (1964), which pioneered other national projects connecting Nagoya, Osaka, and Kobe, the Tokaido Shinkansen (1964), the world's first high-speed railway system connecting Tokyo and Osaka in three hours, and the Tomei Highway (1969), which finally expanded the Meishin highway to the capital. Inter-regional infrastructures are considered crucial for reinforcing urban–rural communication, which is a prerequisite for desakota growth [52].

However, whilst the growing economy had shown a substantial shift towards non-agricultural activities, agricultural land use remained quite persistent in Japan, mainly due to the political presence of farmers. Because of several factors, including cultural ones, Japanese farmers "tended to hold on to their land as long as possible, often maintaining land in active agricultural use long after it is ripe for development" [79], which eventually led to increased participation of farmers in the urbanisation process. Furthermore, weak regulations "allowed farmers to subdivide and sell land on a plot-by-plot basis" [79], which resulted in an extreme fragmentation of land ownership, especially in urban and peripheral areas. This prevented significant urban interventions, giving way to intense mixed-use settlements with tiny plots. Therefore, the Japanese Desakota has a great extent, even undermining the city core of Tokyo with its diffused farming plots [39].

6.2.4 Postwar Konjuka

As a parallel debate to McGee's Desakota concept, particularly in Japan, one cannot avoid touching upon the *konjuka* (coexisting) phenomena. The concept of *konjuka* was recognised amongst Japanese planners as an emerging issue due to rapid urbanisation during the postwar period, especially discussed in the '80s and '90s. It shared certain aspects with McGee's conception [51], that quotes the Japanese term as an alternative for his Desakota type. However, lacking a general and global view of the phenomena, the megalopolitan perspective, as well as its relationship with the water management of the territory brought by dominant wet-rice productions. Instead, *konjuka* studies had a more pragmatic aim. They focused in detail on the demographic/economic mixture of agriculture and non-agriculture, particularly linking it to spatial development issues on the fringe of Tokyo.

Kamata's series of studies on the phenomena attempts to identify the factors and categorise mixed settlements, concluding with an analysis of the spatial distribution of the types in metropolitan Tokyo [42, 72].

The study identifies two factors for the mix of inhabitants: internal and external. These are evaluated through the following indices:

- internal rate of non-agricultural households (INA),
- total rate of external non-agricultural groups (ENA),
- population growth rate.

INA expresses the rate of non-agricultural households within the original population of the settlement, excluding the effect of external non-agricultural population groups, referring to the inhabitants of large postwar housing developments that significantly altered the demographic characters, where applicable. Thus, the index supports the internal factor of the *konjuka* phenomenon. On the contrary, ENA refers to the rate of new non-agricultural inhabitant groups over the total number of households in the settlement. The index qualitatively attests to the external factor. Finally, the population growth rate was employed to evaluate the external factor quantitatively.

Utilising the three indicators, Kamata identifies five typologies of settlements: urban type, agrarian + new development type, relatives/U-turn type, internal non-agriculturalisation type, and agrarian village type, in order of degree of urbanisation. The three typologies in the middle are the mixture of the urban type and the agrarian village type, which can be grouped as konju type.

The spatial analysis of the konjuka phenomenon was conducted by mapping these five categories in the Kanto region. The 1987 analysis concludes that:

- the urban type has a relatively continuous and dominant distribution up until the 40 km border from the city centre of Tokyo, beyond which it stretches along major infrastructural corridors (i.e. railways) in a linear manner, reinforced by secondary city cores.
- outside the 40 km border, the three konju types distribute in a discontinuous and mixed manner. Neither do they form a large patch as the urban type nor distribute in a concentric pattern, although the internal non-agriculturalisation type has a slight tendency of continuity [42].

This spatial distribution character is somewhat relatable to the principles of urban-influenced change developed in the '50s in the United States. The twofold principles by W.T. Martin were defined as follows. The gradient principle illustrates that "the extent of urban-influenced changes in rural areas varies inversely with distance to the nearest city and directly with the size of that city" [49]. He adds that the gradient slope will likely become less steep in the future as communication technology develops, which supports the formation of the extended territories of the megalopolis. The second is the principle of differentiation, which explains that rural areas most under the city's influence would show the most significant differentiation of functional specialisation, and rural areas most isolated from cities would show a minor differentiation [49]. The concept challenged the widespread idea that rural areas have a homogeneous character. Consequently, the typological mixture in the *konjuka* area can be understood as a result of functional differentiation under the influence of the urban area of Tokyo.

Lastly, by observing these areas' transformation, Sakamoto and Kamata detect *konju* areas that exhibit a dynamic process of change, static or stable [72]. This demonstrated their opinion that the *konjuka* phenomenon could not be contextualised as a stage of urbanisation but as an autonomous process under the influence of urbanisation. Meanwhile, it is noteworthy that the study suggests that this Japanese process of agriculture and the non-agriculture mixture does not necessarily result in a decline in the area's agricultural economy, against the common notion that urbanisation of rural regions relates inversely to agricultural activities.

6.2.5 The Tokaido Megalopolis (Gottmann)

The idea of megalopolis was first coined by the French geographer Jean Gottmann who developed the concept during the period that coincided with the Japanese postwar reconstruction and rapid economic growth. Although his first observation of this newly emerging typology was in the United States, explicitly referring to the urbanisation of the Northeastern seaboard stretching from Boston to Washington [31, 32], he later found many of his concepts derived from the American megalopolis much recognisable in the Japanese one, that was quickly expanding, becoming the largest of its kind in the world. It was named the Tokaido Megalopolis after the millennium-old feudal roadway that eventually connected the country's two 'capitals': Kyoto and Tokyo (Edo).

The megalopolis primarily appeared as a spatial continuity of metropolitan economy that chained together expanding metropolises, functioning as the 'main street' of the nation [31]. This resulted in a polynuclear system that spread along a central axis of mobility and communication [33]. The concentration of the population in this corridor was argued to be at least one-fifth of the entire population and required a minimum of 25 million inhabitants to be considered a megalopolis [33]. However, this definition must change according to its era. Such were the fundamental characteristics of the megalopolis that played an increasing role in a given nation. Gottmann described it as an economic hinge that linked "the countries they belonged to and the outside world they participated in" [33]. However, its significance was not only in economic terms but also cultural, and as a whole, expressed as a 'social and economic mosaic' [33].

The Tokaido agglomeration served as a megalopolis par excellence for the French scholar. Its population in 1975 was already over 60 million, meaning that most Japanese lived in the area [56]. The foundation of the concentration along this ancient road was laid long before it was identified as a megalopolis. Still, the postwar period saw a deliberate reinforcement of this axis, starting from constructing prominent transportation infrastructures that connected Tokyo and Osaka (see Chapter about the Japanese Desakota) through direct public investments [56]. Along this 'main street' [31] aligned several nuclei constituting the polycentric system [63]. These metropolises worked cooperatively (*kyodai-Toshi*) "with government and finance in Tokyo, heavy industry and manufacturing in Nagoya, trading and manufacturing in Osaka, high-end cultural production and high-tech ceramics in Kyoto" [78].

Whilst Japan geographically stands between the Pacific Ocean and the Sea of Japan, the Tokaido Megalopolis faces the Pacific, almost neglecting the continental façade. This was largely due to "the reorientation of Japan's international trade after the war towards trans-Pacific markets rather than continental ones, and with the increasing dependence on overseas sources of raw materials and on foreign markets" [56]. Thus, it became the hinge [31] that links the Pacific economy and Japan.

Indeed, Gottmann's conception was largely accepted in Japan, such that he himself was "regularly invited to comment on the Japanese government's long-term plans for the management of the megalopolis" [79]. His followers included the prominent architect Kenzo Tange, who created a growth plan that conceived the whole archipelago as one megalopolis [34], whilst similar but more influential plans came from the politician Kakuei Tanaka, who became the Prime Minister of Japan right after publishing his book, Building a New Japan; A Plan for Remodeling the Japanese Archipelago [80]. His plan recommended "decentralising economic activities to selected locations distributed throughout the territory" [32], connecting them with an efficient network of high-speed infrastructures.

However, all of these were of the twentieth century. Today, the Tokaido Megalopolis is seen as dying, with its population shrinking in most places [78]. It must be recalled that Gottmann had already implied an end of megalopolitan growth in the last century, stating that after the completion of rural-to-urban migration, "ultimately, megalopolitan systems will mainly grow through natural increase and international immigration, if allowed" [33]. Where neither of these factors is active, it is doomed to shrink.

Although a limited number of studies focus on its shrinkage, and the end phase of a megalopolis is unknown, observation of today's Tokaido suggests specific characteristics. Most noteworthy is the extreme concentration in the largest metropolis in the network. Japanese shrinkage parallels Tokyo's one-point concentration of population, economy, culture, and industry. This process not only weakens the structure of the country but also drains the megalopolis of its primary feature of being a product of "deconcentration from the main original nuclei" [33] and its 'polynuclear origin' [31]. Thus, the final phase of the megalopolitan procedure is perhaps the undoing of the 'Gottmann phenomenon'. However, it is not a reverse of the process, but the emergence of the megalopolis has changed significant aspects of the territory once and for all. For instance, the qualitative shift to an urbanised economy and population cannot be easily undone. The hyper-metropolitan region of Tokyo is a product of post-megalopolis (Fig. 13).

7 Case Study—The Filipino Balanced Urban Form

This chapter deals with a specific idea of the Filipino Balanced urban form. It is called 'Urban Adobo' and it deals with the Emergence of the Filipino Balanced Urban Form, focusing on the continuous green-gray and blue infrastructure within the paradigm of the Contemporary Metropolitan Planning Discipline in Central Luzon (Region III).

Adobo is a local garlicky meat dish, cooked with soy and vinegar, that had existed pre-colonial times in the Philippines. Various versions of this dish are expressed differently depending on the region or province you are in. Local food experts claim that it only adopted its name because of its similarity to the Spanish dish known to the colonisers, a case of 'lexical imperialism' [76]. But its uniqueness to the islands is evident as far back as the Austronesian Period [44]. Many variations of the dish, or the sauce [26] are seen as expression of context and cultural identities distinct to the regions where they are found. Albeit studies and articles on this topic, by popular belief, this dish still remains a legacy of colonisers in the minds of the Filipinos.

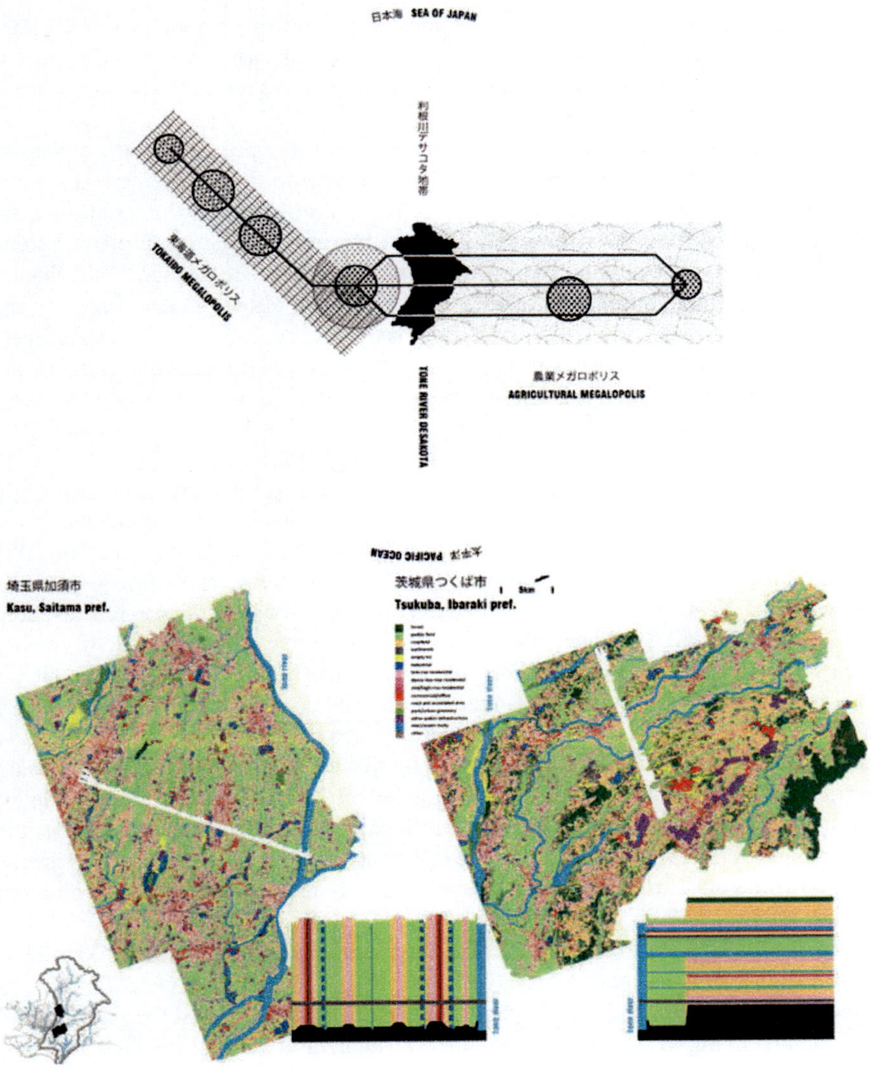

Fig. 13 Addressing the agricultural regions that are shrinking faster than cities (Credits: R. Koike)

Similarly, the Philippines have not escaped the colonial influence on how cities are designed. Through its history, Urban Planning and Metropolitan Architecture has been the channel of development, modernising social, economic, cultural institutions and beliefs that follow patterns of behaviour conforming to the Western world [81], but at the turn of the century, the prominence of the contemporary metropolis, much complex than existing urban centres, is characterised by efficiency in scale, productivity, wealth accumulation, and consumption. Some of these metropolises, including

Metro Manila, are already representing more than half of its country's GDP [60]. This growth explosion is creating gaps and greater disparity in accessing various services, moving towards an unsustainable growth in developing territories, more to those that are part of the global south.

The planning framework needs restructuring. There is a need to inquire, at various levels, for the management framework that is inappropriate to the new scale and context of territories. A framework that is rooted primarily to the local environmental infrastructure of territories, creates the bottom-up synergy in planning that is replicable in other regions and addresses issues on climate change. This chapter is focused on the identification of these processes that allow the emergence of the new Filipino Urban Form that is based on design approaches that assume climate resilience as the main driver for urban development: the Urban Adobo. It is the Urban Centrality, which is characterised by the vernacular mood and identified by locals within the matrices of the contemporary metropolis that surpass traditional local boundaries that espouse public participation through digitation.

In recent years, the Philippines has been touted as having the most promising economic and development growth trend within its region. It has rebounded positively in 2022, even after its last bout with the COVID-19 Delta surge in 2021 (*SPGlobal.com*, Jan 2023). This positive trend is being forecasted to have a sustained growth until 2034, joining a group of countries in the Asia–Pacific region reaching and exceeding a GDP of one trillion dollars. Indicators of this growth are readily seen at the National Capital Region (NCR)—Metro Manila (MM), the country's biggest metropolis, as various infrastructure and development projects are currently directed outwards to nearby provinces with new growth centres emerging and new major mass transport infrastructures being established [58]. Immediate neighbouring regions at the north and south of NCR are affected by this impetus where several major projects are being built. That is why we can speak of Urban Adobo: it represents the Emergence of the Filipino Balanced Urban Form, focusing on the continuous green-gray and blue infrastructure within the paradigm of the Contemporary Metropolitan Planning Discipline.

Region III, the Central Luzon Region, immediately north of Metro Manila is known as the food basket of Luzon. Its land spans to an approximate 22,000 square kilometre area, hosting 12.4 million, as highlighted by the national Census in 2020, with people living in its seven (7) provinces. The natural form of the land envelops an immense flatland that is utilised mainly for rice production, livestock, and other important agricultural crops. Bordering this agricultural plain are two main development corridors protected by mountain ranges on east and west. Several urban *centralities* form a network of cities forthcoming to an emerging *contemporary metropolis* [57] (Fig. 14).

In the last thirty years, a rapid increase of built areas has been observed in the region. Major transport infrastructures are being expanded and a handful of residential, commercial, and industrial projects of significant scale are rising. Just recently, an international airport at the highly urbanised emerging urban centrality at the region's Western front, Clark City in the province of Pampangga, has become operational whilst, another international airport is being built and on its way at the eastern part in

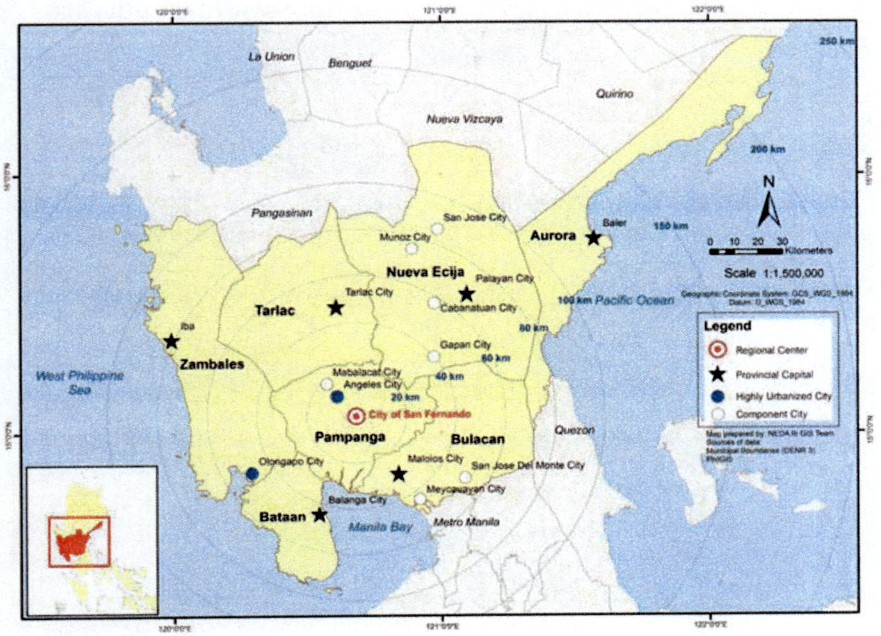

Fig. 14 Central Luzon Map (*Source* Central Luzon Regional Development Plan 2017–2022)

Malolos, Bulacan. These two airports are targeted to service the exodus from NCR at the same time cater and provide more chances for international linkages. At the same time, the expansion and rehabilitation of the major commuter-product railway (Philippine National Railway Rehabilitation Project) coming thru Southern Luzon, Metro Manila, terminating at the New Clark City will provide direct access of people and goods to this region. Furthermore, the new Metrorail-Airport link is anticipated to support the metropolitan linkage in Luzon.

Whilst there is a general positive sentiment from economic managers, reports from the National Economic Development Authority of the Philippines (NEDA-Ph) have identified various issues consequent to this growth trend. First is the issue in food security. 39% of the almost 2.2 million hectares of land of this region is designated and classified for agriculture. The land used for crop-livestock-fisheries and rice, being the main produce in this region, is diminishing due to urban development (Fig. 15).

Even as gross regional domestic product increased from 5.9 in 2017 to 7.3 in 2022 (cfr. RDP-NEDA, 2017–2022), it is undeniable that emerging development concerns are causing multiple problems at various scales. Balancing urban efficiency and capacity, mitigating land-use conflicts and vulnerability brought about by the uncontrolled growth and expansion in urban settlement are already causing the depletion of agricultural land. This will eventually affect food production and security in the

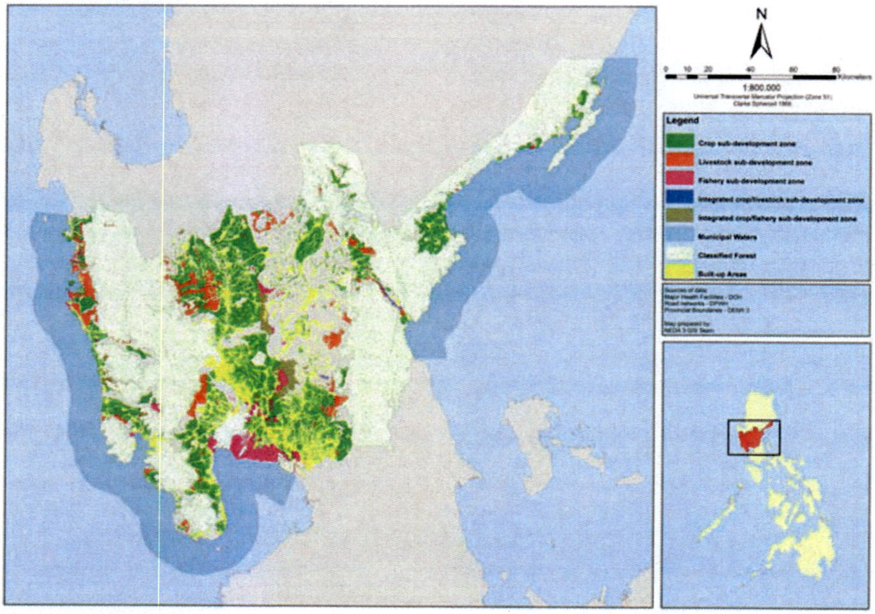

Fig. 15 Land Use classification from Regional Development Plan (*Source* Central Luzon Regional Development Plan 2017–2022)

region causing the need for better strategies in achieving sustainable and balanced development.

Second is the looming threat of climate change on coastal territories of which, the Philippines—an archipelagic land, is regarded as one that will be the most vulnerable and critical in the coming years. Increased occurrence of higher intensity tropical cyclones and long intense droughts have been observed in recent years. An average of fifteen (15) out of the twenty (20) typhoons that pass the country make landfall in Central Luzon, causing extreme hazards such as flooding, landslides, and displacement of families from their homes. Likewise, loss in harvest due to crops and livestock failure, forest fires, and water shortage due to droughts is also recently experienced in the region even though climate simulations show a general increase in rainfall for 2020, 2050, and beyond (NEDA-PH) (Fig. 16).

The higher variability in volume during wet season and longer dry periods during the dry season may change the water supply dynamics in different areas of the territory, year to year contributes to the complexity of issues in the region. Monitoring of variations in river flows and the rate of replenishment of groundwater sources are going to be critical in projecting various strategies in alleviating water problems in the coming years (*NEDA-PH. Regional Development Plan – Region III, 2017–2022*).

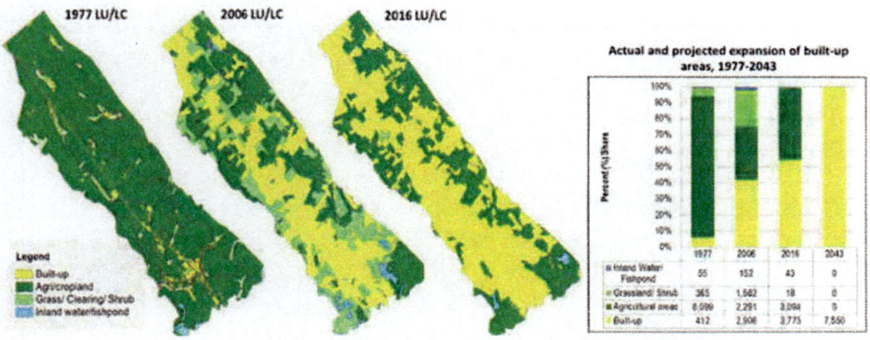

Fig. 16 Actual and projected expansion of built-up areas, 1977–2043 (*Source* Central Luzon Regional Development Plan 2017–2022)

7.1 Economic Aspirations and Planning Prospects

Current regional strategic plans for Central Luzon focus on the expansion of the territories based on socioeconomic indicators. The direction of development is mainly informed by economic aspirations. For example, the *Luzon Spatial Development Framework* although focuses on *concentration, connectivity, and reduction of vulnerability (RDP-RIII-NEDA.2017–2022).* The principle of concentration prioritises the distribution of fiscal resources on tiered urban hierarchy that benefit those that are in the upper tiers (Fig. 17).

Connectivity, on the other hand, provides enhancement of linkages through increase in infrastructure such as communication and transport which will increase efficiency of markets. These priorities aim to reduce vulnerability during emergency situations by having redundancy of transport, goods, and communication routes. Although these are very good responses to improve the market activities and possible calamities, it still creates constant gaps in provision of services and development targets. The design of the plans is merely reactive than proactive.

Some examples of gaps in policies are evident in addressing food security where massive importation of food products is prioritised rather than improvement of agricultural systems in the region. Issues in food security are evident in the policies on massive importation of food products, whilst water management can appropriately be described as *murky and undeveloped.* Based on a study on water management [71], there are various challenges in governing water resources in the Philippines. The lack of financial resource for local water agencies, lack of water data for planning, confusion on various legal documents for governing water resources, and uncoordinated efforts from water agencies sum up its complicated situation. The authors endorsed a review on the institutional framework for water to *create strategies that allow public participation in water governance,* positing that, this will improve planning decisions and policy mandates. It was also mentioned that, although currently,

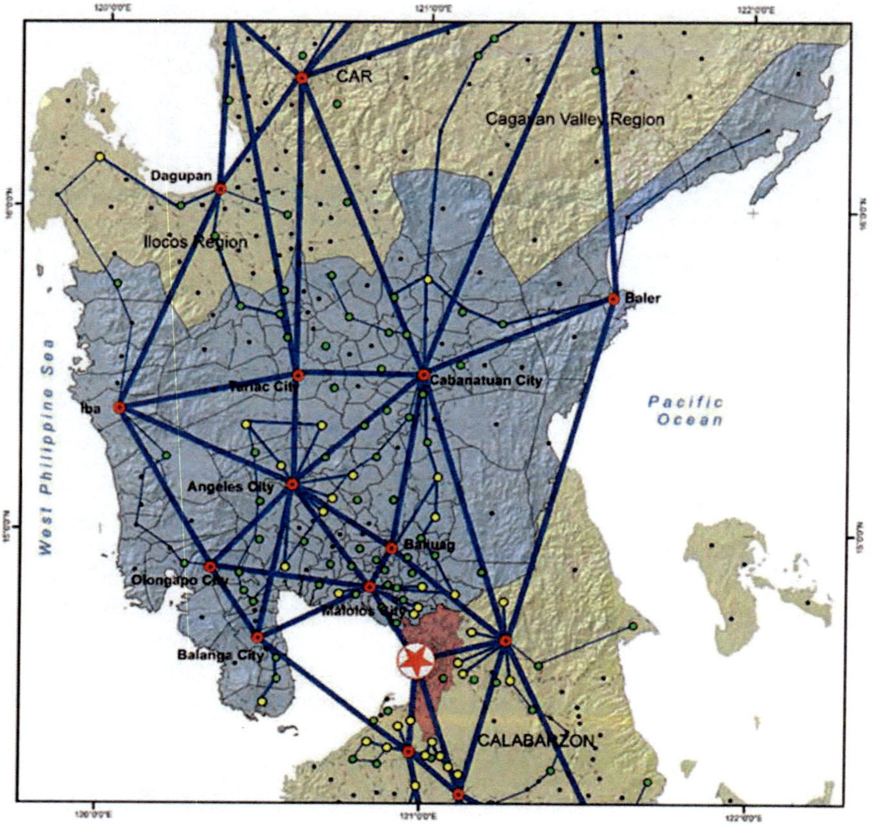

Fig. 17 Five-Tiered Hierarchy Settlement of Central Luzon (*Source* Central Luzon Regional Development Plan 2017–2022)

water governance in the country is hierarchical, *a poly-centric structural mechanism may be contextually relevant* [71].

The *Balik Probinsya Program (Back to the Province)*, the prominent policy program currently being advocated by legislators, is embodied by a *Regional Action Plan* having three strategic programs which are the development of seven selected fifty-hectare (50 ha) township projects, agro-industrial development projects, and integrated transport and logistics development projects within Central Luzon. The township projects aim to provide inclusive access to basic social welfare services in health, employment opportunities, affordable education, and decent housing. The Agro-Industrial projects aim to provide programs for public and private sector investments that promote expansion of medium and large businesses to provincial economic hubs and regional development. Lastly, the integrated transport and logistic projects enforce further the infrastructure development in the region including

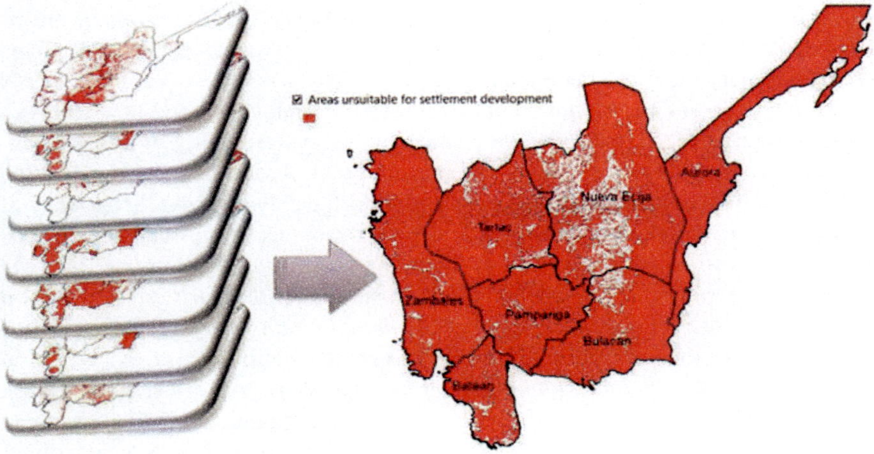

Fig. 18 Settlement Suitability Map (Source: Regional Action Plan – [Balik Probinsya 2] DEC 2020)

public markets, development of logistic chains, cold storage communities, and other facilities that enhance the productivity of the region. Again, as these are very positive steps towards achieving greater goals for economic aspirations, it still lacks a proactive framework that is embedded within the environmental considerations of development. The projects for this action plan are mostly based on planning considerations that can be characterised as a 'firefighting' approach. The strategies were derived mainly on the development of a *settlement suitability map* that considers reactive responses to future calamities such as flooding, landslides, remobilisation in cases of volcanic ash flow. Moreover, classified forestlands, public and communal forests, and strategic agriculture development zones are only designated as restricted areas that can eventually be overridden when the 'right' purpose for these lands is advocated. Programs on restoration, revival, or regeneration of environmental infrastructure, consideration on depletion of production areas are not within the discussions (Fig. 18).

7.2 The Metropolitan Paradigm - Methodology

The emerging growth of urban centralities has created the absolute rapid expansion of contemporary metropolises. If not managed well, will continue to bring about unsustainable practices that are heightened by factors induced by climate change. These directly affect the substantive needs like water and food within territories. There is urgency and need for a paradigm shift on strategies that put *green and blue infrastructure* at the forefront of contemporary metropolitan planning. With the fragmented and multi-layered water governance in the country, the challenge henceforth is the creation of a structural framework that will help consolidate efforts that will put the climate change affecting water resource management in the forefront of contemporary metropolitan planning.

The *contemporary metropolis* is characterised by innovative and new forms of economic activities, having dwellers clamouring for better quality of living conditions [61]. The greater drive for physical consumption and concentration of habitation in urban centralities is currently on high and unfortunately tune out the urgency of careful integrative planning of natural and water resources. *The proposition of this study is to investigate the effectivity of developing a structural framework that can consolidate possible water resource management and food security strategies as it relates to climate change. Driving towards a mindful expansion of a balanced urban development,* the theoretical framework for this study is spurring from the *structural approach* of the *metropolitan planning discipline* characterised by a visual framework of identifying the *Metro-Matrix* that allows better appreciation of urban elements in space [60]. The main goal is to develop tools or mechanisms in driving a clear direction of urban expansion within a territory that considers primarily, the importance of its territorial water resource and consequently uphold the conservation of agricultural productive areas of Central Luzon. To determine a mindful expansion in a balanced urban development, the *Metro-Matrix* is a graphical tool that enables planners to understand developmental strategies [62] in metropolitan and regional scale with ease (Fig. 19).

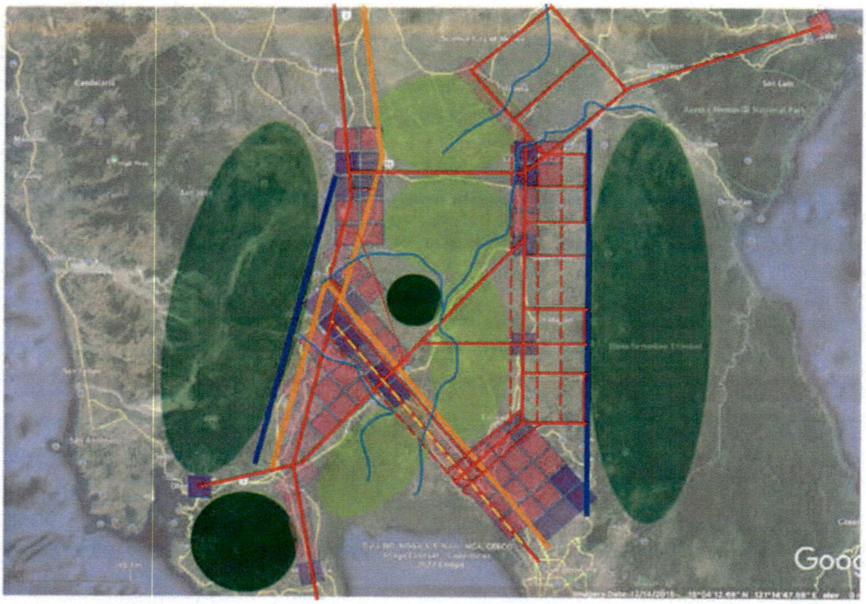

Fig. 19 Metro-matrix of Region III (Credits: K. De Chavez)

The matrix is formed via consideration of the local topographical advantages that dictate the directionality of the physical component of territories. The visual diagram identifies urban centralities that guide the *Balanced Urban Development (BUD)*—system of matrices, characterised by continuous green, gray, and possibly blue infrastructure systems within the urban fabric. The development of the matrix will allow a structured production of detailed *metropolitan cartography* of the territories developed thru integration of big data. This may enforce the concept of *public participation* as data will possibly come from mobile technology usage. A very good example of this was observed in the study on *the Zanjeras Tribe and Ilocos Norte irrigation project.* Although this study was done pre-digital age, there were lessons on environmental sustainability derived from traditional resource management practices in the grassroots level. Exemplifying people's participation in sustainable development [87], these types of activities can now be enhanced by digitation and creation of platforms that identify important parameters creating a 'citizen lead' governance and intervention. In addition to this, traditional surveys, digital check-ins, and digital surveys could be employed in gathering data that will help provide clearer indicators pertaining to water usage, flood levels, river flows, etc. Likewise, the mathematical matrices, a rectangular array of rows and columns, allow a manageable and quantifiable approach in mapping various elements within the territories, going beyond the traditional local governmental boundaries and zones for production and development. The matrices can be a testing ground for the poly-centric structural mechanism that may be relevant for the planning of various *continuous infrastructure* in cities. Application of climate service tools such as *UMEP (Urban Multi-Scale Environmental Predictor)* is also useful to create simulations that may propose regeneration scenarios and reversal of climate change issues. Morphological studies can also be simulated to identify the possible scenarios that will be beneficial for the region.

7.3 Addressing the Gaps and Future Developments

In addressing the gaps, it begs to answer the various questions on this study:

- Whether one can develop a framework that can espouse a paradigm shift on urban development that is based primarily on a substantive ecological approach that focuses on the vernacular practices found within territories?
- Can these practices be identified and enhanced by new technologies to allow bridging of gaps?
- What environmental and ecological parameters are viable for technological approaches that will set up a *digital venue* for public participation?

Through the strategies mentioned above, this study aims to identify and possibly quantify the following factors pertaining water resources:

- Identify significant water resources that are important to various activities within the various matrices, representing urban centralities.

- Identify various water policies implemented across the various matrices, representing urban centralities.
- Identify issues on access, distribution, and supply of water resources within the various matrices, representing urban centralities.
- Identify existing natural and man-made water infrastructure within the various matrices, representing urban centralities.
- Identify risks about water resource management within the various matrices, representing urban centralities.

Likewise, through the strategies mentioned above, this study aims to identify and possibly quantify the following factors pertaining usage of agricultural land, addressing its depletion and managing land use sustainably:

- Measure actual ratio of land use between open-agricultural and built space within various matrices, representing urban centralities.
- Identify and quantify current physical development projects within various matrices, representing urban centralities.
- Measure existing green-agricultural-park open spaces that contribute to the enhancement of water resources within various matrices, representing urban centralities.

Consolidating data and assessment of verifiable information emerging from mapping exercises will allow the ability for better decision-making processes. It also opens doors for a bottom-up approach in governance, where research opportunities in various sectors can be available. The advantage of working within the framework of the matrices, allows flexibility in dealing with various issues and could be a precursor to a paradigm shift on practices in management and governance inducing innovation and practice of open resource can also be achieved.

The result of this study may not be utilised immediately, but the vision is to develop the venue for future studies in contemporary metropolitan planning. Metropolises are becoming the main players in sociocultural development of nations. Its economic strength eventually will dictate the direction of the global futures. This study fundamentally hopes to create development synergies focused on sustainability that will be the basis of the physical quality of our built environment in future cities and metropolises (Fig. 20).

8 Case Study—The Yunnan Province

Our research focuses on Indo-Pacific metropolitanisation and its impact on territorial divisions, development dynamics, and ways of life at the intersection of space and society. Specifically, we are interested in the effects of metropolitan expansion on medium-sized cities, peri-urban and widespread urbanisation, and the informal or illegally inhabited countryside where many people live as invisible individuals, and

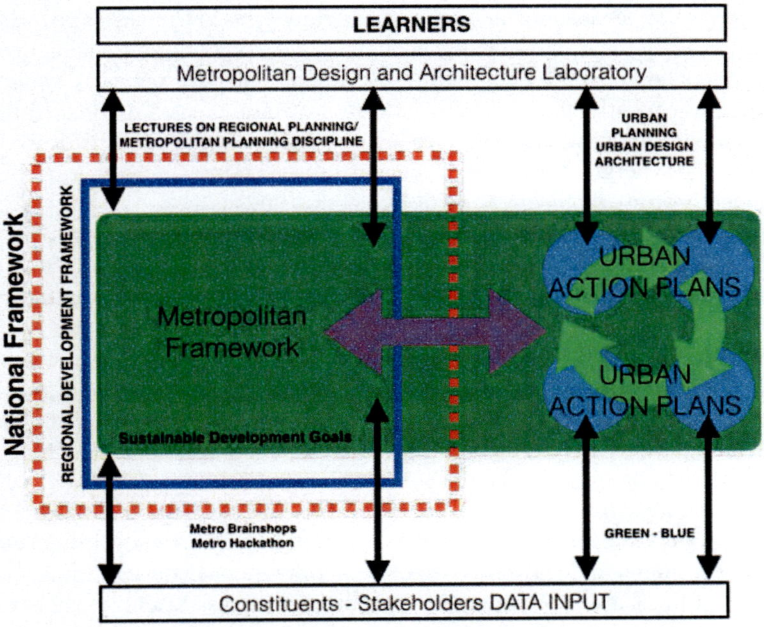

Fig. 20 Conceptual Framework (Credits: K. De Chavez, 2023)

particularly, the abandonment of the countryside territories. We aim to study the physical, social, and spatial transformations in these territories and the resulting economic, social, and cultural inequalities and differences in capital, and we discuss the tourism as the only opportunity for development. Agriculture can play a crucial role in addressing these challenges by providing opportunities for agro-farm development, which can promote social and political stability and private sector growth.

To address these challenges, we propose investing in developing rural and peri-urban areas in metropolitan cities to increase the adaptability of poor communities. This investment should focus on creating new production processes for collaborative metropolitan development that leverage the heritage and culture of local inhabitants and migrants from other territories. We envision a project that enables the regeneration and accumulation of local values, connecting them to the socioeconomic sphere through innovation and participation.

To achieve this goal, we need to think about the leap in scale, capacity building, new open-source mapping methodologies, improved accessibility, good infrastructure for new markets, and new access to credit and land, and most of all welfare. By investing in long-term development that is rooted in the local community, we can achieve endogenous energy and sustainable growth.

In the aftermath of China's reform and opening, rapid industrialisation, marketisation, and urbanisation have brought about significant changes not only to the development process of cities but also to the economic and spatial structure of rural areas

[29]. Agriculture, the leading industry in many rural areas in China, has been dramatically affected by the decrease in the economic efficiency of traditional agriculture, loss of labour force, and reduction of agricultural land since the early twenty-first century. Consequently, the vulnerable situation of rural areas has widened the gap between urban and rural areas and given rise to various social conflicts. Rural transformation and synergic development between urban and rural areas have become the government's primary objective and task to address this issue.

Agro-industry transformation is one of the most crucial methods to promote rural revitalisation, and the Chinese government has placed it at the forefront of the rural renewal strategy. In 2005, the 'New Socialist Rural Construction' strategy was published, supporting agriculture's modernisation through reorganisation, industrialisation, and mechanisation. The 'One Village, One Product' strategy was implemented the same year to promote regional agricultural and economic growth by developing local brands of speciality agricultural products. Whilst over 3,000 model brands have generated over 700 billion yuan in economic spinoffs, many rural areas still need better conditions and outdated infrastructure.

The government released the 'Precise poverty alleviation' strategy to address poverty in remote areas, which lifted many people from poverty by developing specific agricultural industries and advanced technological means. Implementing these policies has transformed poverty in remote areas, leading to the newest 'Rural Revitalization Strategy' [45], which aims to accelerate the modernisation of agriculture, establish a modern industrial system, and promote green agriculture development spinoffs [38].

In addition to supporting the agricultural sector, the Chinese government has also proposed a 'Three-industry integration' development approach, actively promoting the development of secondary and tertiary industries in rural areas and combining the three types of industries [25]. The modernisation of agriculture has increased the value of agriculture itself and created significant economic benefits through the extension of the industrial chain. The diversification and integration of rural industrial sectors, along with the adjustment and transformation of the agricultural structure, have facilitated the development and growth of the non-farm economy, which has gradually become an important development strategy for rural areas. Together with the farm economy, they have become two major driving forces for rural revitalisation, allowing rural areas to establish a relatively independent and complete development system. These efforts have significantly contributed to reducing the disparities between urban and rural areas, enhancing the living conditions and quality of life in rural areas, and promoting sustainable countryside development.

Promoting a sustainable rural area improvement in the metropolitan complexity perspective, though, we need to consider the entire territory system as a Desakota region. Main city, peri-urban zone, agricultural and natural areas must be considered together. The aims of the metropolitan architecture project are to identify the interphase zones amongst these landscapes or *collisive sites* [23] to design linkage patterns amongst them.

Nevertheless, one of the major issues to address from the NUA is *the power of nature in our Metropolises* fostering spatial sustainability as opposed to metropolis construction's fast temporal sustainability. The city is no longer a sustainable entity within itself. However, the metropolis as a system of different places in relation has the capacities to provide for sustainability again. A second issue to be addressed is that of territory fragmentation, easy to recognise if we consider the net-city structure of the nowadays urbanisation and the many territories neglected that exist amongst the networking grey infrastructures, which we must reconceptualise giving them a new function and figure. A systematic design strategy is mandatory in redefining the metropolis as a system amongst intermediate and small cities scales, and the space in-between. We can define four strategies, which support the interrelation amongst Nature and Metropolitan construction.

8.1 First Thesis. Nature and Metropolis. Metropolitan Architecture Project as Collective Landscapes Futures

Metropolitan Architecture projects aim to create livable cities by redefining urban morphotypes and introducing green-grey infrastructure, which connects old and new architectural forms, urban cultivated areas, and their landscapes. The project builds a gradation between related scales using decipherable physical signs, introducing new forms of land use and new built form types capable of introducing urban agriculture and nature by wireless and physical network issues.

In the past, natural places such as volcanoes, hot springs, and mountains held great cultural significance and represented a shared reality in collective memory. However, the Anthropocene Age has brought an epistemological crisis, challenging the traditional understanding of nature and human impact on it. Landscape Urbanism, Metropolitan Architecture, and Metropolitan Cartography [21] are needed to build hospitable cities, creating a delicate balance between the environment's life cycle and human consumption. Metropolitan Architecture aims to create a common temporal space between the city and countryside using technology that can be experienced in the new metropolis' physical locations [74]. The project creates a gradation between related scales to connect old and new places, networks, and landscapes, natural and cultivated, forming a mental image (map) of the metropolitan region. The architectural project of a pattern of linkage urban–rural–rurban–natural spaces at the metropolitan scale determines the passage from one scale to another, from urban, rurban, rural, and natural, averaging them through forms of introduction, such as new built form types and new forms of land use, such as the pattern of linkage. The metropolis is characterised by multipolarity, and a new diffuse metropolitan centrality is emerging, extending into the surrounding territory, mediating amongst built and natural, wet and dry landscapes. This shift towards a sustainable and resilient metropolis is essential for the environment's preservation and reconciliation with urbanisation.

8.2 *Second Thesis. Nature and Metropolis. The Crucial Role of Nature in Shaping the Relationship Between Humans and Technology: A Call for Collaborative Design and Cultural Integration*

Nature plays a critical role in shaping the relationship between humans and technology. As we strive for a more collaborative relationship with nature, technology must be designed to align with nature's intentions. Culture serves as the soul that integrates technology and nature, providing a framework for understanding the relationship between humans, nature, and agriculture. The Metropolitan Architecture Project embodies the idea of a shared experience of time amongst a diverse group of individuals who feel a sense of unity through technology. The project aims to create a common metropolitan temporal space using technology that can be physically experienced in the new metropolis through notable physical locations as new built form types. The project also applies the twin transition of a new urban–rural pattern to introduce a set of scales and landscapes that allow for regenerative agriculture and ensure its survival. Overall, the Metropolitan Architecture Project is a precise and detailed action towards creating a sustainable and regenerative urban environment.

8.3 *Third Thesis. Nature and Metropolis. Promoting Sustainable Green Connections in Metropolitan Areas: Integrated Policies and Adaptive Management for a Resilient, Inclusive Future*

Recognising the metropolitan nature as the bioregion [47] constitutes a new approach to the metropolis from the systems theory and adaptive management as a mechanism that emphasises the need for metropolitan planning to promote understanding of the causes and drivers of territorial processes linked with the deep natural structure of the territory [48]. The new models of metropolises must be based on recognising the bioregion, connecting cities to the territory in which they are situated. To achieve this transformation, we must initiate an eco-social transition that allows us to rethink our metropolises from a systems theory and complexity paradigm [54, 22].

Solutions and policies for creating sustainable connections between various sectors in metropolitan areas, such as water, agriculture, food, and transportation, are needed to reduce negative socioeconomic impacts and promote corrective policies that address existing inequalities. These integrated sectors are crucial in addressing urban–rural linkages outlined in the New Urban Agenda.

To achieve the 17 Sustainable Development Goals, focusing on Goal 11, increasing the number of cities and human settlements that adopt and implement integrated policies and plans is necessary. These policies and plans should promote inclusivity, resource efficiency, climate change mitigation and adaptation, and disaster resilience. Furthermore, it is vital to develop and implement comprehensive disaster risk management strategies at all levels in accordance with the Sendai Framework for Disaster Risk Reduction 2015–2030. The emphasis on integrated policies and plans is also reflected in the European Union's recovery and resilience mechanism, which aligns with the European Green Pact and the Digital Agenda and supports the transition towards a green, digital, and resilient economy.

Assuming ecosystems as complex systems organised structurally and functionally according to a hierarchical configuration, the metropolis ecosystem comprises a set of singular subsystems, with different levels of organisation: metabolic, functional-sensory, and immunological-identity subsystems, which must be kept in balance—functional geography from the nature point of view. The Territorial Heritage System, comprising green infrastructure, blue infrastructure, and agricultural spaces, is crucial in preserving a region's character in the face of external pressures. This subsystem is essential for maintaining the character of a region and can be studied across interconnected spatial and temporal scales [73].

It is not enough to achieve this model change to have a well-designed flow system; inventive adaptive management must be the basis for achieving this transformation. The emphasis should be on flexible institutions and human organisations (*ibidem*) that can build adaptive capacities in synergy with ecosystem dynamics (*ibidem*). This approach represents the capacity and opportunity for urban managers to confront crises, adapt to new situations, and rebuild processes to envision new scenarios. Architects, urban designers, and landscape urbanists can explore a governance implementation approach. It introduces the anti-fragility concept as a category to connect the physical project with the sustainable practices it can provide.

8.4 *Fourth Thesis. Nature and Metropolis. Reimagining the Metropolis: Anthropological Perspectives on Natural and Cultural Paradigms, Economic Futures, and Inclusive Spaces*

To better understand the current relationship between humans and nature, we must adopt an anthropological perspective and critique the Homo oeconomicus paradigm that only focuses on objective observations. We must create new inclusive spaces for the city's productive and economic future and analyse metropolitan planning through an anthropological lens. The concept of entanglement (NEB) highlights the connection between culture, technology, and the biosphere and calls for a deeper understanding of the interconnection between humans and nature.

The globe is anthropocentric, seeing the planet as a resource to be exploited. In contrast, the planet Earth is a place of interdependence and connection. We live in two intertwined worlds, the globe and the planet, with different time logics, one anthropocentric and the other non-anthropocentric. The Anthropocene era, where human actions impact the earth system, has brought changes in earth processes, making it necessary to reestablish our connection with nature.

The distinction between the globe and the planet lies in the fact that the planet is indifferent to human history, despite being intertwined with it. Human action is a geophysical force out of control, which requires understanding the collision between human society's time and the planet's time. We live in both human history's time and geological and biological time, which changes our experience's coordinates. Capitalism and technology's changes in the earth system processes remain open-ended in their outcomes and combined with independent agents, making humans compete for survival with other species.

Standing at the centre of the contradiction between the two worlds requires courage. Humans occupy a delicate zone of a present immersed in change, where invisible forces continuously affect us. Therefore, we must govern the relationship between humans and nature, reestablishing the connection between the two with a non-anthropocentric perspective.

Therefore, an analysis that tends to be anthropological rather than sociological is necessary to explain how a Metropolis is inhabited. Metropolitan Anthropology allows us to inhabit and build Geography today.

In short, the contemporary Metropolis, with its form, modes, and dialectical structure, can be where the life of the modern spirit again expresses itself as the city's destiny if we can connect technology, culture, and design. Nevertheless, it is a question of hypothesising a change of cultural paradigm by taking a stand, choosing the point of observation from which to understand reality, and exposing oneself to a specific anthropological option defining new ways to practice and use the metropolitan territory. A new capacity is needed to imagine the economic-productive future of the city. Still, this Metropolis will also require new spaces where inclusion can occur, and the city can be recreated as a repository of images.

8.5 Heritage in Metropolitan Complexity

Preserving local heritages, particularly those in metropolitan areas, is an urgent problem of importance. This complex topic requires comprehensive understanding. Preservation efforts should strive to maintain a cultural identity whilst allowing integration into the local community. Rather than isolating heritage from its social environment, preservation activities and policies should foster a stronger relationship between heritage and the growth of the local community. This intricate and multi-faceted issue comprises governance, policy-making, economic development, and sociocultural identities.

Heritage preservation is commonly regarded as a complex issue within metropolitan areas, encompassing a range of systemic problems that manifest at different scales. Due to the intricate nature of these issues, more than traditional research methodologies may be needed to understand them fully. The complexities within the metropolitan area give rise to numerous topics, with our primary focus being identifying vulnerabilities across various scales.

8.6 *Shrinkage from the Preservation, a Case of Wengding Village*

Wengding Village is a primitive village located in Yunnan Province, southwestern China. Due to its rugged terrain and mountainous geographical environment, the village of Wengding was isolated from the major Chinese society. The geographical isolation has preserved the local culture; hundreds of unique dry-pen style dwellings and ethnic customs has gained the village of Wengding the acknowledgement of important cultural heritage since 2006.

The village of Wengding has been subject to repeated devastation by fires, with a major one occurring in the 1980s. Since the turn of the Millennium, the population has seen a marked increase which has encouraged a drive to bring the area back to life, leading to a large amount of new housing being built between 1990 and 2004. In the last five years, the village has seen an exponential increase in visitors. The lack of public spaces began to become increasingly apparent. In 2018, due to the implementation of a new preservation policy and the rapid growth of the tourism industry in the area, the majority of inhabitants of the village relocated to the newly built Wengding New Village. Only 17 villagers remain in Wengding New Village, most commuting between this village and their original homes to support local tourism. Thus Wengding has become a workplace instead of a home. On February 14, 2021, a severe fire broke out in Wengding Village again, of 105 unique dwellings in the village, only three remain intact.

The stringent heritage preservation laws, the monolithic industry, and the overdeveloped tourism have all contributed to the dwindling of Wengding Village. These policies have blocked the improvement and development of the village, and the poor administration has further distanced the villagers from their habitat and culture. Moreover, the influx of tourists has caused a shift from the traditional agricultural economy to the tourism industry, decreasing employment opportunities and diversity of jobs. This can be seen at the spatial, economic, and sociocultural levels; the lack of public space and infrastructure has drastically diminished the quality of life for the villagers, and the exodus of the younger generations has caused a shrinkage in population and labour. Intangible shrinkage has also been experienced, as the tourism sector has replaced the traditional agricultural industry and caused a decline in the local economy and social structures (Figs. 21, 22 and 23).

Fig. 21 Bird view image of Wengding Village

8.7 Vulnerabilities Through Multiple Scales

8.7.1 Fragmentation: Wengding in the Metropolitan Scale

Wengding Village is situated approximately 30 km away from the Cangyuan City. The village is administered under the jurisdiction of Cangyuan County and is a popular tourist destination. Visitors from all over China come to explore the picturesque landscape and unique culture. Currently, the only transport available to reach the area is by car, as it does not have a railway or airway connection until 2023.

Cangyuan City is situated in the most expansive valley in the region, encircled by mountains and pristine forests. The village of Wengding and other villages are located in the second largest valley in the area. The Nature Reserve, located between two valleys in southwestern China, is an essential protection for the region's biodiversity. This large area has been preserved to keep the delicate balance of the local environment and promote the sustainability of the native species living there. The Nature Reserve is an effective barrier, delineating the two valleys and leading to a distinct division in the settlement patterns. The urban area was confined to one valley, whilst the others were mostly rural. This division created four clusters of habitation, each with its unique characteristics (Fig. 24).

Geography has been a decisive factor in shaping the divisions between rural and urban areas. Topographical features, including mountains and rivers, as well as policies, and regulations, have all played a role in the distinct separation of the area. This zoning has profoundly affected the population, and its effects can be seen across various areas, from economic disparities to social differences. The fragmentation of rural and urban zones' fragmentation results from a combination of geography, policy, and history, and its effects are still long-lasting and significant (Fig. 25).

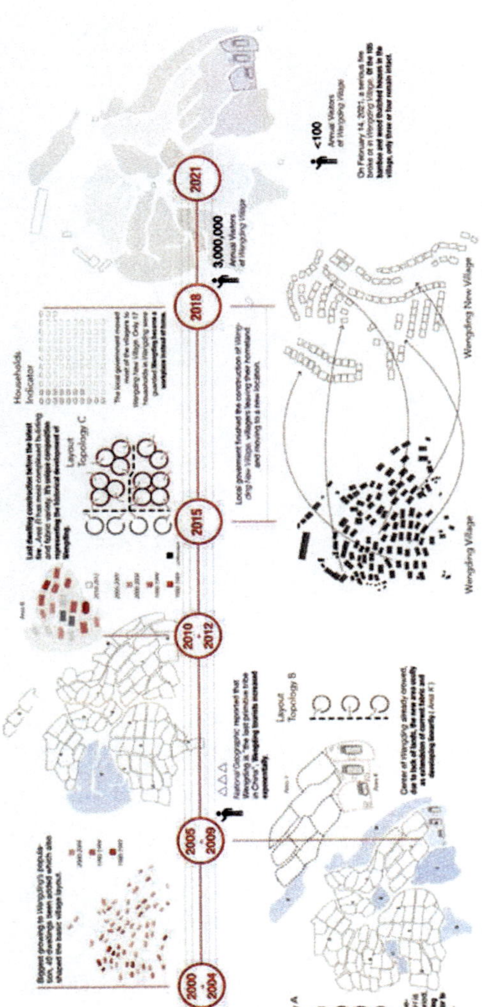

Fig. 22 Timeline of Wengding Village (Credits: D. Yang)

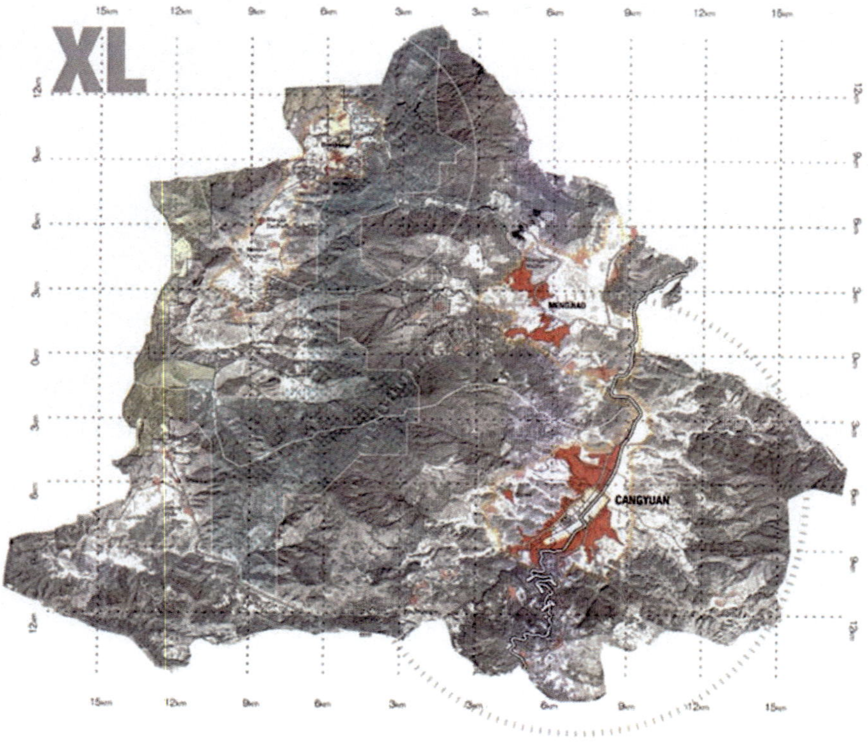

Fig. 23 Map of Administrative Area of Cangyuan County, Yunnan, CHINA

8.7.2 Shrinkage in the Urban–Rural Linkage

To further investigate the vulnerabilities behind the urban–rural fragments, we combined the Wengding-Cangyuan settlement pattern and zoomed in as a focusing area, in order to gain a deeper understanding of the vulnerabilities associated with urban–rural linkage on a more detailed scale. Through this analysis, we aim to gain a clearer insight into the potential issue within this linkage under the context of shrinkage. In this scale, we specifically investigated the connection between rural and urban areas. We sought to understand the way these two areas interact with each other and how the resources and services flow within the rural–urban linkages. Given the available data sources, it is important to consider the rural–urban linkages when examining transportation and population aspects. Such linkages can include both financial exchanges to population flows. It is essential to gain a better understanding of these linkages in order to improve infrastructure and policy-making progress. Ultimately, this could lead to a balance that erases rural–urban inequalities (Fig. 26).

The Wengding-Cangyuan rural–urban linkage faces a significant challenge due to the limited transportation infrastructure connecting the two locations. The only

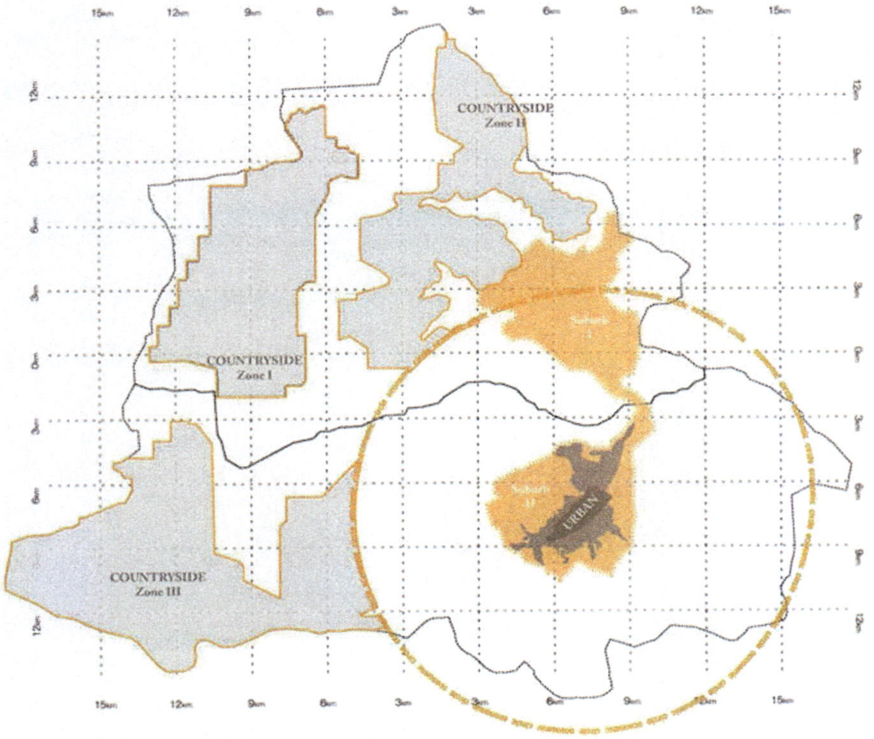

Fig. 24 Map of Boundaries in Research Area (Credits: D. Yang)

transportation path between the two locations is a narrow secondary road, which is roughly 30 km long and takes approximately 40 min by automobile. The absence of efficient transportation networks increases the likelihood of two locations becoming disconnected, making exchanging goods, services, and resources harder for them. This can harm both the local and regional economies, making it difficult for businesses to operate and grow and for people to access. Additionally, this disconnection can lead to decreased social interaction between the two locations, resulting in the overall shrinkage of both places (Fig. 27).

Wengding Village is located in one of China's least developed areas, and the long-term shrinkage in the local population is a trend. On the other hand, Wengding Village, next to the China–Myanmar border, also provides an avenue for villagers to leave, unlike other famous Chinese villages. Both options of migrating to major Chinese cities or Myanmar have increased pressure on the local community. Another critical issue is that the rapid development of tourism has meant the local government has had to restrict the villagers from their homes for preservation reasons, further weakening the cultural and social ties between the villagers and Wengding Village.

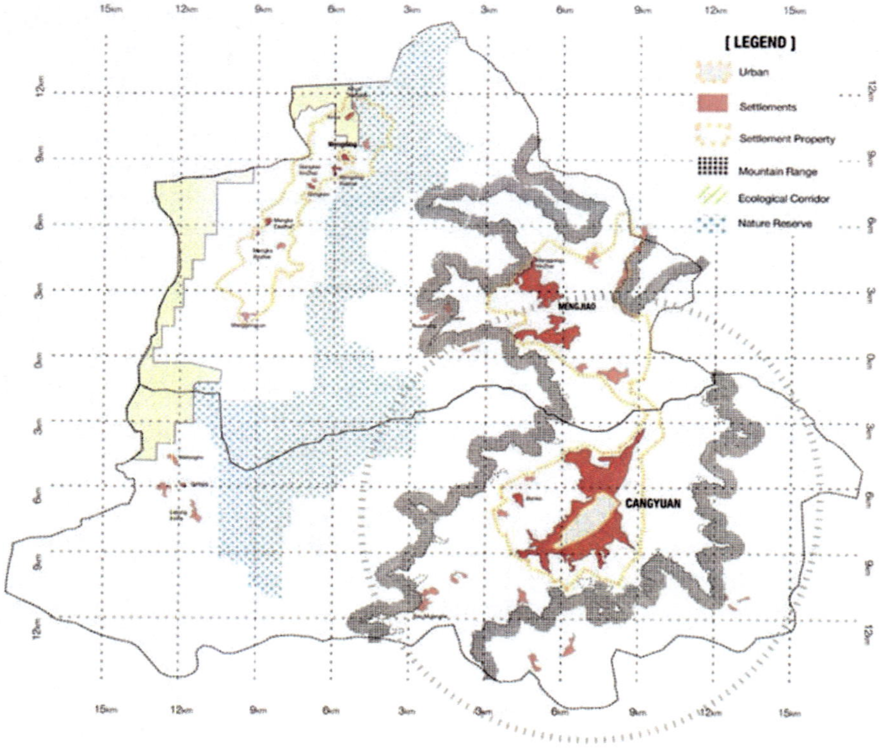

Fig. 25 Map of rural–urban zones (Credits: D. Yang)

8.8 *Solutions*

8.8.1 Breathing Zone: Functional Buffer Spaces

Due to the impacts of local heritage preservation policies and the tourism-centred industry developments, the remote village, already suffering from poor transportation links, has been shrinking into a ghost town. We aim to explore the existence of traditional villages and optimise their quality of public spaces in the context of heritage conservation. We are focusing on finding a balance between the needs of populations and the development of their fragile heritages. A strategy that opens up more spaces in a sensitive area is the best way to reconcile these tensions. With more usable spaces in a protected area, we could improve inhabitants' living qualities and more possibilities for heritage sustainable developments.

We took a vital reference from a traditional Chinese garden. The Lingering Garden has been renovated and enlarged several times over the past four centuries and is one of the most well-known traditional Chinese gardens. In the Lingering Garden, a tiny space known as The Conner Garden has been created to separate buildings

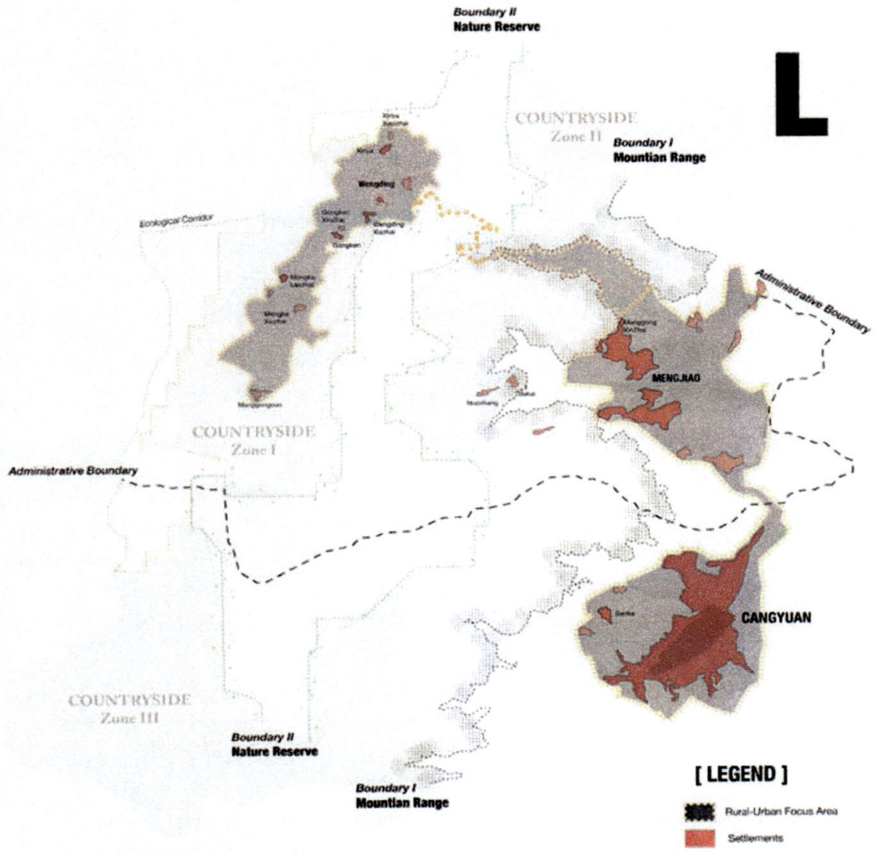

Fig. 26 Map of Focusing Area (Rural–Urban Linkage) (Credits: D. Yang)

and spaces with different themes and priorities. The Conner Garden, serving as a buffer zone to the significant themes, offers separation from various surrounding environments and public services for relaxation and recreation. This scenario is highly conducive to preserving and enhancing Wengding's sustainable development. Based on this reference, the concept becomes opening buffer spaces between different built environments, putting public services into those spaces and ideally offering breathing spaces for the village and the villagers.

8.8.2 Methodology: Open up Boundaries

Through the study of the Wengding and the surrounding area, we have identified six main existing elements, they are farmland, forest, peak, village, river, and road. Due to the sensitivity of Wengding as a cultural heritage site, we decided not to

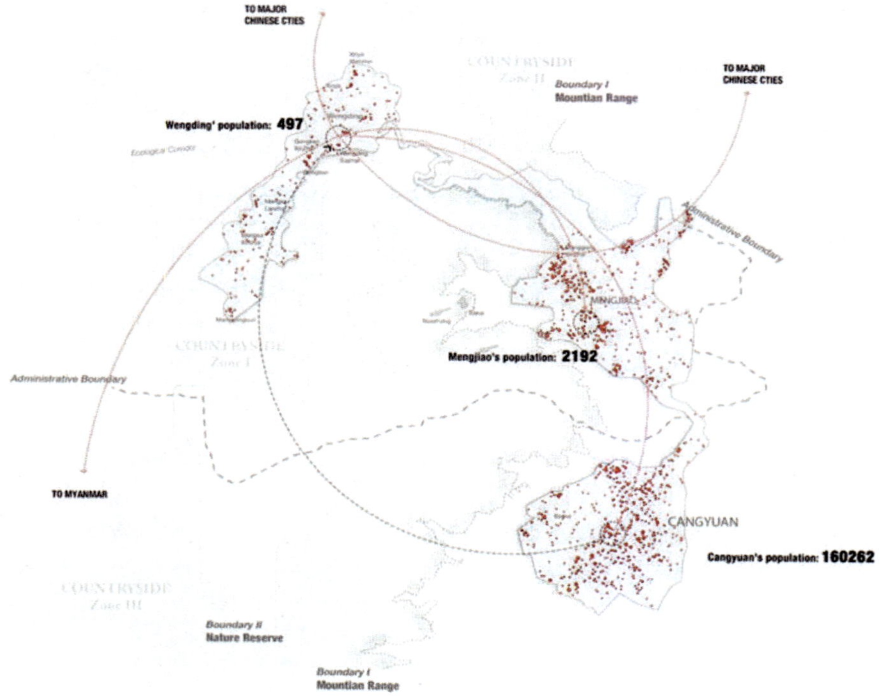

Fig. 27 Population Density and Flows in Rural–Urban Linkage (Credits: D. Yang)

make systematic changes to the local natural landscape, road system, and settlement patterns. Instead, we are investigating the possibility of using the different elements' boundaries to design a protective and spatially functional space. This will provide the fundamental basis for further strategic planning of the area and the metropolitan architecture project. In order to map the specific geographic location for our project, we are utilising the intersections of various elements. These locations are the most effective points for building a network for the sustainable transformation of the area (Fig. 28).

8.8.3 Methodology: Boundaries Interventions

Once the locations of intersecting boundaries that can be used for design were identified, five special interventions were developed based on the intersection of the boundaries of each element. These interventions are intended to provide a unified and harmonious design solution that takes into account the different elements and their boundaries, The goal is to create buffer spaces that are both functional and protective.

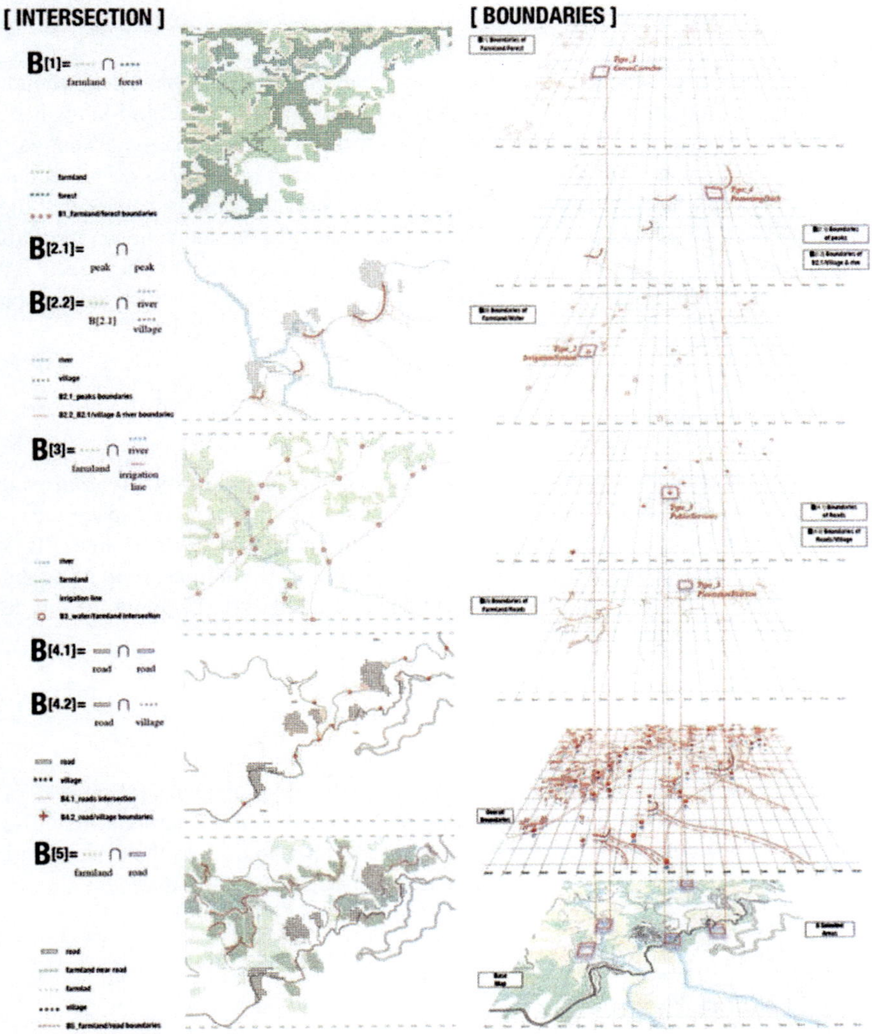

Fig. 28 The Intersections between Different Elements (Credits: D. Yang)

8.8.4 Green Corridor

The Green Corridor is an innovative initiative that seeks to protect and preserve biodiversity by creating a continuous space that overlaps with current farmlands and forest boundaries. It provides an interconnected network of pathways that connects these habitats, allowing for the exchange of species, nutrients, and natural resources. This intervention offers a unique opportunity to protect our biodiversity and promote ecological connectivity and sustainable resource management.

8.8.5 Protecting Ditch

The Protecting Ditch has been constructed as a barrier between villages and mountains, providing additional protection from the risks of flash floods and landslides. Constructing a series of interconnected ditches forms a buffer zone that effectively controls erosion risk from the surrounding mountains. This system of protective ditches works to limit the damage caused by soil runoff, helping to ensure that the land remains stable and secure. The buffer zone also absorbs the shock of water runoff during heavy rains, thus reducing the risk of flooding in the surrounding areas. By providing a safe, stable environment, these ditches help to protect the local landscape and the people who inhabit the village.

8.8.6 Public Services

The current transportation infrastructure has the potential to be drastically improved by lifting the boundaries between roads and villages. This would create a space accessible to the public, allowing for easier access to public services in rural areas. This initiative could revolutionise transportation within rural communities, promoting the free flow of resources and people. This would be a major step to bridging the divide between urban and rural life.

8.8.7 Irrigation System

The boundaries of rivers and farmlands were connected through an irrigation system that utilised the height differences to transport fresh water to the farms efficiently. This system allowed for the optimal allocation of water resources to the agricultural areas. As a result, the valley could sustain a higher population and maintain a more prosperous economy from agricultural development.

8.8.8 Plantation Tourism

Plantation Tourism is an innovative art installation that has been designed to provide a visual representation of the boundaries between roads and farmlands. It provides an educational experience for those interested in exploring the history and culture of the area, as well as a unique perspective on the landscape. Through this art installation, viewers can gain insight into the agricultural history of the region, as well as appreciate the beauty of the natural environment. Creating an agritourism industry in this region can act as a temporary separation between the traditional agricultural industry and the tourism industry. It can also serve as an opportunity to attract tourists to the area and provide them with a unique experience. By providing visitors with insight into the local agricultural industry, agritourism can generate revenue for local businesses and provide a valuable educational experience to those who visit (Fig. 29).

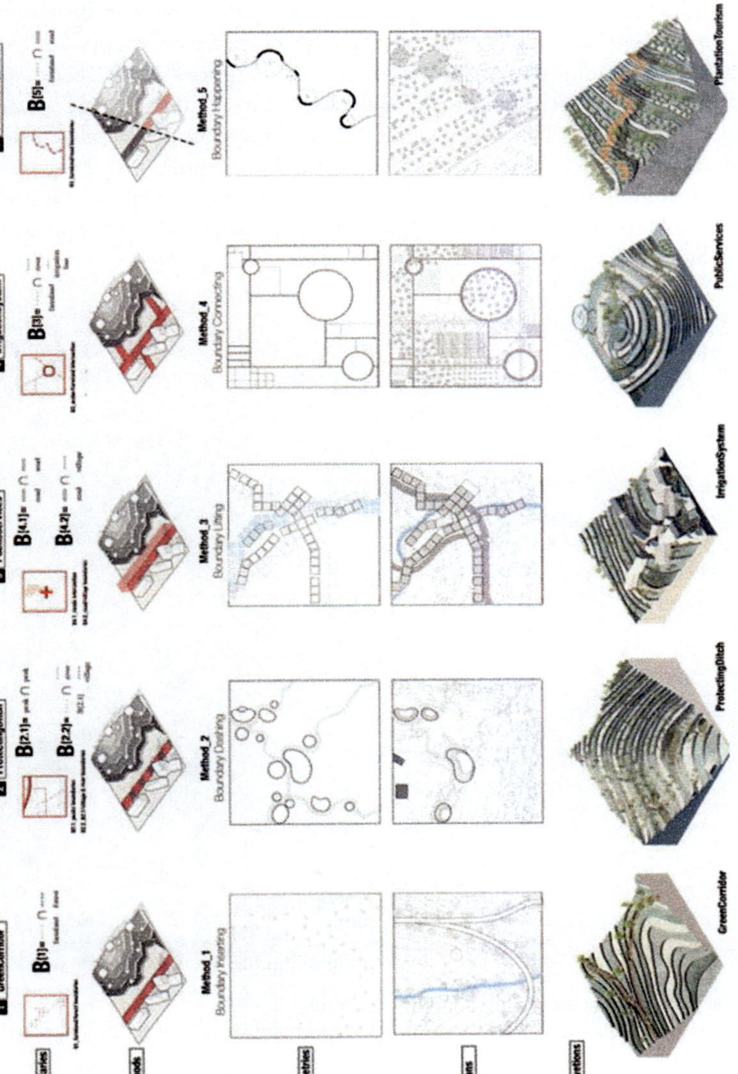

Fig. 29 Boundaries Intersections-Interventions Matrix (Credits: D. Yang)

8.9 Master Plan Proposal

This project aims to provide a new angle to using metropolitan architectural projects to meditate on the conflicts caused by simple preservation policies. Also, it proposes a vision to deal with the shrinkage brought along by the rapid tourism development in remote areas. To answer the Metropolitan Complexity, the project is not merely the construction of buildings but rather a process that integrates diverse ecosystems, rural landscapes, cultural heritage, and local infrastructure. Different elements intersected a new presentation of Rural–Urban Linkage. The methodology also provides a systematic approach to understanding the context and complexity of the site and assists in the area's sustainable development. The methodology that opens breathing spaces in dense land fabrics could also have practical usage in other rural or even urban environments (Fig. 30).

Fig. 30 Master Plan of Wengding Village (Credits: D. Yang)

9 Closing the Knowledge Gap

Through spatial analysis, we aim to generate a strategic vision of territorial shrinkage for Japan, a world-leading shrinking country in Asia. This was positioned in a larger picture of shrinkage worldwide, especially identifying that the Asian population decline in the second half of the twenty-first century needs careful observation and strategy. The particularity of Asian territory and cities, often forming a megalopolitan region with extended settlements, made exploring an original shrinkage model for these Eastern megalopolises necessary. Thus, the primary knowledge gap was set to the need for more methodology to deal with this unprecedented large-scale shrinking. In addition, it was argued that the transformation should positively impact the territory, improving the quality of a given region.

The knowledge gap identified was accompanied by a leading research question and three sub-research questions:

- What could be the shrinkage model for an Asian megalopolis that can enhance territorial sustainability and identity?
- What is the Japanese striking population shrinkage mechanism, and how could spatial planning affect the trend?
- What are the characteristics of the 'in-between' territories of Japanese and Asian megalopolises?
- How is it possible to maintain/enhance the shrinking agricultural regions faster than cities?

The primary cause of the population decline in Japan is demographic change, mainly due to a low total fertility rate. However, the spatial distribution of population and economy in the country is also a contributing factor, with the concentration of population and economic activity in Tokyo being a major driver of the lowering fertility rate. Spatial planning strategies can potentially address this issue by directly confronting the consequences of population shrinkage.

Megalopolises in Japan and Asia have characteristic 'in-between' territories, known as Desakota zones, which result from the deconcentration from the original urban pole combined with the pre-existing density of rural areas. These zones require sufficient infrastructure and function as major agricultural production centres for the country. The sustainability of shrinking territories in Asia largely depends on the behaviour of these Desakota zones.

The focus needs to shift from urban cores to Desakota regions to maintain and enhance shrinking agricultural regions. By promoting the Desakota phenomenon and fostering the proximity of urban settlements and agriculture, the depopulating farmlands can be reinforced through increased connectivity within self-sufficient regions defined by geographic characters and hydraulic systems. This can strengthen food production and enhance regional landscape identities.

A hypothetical shrinkage model for Asian megalopolitan shrinkage was proposed through a reinforcement of the Tone River Desakota in the Kanto region of Japan. This strategy contributes to infrastructural shrinkage of the region, reinforcement of agriculture, optimisation of existing high-density urban settlements through deconcentration, and increased communication within the geographic region.

The post-megalopolitan procedure that strictly follows the line of Gottmann's theory involves strategically manipulating territorial shrinkage through a concentration of capital in the 'in-between' Desakota regions. The proximity between agricultural supply and demands, urban activities, and food production is of primary concern, and utilisation of geographic features of the territory, particularly water systems, may become significant actors in shrinking territories. This model has the potential for adoption in other future shrinking megalopolises in Asia.

In terms of future investigations, from the discussions made, multiple aspects of this thesis arise as topics that require further studies. Primarily, the Desakota shrinkage model shall be tested in other regions of Asian megalopolises. Simultaneously, it must be stated that this model is highly hypothetical without exact precedent experiences. It will require a detailed investigation to evaluate its feasibility.

Other issues concern the reality of urban and agricultural mixture. As McGee has already pointed out, this juxtaposition is not always environmentally sustainable, where many of the Desakota region in developing countries suffer from pollution and lowered quality of life. However, this can be overcome by an invention of a particular pattern of mixture, and a design of transition between wet-paddy fields and urban settlements.

Lastly, practical methods of the 'urban weight shift' must be pursued, in order to avoid a double thinning out due to expansion of the territory.

Nevertheless, it must be recalled that there is a limited amount of time for investigation as the world, and particularly Asia, is about to reach its peak of population, anticipating a large shrinkage in the near future.

10 Conclusion

According to Fernand Braudel's work (1981–1984; [5]), the towns and cities of early modern Europe can be compared to electrical transformers due to their significant impact on the surrounding areas. These urban centres played a crucial role in producing, transporting, and exchanging goods and services and providing legal, financial, and educational services. They also housed bureaucratic structures and served as communication hubs for various forms of information, whilst offering diverse leisure activities. This chapter focuses on changing patterns of Indo-Pacific urbanisation and explores the unique characteristics and roles of the Desakota model in the Tokyo region, highlighting their specialised economic functions, distinctive relationships between the mother city and countryside, governance, material environment and spatial characteristics, and the metropolitan mentality and culture they foster. The chapter covers topics such as the shrinkage of rural metropolitan areas but instead emphasises their relevance in contemporary urbanisation.

The chapter provides an overview of the Philippines' experience regarding a sustained economic growth trend until 2034, with indicators of growth seen in the National Capital Region and nearby provinces. The emergence of the Filipino Balanced Urban Form, known as Urban Adobo, is being developed within the paradigm of the Contemporary Metropolitan Planning Discipline. However, this growth trend is causing issues in food security and the looming threat of climate change on coastal territories, which may affect food production and security in the region. The chapter demonstrates that better strategies for achieving sustainable and balanced development are necessary, including monitoring variations in river flows and groundwater sources to alleviate water problems in the coming years.

As conclusion, this chapter presents a strategic planning example to deal with the shrinkage of Chinese rural heritage, focusing on the case of Wengding Village. The methodology involves the axis of rural fragments, opening spaces between boundaries, and strategic interventions based on boundary conditions. The metropolitan architecture project aims to reconnect Wengding Village with its rural heritage and create a sustainable future for the community by incorporating regional architecture techniques sensitive to the local culture and environment. This research focused on the role of culture in the development of cities, with a particular emphasis on small and intermediate cities. The study aims to identify practical knowledge-to-action proposals that can be applied to urban development.

We considered the different forms of Eastern religions, such as Confucianism and Buddhism, which can shed light on the principle of Harmony, which underpins Chinese culture. The study aims to explore how Harmony can be integrated into an Operating Culture that encompasses anthropological, social, and economic aspects of cooperation with nature. Moreover, Nature is critical in shaping the relationship between humans and technology. As we move towards a more collaborative relationship with nature, technology must be designed to align with nature's intentions. Culture is the soul that integrates technology and nature, providing a framework for understanding the relationship between human beings and nature. The study aimed to identify the values of this relationship, which will also serve as the basis for future development.

Chinese' significance lies in its technological advancements and its potential to help China overcome its deep-seated tendencies towards particularism and discover new ways of understanding its unique forms of unity. We must be careful not to view these modes of unity as fixed identities, as they are dynamic and constantly evolving. If individuals can move beyond their ties to clan and land and create a new vision for the future, then China could have a major impact on the world as a whole, as we know.

Overcoming particularism is a complex issue involving more than just 'rule by law' or 'rule by people' but the urbanity that can talk about Culture as an innovation operator for a global city in today's post-COVID-19 situation defining culture as a driver of relationships.

The Chinese experience has shown how the community dimension has been effective in an emergency. Community means mutual attention. So, there could be two issues to analyse:

a) what are the words and images through which China can define a flag project to say the relational experience it has lived: how the limits developed a relationship towards the community? We are talking about areas of living; areas of common and public spaces; governance strategies; shared technical skills; communication skills.
b) how the vulnerability of a community was transformed into a resource. It is about highlighting the potential of Chinese collective intelligence that was exposed to a wound caused by the development of a 'healing strategy'.

Chinese culture can teach us a lot about how a community can develop skills in a moment of crisis, especially concerning the relationship between human beings, nature, and 'the unknown' (I don't know what it is) that today for us is COVID-19. How is Chinese thought in an 'unprecedented' situation? With which logical structures did it react?

The development of the Chinese city as an 'instant city' shows today the need to highlight a Chinese approach to 'eventuality', such as an ability to think about the present and the unknown: to be present means not succumbing to the past or the future. That is a thought on the times of the city: from the very fast of the metropolis to the slow and very slow of the Chinese garden.

This is a fundamental challenge that China must grapple with in the twenty-first century. The Chinese's everyday behaviours such as crossing the street or standing in line at the post office, seem to be opposed to just the numbers and statistics of technological innovation and exports. Upon careful observation of public behaviour in China, including Taiwan, it becomes evident that 'rule by law' is not always a practical concept, despite its frequent proclamation by the State. Therefore, it is essential to consider these issues to understand China's trajectory.

Chinese's ability to incentivise individuals to take responsibility for their behaviour in public spaces would represent a significant innovation. However, simply having numerous laws in place is not enough. Realistic contexts that promote responsible behaviour are necessary to achieve the desired outcomes. Chinese societies have demonstrated strongly innovative behaviour. A type of bricolage, for example, is expected as traditional cultures encounter the wider world. However, the challenge remains of harnessing and transforming these energies into productive innovation. The proposed issues aim to promote innovation in a broader sense by addressing how to channel and utilise these energies effectively.

Moreover, is the Desakota model we would like to emphasise. The Desakota model is essential in defining the metropolitan city of the second modernity, as it allows us to reconnect with the natural life cycle and embodies the concept of temporal complexity that occurs between spaces due to technology. It also helps to introduce the issue of commonplace and time within the definition of a metropolis, where technology marks these places. Thus, we can experience the novelty introduced by technology, which raises questions about our position relative to nature and the value and sustainability of our actions. The productivity of the Desakota region also prompts us to question the meaning of economy, which is now understood as a relationship between energy production and consumption, deeply tied to the cost and benefit of production. At the same time, the market is only concerned with the combination of market and distribution.

The Desakota model also raises the question of a paradigm shift—how we can give back to rivers instead of solely benefiting from them. The ancestral Indo-Pacific world used water as a cultural resource rather than just for economic purposes. This allows us to consider a geographic economy built on a chain of values that determines the connection between supply chains and choices that prioritise activities located in specific places. This is a place born from a wealth of knowledge, and it can be innovated through programming and investment opportunities based on a strategy that identifies the region's priorities.

Acknowledgements Antonella Contin is the supervisor of all the chapter contents, plus she authored all the paragraphs. Ryusei Koike is the co-author of chapters III-IV-VI, K. De Chaves of chapter VII and D. Yang of chapter VIII. Alessandra Maria Pandolfi contributed to chapters I, IV, and V.

References

1. Aiba S (2015) 都市をたたむ. Kadensha, Tokyo
2. Alkema L, Raftery AE, Gerland P, Clark SJ, Pelletier F, Buettner T, Heilig GH (2011) Probabilistic projections of the total fertility rate for all countries. Demography 48(3):815–839. https://doi.org/10.1007/s13524-011-0040-5
3. Balducci A, Karimi K, Kunzmann M (2019) The darker side of smart city development. Urbanistica 71(163):90–92. ISSN: 0042–1022
4. Barca F (2022) Disuguaglianze Conflitto Sviluppo.La pandemia, la sinistra e il partito che non c'è. Donizelli Editore. ISBN: 9788855222891
5. Braudel F (1981–1984) Civilization and Capitalism, 15th–18th Century. Harper and Row, New York
6. Brusa G (2007) Metodo e prassi estimativa: dal 16. sec. agli anni '70 del 20° secolo. Maggioli, Bologna. ISBN: 978–88–387–4106–7
7. Cabinet Office, Government Of Japan (2019) まち・ひと・しごと創生長期ビジョン (令和元年改訂版) 及び 第 2 期「まち・ひと・しごと創生総合戦略」(概要). 地方創生. https://www.chisou.go.jp/sousei/info/pdf/r1-12-20-gaiyou.pdf. Accessed June 1, 2022
8. Campi P, Pandolfi A, Terlizzi L (2016) Using the Geographical information systems for the quantitative and qualitative landscape analysis. 9th International Conference on Social Science and Humanities (ICSSH). https://hdl.handle.net/11311/1008156
9. Campi P, Pandolfi A, Terlizzi L (2017) GIS and landscape analysis. PEOPLE: International Journal of Social Sciences, Special Issue 3(1):629–643. Doi: https://doi.org/10.20319/pijss.2017.s31.629643
10. Chapple J (2004) The dilemma posed by Japan's population decline. Electronic journal of Contemporary Japanese Studies, Discussion Paper 5/2004. https://www.japanesestudies.org.uk/discussionpapers/Chapple.html
11. Conconi, M., Pandolfi, A. (2018). Doing Something Scenario: Teaching Students an Incremental Approach to Design Strategies for Development Countries. Int J Innovation and Res Educ Sci vol 5
12. Contin A, Ortiz P (2015) Qualitative Metropolis. A new culturally encoded concept with qualitative regeneration of formal and informal settlements. Rule Model TECHNE-J Technol Arch Environ 10:120–129. https://doi.org/10.13128/Techne-17508

13. Contin A, Ortiz PB, Galiulo V, Siahkali RS, Pandolfi A, Campi P, Lutukurthi S, Sathiwada R, Kumar K, Girgaonkar P (2021) Territorial intelligence project: governance for megalopolis urban-rural linkage pattern. Springer, Singapore. https://doi.org/10.1007/978-981-16-2023-2_3
14. Contin, A., Ortiz, P. Galiulo, V. (2022). The Inclusion of "Unequals:" Hotspot Network Strategy for a Metropolitan Agricultural Revolution Eluding Informality. In: Marinic G, Meninato P (eds) Informality and the city theories, actions and interventions. Springer Nature, Switzerland
15. Contin A (2006) Landmark e paesaggi interni. In: d'Alfonso E (eds) Milano Malpensa. La regione Urbana nello Spazio dei Flussi. Alinea, Firenze
16. Contin A (2015) QUESTO Architettura Metropolitana. Einaudi, Torino Contin A et al (2021) TellMe- Inaugural book. Cippec, Buenos Aires
17. Contin A, Del Bosco M, Frigerio A (2014) SMART.MAP: Smart Sensitive City New Mapping tor 21st Century Citizens. The Electronic Journal of Communication. Communication Institute for Online Scholarship, New York Fujisawasst (20/10/2020). Sustainable Smart Town. https://fujisawasst.com/EN/
18. Contin A, Kim J (2019) Knowing how to grasp the power of the place: the project for the latin american metropolitan centralities. 55th ISOCARP World Planning Congress Jakarta-Bogor, Indonesia International Society of City and Regional Planners. https://dryfta-assets.s3.eu-central-1.amazonaws.com/assets/isocarp2019/eventdocs/1576665851ISOCARP_55th-Congress_Proceedings2.pdf
19. Contin A, Lanfranchi G (2017) The Rise of a new Discipline to Manage Metropolitan Urban system. In: Rajack DGÁR, Moreno EL, Lanfranchi G (eds) Steering the Metropolis Metropolitan Governance for Sustainable Development Goals. Cippec, Buenos Aires. Doi: https://doi.org/10.18235/0000875
20. Contin A, Ortiz PB, Zammataro A (2015) Towards a metropolitan approach for the definition of a network of intermediate cities, in regional development dialogue. United Nations Centre for Regional development, Nagoya
21. Contin A et al (2021) TellMe- Inaugural book. Cippec, Buenos Aires
22. Cook IR, Swyngedouw E (2012) Cities, social cohesion and the environment: towards a future research agenda. Urban Stud 49(9):1959–1979. https://doi.org/10.1177/004209801244488
23. Corner J (2006) Terra fluxus. In: Waldeim, C (eds) The landscape urban Reader. Princeton, New York
24. David PA, Foray D (2001) An introduction to the economy of the knowledge society. UNU-MERIT Research Memoranda, Tokyo
25. Deng Q, Erling L, Yang Y (2022) Politics, policies and rural poverty alleviation outcomes: Evidence from Lankao County. China. Habitat Int 127:102631. https://doi.org/10.1016/j.habitatint.2022.102631
26. Fernandez D, Alegre E (1991) Kinilaw. Makati, Metro Manila
27. Forero R, Sietchiping R (2022) Metropolises overcoming the COVID-19 pandemic: an urgent call for territorializing global agendas at subnational levels. In: Nunes Silva C (eds) Local government and the COVID-19 pandemic. Springer Nature, Switzerland
28. Fujii Y (2005) Shrinkage in Japan. Shrinking Cities 1:96–100. Hatje Cantz Verlag, Ostfildern-Ruit
29. Gao X, Anqi X, Lun L, Ouping D, Min Z, Jing L, Yali W (2017) Understanding rural housing abandonment in China's rapid urbanization. Habitat Int 67:13–21. https://doi.org/10.1016/j.habitatint.2017.06.009
30. Ginsburg N (1991) Extended metropolitan regions in Asia: a new spatial paradigm. In: The extended metropolis: settlement transition in Asia. University of Hawaii Press, Honolulu
31. Gottmann J (1957) Megalopolis or the Urbanization of the Northeastern Seaboard. Econ Geogr 33(3):189. https://doi.org/10.2307/142307
32. Gottmann J (1961) Megalopolis: the urbanized Northeastern seaboard of the United States. The Twentieth Century Fund, New York
33. Gottmann J (1976) Megalopolitan Systems around the World. Ekistics 41(243):109–113

34. Gottmann J, Harper RA (1967). Metropolis on the Move: Geographers Look at Urban Sprawl. John Wiley & Sons Inc: Hoboken (US)
35. Gottmann J (1961) Megalopolis: The urbanized northeastern seaboard of the United States. The twentieth century fund, New York
36. Gu D, Andreev K, Dupre ME (2021) Major trends in population growth around the World. China CDC Wkly 3(28):604–613. Doi: https://doi.org/10.46234/ccdcw2021.160.
37. Haase A, Rink D, Grossmann K, Bernt M, Mykhnenko V (2014) Conceptualizing urban shrinkage. Environ Plan A: Econ Space 46(7):1519–1534. https://doi.org/10.1068/a46269
38. Han Y, Yi C, Yi P, Ke G, Jin Yi B (2022) 'One village one product' can go a step further. Rural Revital 2022–1–5:16–38. http://www.zgxczx.cn/content_35521.html
39. Hebbert M (1994) Sen-Biki amidst Desakota: urban sprawl and urban planning in Japan. In: Planning for cities and regions in Japan. University Press, Liverpool
40. Huguet JW (2003) Can migration avert population decline and ageing in East and Southeast Asia? J Popul Res 20(1):107–124. https://doi.org/10.1007/bf03031798
41. Joint Research Centre (2023) Digital transition. Portfolio 14. ISBN 978-92-76-99490-9
42. Kamata M (1987) Study on the classification of coexisting phenomena and problems on planning in agglomeration. J Arch, Plan Environ Eng 375:104–113. https://doi.org/10.3130/aijax.375.0_104
43. Latz G (1991) The persistence of agriculture in urban Japan: an analysis of the Tokyo Metropolitan Area. In: The extended metropolis: settlement transition in Asia. University of Hawaii Press, Honolulu
44. Lim-Castillo P (2006) Traditional Philippine Vinegar and their role in shaping the culinary culture. In: Authenticity in the kitchen. Prospect Books, London
45. Long H, Yansui L, Xiubin L, Yufu C (2010) Building new countryside in China: a geographical perspective. Land Use Policy 27(2):457–470. https://doi.org/10.1016/j.landusepol.2009.06.006
46. Ma X, Yan Q, Pan Q, Chen X, Li G (2023) Identification and classification of urban shrinkage in Northeast China. Land 12(6):1245. https://doi.org/10.3390/land12061245
47. Magnaghi A (2006) Dalla città metropolitana alla (bio)regione urbana. In: Marson A (a cura di), Il progetto di territorio nella città metropolitana, Alinea. Firenze p. 69–112
48. Magnaghi A (2014) La bioregion urbaine. Petit traité sur le territoire bien commun, Eterotopia, Paris. http://www.eterotopiafrance.com/catalogue/la-bioregion-urbaine/
49. Martin WT (1957) Ecological change in satellite rural areas. Am Sociol Rev 22(2):173. https://doi.org/10.2307/2088854
50. Masuda H (2015) To avoid the emergence of a polarized society. Jpn J R Estate Sci 29(2):28–35. https://doi.org/10.5736/jares.29.2_28
51. McGee, Terry G. (2013) The Emergence of Desakota Regions in Asia: Expanding a Hypothesis. In: Brenner N (eds). Implosions/explosions: towards a study of planetary urbanization. JOVIS Verlag GmbH, Boston pp 121–137. https://doi.org/10.1515/9783868598933-010
52. McGee T (1991) The emergence of Desakota regions in Asia: expanding a hypothesis. In: Ginsberg N, Koppel B, McGee TG (eds). The extended metropolis: settlement transition in Asia. Haxaii Press, Honolulu
53. Ministry of Agriculture Forestry and Fisheries (2022) 食料自給率 - Maff.go.jp.知ってる?日本の食料事情. https://www.maff.go.jp/j/zyukyu/zikyu_ritu/attach/pdf/012-2.pdf. Accessed June 1, 2022.
54. Morin E (1993) Introduzione al pensiero complesso Gli strumenti per affrontare la sfida della complessità. Sperling & Kupfer, Milano
55. Moss T (2008) 'Cold spots' of urban infrastructure: 'shrinking' processes in Eastern Germany and the modern infrastructural ideal. Int J Urban RegNal Res 32(2):436–451. https://doi.org/10.1111/j.1468-2427.2008.00790.x
56. Nagashima C (1981) The Tokaido Megalopolis. Ekistics Japan's Organ Space 48(289):280–301
57. National Economic Development Authority Regional Office 3 (2017) Central Luzon Regional Regional Development Plan 2017–2022. https://neda.gov.ph/ wp-content/uploads/2018/02/3-Central-Luzon-RDP-2017–2022.pdf

58. National Economic Development Authority Regional Office 3 (2021) Central Luzon water supply and sanitation databook and regional roadmap. https://neda.gov.ph/wp-content/uploads/2021/09/03-Region-3-Databook-and- Roadmap_4June2021.pdf
59. Ono J (2014) 森林面積の長 期的変 変遷 - Lij.jp. 土地総研リサーチ・メモ. — 般財団法人土地総合研究所. https://www.lij.jp/news/research_memo/20141128_3.pdf
60. Ortiz P (2013) The art of shaping the metropolis. McGraw-Hill Education, New York
61. Ortiz P (2020) Megapolis governance new scale, new challenge, cautious solutions. http://www.pedrobortiz.com/
62. Ortiz PB, Kamiya M (2016) The political economy of the metropolis in the Global North and South: Connecting, financing, and ruling. In: Governing the metropolis. World Bank, UN-HABITAT, IADB
63. Ostrom E (2010) Beyond markets and states: polycentric governance of complex economic systems. Am Econ Rev 100(3):641–672. https://www.jstor.org/stable/27871226
64. Oswalt P, Rieniets T (2006) Atlas der schrumpfenden Städte/Atlas of Shrinking Cities. Hatje Cantz Verlag, Ostfildern-Ruit
65. Oswalt P (2005) Introduction. In: Shrinking Cities, Volume 1 International Research. Hatje Cantz Verlag, Ostfildern-Ruit
66. Pagano U, Rossi M (2009) The crash of the knowledge economy. Camb J Econ 33(4):665–683
67. Pandolfi A (2019) The importance of including the value of soil in metropolitan planning strategies. Int J Eng Tech Res 8(12):664–669. https://doi.org/10.17577/IJERTV8IS120359
68. Paolillo PL (2002) Problematiche del parametro suolo. Uno sguardo preoccupato alla situazione regionale. Franco Angeli/Urbanistica, Milano
69. Prigge W (2005) On the origins of shrinkage. In: Shrinking Cities, Volume 1 International Research. Hatje Cantz Verlag, Ostfildern-Ruit
70. Ripple WJ, Wolf C, Newsome TM, Barnard P, Moomaw WR (2019) World scientists' warning of a climate emergency. Bioscience 70(1):8–12. https://doi.org/10.1093/biosci/biz088
71. Rola AC, Pulhi JM, Tabios GQ, Lisada JC, Dayo MHF (2015) Challenges of water governance in the Philippines. Philipp J Sci 144(2):197–208. https://philjournalsci.dost.gov.ph/images/pdf/pjs_pdf/vol144no2/pdf/challenges_of_water_governance_in_the_Phils_FinalCopy_05_April_2016.pdf
72. Sakamoto J, Kamata M (1996) Study on Tendency of the coexisting phenomena in Tokyo Metroporitan Area: through the comparison between 1980 and 1990. J Arch Plan 61(479):149–158. https://doi.org/10.3130/aija.61.149_1
73. Sancez Fuentes D, Mascort-Albea EJ (2021) Metropolitan metabolism: the ecological footprint. In: Contin A, Giordano P, Nacke M (eds) Training for education, learning and leadership towards a new metropolitan discipline Inaugural book. Cippec, Buenos Aires
74. Shane DG (2005) Recombinant urbanism. Conceptual modeling in architecture, urban design and city theory. Wiley-Academy, West Sussex
75. Siddiqui MH (2013) Industrial and agriculture in urban land use planning. Random Publications, New Delhi
76. Sokolov R (1991) Why we eat what we eat: How Columbus changed the way the World eats. Atria Books, New York
77. Sona G, Pandolfi A (2021) Metropolitan cartography, remote sensing and geographic information systems. In: Contin, A. (eds) Metropolitan landscapes. Springer Nature, Singapore
78. Sorensen A (2019) Tokaido megalopolis: lessons from a shrinking mega- conurbation. Int Plan Stud 24(1):23–39. https://doi.org/10.1080/13563475.2018.1514294
79. Sorensen A (2004) Megalopolitan Development and the transformation of rural Japan: sustainability implications of extended metropolitan regions in Asia Human Settlement Development. In: Sassen S (eds) Encyclopedia of Life Support Systems (EOLSS). http://www.eolss.net
80. Tanaka, K. (1973). Building a New Japan: a plan for remodelling the Japanese archipelago (Simul International, Inc., Trans.). Simul Press.
81. Torres CE (2000) The Americanization of Manila 1898–1921. QC University of the Philippines Press, Manila

82. UN-Habitat (2019) Urban-rural linkages: guiding principles and framework for action to Advance Integrated Territorial Development. UN-Habitat, Nairobi
83. United Nations Department of Economic and Social Affairs, Population Division (2001) Replacement migration: is it a solution to declining and ageing populations? (ST/ESA/SER.A/206)
84. United Nations Department of Economic and Social Affairs, Population Division (2019a) World Population Prospects 2019: methodology of the United Nations population estimates and projections (ST/ESA/SER.A/425)
85. United Nations Department of Economic and Social Affairs, Population Division (2019b) World Urbanization Prospects: the 2018 revision. https://doi.org/10.18356/b9e995fe-en
86. Uto M (2018) Examining environmental risks from the viewpoint of de- population and infrastructure issues. J Natl Inst Public Health 67(3):306–312
87. Yabes R (1995) Zanjeras and the Ilocos Norte Irrigation Project: lessons of environmental sustainability from Philippine traditional resource management systems. In: Strong M, Ghai D, Vivian JM (eds) Grassroots Environmental Action, 1st Edition. Routledge, Abingdon. ISBN: 9781315832166

Antonella Contin is Coordinator of MSLab—Measure and Scale of the Contemporary city. She is scientific and full organizational responsible of international workshop, seminar and projects. Her research deals with evaluation methods and design strategies of metropolitan dimension, urban sets and landscapes. In particular her recent work is related to developing "a Metropolitan Discipline" that integrates scales of territorial development, institutional levels and disciplines, including new approaches to mapping. Since 2017 she takes part to the activities of the UN-Habitat MetroHUB.

Valentina Galiulo is an architect whose research focuses on Metropolitan Cartography as a methodological tool for sustainable and integrated development in contemporary metropolitan cities. She holds Master of Science in Architecture and Architectural Engineering at Politecnico di Milano, and a Bachelor's Degree in Science of Architecture at Politecnico di Milano. She worked at the University of Seville in Dep. Urbanism and Territory of Universidad de Sevilla for Traineeship Erasmus +. She followed experience at CONICET National Research Center and in University of Cuyo of Mendoza (Argentina). She is a research fellow in DASTU Polimi for TELLme European project; she is a Ph.D. student in Architecture ETSA (Seville) for City, Territory and Habitas research line.

Alessandra Maria Pandolfi holds a double degree MSc in "Environmentally Sustainable Urban Planning" at the Politecnico di Milano (2006, Best Graduate of her course) and Torino (2008). She graduated (2008) with merit in the 1st cycle of Alta Scuola Politecnica. After two specialization courses in GIS (2007) and Real Estate (2008–2009), in her Ph.D. (2009–2012), she studied the economic values of development rights and their transfers in Italy and in other countries. Since 2010, she is adjunct professor of "Project Appraisal and Evaluation" at the Politecnico di Milano and author of about 60 international publications.

Ryusei Koike is an M.Sc. student at Politecnico di Milano, specializing in Sustainable Architecture and Landscape Design. He holds a B.Sc. degree with honor in Architectural Design from the same university. His graduation thesis entitled "Desakota Shrinkage. The Asian post-megalopolitan shrinkage model through the case of Tokyo, Japan" (2022) called for a distinguished spatial approach to the upcoming population shrinkage of Asian landscapes. He has special interest in the transformation of contemporary rural landscapes.

Dicheng Yang is a PhD candidate of Politecnico di Milano with eight years study experience of Architecture Design and Urban Planning. He focus on Chinese rural studies. In collaboration with Alibaba Group and the International Forum of Urbanism (IFoU), he has conducted research on the challenges arising from E-commerce and the urban-rural relations in east China (2020). After graduating with honors from Politecnico di Milano in 2022, his research centers on urban studies, incorporating perspective from the neuroscience discipline. His research specializes in people's well-being and perception related with urban complexities.

Kelvin C. de Chavez is a practicing Architect and Environmental Planner. He is also a tenured Faculty at the University of the Philippines, College of Architecture. He holds a MSc. degree in Urban Planning and Policy Design from Politecnico di Milano, Italy. He currently is doing his PhD in Sustainability and Urban Regeneration at the Universidad Politecnica de Madrid, Spain. His research is focused on metrics for climate change during the contemporary metropolitan age.

Japan

Japanese Methods for Promoting Harmonious Regionalization in Asia

Chris G. Pope and Aki Suwa

Abstract This chapter examines the ways in which Japanese leaders in the lithium industry have sought to adapt to the changes in the international political economy, particularly with respect to the varying states of Sino-Japanese relations. To do so, it first provides an overview of the ways in which regional trade systems are transforming and the impacts this exerts on the Japanese economy. Following this, the chapter provides a qualitative examination of the methods and considerations of adaptation among Japanese businesses in the lithium industry, particularly with respect to businesses operating within Osaka as a major industrial center. By way of conclusion, the chapter offers a broad discussion of the potential avenues for further development for Japanese businesses with a view on how to develop a win–win dynamic based on economic cooperation and a harmony of interests across central and east Asia.

Keywords Japan · China · Lithium · Development · Globalization · International political economy · Trade

1 Introduction

This chapter examines and assesses the ways in which Japanese corporations have adapted to globalization, taking up the companies in Osaka involved in the lithium industry as a case study. This is because this chapter also aims to contemplate the ways in which Japanese corporations can continue to adapt to the profound developments taking place on the Asian continent in terms of international trade. While there are legitimate concerns over the efficiency of lithium batteries and the environmental damage caused by lithium mining [1], the chapter focuses on the lithium industry

C. G. Pope (✉)
Department of Commerce, Fukuoka University, 8-19-1, Nanakuma, Jonan-ku, Fukuoka-shi, Fukuoka-ken 814-0133, Japan

A. Suwa
Department of Contemporary Society, Kyoto Women's University, 35 Imakumano Kitahiyoshi-cho, Higashiyama-ku, Kyoto-shi, Kyoto-fu 605-8501, Japan

T. M. Vinod Kumar (ed.), *Indo-Pacific Smart Megacity System*, Advances in 21st Century Human Settlements, https://doi.org/10.1007/978-981-99-6218-1_5

because it is nonetheless considered a competitive driver of new areas of economic growth, and there are significant reserves located in Asia, particularly in China. It therefore constitutes an appropriate case study to examine the ways in which economic integration in the region may develop in the future.

Lithium is of course used in lithium-ion batteries, which are installed in an array of so-called smart technologies linked to a new wave of technological development [2]. However, its uses in fact apply to a wide array of goods from lubricant greases to pharmaceuticals to being used as an alloying agent for the production of metals in the aviation industry, and more [3]. It is even used to produce collection lenses employed by intergovernmental organizations such as CERN and the U.S. Fermi National Accelerator Laboratory [4–6]. As a result, the lithium industry constitutes an appropriate case study to both elaborate on and assess the structure and mechanisms of regional economic integration between Japan and other Asian countries. The position this chapter adopts is that development can be achieved in a "win–win" dynamic as advocated by many transcontinental economic development programs in the past [7, 8]. Therefore, it is of paramount importance to understand how Japan will be able to adapt to new systems of trade and potentially realize jointly held interests in the region centered on economic development which elevates public welfare. Elaborating on this is the principle goal of this chapter.

To do so, the text is separated into four sections as follows: The first section summarizes recent developments in the formation of new systems of international trade and finance in order to suitably contextualize the problems and opportunities that the Japanese economy presently faces internationally. The second section then details the course charted by the Japanese economy in the so-called postwar period, from the end of the Second World War to the present day. This section explains in more detail exactly the issues the Japanese economy faces and its causes. Having suitably situated the issue at hand in its broader context, the third section then discusses the ways in Japanese companies have, up to now, responded to the changes in the international political economy (IPE) with respect to the lithium industry. Based on this, the fourth section discusses how lithium battery manufacturers in Japan might navigate changes in the IPE, including potential changes in Sino-Japanese relations, as regional neighbors, and its potential impact on their operations and the industry as a whole. Following this, a summary of the chapter as a whole is provided in the conclusion.

2 A New Economic Paradigm in the Global South

In recent years, developing countries have begun to implement new methods of trade, portending seismic changes to the IPE and therefore the mechanics of economic regionalization and development. The BRICS nations (Brazil, Russia, India, China, and South Africa), comprising some of the world's leading growth economies, are pioneering new means of cross-border trade. For example, they recently discussed the expansion of new mechanisms and frameworks of international trade as an alternative to the U.S. dollar at the summit in Johannesburg, South Africa on August 22, 2023 [9]. This is nothing new, but is in fact part of a larger discussion that has been ongoing since at least 2018 [10].

Of course, this does not pertain solely to the BRICS nations but is extended out to include as many countries that wish to participate. Both China and Russia have already launched new cross-border interbank payment systems and rapidly expanded trade in non-dollar currencies across the world in order to strengthen their own global commodities trading systems. Brazil and China agreed to carry out bilateral trade in their own currencies at the Brazil-China Business Seminar on March 29, 2023, while Saudi Arabia is considering accepting oil sales to China in yuan [11, 12]. This would not be a surprise, as most oil exporters to China already accept yuan, with China both having launched the yuan-denominated crude oil futures market in Shanghai in 2018 and allowing exporters to convert revenue into gold on the Shanghai and Hong Kong exchanges. With the onset of viable alternatives to the dollar for international trade, the current international financial system based primarily on international trade in U.S. dollars is at an impasse.

The architecture of international trade is indeed changing. For example, India, Iran, and Russia are involved in expanding energy trade across the International North South Transportation Corridor (INTSC) which spans 7000 km from India across the Arabian Sea through Iran and passed the Caspian Sea to Russia, while both the Iranian and Russian central banks are studying the adoption of a stable coin for foreign trade settlements [13]. Furthermore, India has stepped up plans to internationalize the rupee, with the Reserve Bank of India allowing "authorized dealer" banks from, up to now, 18 countries across the world to open up Special Rupee Vostro Accounts for settlement in rupees, amidst plans to link its domestic financial messaging system with financial systems alternative to SWIFT [14, 15]. Meanwhile, President Widodo of Indonesia has urged the country to ditch foreign payment systems such as MasterCard and Visa in favor of Indonesia's domestic alternative, the National Payment Gateway [16]. In the fourth quarter of 2022, the IMF recorded a drop in the share of U.S. dollars held as allocated reserves by central banks to its lowest level on record (from 1995), while Algeria, Argentina, and Iran have applied for membership to BRICS, with Indonesia, Saudi Arabia, Türkiye, United Arab Emirates, and many others also expressing an interest [17, 18]. This is because new systems are being designed to facilitate a transition away from international trade based primarily on the U.S. dollar.

For Japan, this change could not be more portentous. As outlined in the following section below, Japan's manufacturing bases are located in Asia, where these changes are taking place, while their chief export market is the United States., the home of the dollar. However, the United States and other Western economies are suffering from an illiquidity crisis—an effusive wound which their respective central banks tried to bandage up by implementing quantitative easing (QE) for fifteen years following the financial recession of 2008. QE was Western economies' attempt to drive up liquidity in order to increase lending and hence credit-based growth in their economies. This was the alleged reason and the means involved offering ultra-low (sometimes minus) interest rates on both reserves (bank money) held as deposits at central banks and on lending, and repurchasing government bonds (sovereign debt) held by financial institutions to control bond prices for budgetary purposes and greatly increase central bank reserves.

Crucially, however, this did not involve any rigorous mechanism for the guidance of credit flow, unlike growth states such as China, India, Vietnam, and others. Thus, Western financial institutions funneled credit into non-productive assets, causing an enormous financial bubble. Financial institutions were able to profit from arbitrage on interest rates by taking out loans at low cost, speculating on/investing in securities with higher yields and then pocketing the difference, all at the expense of the real economy which, without rigorous investment, stagnated. Due to this, commercial banks began to make loans of regressively worse quality so that they could be used as a springboard for more lucrative shadow banking activities and other forms of financial speculation. Meanwhile, other countries had to sell goods with intrinsic value—in the sense that they contribute to elevating people's welfare—for a fiat currency with diminishing purchasing power or the promise thereof in the form of U.S. Treasuries, for instance. This speculative activity also pushed up asset prices in developing countries, while trade surplus countries saw much of their surplus recycled back into the U.S. financial sector as central banks bought up U.S. securities in order to recalibrate exchange rates with the dollar to ensure that exports would not lose competitiveness. In other words, financialization has sucked out resources that could have been used for revitalizing the real economy and realizing robust, sustainable growth. This is why developing countries such as the BRICS nations are seeking an alternative.

However, as a result of this, Western economies have become increasingly reliant on financial markets and now must ensure that the music keeps on playing. Therefore, there are frequently new rounds of QE to make sure newcomers do not miss their margin calls and thus trigger a financial panic. Owing to the issue of rehypothecation and counterparty risk, there is no real knowing of exactly how bad a crisis would be and exactly what would trigger it. However, according to the Bank of International Settlements (BIS), there are more than $80 trillion in outstanding obligations to pay in U.S. dollars in short-term, "off-balance sheet" foreign currency swaps/forwards. This collectively exceeds the stocks of all U.S. Treasury bills, repo, and commercial paper combined, and is also well in excess of global trade which allegedly stands at only $29 trillion [19]. With demand for the dollar declining, adducing that there is potential here for a catastrophic liquidity squeeze and economic downturn does not constitutea a leap of the imagination. Indeed, historians may come to view the music of the financial sector as an historical parallel to Emperor Nero playing his fiddle as Rome burned to the ground.

As for Japan, it is unable to excise itself from the risk of economic collapse, particularly if it emanates from Western economies. Its production base has relocated principally to mainland China and Southeast Asia where it has relied on its well-developed supply chains across the region, exporting products to North America, Europe, and, increasingly, China, as the country looks to become a consumption hub in Asia to avoid the so-called middle-income trap [20]. Japanese investors and the Bank of Japan are collectively the largest foreign holder of U.S. sovereign debt while Japanese investors have also invested heavily in Western equities and securities in order to profit from the higher yields offered as the Bank of Japan committed the country to QE. Thus structural inflation brought about by the stagnation of the real economy and financialization in Western economies as well as economic sanctions poses a risk for Japanese exporters as it diminishes the purchasing power of consumers in export markets. However, quantitative tightening and monetary tapering in response is not much better as it will cause Japanese investors to suffer unrealized losses on financial assets held on their books, and thus pose a substantial risk of triggering a financial panic as investors look to hedge against currency risk and the exposure of asset impairment across financial markets.

Traditionally, Japan has relied on playing a "bridge-building" role between East and West for decades with varying degrees of success. Although it is likely to continue to play this role, it is a fair assessment to suggest that because one economic bloc is apparently in the process of sinking and the other is charting a course to new waters, Japan is unlikely to be able to continue straddling two different boats as it has done while the world relied on U.S. dollars for international trade. Rather, economic development in Asia presents an opportunity for Japan to acquire foreign demand for Japanese goods and services, while Japan's banking sector can continue to build bridges across different regions as new infrastructure projects across Asia develop more markets from which to profit. What must take place, however, for this to be successful is an adaptation to the burgeoning systems of international trade and finance in the region. However, given that these systems are still in the process of development, there is no clear answer or blueprint on which to rely, regarding the question of how to do so.

The following section, therefore, addresses and details the challenges that beset the Japanese economy, including the considerations of Japanese companies in how to manage risk and uncertainties in IPE. It does so by outlining Japan's course of economic development in the postwar period in order to assess in the following sections the ways in which it may adapt to the developments outlined above. It is to this discussion that the chapter now turns.

3 Postwar Japanese Economic Development

While Japan has its own unique economic systems, institutions, and customs, its lack of natural resources, particularly as it pertains to energy, means that it must import from abroad—making it, in a sense, highly internationalized. While this section details Japanese postwar economic development in broad overtures, this fact is of foundational importance to understand Japanese economic strategy and must not be overlooked.

In the postwar period, Japan reconfigured its war-time economy to an export-intensive economy based on credit-led growth. Benefiting from the "special demand" (*tokuju*) from U.S. troops for the direct procurement of goods and services during the Korean War (1950–1953), the Japanese economy acquired a foreign source of demand necessary to rapidly redevelop but at the cost of becoming highly dependent on exporting products to the United States. In this way, Japan played a bridge-building role between West and East as Japanese economic development helped to entrench the new Bretton Woods monetary system where the dollar was convertible to gold at a guaranteed rate and the currencies of all other member states to the IMF had to stay within a 1% range to the dollar. The benefits of Japanese economic growth for the Asian region were felt also first by the "Asian Tiger" economies (Hong Kong, Singapore, Thailand, and Taiwan), and then others. As the economy industrialized, Japanese mass production began to shift its focus on cheap goods, metals, and heavy chemicals (many of which relocated throughout Southeast Asia and then China in the ensuing decades), to technologically sophisticated products. Nonetheless, the bulk of products manufactured were destined for foreign markets, mostly American and European, where the revenue earned would be recycled into the Japanese economy to further the drive for postwar reconstruction and becoming an "advanced" economy again.

The relocation of Japanese industries was expedited by the change in the international monetary system from its foundations fleshed out during the 1944 Bretton Woods conference to the floating exchange rate. It caused the internationalization of the Japanese economy to take on another form, however, as rather than being reinvested in Japanese industries, much of the surpluses earned were used to purchase U.S. sovereign debt and financial assets instead. Firstly, in response to the oil shocks of the 1970s, Japanese businesses underwent radical changes. Industries that are easily moved abroad began to relocate, while others had to rely on innovations in the production process or otherwise face stagnation and eventual bankruptcy. As Western economies began to financialize, manufacturers moved abroad where production costs were lower and Japan followed suit so as to maintain its competitive edge, particularly given the large discrepancy in the stages of economic development between Japan and its regional neighbors. This process was facilitated by the Plaza Accords of 1985, where the Japanese yen was artificially lowered to the U.S. dollar, because Japanese exporters lost their competitiveness and so relocated operations overseas, particularly to Southeast Asia. Consequentially, developing countries were suffering from debt crises and so, Japanese economic activity within these regions would likely have been welcome for nations struggling to meet their international financial obligations.

However, Japan had long since sought to design methods to insulate both Japan and Southeast Asia from the heightened currency risk brought about by the onset of the floating exchange rate system with, for example, the development of the Asian Monetary Fund and the New Miyazawa Initiative [21]. Nonetheless, the collapse of the Japanese bubble economy in 1991 ultimately portended a decline in Japanese politico-economic influence in the region. Instead, Japan has had to contend with decades of deflation and the slow bottoming out of domestic demand for credit owing both to financial crises and a declining population base. Japan's numerous city banks have now merged largely into to five highly internationalized megabanks, while Japanese regional banks, which had traditionally supported Japanese regional communities and SMEs, have slowly declined with the so-called shutter economy of closing businesses and a diminishing customer base due to low birth rates [22].

Unlike Japanese smaller businesses, the large corporations and financial institutions have largely profited from low-interest rates in Japan by investment in long-term securities, particularly OTC derivatives, and equities with higher yields abroad. Small-to-medium-sized enterprises traditionally benefitted from Japan's developmentalist model of growth including the use of deposits held at Japan Post, Japan's public bank. This was part of Japan's infamous Fiscal and Investment Loan Program which provided a means of financing public works throughout Japan without overburdening citizens with tax and has been emulated and developed upon in today's growth economies such as China also. However, with its privatization finalized in 2015, SMEs and Japan's regional economies have had to adapt to a new reality with less state guidance and support when it comes to the use of credit. At the same time, larger companies have relied on retained earnings to finance projects and not Japanese banks per se, causing a divide in the Japanese economy which has become more pronounced as globalization progressed.

Given this transformation in the economy and the shrinking of the population in Japan, large businesses have opened up shop abroad. For instance, Japan invests approximately 5.1 trillion yen in Asia, 4.6 trillion yen in Europe, and 8.2 trillion yen in North America [23]. Roughly 60 percent of investment in Europe and 75 percent in North America is in non-manufacturing industries. However, non-manufacturing industries such as finance and insurance, and wholesale and retail, increasingly comprise a large portion of Japanese direct investment also in Asia, representing 56% of investment overall [23]. This is due to the increasing purchasing power of developing countries such as China, whose private consumption is believed to match the United States at current levels by 2030 [24].

Japan is indeed largely dependent on its primary income as opposed to a trade surplus to balance its current account owing to the slow decline of its domestic economy. The question, therefore, concerning which region will be the most profitable to invest in in the years ahead, is crucial to the country. Nonetheless, since the collapse of the bubble economy in 1991, Japan has implemented monetarist methods to simulate market conditions favorable to credit creation in order to prop up the economy. The Bank of Japan has maintained low-interest rates, provided financial institutions with a large amount of central bank reserves, bought up a large proportion of stock and equities from Japanese markets, controlled the price of 10-year government bonds—largely taken to be an indicator of future interest rates—and both promised and carried out economic liberalization measures, all, ostensibly at least, to convince its large banks to invest in the real economy. Under Abe Shinzō (2012–2020), in particular, this was combined with an expansive fiscal policy, in an attempt to kick-start investment in Japan, while other administrations such as Jun'ichirō Koizumi's (2001–2006), relied on austerity so as to not crowd out private investment. Nonetheless, all these efforts were carried out under the pretense of convincing financial institutions that conditions are good for lending and that the central bank would not ruin the party any time soon by causing lending costs to rise. Unlike the many growth-centers in Asia today, however, this came with little-to-no guidance from the state as to how credit should be used. Thus, we see that credit creation in Japan has not been lower in over seven decades and the country is in need of demand for credit from its commercial banks [21], while government debt has never been higher. The central bank owns most of the bonds it has issued, including for a time over 100% of newly issued 10-year government bonds. As a result, the Bank of Japan is hamstrung because it needs to buy bonds in order to mark up their price so as to cover expenses in the following year's budget with bond issuance, but simultaneously it faces a serious and seemingly chronic illiquidity issue within the Japanese economy. This has caused the Japanese yen to plummet to historical lows [25] and has fomented concerns among investors that at some point, the Bank of Japan will have to relax its yield curve control and allow lending costs to rise after all which will place pressure on Japanese citizens and its export industries.

Needless to say, one crucial thing that is necessary to fix this issue for Japan is demand for Japanese credit for productive investment whether domestic or from abroad. The new system of international trade pioneered by developing countries and emerging economies will likely focus on infrastructural improvements and assessing economic performance based on metrics of public welfare as opposed to the appreciation of financial assets designed for speculative activities. In such a scenario, Japan can utilize its position as a nation that exports highly sophisticated technologies to help developing countries within the region increase their productive power and hence would benefit from the gains achieved therein. This is not a new concept for Japan. As outlined above, it comprised the basis of its postwar economic reconstruction also. In other words, there is opportunity for the realization of an harmony of interest between various actors within Japan and Asia which would result in the elevation of public welfare for all concerned. What is of crucial importance here, however, is assessing the ability and means by which Japanese businesses are able to adapt to the challenges presented with due to such developments in the IPE. To provide a tentative assessment thereof, the following section presents the case of study of how companies in Osaka, Japan, have responded to the emergent and developing lithium industry on the Asian continent.

4 Cooperative Adaptation of Lithium Industries in Osaka

In order to assess the ability and means for Japanese business to adapt to changes in the IPE, the battery industry can provide an indicative example, with its resource trade and web of network for supply chains. Lithium-ion batteries are installed in a wide range of electronic and electronic devices, including smartphones, mobile phones, notebook PCs, digital cameras, video and tablet terminals, and music players. In addition, practical use is progressing as a power source for electric and hybrid vehicles. Furthermore, research and development is progressing as a power storage device for power leveling and smart grid [26].

Growth is expected particularly in the automotive lithium-ion battery market. The shift to environmentally friendly electric vehicles is intensifying, and demand is expected to expand rapidly in the future due to the expansion of production of eco-friendly vehicles by automakers and the reduction in the cost of lithium and ion batteries for automobiles. Lead-acid batteries, nickel, hydrogen, batteries, and lithium-ion batteries are mainly used in automotive secondary batteries, but lithium-ion batteries have the highest performance, and in terms of energy density, they are about four times that of lead batteries and about twice that of nickel-metal hydride batteries. For this reason, it is indispensable for EVs, which require particularly high energy density. In the past, there was a limit to the installed capacity due to the high cost, but in recent years, the cost of lithium-ion batteries has been dramatically reduced, and in some cases, they are equivalent to lead-acid batteries.

Lithium-ion batteries were put into practical use by Asahi Kasei and Sony in 1991. In 1994, Sanyo Electric commercialized a lithium-ion battery made of graphite carbonaceous material. In 1998, a lithium-ion polymer battery that uses polymer as the electrolyte was introduced to the market. It means there was a time when Japan manufacturers had a high market share of lithium-ion batteries. Panasonic (having Sanyo Electric under its umbrella), Sony, Hitachi Maxell Energy, NEC Energy Devices, etc. were known as the main manufacturers.

On the other hand, production in South Korea (Samsung SDI, LG), China (BYD, etc.), and in Taiwan, etc., increased since 2010s, and Sanyo Electric, which had been the top until then, was overtaken by Samsung SDI on a shipment basis. Thus, the Japanese lithium-ion battery industry used to be seen as a significant player in the global market, but now faces challenges in the context of international lithium trade. One of the primary challenges is the dependence on imported lithium resources. Japan has limited domestic lithium reserves and relies heavily on lithium imports to meet the growing demand for batteries. The majority of lithium resources are concentrated in Argentina, Australia, Bolivia, and Chile, but there are reserves also in China and the United States. This concentration of reserves, however, creates a potential vulnerability in the supply chain, as any disruptions in the international lithium trade could impact the availability and cost of lithium for Japanese battery manufacturers.

Sino-Japanese international relations are indeed complex and multifaceted, and the relationship between Japan and China specifically in the context of lithium-ion battery development and manufacturing is no exception to this. Both countries are significant players in the global battery industry, and their relationship is characterized by cooperation as well as competition (Fig. 2) [27]. Competition between Japanese and Chinese battery manufacturers is intensifying. China has made significant strides in battery production, becoming the largest global producer of lithium-ion batteries. Its manufacturing capabilities, economies of scale, and cost advantages have allowed Chinese companies to dominate the battery supply chain. This has posed challenges for Japanese battery manufacturers to compete on cost and scale.

Another aspect of the Japan–China relationship is the supply chain for lithium and other raw materials. China is a major producer of lithium and controls a significant portion of the global supply. Japanese battery manufacturers heavily rely on imported lithium, and China's dominance in the supply chain creates a potential vulnerability for Japan's battery industry. Ensuring a stable and diversified supply of raw materials, including lithium, is crucial for the development and growth of the Japanese battery sector. In recent years, both countries have recognized the strategic importance of the battery industry and have taken steps to strengthen their domestic supply chains. Japan has been investing in research and development to advance battery technologies and reduce its reliance on imported materials. China, on the other hand, has focused on securing lithium resources and developing a vertically integrated battery industry.

Furthermore, the increasing global demand for lithium-ion batteries, particularly for electric vehicles (EVs) and renewable energy storage systems, has led to a surge in lithium prices. The competitive nature of the international lithium market, together with limited supply, has pushed the costs of raw materials up. This sets a challenge for Japanese battery manufacturers to remain competitive while ensuring a stable supply of lithium. In response to these challenges, Japan has been actively exploring strategies to secure a stable lithium supply. One approach is to diversify the sources of lithium imports by engaging with a broader range of countries that supply them.

Although Japan is investing in research and development efforts to improve lithium-ion battery technologies, with the aim to reduce its dependence on lithium resources and mitigate the challenges associated with international lithium trade, collaboration, and partnerships with lithium-producing countries are also being pursued to ensure a stable supply of lithium. Japan has been involved in various initiatives and agreements with resource-rich nations to secure long-term lithium supply contracts. These collaborations are imperative in thriving in the international market.

The lithium-ion battery industry in Japan is, as a result, undergoing accelerated restructuring. Sony transferred its lithium-ion and battery businesses to Murata Manufacturing in 2017. In 2019, Nissan Motor sold its joint venture subsidiary, Automotive Energy Supply (AESC), to the China Envision Group. In addition to electric vehicles, the use of lithium-ion batteries is expanding in a wide range of fields such as digital devices and power supply facilities, and related industries such as automobiles and electrical manufacturers are expanding their businesses, where CATL and LG Group accounting for half of the shipments. The presence of Japanese manufacturers is getting thin [27, 28] (Fig. 1).

To adapt to the changing global economic conditions, Japanese manufacturers have sought to establish relationships with international players through collaborations, partnerships, and market expansion efforts. One prominent Japanese lithium-ion battery manufacturer is Panasonic. Panasonic has been a key player in the industry, supplying batteries for a wide range of applications, including consumer electronics, electric vehicles (EVs), and energy storage systems. Panasonic's partnership with Tesla, an American electric vehicle manufacturer, is one of the notable collaborations in the industry. Panasonic supplies cylindrical-shape battery to Tesla's EVs and square-shape battery to EVs, HVs, and PHVs to other automakers. In addition to domestic production bases such as the Suminoe Plant and Kaizuka Plant in Osaka Prefecture for cylindrical-shape battery, the Kazo Plant in Hyogo Prefecture, the Tokushima Plant in Tokushima Prefecture, and the Sumoto Plant in Awaji Island for square-shape battery, a new square type plant has been built in Dalian, China, and

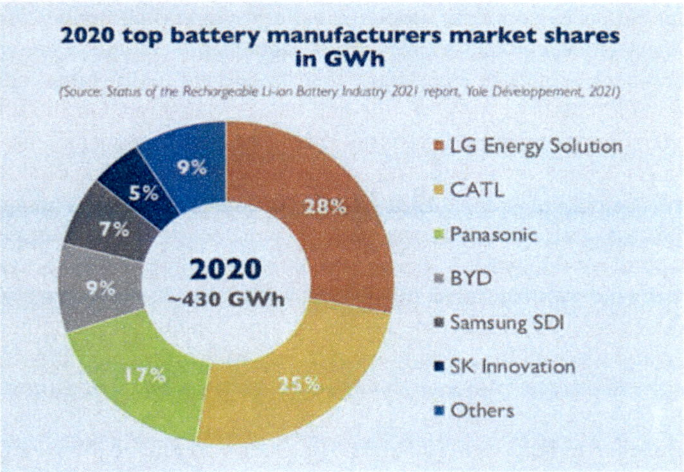

Fig. 1 2020 top battery manufacturers market share in GWh. *Source* Status of the Rechargeable Li-ion Battery Industry 2021 Report

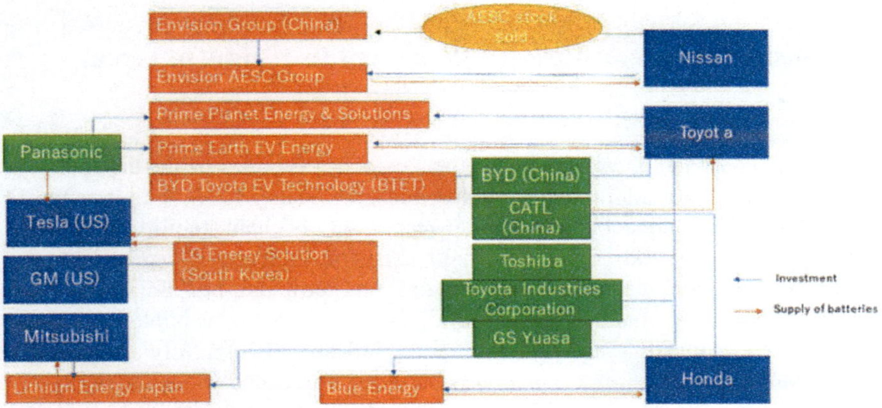

Fig. 2 Lithium-ion battery investment and supply relations [27]

mass production has begun. In North America, the company built one of the world's largest lithium-ion and battery factories with Tesla in Nevada [26] (Fig. 2).

Japanese business in local areas is no exception to adapt to this global challenge. The Osaka region in Japan, for example, has emerged as a significant hub for the lithium-ion battery industry, playing a pivotal role in the development and production of advanced batteries. The region's battery industry has witnessed significant growth and faces both opportunities and challenges. As above, companies like Panasonic and GS Yuasa, among others, have established manufacturing facilities and research centers in the Kansai region. These companies have been at the forefront of battery technology development, supplying batteries for various applications such as consumer electronics, electric vehicles (EVs), and energy storage systems.

Large companies, such as Panasonic, tend to survive international competition through strategies such as forming international partnerships. On the other hand, SMEs provide different picture in terms of their strategies for the changing competitive context. For example, this paper interviewed one of the SMEs in Osaka, having a long history of supplying copper mechanics for battery production to manufacturers, such as GS Yuasa, and the company provided us with the limited sense of global risks and the adaptation strategy for it. Since the company has been incorporated into the supply chain of large manufacturers for many years, it does not have any intention to explore its own sales channels domestically or globally. Also, the company receives specific requests from manufacturers for product design, it is not willing to develop any new product in-house. In addition, the company is even not aware of the origins of the resources they use, as they purchase the necessary resources from trading companies in long-term relations. It does not automatically mean that they lack sense of risks for losing the business facing the international challenge, but the company responded to us that they will seek another business frontier, if necessary, only when the battery business goes further down.

5 Considerations on Risk Management in a Time of Great Change

Lithium battery manufacturers navigate the tension between Japan and the emerging challenge, especially addressed by China and other competitors with a range of perspectives. While concerns over market access, supply chain stability, intellectual property protection, and geopolitical risks exist, manufacturers also seek opportunities for collaboration, diversification, and long-term stability. Balancing these considerations allows manufacturers to mitigate risks, leverage technological expertise, and drive innovation in the lithium battery industry. By carefully assessing the implications of the tension and adopting strategic approaches, manufacturers can navigate the complexities of the Japan–China relationship and the changing nature of the IPE *inter alia*, while contributing to the growth and advancement of the global lithium battery market.

The tension between Japan and China has been a longstanding issue, rooted in historical and geopolitical factors and thus changes in the IPE present both the opportunity for resolution and the risk of aggravation with respect to these tensions. As both countries play a significant role in the lithium battery industry, it is essential to examine how manufacturers in this sector perceive and navigate this tension as a means for understanding how it is that they might develop in the future. This section aims to provide insights into how lithium battery manufacturers view the Japan–China tension and its potential impact on their operations and the industry as a whole.

1. Market Access and Supply Chain Concerns:
   For lithium battery manufacturers, market access and supply chain stability are vital considerations. The Japan–China tension can create uncertainties regarding trade relations and the ease of doing business between the two countries. Manufacturers may perceive potential challenges in exporting products, accessing critical raw materials, or collaborating with Chinese suppliers or customers. Disruptions in the supply chain can have significant consequences on production schedules, costs, and overall business performance.
2. Diversification Strategies:
   To mitigate risks associated with the Japan–China tension, lithium battery manufacturers often adopt diversification strategies. This includes seeking alternative markets for exports and diversifying their supply chains by reducing reliance on a single country. Manufacturers may explore opportunities in other Asian countries, such as South Korea, Taiwan, and Southeast Asian nations, to ensure a more balanced distribution of their products and mitigate potential disruptions caused by the Japan–China tension. However, given the geographical limitation on lithium resources and proximity of China to Japan, there is a limit on the extent to which diversification will mite risk.
3. Technological Collaboration and Intellectual Property Protection:
   Technological advancements and innovation are crucial drivers of the lithium battery industry. Manufacturers often engage in collaborations and partnerships to accelerate research and development efforts. However, the Japan–China tension can impact such collaborations due to concerns over intellectual property protection. Lithium battery manufacturers may approach collaborations cautiously, evaluating potential risks associated with sharing proprietary technologies and trade secrets. They may prioritize partnerships with countries or organizations that offer strong intellectual property protection frameworks to safeguard their innovations.
4. Geopolitical Considerations and Risk Management:
   The Japan–China tension has geopolitical implications that can impact the overall business environment for lithium battery manufacturers. Heightened political tensions can create uncertainties, affecting trade policies, investment regulations, and business operations. Manufacturers may evaluate the geopolitical risks associated with operating in either Japan or China, considering factors such as political stability, diplomatic relations, and the potential impact of any escalation in tensions on their business continuity. Robust risk management strategies, including scenario planning and diversification of business operations, can help mitigate these risks.

Of course, none of this is to suggest that the Japanese lithium industry would seek to decouple or separate from China's completely, as it is both unfeasible and unnecessary. There are scenarios in which cooperation and collaboration could be of great benefit to industrial development of both countries. For instance, China's Belt and Road Initiative (BRI) potentially provides partners with access to a large majority of rare earth minerals and precious metals essential for the production of high-tech goods including end-products based on lithium technology. The BRI is an investment project ostensibly based on the China's "Five Principles of Peaceful Coexistence," where the fourth principle is "equality and cooperation for mutual benefit," first laid out in the Panchsheel Agreement between China and India in 1954 and included in the preamble of China's constitution. For this reason, the BRI is considered to mutually combine with other regional developmental initiatives such as the Russian-led Eurasian Economic Union.

There can be little doubt that the development of the regions across the Eurasian continent could create demand for Japanese credit and development opportunities for the Japanese lithium industry. The development of new systems of international trade including new trade corridors, systems of international settlement, and even innovation and breakthroughs in technological progress all provide Japanese manufacturers with both risk and opportunity when it comes to pursuing avenues of sustainable, regional economic development. Nonetheless, despite the issues that exist today and the clear, rational need for Japanese manufacturers to diversify their partners simply for the purposes of maintaining economic security amidst high levels of uncertainty with respect to the geopolitical developments in the IPE, the opportunities for a peaceful co-existence and cooperation for mutual benefit should not be dismissed given the import with which it has been upheld in both Chinese law and diplomacy.

Indeed, despite the tension between Japan and China, lithium battery manufacturers also recognize that both countries possess significant technological expertise and resources in the industry also. Rather than solely focusing on the negative aspects of the tension, some manufacturers seek opportunities for collaboration and cooperation. This may include joint research projects, technology transfers, or mutually beneficial partnerships to leverage each country's strengths and drive innovation. While not losing sight of the geopolitical tensions, manufacturers should recognize that there may be avenues to foster collaboration that benefits the industry as a whole within the economic region. This would be consistent with China's principles of peaceful co-existence and indeed the Japan–China Joint Communiqué in 1972 which directly reiterated these principles agreed upon by both parties.

6 Conclusion

This chapter has examined the ways in which lithium battery manufacturers in Japan have sought to respond to the various challenges posed by globalization. Overall, lithium battery manufacturers operate with a long-term perspective, considering market stability and predictability. Thus, while the Japan–China tension may introduce short-term uncertainties, manufacturers may perceive it as a part of a larger geopolitical landscape that can evolve over time. They may focus on building resilient business models that can adapt to changing circumstances and maintain stable operations despite geopolitical challenges. This long-term outlook allows manufacturers to navigate through periods of tension and plan for growth in the industry, potentially contributing to inclusive and sustainable economic prosperity in the Asian region as the world economy undergoes change.

On the other hand, smaller companies have found it hard to adapt to globalization owing to a lack of flexibility from the rigid supply chain established over a long period. Although these SMEs have played a vital role in determining the prospects of the Japanese economy, at least looking at the lithium industry, there is a gap with the large-scale player in terms of explicitly addressing the Asian harmonious regionalization, particularly in China. In order for Japan to adapt to new systems of trade and mutually held interests in the region, the economic integration, both by large and smaller players, in the region may need to be developed further in the future. Thus, in this sense, the future may appear very much uncertain for many of these industries. Nonetheless, despite the emphasis on new developments within the IPE, there are diplomatic frameworks for co-existence in the region upheld by Japan's regional partners that may provide space for collaboration and cooperation in the future that benefits the region and thus its people as a whole.

References

1. Kaunda RB (2020) Potential environmental impacts of lithium mining. J Energy & Nat Resour Law. https://doi.org/10.1080/02646811.2020.1754596
2. Glazyev SY (2012) Современная Теория Длинных Волн В Развитии Экономики [The modern theory of long waves in the development of the economy]. ECR-RAS 2(57):27–42
3. Geoscience Australia (2023) Lithium. Commonwealth of Australia, June 7. Available at: https://www.ga.gov.au/scientific-topics/minerals/mineral-resources-and-advice/australian-resource-reviews/lithium. Accessed 30 June 2023
4. Bayanov B, Belov V, Chernyakin A, Eschenko V, Karasuk V, Petrichenkov M, Silvestrov G, Vsevolozhskaya T (1999) Liquid lithium lens with magnetic fields. Proceedings of the 1999 particle accelerator conference, New York, pp 3086–3088. Available at: http://cds.cern.ch/record/552773/files/thp58.pdf. Accessed 1 July 2023
5. Tariq S, Ammigan K, Hurh P, Schultz R, Liu P, Shang J (2003) Li material testing—Fermilab Antiproton Source lithium collection lens. Proceedings of the 2003 particle accelerator conference. https://doi.org/10.1109/PAC.2003.1288558
6. Clarke J, Bailey I, Baynham DE, Bharadwaj V, Bradshaw T, Brummitt A, Bungau A, Carr F, Collomb N, Dainton J, Dollan R, Gai W, Gronberg J, Hartin A, Hesselbach S, Hock KM, Ivanyushenkov Y, Jenner LJ, Laihem K, Zang L (2008) The design of the positron source for the International Linear Collider. Proceedings of EPAC08. Genoa, Italy, pp 1915–1917
7. Carey HC (1851 [2010]) The harmony of interests: Agricultural, manufacturing, and commercial. Kessinger Publishing, Whitefish, MT
8. Sun Y-S (2011) The three principles of the people (trans: Price FW). Soul Care Publishing, Vancouver
9. Vecchiato P (2023) South Africa urges careful debate on option of introducing BRICS common currency. Bloomberg, May 10. Available at: https://www.bloomberg.com/news/articles/2023-05-10/south-africa-urges-careful-debate-on-option-of-introducing-brics-common-currency. Accessed 30 June 2023
10. Lassovolik, YD (2018) Boosting the USE of national currency among BRICS. Russia in Global Affairs, 2002–2023, September 9. Available at: https://eng.globalaffairs.ru/articles/boosting-the-use-of-national-currencies-among-brics/. Accessed 30 June 2023

11. Betz B (2023) Brazil, China strike trade deal agreement to ditch US dollar. FOX News Network LLC, March 29. Available at: https://www.foxbusiness.com/markets/brazil-china-strike-trade-deal-agreement-ditch-us-dollar. Accessed 30 June 2023
12. Yu J, Xiaojing X, Weiduo S (2022) Strengthened China-Saudi ties raise prospects for us of yuan in oil settlement. The Global Times, December 8. Available at: https://www.globaltimes.cn/page/202212/1281416.shtml. Accessed 30 June 2023
13. Escobar P (2023) Global South: Gold-backed currencies to replace the US dollar. TheCradle, January 19. Available at: https://thecradle.co/article-view/20532/global-south-gold-backed-currencies-to-replace-the-us-dollar. Accessed 30 June 2023
14. Reserve Bank of India (2022) Chapter III: regulatory initiatives in the financial sector. Reserve Bank of India, December 29. Available at: https://rbi.org.in/scripts/PublicationReportDetails.aspx?ID=1226. Accessed 30 June 2023
15. Khan A (2023) Dedollarization trend India's strategy to boost its economic power. South Asia Journal, April 30. Available at: https://southasiajournal.net/dedollarization-trend-indias-strategy-to-boost-its-economic-power/. Accessed 30 June 2023
16. Marchel E (2023) Indonesian government encourages the use of GPN to replace Visa, Mastercard. Indonesia Business Report, March 30. Available at: https://indonesiabusinesspost.com/insider/the-indonesian-government-encourages-the-use-of-gpn-to-replace-visa-mastercard/. Accessed 30 June 2023
17. IMF (2023) Currency composition of official foreign exchange reserves (COFER). International Monetary Fund, March 31. Available at: https://data.imf.org/regular.aspx?key=41175. Accessed 30 June 2023
18. Devonshire-Ellis (2022) The new candidate countries for BRICS expansion. Silk Road Briefing, Dezan Shira & Associates, November 9. Available at: https://www.silkroadbriefing.com/news/2022/11/09/the-new-candidate-countries-for-brics-expansion/. Accessed 30 June 2023
19. Bank for International Settlements (2022) BIS quarterly review: international banking and financial market developments. Bank for International Settlements, December. Available at: https://www.bis.org/publ/qtrpdf/r_qt2212.pdf. Accessed 30 June 2023
20. Koo R (2015) The escape from balance sheet recession and the QE trap. Wiley, Singapore
21. Yamamoto K, Toritani K (eds) (2023) 世界経済論: 変容するグローバリゼーション [World economics: The transformation of globalization], 2nd edn, Minerva, Kyoto
22. Ono S (2016) メーンバンク制と金融産業の今後~新しい金融システムを目指して~ [The megabank system and financial industries from now on: towards a new financial system], Planning and Coordination Office, Secretariat of the House of Councilors, Keizai-no purizumu 151:1–24. Available at: https://www.sangiin.go.jp/japanese/annai/chousa/keizai_prism/backnumber/h28pdf/201615102.pdf. Accessed 14 April 2023
23. Ministry of Finance, Japan (2022) 対外・対内直接投資 (地域別・業種別 [Outward/inward direct investment, breakdown by region and industry]. Ministry of Finance Japan. Available at: https://www.mof.go.jp/policy/international_policy/reference/balance_of_payments/bpfdii.htm. Accessed 30 June 2023
24. Cheng E (2021) Chinese consumer spending is set to double by 2030, Morgan Stanley predicts. 2021 CNBC LLC, January 29. Available at: https://www.cnbc.com/2021/01/29/chinese-consumer-spending-to-double-by-2030-morgan-stanley-predicts.html. Accessed 30 June 2023
25. Iwamoto K (2022) Yen weakens to 32-year low, declining past 147 per dollar. Nikkei Inc. October 13. Available at: https://asia.nikkei.com/Business/Markets/Yen-weakens-to-32-year-low-declining-past-147-per-dollar. Accessed 30 June 2023
26. Japan Economic Center (2020) Lithium-ion battery market facts and future prospects. Japan Economic Center Co., Ltd.

27. Nikkei Cross Tech (2021) 出資・供給関係が一目で分かる、リチウムイオン電池の業界地図 [Lithium-ion battery industry map showing investment and supply relationships at a glance]. Nikkei Business Publications. Available at: https://xtech.nikkei.com/atcl/nxt/column/18/01770/090900012/. Accessed 30 June 2023
28. Agarwal S (2021) Status of the rechargeable Li-Ion battery industry 2021. Market and Technology Report-July 2021. July 2021. Yole Dévelopement. Available online at: https://s3.imicronews.com/uploads/2021/07/YINTR21186_Status_of_the_Rechargeable_Liion_Battery_Industry_2021_flyer.pdf. Accessed: October 6, 2023.

Chris G. Pope is a lecturer at Fukuoka University where he teaches International Political Economy. He teaches a variety of courses on international political economic theory and the rise of new currency systems and systems of international trade and finance in Japanese. Prior to this he was an assistant professor at Kyoto Women's University, having written an award-winning doctoral dissertation at the University of Sheffield's School of East Asian Studies. His research focuses on the International Monetary System, international trade and East Asian international economics and international relations, particularly in the context of multipolarity pioneered by emerging nations. He speaks English, Japanese and Mandarin Chinese.

Aki Suwa is a professor at Kyoto Women's University, and a Lecturer at the College of Policy Science, Ritsumeikan University, Japan. Prior to her appointment she was a research fellow at the United Nations University Institute of Advanced Studies (UNU-IAS). Her research focuses on how environmental understanding facilitates policy formulation, and how this subsequently influences the degree of the policy implementation on ground. Prior to working in UNU-IAS, she completed a doctorate from the University College London, the University of London. She also holds a MSc degree in the Environmental Technology and Policy from the Imperial College of Science, Technology and Medicine.

Kolkata

Indo-Pacific Smart Megacity System Megacity Studies: Is Kolkata Ready to Be the Gateway to Indo-Pacific Region?

Prabh Bedi and Mahavir

Abstract Locational factors along with geopolitical conditions are contributors to the growth or decline of settlements since historical times. Kolkata region has gained from these factors and has experienced phenomenal growth leading the country not only in socio-cultural and political ethos, but also in economic and urban planning as well. The authors consider the factors that have led to shaping the region to be causal to its growth as well as its decline. This research delves into the pertinent aspects in current times to understand the benefits that can be derived from learnings so as to push Kolkata into the next phase of leading the nation by taking advantage of the current geopolitical environment.

Keywords Kolkata · Indo-Pacific · Industry · Geopolitics · Megacity

1 Introduction

Kolkata has been at the helm of industrial activity in India from medieval times. The city and the area around it have reinvented itself from time to time under various regimes and have maintained the continuity of various economic activities in varying phases of economically successful and vibrant decades with some declining periods of growth. Each decline has, however, seen a revival mainly due to the resilience of its people.

The last few decades have been that of declining industrial growth, which has impacted the socioeconomic and cultural aspects of the society. The city region is currently at the juncture of the next stage of success in the form of the opportunity provided by the role being played by the country in the Indo-Pacific region. Recent policies at the national level to liberalize the economy and to promote industrial

P. Bedi (✉)
Resonance Integrated Solutions Canada Inc., Victoria, Canada
e-mail: prabhb@gmail.com

Mahavir
School of Planning and Architecture, New Delhi, India

T. M. Vinod Kumar (ed.), *Indo-Pacific Smart Megacity System*, Advances in 21st Century Human Settlements, https://doi.org/10.1007/978-981-99-6218-1_6

development, need to be adequately capitalized at the state level to erase the deterrents of industrial growth.

The geopolitical developments in the eastern part of undivided India have gravely impacted the societal ethos which has over the years impacted the political spirit and in turn the development policies of successive state governments. It is high time that the learnings from past events be appropriately shaped in policies for the overall development of the region.

This research traces the history and growth of Kolkata since medieval times and highlights the problems faced currently, the emphasis being on the industry and the port of Kolkata on the premise that manufacturing or secondary sector is the key to a long-lasting growth of the regions and the sector needs to be substantial contributor to the regions gross domestic product. The port holds importance as it is the interconnecting link between the hinterland and the overseas world and has the capacity to provide leapfrog jump in the growth of the economy. The study concludes with discussion and puts forth ideas such that Kolkata can successfully emerge as the gateway to the Indo-Pacific.

The words Kolkata and Calcutta have been used interchangeably, as these appear in the timeline of the growth of city. The state government adopted the name Kolkata in 2001 [1]. Attempt has been made as best possible to use the two names as these appear on the time scale of city. Similarly, Hoogli and Hooghly have been used interchangeably, Hoogli being in use mainly during the British era.

2 Indo-Pacific Region

In the widest sense, Indo-Pacific region comprises of the waters of Indian Ocean and Pacific Ocean and all the nations surrounding both the oceans encompassing east Africa, Asia, Australasia, Western Americas, and the island nations in these oceans. Ichthyological studies based on the occurrence of number of species in the region define the Indo-Pacific region as the tropical waters of the Indian Ocean, the western and central Pacific Ocean, and the seas connecting the two [2]. Based on the variation in the presence of species, it does not include the temperate and polar regions of the Indian and Pacific oceans, nor the Tropical Eastern Pacific.

Indo-Pacific identity did not exist until the eighteenth century, an era when India and China had started to have an impact on the world, with both collectively contributing more than half of the global gross domestic product [3, 4]. This prominence of the two countries on economic front on the global scale had become the basis for Karl Haushofer's identification of Indo-Pacific space [5].

The term Indo-Pacific combines the Indian and the Pacific oceans into a singular regional construct as stated by [4] the two regions are vastly dissimilar in geo-economics as well as security environment which largely shapes the geopolitics [4]. In contemporary times, the term Indo-Pacific was first described in a geopolitical context by Japanese Prime Minister Shinzo Abe's speech entitled [5]. Since the late 2010s, the term has been increasingly used in geopolitical discourses.

The recent prominence of the region in geopolitical, research, and academics discourses is nested in the feelings of political insecurity experienced by the countries in the region due to One Belt One Road initiative of China. Through this initiative, China plans to economically dominate over much of the Indo-Pacific space across Central and Southeast Asia, to Africa, and Europe. China's military buildup in the region, particularly its island-building strategy in the South China Sea as well as the building of foreign bases in East Africa and along the Indian Ocean coast [6, 7] and its strategy of building and maintaining hubs in countries like Pakistan, Oman, Yemen, Djibouti, Sri Lanka, Bangladesh, Singapore, and Myanmar [8] has been a step toward destabilizing Asia. The countries in the Indian and Pacific Oceans feel a common threat by China and are coming together to build a common identity to establish an alliance. The term Indo-Pacific is a response to these developments, as countries are concerned for their survival based on free and open trading routes in the Pacific and Indian Oceans.

China's expanding military footprint in Indian Ocean made the United States realize the inadequacy in its use of term Asia–Pacific to meet the geopolitical objectives in Asia. The term Asia–Pacific was used by United States but excluded South Asia and the Asia–Pacific Economic Cooperation excluded India as a full member. With the remarks by Hillary Clinton, US Secretary of State in 2010, United States officially recoganised Indo-pacific for the first time [9]. As mentioned by Khurana [4], in 2012, the US PACOM's command was renamed from Asia–Pacific to Indo-Asia–Pacific. President Trump, in 2017 further altered the Indo-Asia–Pacific formulation to Indo-Pacific contextualizing it to partnership with India. In 2018, Secretary of Defense James N. Mattis officially announced the US Pacific Command would become US Indo-Pacific Command, recognizing the increasing connectivity between the Indian and Pacific [10]. Within days of President Trump's Indo-Pacific articulation, India, Australia, Japan, and United States of America held their first joint secretary level meeting of the Quadrilateral Dialogue (QUAD) [11]. Current US leadership under President Joe Biden regards China as the main challenger to US global primacy and is building a coalition of democracies aimed at outcompeting China on a wide range of issues, mainly economic [12].

The Indo-Pacific region, pivotal to defining the course of the twenty-first century, is the world's fastest growing region. It offers innumerable opportunities in sectors of trade, research, manufacturing, and education. The stability of this region is an assurance to the development of the global economy and international security.

In the Indo-Pacific, the focus of India is to bring countries together in the spirit of cooperation. India revised its geopolitical strategy in 1990s with the 'Look East' policy of the P.V. Narasimha Rao government in 1991 [13], which was an effort to cultivate economic and strategic relations with Southeast Asia. As stated by [12], this was seen as an important component of India's decision to open its economy and take advantage of the dynamic East Asia region. The guiding force for the change in the strategy was disputes with Pakistan in the west and China to the north which were limiting its overland communications and trade options. US policy of rapprochement with India as an aftermath of 1998 nuclear tests leading to closer Indo-US ties and later formation of coalition between United States, India, Australia, and Japan to provide assistance to those affected by the tsunami of December 26, 2004, though only for a week [12], formed the basis of notion for quadrilateral grouping. In the Info-Pacific geography, India may not be a significant player east of Malacca Straits, but to its west, India dominates owing to its maritime and land borders with four of the eleven ASEAN nations. India recognizes the centrality of ASEAN to its Indo-Pacific strategy. India regards the Indo-Pacific as a geographic and strategic expanse, with the 10 ASEAN countries connecting the two great oceans.

India has been active in advocating the concept of free and open trade in Indo-Pacific introduced by Prime Minister Shinzo Abe in 2016 [14]. The United States, Australia, and the members of the ASEAN have all expressed a common view that India should play a greater role in the region. India does not see the Indo-Pacific Region as a strategy or as a club of limited members and maintains a stance that security in the region must be maintained through dialogue, a common rules-based order, freedom of navigation, unimpeded commerce, and settlement of disputes in accordance with international law. India's view is to work with other like-minded countries in the Indo-Pacific region to cooperatively manage a rules-based multi-polar regional order and prevent any single power from dominating the region or its waterways [14].

Taking this stance of India as the basis, this research delves into the plausibility of Kolkata megacity as being the gateway to Indo-Pacific for India.

3 Constituents of a Megacity

A megacity as defined by the United Nations [15] is a city with more than 10 million inhabitants. The United Nations Department of Economic and Social Affairs in its 2018 World Urbanization Prospects reported urban agglomerations having over 10 million inhabitants as megacities. The term megacity came into common use in the late nineteenth or early twentieth centuries when it was documented by the University of Texas in 1904 [16]. In the mid-1970s, the term was coined by urbanist Janice Perlman referring to the phenomenon of very large urban agglomerations [17]. The definition of megacity has evolved with increasing concentration of people in large urban centers as cities having a population of more than five [18], eight [19, 20], or ten million people [18]. Initially, the United Nations used the term to describe cities of 8 million or more inhabitants, but now uses the threshold of 10 million [15]. A megacity is an outgrown metropolitan city that is a contiguous spatial spread of the leading and subsidiary city along with the peripheral rural area with a population above 10 million [21].

Concentration of population in urban centers is an increasing trend and a recent phenomenon that has occurred since the first industrial revolution. In 1800, only 3 percent of the world's population lived in cities, which increased to 47 percent by the end of the twentieth century. As per the United Nations [22] estimates, 60 percent of the world's population will be living in urban areas by 2030.

While the big cities that emerged due to concentration of people in big cities for work in the now developed world because of the industrial revolution, the trend of emergence of megacities is predominantly in the less developed parts of the world. In 1950, there were only two megacities, New York and Tokyo. This number was 14 in 1955 and currently is 34. Of these, only 4 are in the developed world and 6 are in India, namely Mumbai, Delhi, Chennai, Bengaluru, and Kolkata. Being the country's first metropolitan center, having a population of above 1 million in 1901, Kolkata has been among the largest cities of India since before Independence. India, recoganises the definition of United Nations and considers those metropolitan cities that have a population above 10 million as megacities.

4 Kolkata's Role in Indo-Pacific Region

Kolkata (erstwhile Calcutta) that was initially a village and was developed as a trade post by the East India Company of British is now one of the five megacities of India. It consists of a complex set of administrative entities—4 Municipal Corporations, namely Kolkata Municipal Corporation, Howrah Municipal Corporation, Bidhan Nagar Municipal Corporation, and Chandan Nagar Municipal Corporation, 37 Municipalities, and 23 panchayat samities[1] [24]. Kolkata Municipal Corporation, spread over an area of 185 sq km and consisting of 141 wards, grew out of villages of Gobindpur, Kalikata, and Sunatuni and developed into a trade post and later into the city of Calcutta.

Calcutta has continuously expanded over the years and is now the core of 1851 sq kms Kolkata Metropolitan Area. Kolkata Metropolitan Area lies between 22° 19′ N to 23° 01′ North latitude and 88° 04′ E to 88° 33′ East longitude and within this Kolkata Municipal Corporation is located between 22° 37′ N to 22° 30′ North latitude and 88° 23′ E to 88° 18′ East longitude (Ref). Kolkata Metropolitan Area stretches in a linear pattern along the east and west bank of the river Hooghly. The present-day Kolkata Municipal Corporation was once surrounded by marsh and swamp, which has been reclaimed and developed upon for urban use in the last over 300 years. The urban local bodies of Bidhan Nagar, Rajarhat, Sonarpur, and Maheshtala [25] are a few that have been developed on the wetlands [26] in recent years primarily for decongesting of the urban core of Kolkata.

While Kolkata is often perceived as a coastal city, it is about 145 km inland from the coast of Bay of Bengal but within tidal reach of Hooghly River. Kolkata is home to one of the oldest airports in the country. It is a major domestic hub of not only East and North-Eastern India but also neighboring countries in Southeast Asia. Kolkata Airport has non-stop passenger flights to 63 destinations in 10 countries and 49 domestic flights to major and small cities in the country [27]. In the last few years, cargo operations have seen a steady rise [28], especially from the North-Eastern

[1] Rural local government at grassroots level [23].

India. The growth of the airport and seaports in domestic and international space establishes the primacy of Kolkata in the region.

4.1 History of Kolkata

Kolkata the capital of West Bengal and a primate city is located on the eastern bank of the Hooghly River. It is the business, commercial, and financial hub of Eastern India. As per the Census of India, 2011 [29] Kolkata was the seventh-most populous city in India, having a population of over 4 million within the municipal limits forming the core of the Kolkata Metropolitan Area which had a population of over 14 million making it the third-most populous metropolitan area in the country. In 2021, it is estimated (Census 2021 is deferred due to COVID-19 pandemic [30]) that Kolkata metropolitan area had about 15 million people living in the region [31]. The Port of Kolkata is India's oldest and only riverine operating port.

The word Kolkata derives from *Dihi Kolikata* in the local language, Bengali, and was the name of one of the three villages, where the root of the present-day city lies [32]. Before the British took control of the area, the region was ruled by the Nawab of Bengal under Mughal suzerainty. The British trading company, East India Company had obtained trading rights in 1690 and the area was developed by East India Company into an increasingly fortified trading post. In the next hundred years, East India Company was able to abolish the native rule and assume full control of the region, which primarily included the villages of Kolkata, Sutanuti, and Gobindpur [33]. Under the East India Company and later British rule, Calcutta served as the capital of British held territories in India until 1911, when the capital was shifted to a more central location at New Delhi. Calcutta continued to be the capital of the Bengal Presidency. Partition and Independence in 1947 resulted in political violence and economic stagnation bringing much loss to Kolkata and the areas surrounding the city, which was once the cultural, political, and economic hub of the country.

4.2 Demography, Land Use and Planning

The metropolitan region of Kolkata lists among the 34 largest megacities of the world having a population of more than 10 million [15]. The population 2011 Census [29], the population of Kolkata Metropolitan Area was 14.11 million, of which Kolkata Municipal Corporation accounted for 4.49 million. The population density in Kolkata Metropolitan Area was 7,950 persons per square kilometer. The municipal core of Kolkata, which is the heart of the region, is denser with a population density of 23,149 persons per square kilometer accounting for 31 percent of the region's population within 10 percent of its area. Since the turn of the century, the population has been declining in the region. The population growth rate of Kolkata Metropolitan Area was 7.6 percent between 2001 and 2011 and has declined to 6.1 in 2011–2021. Kolkata

Metropolitan Area population was projected to increase to 20 million in 2021, and 21.1 million in 2025 [24].

Owing to the COVID-19 pandemic, the census as scheduled in 2021 has not been conducted [30]. It is estimated to be 14.97 million in 2021 which is much lower than the projected population. However, there has been a 28 percent increase in the Kolkata Municipal Corporation population [31] after a negative growth rate in the previous decade which needs to be ascertained after the actual census count. Although the population growth rate in Kolkata Metropolitan Area shows a declining trend, the absolute size of the population and its high density continue to pose challenges to urban growth.

Kolkata's demographic growth trend has been irregular over the decades, primarily due to its conspicuous socioeconomic and political trajectories. Kolkata's population has had a declining growth rate since 1981. The first decline in the growth rate was seen in 1960s which reversed in the next decade, but since has been steadily declining and in 2001–2011 has experienced a negative growth rate [29]. The population in the Kolkata Metropolitan Area, however, has been growing at a much higher rate than Kolkata Municipal Corporation (Refer Table 1). Reasons for the decline can be a saturation to absorb population in terms of available built-up space, poor and depleting infrastructure, a stagnant economic structure of the city, and an increased cost of living. Further, transportation systems developed in recent decades have facilitated daily commute allowing individuals to move to residential locations developed in peripheral areas, where land and low-cost housing is easily available, a reason for rapid growth of peripheral areas. Though this has exerted excessive strain on economy, infrastructure, environment, and land, in turn leading to rise in poverty and deteriorating the overall quality of life.

The measures adopted by the government have been in terms of establishing new townships through financial deregulation and public–private partnership development model which has emerged to be a real-estate-driven model benefitting the housing needs of a better-off section of the society. Unfortunately, both the local and state governments have been ineffective in resolving the urban issue.

4.3 Land Use

The city's role as the nucleus of labor employment and economic opportunities has attracted both short and long-distance inter and intra state migration. Efforts have been made to divert the migration to other areas of the state by developing some secondary cities such as Asansol, Durgapur, and Siliguri. However, Kolkata continued to remain the city of attraction for livelihood leading to horizontal and vertical expansion of the city. Unplanned expansion led to encroachment of agricultural lands and wetlands. The urban pattern as it exists today is of a highly dense core which is linear residential development.

Urban built-up land constitutes 54.2 percent of the total area of Kolkata Metropolitan Area with the remaining under non-urban use. Among urban uses,

Table 1 Population and growth rate of Kolkata Municipal Corporation and Kolkata Metropolitan Area

Census year	Kolkata Municipal Corporation population	Growth rate	Kolkata Metropolitan Area population	Growth rate
1901	1,009,853	–	–	–
1911	1,117,966	10.71	–	–
1921	1,158,497	3.63	–	–
1931	1,289,461	11.30	–	–
1941	2,352,399	82.43	–	–
1951	2,956,475	25.68	–	–
1961	3,351,250	13.35	–	–
1971	3,727,020	11.21	–	–
1981	4,126,846	10.73	9,194,000	–
1991	4,399,819	6.61	11,021,900	19.88
2001	4,572,876	3.93	13,114,700	18.99
2011	4,496,694	(1.67)	14,112,536	7.61
2021	5,800,000	28.98	14,974,073	6.10

Source Primary Census Abstract, West Bengal, Census of India, 2011 [29]

residential use is pre-dominant (31.2 percent of total area), followed by industrial, transportation, public, and recreational uses (14.7 percent of total area), and the remaining is under mixed built-up land use (8.3 percent of total area) [34].

The land use pattern in Kolkata Municipal Corporation is even more urban, reflecting 300 years of organic growth. The white town areas were planned by the British, whereas the rest of the areas developed organically and was later given a piecemeal and surgical treatment to make the living conditions better. The land over which urban expansion occurred was mostly wetlands requiring reclamation. Various organizations have progressively converted the wetlands over time for urban use. The Calcutta Improvement Trust reclaimed considerable area to the east and south and developed planned colonies for residential purposes [26]. With almost nonexistent land use planning or control, the bulk of land use reflects mixed use. Both residential and non-residential uses co-mingle in most areas with little or no demarcations. In the Kolkata Municipal Corporation, residential or mixed residential accounts for 68 percent of land use followed by 9.5 percent under open space and parks, 6 percent under industrial, 5 percent under commercial, 4 percent under agriculture, and 3 percent under transportation and storage [34].

To comprehend the issues in Kolkata region, an intricate composition of the urban space needs to be understood. Kolkata Municipal Corporation has a large proportion of its population living in slums and squatters, constituting 33 percent of the total population, in 2,011 registered, and 3,500 in unregistered such settlements [35], which are spatially concentrated in six of 141 wards. Slums in Kolkata Municipal Corporation are the hub for many informal manufacturing units, some of which are producing highly toxic acids and chemicals [34]. The increasing population pressure, dense urban environment, and low coping capacity of population living in slums make these the most vulnerable parts in the city. The infrastructural facilities in terms of water supply, roads, electricity, sewerage, health, and education remain sub-standard.

4.4 Urban Planning and Development

Kolkata became a corporation on September 4, 1726, by a royal charter of the then British colonial system. Since then, several laws have been passed and amended to improve municipal governance of the city. Calcutta Municipal Act, 1923 provisioned the election of a mayor in the municipal Government of Kolkata and the enfranchisement of women. Municipalities, namely Cossipore, Maniktala, Chitpore, and Garden Reach were merged with Kolkata though later Garden Reach was separated. Kolkata Municipal Corporation was envisioned as a policymaking, directive, and rule-making civic institution with the enactment of Calcutta Municipal Act, 1951 [36]. Despite several amendments over the next two decades, the provisions of Calcutta Municipal Act, 1951 were, however, inadequate to provide requisite guidelines for the better governance of a problem-ridden city.

In response to the demands of the expanding city and the depleting civic conditions of Calcutta, an Improvement Trust was established in 1911, which continued to be the agency responsible for expansion and improvement of Calcutta and its urban surroundings. The trust was responsible for development of urban infrastructure through maintenance and building lakes, housing, roads, rehabilitation projects, and commercial complexes.

Multiple urban development plans have been prepared since the 1950s. The Calcutta Metropolitan Planning Organization was created in 1961. Along with a team of eminent international and Indian planners, and supported by the Ford Foundation, Basic Development Plan for 1966–1986 for the urban development of Calcutta Metropolitan District was prepared [34].

Based on the analysis of land use and population growth patterns, the plan recommended the development of a counter magnet, Kalyani-Bansberia in the northern-most periphery of Calcutta Metropolitan Area [37]. In 1970, the Calcutta Metropolitan Development Authority was created by a state legislation and was given the responsibility of implementing the plan. However, substantial changes had taken place in the urban structure during the intervening two decades. Population growth in Calcutta had declined whereas it had grown considerably in the suburbs. Addressing these changes, another plan was developed in 1976, the Perspective Plan which recommended a multi-modal development with several growth centers [37]. The plan was further revised in 1982 and has also witnessed additional changes over the years based on ground realities and current situation.

From 1970 to 1981, Kolkata Metropolitan Development Authority worked as the development agency in the Calcutta Metropolitan Area primarily because most of its constituent Urban Local Bodies were under suspension, with no elected representative. The Kolkata Metropolitan Development Authority has since functioned in collaboration with the elected Urban Local Bodies, working toward efficient governance and holistic development of the region. With the enforcement of the 73rd and 74th Constitutional Amendment Act, 1992 and the West Bengal Metropolitan Planning Committee (WBMPC) Act, 1994, the process of decentralized planning and development gained momentum in the Kolkata Metropolitan Area. Kolkata Metropolitan Planning Committee was formed in 2001, following the provisions of the WBMPC Act, 1994. The WBMPC mandates the Kolkata Metropolitan Planning Committee to prepare the development plan for the Kolkata Metropolitan Area [24]. Kolkata Metropolitan Planning Committee is also provisioned to oversee the developmental activities launched by various state and central government agencies within the administrative jurisdiction of the Kolkata Metropolitan Area, to ensure that such activities are in consonance with the development plans.

The Kolkata Metropolitan Development Authority's functions broadly cover three domains: planning for development, project implementation, and regulatory functions. In addition to this, the Kolkata Metropolitan Development Authority acts as a nodal agency, providing consultancy services and implementing projects on behalf of various government departments and institutions. Over time, the Kolkata Metropolitan Development Authority has become the technical secretariat of the Kolkata Metropolitan Planning Committee. Thus, the Kolkata Metropolitan Development Authority has carried out several projects in diverse sectors of civic infrastructure, such as water supply, sewerage and drainage, traffic and transportation, township and area development, housing and slum improvement, commercial complexes, and parks and playground [36, 38].

According to the guidelines of the West Bengal Town and Country (Planning and Development) Act, 1979, the Kolkata Metropolitan Development Authority must prepare existing Land Use Maps and Registers (LUMRs) for various ULBs within the Kolkata Metropolitan Area to help them prepare Land Use and Development Control Plans (LUDCPs) for those sub-areas. The responsibility of enforcing LUDCPs and DCRs is delegated to the individual ULBs, while the KMDA retains enforcement power in particular areas and zones [36]. The KMDA executes various planning exercises, ranging from perspective plans to the formulation of investment programs; it must continually upgrade perspective plans and sectoral development plans to keep up with changing socioeconomic, physical, and other contexts [24].

In 2017, the state government merged Kolkata Improvement Trust and Kolkata Metropolitan Water and Sanitation Authority with Kolkata Metropolitan Development Authority under the aegis of West Bengal Town and Country Planning Act, 1979 which resulted in repeal of Kolkata Improvement Trust Act of 1919 and Kolkata Metropolitan Water and Sanitation Authority Act of 1966 [39]. After the merger, Kolkata Metropolitan Development Authority is responsible for the planning and infrastructure development of 125 municipalities and the intermediate rural area across either side of Hooghly [24].

In order to improve the infrastructure of the city, City Development Plan for Kolkata Metropolitan Area was prepared under JNNURM, a Government of India program launched in 2005 [40], which aimed at creating economically productive, efficient, equitable, and responsive cities, with a focus on -

- improving and augmenting the economic and social infrastructure of cities;
- ensuring basic services to the urban poor including security of tenure at affordable prices;
- initiating wide-ranging urban sector reforms with a primary objective of eliminating legal, institutional, and financial constraints that impede investment in urban infrastructure and services;
- strengthening municipal governments and their functioning in accordance with the provisions of the 74th Constitutional Amendment Act of 1992.

The metropolitan region of Kolkata has been dotted with small towns at the periphery of the dense core with the intent of decongestion. These small towns have been treated as experimental grounds to de-densify the metro region [41] The development of Rajarhat on the eastern fringe of Kolkata is one such case. At a distance of 17 km from Kolkata, the area once comprised of water bodies, salty marshes, and extensive areas of wasteland. Rather than being a city-for-all as planned and initially envisioned, Rajarhat's on ground development has been exclusionary favoring high-income consumerist inhabitants. As stated by Chowhan et al. [41], the idea to develop an eco-friendly, inclusive township with minimal dispossession turned into a nightmare for fairly developed self-reliant villages of Rajarhat. Instead of blindly valorizing economic growth or technological advancement, the need to situate sustainability within the rubrics of urban planning by anchoring down justice, equity, and inclusivity at the heart of city-making projects [41]. The case showcases the existing situation of the majority in the city in varying measures.

In 2015, the union government launched its new urban agenda, aiming to develop 100 smart cities across the country. According to Smart City Mission's (SCM) guidelines, these cities would have adequate civic amenities and services; affordable housing; smart transport solutions; efficient governance and participation of citizens; safe, secure, and sustainable city environment for all, with particular emphasis on the elderly, children and women; and accessible healthcare and education [36]. While the SCM's agenda and methods of implementation have triggered a debate in the academic and political spheres, it is crucial to examine critically the extent to which smart cities can help achieve a sustainable and inclusive city (SDGs Goal 11, put forth by the UNDP) [36, 42]. However, as stated by Tiwari, Sarkar, and Jana [43], integration of the SCM with the development plan of the city will be pertinent as it will ensure efficiency and facilitate the completion of projects.

Despite various urban development and welfare policies, programs, schemes, and missions under the aegis of both national and state governments and other development agencies, Kolkata is yet to become a sustainable and inclusive city. While some sectors like real estate, information technology, tourism, and physical infrastructure have progressed significantly since the economic liberalization, at the micro-level, spatial inequalities persist in terms of adequate housing, access to essential amenities and civic services, and the financial well-being of residents. Moreover, issues continue to pose challenges to making Kolkata a sustainable and inclusive city: unregulated built-up growth, planning violations, mismanagement of municipal waste, environmental degradation, traffic congestion and restricted spatial mobility, unsafe public spaces, exclusionary social practices, and unaffordable technology-driven infrastructure. Considering these issues, the question persists if Kolkata is ready for being the gateway under the Indo-Pacific Policy of Government of India.

Kolkata, a mega agglomeration, and a center of economic, political, and business activities in East India, has held geopolitical importance since colonial times. In east India, it is a popular livelihood destination for migrant labor from the neighboring states of Uttar Pradesh, Bihar, Odisha, Sikkim, and seven North-Eastern states which along with cross India migration has probably pushed Kolkata into a premature megacity.

4.5 Industry and Economy

In the pre-Independence period, Bengal experienced a very long and economically rewarding industrial history since ancient times. The cottage industries were dominant comprising mainly of silk weaving in Murshidabad and Malda areas, metal work in Bishnupur, embroidered clothes in Dacca, conch-shell bangles and pearl button of Kharagpur and rural area of Dacca (now Dhaka, capital of Bangladesh), are significant examples of industries that existed in Bengal [44]. When the British arrived first in India, Bengal's cotton handloom industry was the leading in the production of highest quality of cotton [45]. However, its decline started in the early nineteenth century due to British economic and political policy and resultant deindustrialization [46]. The focus of the British in India shifted from industrial production to agricultural production. The agricultural produce was exported to Britain and factory manufactured finished goods were imported to India, which severed as a stiff competition to the indigenous produce of the cottage industries. There was some industrial growth from 1854, when the British established cotton and jute mills in Calcutta and Bombay areas which too served as a source of competition to the local cottage industry setups. During the First World War, Indian industries got a boost as India became the key supplier of cotton and woolen textiles. During the Second World War, due to the paucity of raw materials, industry experienced a decline.

After Independence, India adopted a system of Five Year Plans, under which each period focused on one key sector's development. While the first Five Year Plan focused on the agricultural sector, the second plan laid emphasis on industrial growth. The period from 1950 to 1970 was of the first three national Five Year Plan. There was a compounded growth rate of industries in West Bengal at 5.7 percent during the first Five Year Plan which increased to 7.2 percent during the second plan and rose to 9.0 percent during the third Plan period. During the period 1951–1965, West Bengal's industries prospered keeping pace with the all-India industrial growth rate [47]. In the period 1951–1965, the value of industrial output in West Bengal increased by 28.7 percent. Registered factory employment in West Bengal during this period increased by about 35.02 percent [48]. West Bengal during his period was among the leading states in the country having concentration of industrial units in and around Calcutta.

The period from 1970 to 1980 having two Five Year Plan saw industrial recession primarily due to the Indo-China war (1962) Indo-Pak wars (1965 and 1971), which resulted in diversion of public investment to unproductive uses leading to the decline in the growth of the industrial sector. Successive draughts, from 1965 to 1966 and 1966–1967 and later 1971–1972 and 1972–1973 reduced agricultural production and in turn adversely affected industrial development. The oil crisis in 1973 further led to a significant imbalance in industrial production. These collectively caused an industrial recession in India and the state as well [44, 49].

During the next two Five Year Plans in the 1980s decade, relatively buoyant economic conditions prevailed in India as well as in West Bengal, though lack of interest was shown by the big investors to invest in the large-scale industrial sector in the state [50] due to persistent labor issues and not so favorable industrial policy of the state.

The 1990s saw a positive industrial growth both in India and West Bengal. There were two phases of industrial growth, though first four years faced prolonged recession, in the second phase (1992–1996), steady industrial growth had been seen after the extended industrial deceleration [51]. In 1997, the State Government announced an investor-friendly industrial policy which was a turning point in the industrial sector and opened the private capital investment in the still not so vibrant economic backdrop of the state [52]. The state government introduced schemes for developing the infrastructural segment as well, like road transport and development of the power sector [53]. This resulted in an increase in industrial output in West Bengal. It gained further momentum at the turn of the century as the state government took many steps for development in Information Technology [54]. Traditionally, an agrarian economy, nearly 70 percent of West Bengal's population continues to live in rural areas [55]. In recent years, the tertiary sector has started to dominate the state GDP, currently at around US$ 85 billion [56].

As with all metropolitan regions, Kolkata too has seen a rise in the secondary and tertiary sectors and a decline in the share of primary sector (largely agriculture). Manufacturing industries and construction activities constitute the core of secondary sector activities while the banking, insurance, and transport sub-sectors constitute the core tertiary sector. The jute textile industry has been traditionally dominant in Kolkata Metropolitan Area making the region the largest hub in the country. The growth rate of state domestic product reflects the recent trends of structural changes in the economy of Kolkata Metropolitan Area. Contribution of Kolkata Metropolitan Area to state domestic product and variation in the per-capita income over the years. The average per-capita income for 2001–2002 in Kolkata Metropolitan Area was Rs. 15, 281 (at 1993–94 prices) (Refer Table 2).

In the Kolkata Metropolitan Area industries are concentrated in Kolkata Municipal Corporation (68 percent) and Howrah Municipal Corporation (14 percent) that of jute manufacturing, light and heavy engineering, leather products, textiles, paper, pharmaceuticals, chemicals, tobacco, food products, glass products, and electrical

Table 2 Annual average growth rate of state domestic product for Kolkata Metropolitan area

Sectors	Kolkata Metropolitan Area		West Bengal	
	1985–1986 to 1993–1994	1993–1994 to 2001–2002	1985–1986 to 1993–1994	1993–1994 to 2001–2002
Primary	1.58	4.79	5.62	4.18
Secondary	1.81	5.71	4.92	6.36
Tertiary	3.80	11.06	8.42	9.50
Total	**2.96**	**9.04**	**6.51**	**7.13**

Source JNNURM, 2005 [34, 57]

and electronic products. Among the 9,000 industries in Kolkata Municipal Corporation about 1 percent are categorized as large, 3 percent as medium, and the remaining 96 percent as small-scale industries [34, 57]. Highlighting the dominance of small-scale industries since historical times (Refer Table 3), the concentration of number of units is highest in Kolkata though area-wise Murshidabad has highest area under industrial estates.

There has been an economic resurgence in the last decade due to the information technology industry and business parks in areas like Bidhan Nagar and Rajarhat-New Town having a trickledown effect, furthering dichotomy between the haves and have nots. The city's IT sector has grown at a rate of 70 percent per annum, twice that of the country average [36]. There has also been a surge in investment in Kolkata's booming real-estate sectors. Despite the development in the IT sector, the city continues to struggle with endemic poverty, traffic congestion, and pollution.

Independence and Partition resulted in the decline of industry and commerce in Kolkata. Two-thirds of the province, both area and population-wise, had become part of West Pakistan. Industry in Bengal had been centered around Kolkata. However, its primary and commodity hinterland that was rich and fertile was in East Bengal (now Bangladesh). It was the source of food and other agricultural produce required both by the city's growing population as well as for industries as raw material. With the partition, a greater part of this hinterland, industry, particularly jute, was gravely impacted. Most of the jute growing areas were in East Bengal, while the majority of the jute mills were in West Bengal [50].

Partition had an adverse effect on the investible surplus too as most of the capital invested by the Bengali entrepreneurs came from the large surpluses of the big landlords who were based in East Bengal and had been left bereft of land and money due to the partition impacting the loss of surplus for industrial investment. The continuous influx of refugees and cross-border illegal migration was mainly to Calcutta that increased the stress on the city's civic amenities, employment, food supplies, and its living space which overwhelmed the government machinery. This was something which the government could hope to tackle for many decades.

The distance between Kolkata and the North-Eastern hinterland increased and Assam, Tripura, Manipur, and other North-Eastern states suddenly became far-off places. Transportation between these places and Kolkata became much more time

Table 3 Distribution of industrial estates and key industries in West Bengal

District name	No. of industrial estates	Major industries	No. of Units	Area (hectares)
Kolkata	12	Leather goods, printing, binding, Engineering, Fabrication Job, Transformer mfg., Pharmaceuticals, Air-Conditioning, Synthetic fishing nets, food manufacturing, offset printing, handicraft, detergent powder	581	29.06
Birbhum	1	Tiles, Plastic packaging	32	8.47
Jalpaiguri	3	Engg., Electrical, Electronics, Food Products, Plastic, Furniture, Chemicals, Sports goods, Transformer	253	63.03
Murshidabad	2	Hotel & Restaurant, garments mfg., Diagnostic center	24	80.19
Nadia	3	Engg. Fabrication job, Woolen fancy garments, Technical Institute, Product of Haldia Downstream Chemical, Plastics, Packaging, Engg. Fabrication job	184	28.53
North 24 Pargana	1	Garments and Chemical products	10	0.73
Purba Bardhaman	5	Fabrication job, Engg. Work, Wooden furniture, Electronics goods	169	26.63
Purab Medinipur	1	Plaster of paris, Engineering materials	54	5.20
South 24 Pargana		Engineering fabrication job, Printing, synthetic fishing net, garments	89	15.69
Uttar Dinajpur	1	Under allotment	2	6.74

Source West Bengal Small Industries Corporation [58]

consuming as now travel and transportation of goods were required to be routed through north Bengal.

Though many Bengalis had an acumen for industry, the commercial and industrial sector of the province had been dominated mainly by the British, and among the Indians, by the Marwaris [59]. This socioeconomic change clubbed with poor government incentives and unfavorable economic conditions led to slow modernization, product development, and diversification causing obsolescence of machinery and shrinking of markets especially in jute, engineering goods, and tea industries.

The industry's problems were compounded by the Government of India's freight equalization policy of 1956 that persisted till 1993 [60], destroyed the locational advantage of Eastern India. For transportation of minerals, freight rates were equalized which became advantageous for industrialists to set up factories in other states as the minerals could be procured at same rates as from within West Bengal, Bihar,

Orissa, and other mineral rich states. In terms of industrial development, this policy shifted the center of gravity to Western India. Discovery of oil on the Arabian Sea seaboard was another factor which gave a fillip to the petro-chemical industry in Western India.

In time, Bangladesh, which had most of the jute growing areas started competing with Indian jute. Internationally, the demand for tea fell and countries like Sri Lanka, Indonesia, and Kenya emerged as major competitors with cheaper tea.

Since before Independence, West Bengal suffered from protracted industrial unrest. The State experienced labor movements which continued in one form or other for many decades after Independence, digressing the industrialists due to lack of political stability in the state. All these factors have collectively put West Bengal into a quandary having a negative multiplier effect on the psyche of the people, especially industrialists. All these factors collectively led to the decline and stagnation of industrial growth of Kolkata and its immediate hinterland.

4.6 Kolkata Port

A port city is a settlement situated at the coast or the bank of a river close to the coast, where land and sea meet, which over time may have grown into a town or a city, where ships load and unload or where ships may take refuge in times of storm. Port cities grow into centers of exchange of cultures, goods, people, and ideas. They essentially are the links between their hinterland and the outside world, across the seas and oceans.

Port cities are not merely large settlements on the shore but are maritime economies [61]. In growing urban morphology and culture, port cities need to be nodes having economic functions based on prominence of sea-based trade [62]. More than being a settlement with a port as an appendage, the port cities are a space on land and a node of amalgamation of goods and people and intermixing of cultures and ideas coming from the maritime space [63]. Port cities function not merely as entry or exit points for goods, labor, and capital but also serve as nodal centers for the reception and transmission of culture, knowledge, and information [64].

Port cities thrive on the interlinkages between internal (rail, road, and air) networks with external trans-national ones. The essential functions of the port city are largely determined by their relative positions between hinterland on the one hand and the ports overseas [65].

Hinterlands are one of the major reasons for the ports to thrive. The concept of the hinterlands, however, is varied, and their typologies include the immediate hinterlands that port area itself, primary hinterland that is the area where port and city assume a commanding role and determine life of area; commodity hinterland which is based on shipment of particular types of commodities; and the inferred hinterland where the port's hegemony over a particular area, to the extent that it satisfies the demand for imports and exports in the area it serves [66]. The influence which the hinterland exerts on the port city itself depends on the economic, political, and social relationships between the geographical entities.

Prior to the three villages being developed as Calcutta, the area was a series of small riverine marts settled by weavers and artisans. It transformed from a small weaver's settlement to a center for India's trans-oceanic trade and a thriving port on river Hoogli, with a large hinterland from the foothills of Assam and the fertile valley of river Brahmaputra in the northeast to the fertile plains of Ganga in the north largely due to the colonial interests of the British. The geopolitical situation of the time and the consolidation of British power initially in Bengal and later whole of the subcontinent led to the Calcutta gain its importance.

The emergence of Calcutta port in the map of Indian subcontinent was preceded by two important ports of the Mughal empire Hoogly and Sonat on the west coast of the river. The then prevalent structure of intra-Asian trade in the western Indian Ocean had been disrupted and eventually collapsed, which became a growth catalyst for the British led trade's reorientation toward China with the final destination in Europe. Eventually, this led to Britain's dominance in India's international trade through the ports of Calcutta in the east and later Bombay in the West.

Apart from the changing political and geopolitical environment on the subcontinent, the navigational aspect of the prior ports or the lack of it, too led to the growth of Calcutta. The dominance of the settlements along the Bhagirathi Hoogli channel of Ganga in the sixteenth century had become unsuitable for navigation due to silting, especially for the bigger vessels. Apart from this, the frequent bends in the river upstream made it very difficult for large vessels to sail through, making it more conducive to the development of Calcutta port.

One of the settlements upstream, Satgaon, which was a flourishing riverine port at the confluence of rivers Saraswati and Bhagirathi had become inaccessible for sea going ships causing the large sea going ships to dock at Betor and be fed by smaller ships from Satgaon, which later became the reason for the shifting of the factories from Satgaon to Hoogli. Hoogli continued to be an important center of trade through seventeenth century under the influence initially of Portuguese and later Dutch and British. The trade from Hoogli led to the development of trade marts along the river downstream up till Sunatuni and Govindpur. These two settlements, later, along with Kolkata became the hub of British East India Company. The setting up of factories in these new settlements gained importance which eventually led to the shifting of key Indian businesses of the time from Satgaon.

Moreover, the region had been well surveyed by the British to gauge the navigability of the area in terms of its capacity to handle large ships. The trade through Calcutta primarily consisted of export of agricultural and semi-finished products. The growth of exports led to the setting up of ancillary and support industry in the service sector like insurance, banks which furthered the siphoning off the savings and profits from India to Britain weakening the economic structure of the society.

Calcutta and the trade from the city got impetus with the setting up of Calcutta Port Trust in 1870 that was responsible for the development of infrastructure facilities at the port. The development of railway network in the subcontinent enhanced trade providing easier means of transportation of goods from remote parts of the rich inland regions to the port. By then the British had gained a stronger political hold in the subcontinent, which furthered the export of raw materials and import of finished products and Calcutta continued to be the gateway for this trans-continental trade.

At the time of formation of the Trust, there were four jetties and a wharf accommodating 52 vessels with a total tonnage of 48,000. During the next decade, the number of jetties increased to 6 accommodating 143 vessels with a total tonnage of 2,22,000 tons. The goods handled were grains, seeds, raw and semi-finished jute products. In 1860, Budge Budge petroleum wharf and a tea warehouse were added as well and by 1892 Kidderpore docks were completed leading to phenomenal growth in export and import through Calcutta which by then included coal. From 1772, when Calcutta became the capital of British India until 1911 when New Delhi was made the capital, the port city thrived as an administrative and cultural port city.

Calcutta's growth all through the British colonial period was due to its vast hinterland that was tapped by the British traders linking the agricultural producers and the merchants in India to consumers in China and Europe [67]. Prior to the British establishment, the Mughals had controlled the river systems and by denying the Europeans trade rights and access to inland markets, the Mughals had been able to keep the Europeans at bay. With the decline of Mughals and their ports, the British were able to gradually develop an alternative system that of railways for supplying goods to the port.

With Calcutta emerging as a nationalists center the British moved their capital to New Delhi which is seen as a deterrent to the growth of Calcutta in the long run. Calcutta continued to grow as a port city, especially after opening up of Suez Canal, which shortened the route to England by 6500 KMs leading to increase in volume of commercial traffic moving through the port which in time required the development of new docks, like King George docks in 1929 [68].

Partition of Bengal in 1905, Bengal famine 1943–1944, Independence in 1947, Partition of India and the near continuous influx of refugees and illegal migration from Bangladesh and Myanmar have had an adverse impact on the port city's growth. The boundary created by the Radcliffe Commission left 42 percent of the total Hindu population of undivided Bengal (12 million) in the newly created state of West Pakistan. There was initially no massive migration as had happened in Punjab, but the refugees and cross-border migrants came in a series of waves. For the first 10 years following partition (1947–1958), over 4 million refugees and migrants came to West Bengal. Between 1959 and 1971, another 1.5 million from West Pakistan sought refuge in India. By the early 1980s, it was estimated that at least 8 million migrants from the region of East Bengal had settled in West Bengal.

This influx of migrants put considerable strain on the resources of West Bengal, aggravating social problems and causing severe over-crowding. Most of the migrants converged around the Calcutta area, and 70 percent of the total refugees and migrants came to concentrate in the city, giving rise to proliferation of squatter colonies in the city [65].

The Partition deprived the Calcutta port of nearly 30 percent of its primary hinterland. The impact of this loss was tremendous as [69] has shown the sharp decline in the volume of cargo handled at the Calcutta port from 95 percent in the eastern

Indian sector in 1951–1960 to 51.4 percent in the decade 1961–1970. A contributory factor leading to the decline of the port was the silting of the approaches to the Calcutta port. By early 1960s, deep-drafted vessels had found it almost impossible to navigate their way into the Calcutta port, prompting the port authorities to develop a new satellite port at Haldia, about 50 miles downstream of Calcutta to overcome the difficulties encountered by the old port. This development which was complemented with an adjoining industrial township and connected by roads and railways to major cities in the region, was major blow to Calcutta's primacy as the major port city of east India [70].

After New Delhi had replaced Calcutta as the political center of India in 1911, Mumbai soon became the financial and economic capital. West Bengal, along with Maharashtra, had been India's leading industrial state until the 1960s, accounting for 14 percent of the country's overall industrial output. By 1980–1981, West Bengal's share had fallen to 9.8 percent, and a decade later had dropped to 5.6 percent. The declining share of Calcutta's port in the country's overall foreign trade showcases the decline. In 1964, Calcutta port handled 92 percent (by tonnage) of India's export and 25 percent of its import; 30 years later, by 1995, this had fallen to 2.69 percent and 5.5 percent, respectively, which in 2019 was 2.65 percent [65, 71]. Political tension between the state and the center complicated attempts at successful planning in Calcutta, leaving the city and state in a state of neglect, a situation that has not been helped by an economic stagnation, decaying infrastructure, and continued political confusion.

Having the longest navigational channel in India, longer than all the major Indian ports combined, Kolkata port today has two dock systems—the Kolkata Dock System (KDS) and the Haldia Dock System (HDS) (Kamath 2009). In 2011, Kolkata Port handled about 43.2 million tons of cargo, 0.6 million TEUs of container traffic, and 3,186 vessels. By 2020, this had increased to 47.09 million tons, KDS' share being 13.34 and HDS' being 33.76 MT. The port has 52 berths, 35 including 6 oil jetties at KDS and 17 including 3 jetties at HDC [72]. Two deep seaports are on the anvil -

- a deep seaport at Sagar is planned as a joint venture of the Government of West Bengal and Ministry of Shipping, Govt. of India in PPP mode, and
- a deep seaport is planned at Tajpur, in Purab Medinipur in PPP mode by State Government.

It is to be noted that the ports on the East Coast of India, like Krishnapatnam, Karaikal, Vishakhapatnam, and Paradeep, even though smaller in size and capacity serve as a competition to the port in Kolkata due to there being good deep-water ports and having sufficient catchment areas to handle large container ships.

Kolkata Port is the gateway to eastern and northern India, including landlocked countries of Nepal and Bhutan. As per the Kolkata Port Trust, it has a vast hinterland, comprising the entire Eastern India including West Bengal, Bihar, Jharkhand, Uttar Pradesh, Uttarakhand, Madhya Pradesh, Chhattisgarh, Punjab, Haryana, Rajasthan, Assam, Northeastern States, and the two landlocked neighboring countries viz. Nepal and Bhutan. The industrial development, commerce, and trade of this vast hinterland are inseparably linked to the development of Kolkata Port and vice versa.

5 Discussion

Decentralization of industries and regional development are interrelated phenomena. The development on the periphery of the city undoubtedly contributes to a more balanced local and regional growth. Decentralized industrialization leads to lesser concentration of population in select cities, resulting in greater dispersal economic benefits and development. The Government of West Bengal has taken some initiatives for industrial expansion—setting up new industries in urban fringe and rural areas; strengthening the linkage between agrarian and industrial economy; and setting up of sector focused industrial parks for sustenance of the MSMEs.

It is imperative that manufacturing units be improved in capacity and quality of products to compete with national and international markets. In order to enhance the level of economic development, government policies needs to ensure higher and continuous channels of investment in secondary sector which needs to be embedded in the policy of decentralization of industries so as to make utilization of raw materials and human resource, thereby leading to balanced regional development in the state.

To summarize, the industrial policy of 1978 had given priority to small and cottage industries and sought to curtail the dominance of big industrial houses and foreign multinational firms. With the intent of reviving the inadequacies and product of small and local units. However, success was not achieved in reviving the industrial situation in West Bengal which can be attributed to poor implementation of the policy and conflict with the central government. This led to the state's ranking falling continuously in industrial production.

The 1990s saw the liberalization of India's economy. The initial response of the Government of West Bengal was to oppose the liberalization policy [73]. However, taking advantage of the central government initiative, the West Bengal government attempted to bring an end to the license raj by formulation of the Industrial Policy in 1994. This was a radical move from the earlier policy as it welcomes foreign investments in the state, with initiatives from successive chief ministers to attract foreign investments. However, despite the new policy and some increase in investments, the state continued to experience the decline in its share of ex-factory value added output in 1990 from 6 percent to 3.9 percent by 2000 [73].

In the last decade, the compounded annual growth rate of gross value added for West Bengal was 2.7 percent whereas that of all India was 6.4 percent. Further, according to the Annual Survey of Industries, the state slipped from 9th rank at all India level to 11th rank in terms of net value addition in the last decade. However, even in the early phase after Independence (1948–1958), West Bengal's manufacturing industries recorded a compound annual growth rate of 3.3 percent whereas, that of all India was 2.8 percent. Standing in the middle of twenty-first century, West Bengal's industrial legacy appears as a talk of antiquity when we see the closed shutters of the jute mill factories and the chemical plants [74].

West Bengal's manufacturing sector is predominantly informal in nature where the firms are not registered with the government. Hence, these firms do not have access to any government securities and support, making them vulnerable to market fluctuations. As of 2015, the informal firms accounted for more than 95 percent of the total manufacturing firms in the state. These informal firms employ 93 percent of the total workers working in the manufacturing sector.

Most of the industrial initiatives of the state since 2013 have been directed toward marketing and advertising self-employed small and medium enterprises through the flagship program of *Bishwa Bangla* and the cash transfer scheme of *Shilpa Shathi.* Though marketing does increase the sales revenue but only if there is capacity to produce. The loans given under the cash transfer scheme do not ensure revenue generation due to the weak follow-up strategies of the scheme. The acquired loans are utilized inappropriately, and the small-scale industrialists are not able to break free from the indebtedness [74].

It was argued that due to industries becoming more capital intensive, the employment in the manufacturing sector has fallen from 23 to 18 percent (Employment and Unemployment Survey and Periodic Labour Force Survey) from 2011 to 2019. This is falsified in case of West Bengal due to two reasons, firstly, the state is lacking in adequate capital generation to be invested in capital intensive production techniques and secondly the informal manufacture employs 93 percent of the workforce in the manufacturing sector. West Bengal has the highest number of micro and small-scale enterprises (MSMEs), comprising of 11 percent of the total country's MSMEs, and these are in the informal sector, which face the twin issue of lack of capital creation and unavailability of loan advances from the government for investment in any kind of industrial improvement and enhancement.

Major reasons for the continuous decline and non-revival of industries are coordination failure between the Center and the state government, lack of a comprehensive industrial policy, and lack of conducive political environment for industrial growth. It should be highlighted that the factors that resulted in a crisis of capital creation, which is the underlying economic reason behind the industrial crisis of West Bengal are predominance of informal firms in the sector who are suffering from access to capital both in terms of fixed assets and loans and decline in the formation of fixed assets for the formal sector firms.

This research recommends and reiterates the need to create a sound decentralized industrial policy with a firm agenda of creation regional capital by empowering the MSMEs and a progressive intervention to boost large enterprises investments. Government initiatives and policies need to be driven toward capital formation, production, and revenue generation.

To conclude, a multi-pronged approach needs to be adopted, at the node level that is the port of Kolkata modernization, and expansion requires augmentation, which needs to be based on digital technology and concept of ease of business. This system should be robust enough to attract business to use Kolkata as a desirable port for export and import. Further, in this whole gamut, as stated by Bedi and Mahavir [21], there is a need to balance the environment, competitiveness, and quality of life of the citizens which can be achieved through better governance based on extensive use of technology.

At the area level, the livability in Kolkata metropolitan region needs redressal. Decongestion and decentralization need to be the basis for future planning based on the principles of ecology of the city rather than ecology in the city. On a linear scale, interconnections with the hinterland need to be reliable, inexpensive, and rapid. This needs to be particularly addressed in relation to the states in the northeast India.

References

1. Dasgupta P. Capital shift. [Online] August 03, 2016. [Cited: December 02, 2022]. https://timesofindia.indiatimes.com/city/kolkata/capital-shift-after-15-years-why-kolkata-is-still-calcutta-for-many/articleshow/53515720.cms
2. Wafar M et al (2011) State of knowledge of coastal and marine biodiversity of Indian Ocean countries. PLoS One 6(1):e14613
3. Maddison A (2007) Contours of world Economy I-2030 AD. Oxford University Press, Oxford
4. Khurana GS (2019) What is the Indo-Pacific? The new geopolitics of the Aisa-Centered Rim Land. In: Berkofsky A, Miracola S (eds) Geopolitics by other means, the indo pacific reality. Ledizioni LediPublishing, Milan, pp 13–32
5. Li H (2022) The "Indo-Pacific": intellectual origins and international visions in global contexts. Mod Intellect Hist 19(3):807–833
6. Dar AI (2022) India's Indo-Pacifc policy: unpacking the underpinnings and challenges. Chinese Political Science Review
7. Shishir M, Sakib NH (2022) How interests and ideas of a dominant actor make a big diference. Chinese Political Science Review
8. Ismail M, Husnain SM (2022) Recalibrating impact of regional actors on security of China-Pakistan economic corridor (CPEC). Fudan J Hum Soc Sci 15:437–462
9. Gnanagurunathan D. India and the idea of the 'Indo-Pacific'. [Online] October 20, 2012. [Cited: May 23, 2023]. https://www.eastasiaforum.org/2012/10/20/india-and-the-idea-of-the-indo-pacific/
10. US Indo Pacific Command. History of United States Indo-Pacific Command
11. Bhattacherjee K (2017) India, Japan, U.S., Australia hold first 'Quad' talks at Manila ahead of ASEA Summit. s.l. The Hindu
12. Joshi M. What's in a name? India's role in the Indo-Pacific. [Online] July 22, 2021. [Cited: April 23, 2023]. https://www.orfonline.org/research/whats-in-a-name-indias-role-in-the-indo-pacific/
13. Mehrotra, L. (2012) India's look east policy: its origin and development. Indian Foreign Aff J 7(1): 75–85
14. Saha P. India's role in the emerging dynamics of the Indo-Pacific. [Online] January 26, 2022. [Cited: May 01, 2023]. https://www.orfonline.org/expert-speak/indias-role-in-the-emerging-dynamics-of-the-indo-pacific/
15. United Nations (2019)World urbanization prospects, the 2018 revision. United Nations, New York
16. Hemsfile (1999) Perspectives on political and economic trends in the Americas, vol 5
17. Caves RW (2004) Encyycolpedia of the City. s.l. Routledge
18. Kraas F, Mertnis G (2014) Megacities and global change. In: Kraas F, Aggarwal S, Mertins M (eds) Megacities—Our global urban future. s.l. Springer
19. Chen NY, Heligman L (1994) Growth of world's megapolises. In: Fuchs RJ et al (eds) Mega-city growth and the future. The United Nations University, Tokyo
20. United Nations (1984) Population and policies in mega-cities: Bangkok
21. Bedi P, Mahavir (2021) Spatial strategies for Kolkata Metropolitan Region. In: Vinod Kumar TM (eds) Smart global megacities. Advances in 21st century human settlements. Springer, Singapore

22. United Nations. Population facts: policies on spatial distribution and urbanization have broad impacts of sustainable development. [Online] December 2020. [Cited: May 21, 2023]. https://www.un.org/development/desa/pd/sites/www.un.org.development.desa.pd/files/undes_pd_2020_popfacts_urbanization_policies.pdf
23. Singh V (2003) Panchayati Raj and village development. Sarup and Sons, New Delhi
24. Kolkata Metropolitan Development Authority. Introduction to KMDA. [Online] January 2022. [Cited: May 03, 2023]. https://kmda.wb.gov.in/page/cms/introducing_kmda_011393
25. Dasgupta S et al (2013) A megacity in a changing climate: the case of Kolkata. Clim Change 116(3–4), 747–766
26. Ghosh D, Sen S (1987) Ecological history of Calcutta's wetland conversion. Environ Conversat 4(3)
27. Direct Flights from Kolkata. [Online] June 29, 2023. [Cited: June 29, 2023]. https://www.flightconnections.com/flights-from-kolkata-ccu
28. Statesman News Service. Kolkata airport sees rise in both passenger & cargo flights. [Online] January 14, 2021. [Cited: May 21, 2023]. https://www.thestatesman.com/cities/kolkata/kolkata-airport-sees-rise-in-both-passenger-cargo-flights-1502946372.html
29. Census of India (2011) Primary census abstract, West Bengal. s.l. Government of India
30. Hindustan Times. Census 2021 deferred till further orders: Centre. [Online] July 27, 2022. [Cited: April 19, 2023]. https://www.hindustantimes.com/india-news/census-2021-deferred-till-further-orders-centre-101658861900488.html
31. Population Census Data. Kolkata population 2023. [Online] 2022. [Cited: June 02, 2023]. https://www.census2011.co.in/census/city/215-kolkata.html
32. Banerjee S (2016) Memoirs of roads: Calcutta from colonial urbanization to global modernization. Oxford University Press, New Delhi
33. Sanjoy C (2023) From colonial city to globalizing city? The far-from-complete spatial transformation of Calcutta: an update after 25 years. In: Bandyopadhyay S (eds) Kolkata—the colonial city in transition: reflections in geographies of Urban India. s.l. Routledge
34. World Bank (2011) Vulnerability of Kolkata metropolitan area to increased precipitation in a changing climate. s.l. World Bank
35. Roy B, Bahdyopadhyay S (2023) Restructuring urban space: how inclusive are neighbourhood transformations? In: Bandyopadhyay S (eds) Kolkata—the colonial city in transition: reflections in geographies of Urban India. Routledge, New York
36. Haque I, Mehta S, Kumar A (2019) Towards sustainable and inclusive cities: the case of Kolkata. s.l. ORF
37. Kumar S, Banerji H (2018) Analysis of transformation of urban planning practice by mapping changes in economic, social, cultural and built environment of the changes in economic, social, cultural and built environment of neighbourhood level urban communities: case Study of Kolkata. 54th ISOCARP Congress.
38. Ministry of Urban Development (2015) Smart city mission guidelines. s.l. : Government of India
39. Bandyopadhyay K. Times of India. [Online] February 28, 2017. [Cited: May 31, 2023]. https://timesofindia.indiatimes.com/city/kolkata/three-planning-organizations-of-kolkata-merging-into-one/articleshow/57394914.cms
40. Ministry of Urban Development (2007) Jawaharlal Nehru National urban renewal mission: overview. s.l. Government of India
41. Chowhan G, Sen A, Mukhejree J (2022) Sustainable and 'smart' restructuring around the making of mega and world-class cities in India: a critical review. GeoJournal 87(4), 897–910
42. UNDP (2019) Hazard profiles of India districts, national capacity building project in disaster management. s.l. UNDP
43. Tiari V, Sarkakr A, Jana A (2022) Integrating master planning with smart city initiative: challenges and opportunities in Newtown, Kolkata. In: Vinod Kumar TM (ed) Smart master planning for cities. Springer, Singapore
44. Ganguly L (2020) Past and present economic status of West Bengal: a review. ARSS 9(2)

45. Saini A, Chaudhary M, Solanky R (2015) Survival of handloom industry duing 19th and early 20th century in India. Int J Trade Commer 4(1):115–120
46. Nandi B (2015) The decline in organized manufacturing—policy and industrial performance of West Bengal.Quest Multidiscip J HumIties Soc Sci 2(2):1–21
47. EPW. Economic and Political weekly. [Online] January 26, 1955. [Cited: June 27, 2023]. https://www.epw.in/system/files/pdf/1955_7/4-5/problems_of_west_bengal_industries.pdf
48. Lahiri A. Ideas for India. [Online] 2016. [Cited: February 20, 2023]. https://www.ideasforindia.in/topics/moneyfinance/west-bengal-from-an-external-perspective.html
49. Ray SC (2011) The political economy of decline of industry in West Bengal: experiences of a marxist state within a mixed economy. University of Connecticut Working Paper 2010–2011
50. Chaudhari B, Sarkar S, Panigrahi AK. Chronic problems in industrialization in West Bengal. [Online] 2014. [Cited: February 22, 2023]. https://www.theigc.org/sites/default/files/2015/01/Chaudhuri-et-al-2014.pdf
51. Karak A (2017) Pattern of industrial growth in West Bengal during 1980–1991: structural demand and agricultural industry relations. J South Asian Dev 12(1): 65–88
52. Sen R. The Evolution of industrial relations in West Bengal. [Online] December 15, 2009. [Cited: January 22, 2023]. https://www.ilo.org/newdelhi/whatwedo/publications/WCMS_123342/lang--en/index.htm
53. Power for All, West Bengal. [Online] 2016. [Cited: May 23, 2023]. https://powermin.gov.in/sites/default/files/uploads/joint_initiative_of_govt_of_india_and_West_Bengol.pdf
54. Burange LG, Yamini S. A review of India's industrial policy and performance. [Online] January 2011. [Cited: January 10, 2023]. https://www.researchgate.net/publication/280727114
55. Roy BC, Khatun D, Roy A (2018) Rural livelihood diversification in West Bengal. s.l. Ministry of Agricultural and Farmers Welfare
56. ASSOCHAM India. Rising Bengal: The land of opportunities. [Online] October 2011. [Cited: June 23, 2023]. https://www.pwc.in/assets/pdfs/industries/rising-bengal-151211.pdf
57. Government of India. Status of city development plan: Kolkata. [Online] 2005. [Cited: December 20, 2022]. http://jnnurm.nic.in/nurmudweb/toolkit/kolkata.htm
58. Government of West Bengal. Know your industrial estate. [Online] 2023. [Cited: May 21, 2023]. https://www.wbsidcl.in/estate/industrial_estates
59. Dasgupta S. Bengal's long lost entrepreneurial spirit. [Online] May 26, 2019. [Cited: March 20, 2023]. https://medium.com/@soumyadasgupta.91/bengals-long-lost-entrepreneurial-spirit-4ea63a5d0a5d
60. Roy SN (1996) West Bengal's industrial development policy. Econ Polit Wkly 31(18):1094–1096
61. Reeves P, Broeze F, McPherson K (1989) Studying the Asian Port City. In: Broeze F (ed) Brides of the sea: port cities of Asia from 16th–20th centuries. s.l. University of Hawaii Press
62. Driessen H (2005) Mediterranean port cities: cosmopolitanism reconsidered. Hist Anthropol 16(1):129–141
63. Kidwai AH (1992) Conceptual and methodological issues: ports, port cities and port hinterlands. In: Banga I (ed) Ports and their hinterlands in India 1700–1950. Manohar, New Delhi, pp 7–43
64. Frost MR (2002) Wider opportunities: religious revival, nationalist awakening and the global dimension in Colombo 1870–1920. Mod Asian Stud 36(4):937–967
65. Tan T-Y (2007) Port cities and hinterlands: a comparative study of Singapore and Calcutta. Polit Geogr 26, pp 851–865
66. Bird JH (1971) Seaports and seaport terminals. Hutchinson and Co., London
67. Bayly CA (1985) Inland port cities in north India: Calcutta and the Gangetic Plains 1780–1900. Basu DK (ed)The rise and growth of the colonial port cities in Asia. University Pres of America, Berkeley, pp 13–17
68. Ray, A. The Calcutta port. [book auth.] S. Chaudhari. *The present and future. Calcutta: The living city, Vol. II.* Calcutta : Oxford University Press, 1990.
69. Kidwai AH (1989) Port cities in a national system of ports and cities: a geographical analysis of India in the twentieth Century. In: Broeze F (ed) Brides of the sea: Port cities of Asia from the 16th–20th centuries. University Of Hawaii Press, Honolulu, pp 207–222

70. Chaudhuri MR (1964) Geographical notes: Haldia—a major port in making. Geogr Rev India 26(1): 40–41
71. Pal A (2006) Scope for bottom-up planning in Kolkata: rhetoric vs reality. Environ Urban 18(2):1–21
72. Ministry of Shipping. Annual report 2019–20. s.l. Government of India
73. Bhattacharya D. Why the lack of industry in West Bengal? [Online] May 22, 2019. [Cited: May 22, 2023]. https://www.downtoearth.org.in/blog/economy/why-the-lack-of-industry-in-west-bengal--64679
74. Nataraj M. West Bengal's industrial decline: a crisis of capital creation. [Online] March 22, 2022. [Cited: June 03, 2023]. https://indianresearcher.in/west-bengals-industrial-decline-a-crisis-of-capital-creation/

Dr. Prabh Bedi is a geographer, regional planner and geospatial technology expert. She has been engaged in the field of planning and geospatial technologies both in industry and academics for over 25 years. She has published research on NUIS, Smart Cities, urban greens and urban hydrology. Her research interests include information systems, especially geospatial technology, and its applications in urban planning.

Prof. Dr. Mahavir was engaged in teaching at School of Planning and Architecture, New Delhi for over 35 years. Planning Techniques, Geo-Informatics, Regional Planning and Environmental Planning have been his areas of interest. He holds degrees in Architecture, Urban and Regional Planning, PG Dip. in Remote Sensing and a Ph.D. jointly from ITC and the Utrecht University, The Netherlands. He contributed several papers and book chapters in National and International publications. Besides teaching, he has held the positions of Dean (Academic) and has headed the Departments of Physical Planning, Environmental Planning, Regional Planning and Centre for Remote Sensing and GIS at the School. He has been on several committees of the Government of India related to Urban and Regional Planning and Planning Information Systems.

Mumbai

Positioning of Mumbai in the Indo-Pacific Megacity System: Economic and Transport Opportunities

Amit Chatterjee, Mitali Mondal, Jilik Dhar, and Md. Imran Nazir

Abstract Mumbai is the largest megacity in India and one of the world's largest urban agglomerations, with a population of approximately 20 million in 2018 and is expected to increase to around 24.5 million by 2030, according to UN-Habitat estimates. Mumbai has emerged as an integral part of the Indo-Pacific megacity system in geopolitics, global economy, international trade, and inter-regional transport, art, and culture (Hindi film industry, popularly known as Bollywood). Being a megacity (with more than twenty million people) in the Indo-Pacific region, rapid urbanization and economic development are resulting in changing patterns of city functions. Mumbai's significance in the world is continuously expanding due to the juxtaposition of its locational and functional advantage, i.e. flourishing economy, technological advancement, and its connection with the outer world. Along with these economic advantages, Chhatrapati Shivaji Maharaj International Airport in Mumbai is an important global hub of the air transportation system. Once operational, Navi Mumbai International Airport will boost capacity and improve connection to Mumbai's air transport system with the rest of the world. The Jawaharlal Nehru Seaport, India's largest container port, is essential to India's foreign trade. Further, the International North-South Transport Corridor, consisting of rail, road, and maritime transportation networks, starts from Mumbai and is a prime economic growth corridor for promoting economic cooperation among the participating countries. Positioning of Mumbai city in the Indo-Pacific setting can be conceptualized by reviewing the city's economy and transport opportunities.

Keywords Mumbai · Indo-Pacific · Megacities · Economy · Bollywood · Transport · North-South Corridor

A. Chatterjee (✉) · M. Mondal · J. Dhar · Md. I. Nazir
Department of Geography, Visva-Bharati, Santiniketan 731235, West Bengal, India
e-mail: amit.chatterjee@visva-bharati.ac.in

T. M. Vinod Kumar (ed.), *Indo-Pacific Smart Megacity System*, Advances in 21st Century Human Settlements, https://doi.org/10.1007/978-981-99-6218-1_7

1 Introduction

Change in the term from 'Asia-Pacific' to the 'Indo-Pacific' indicates India's emergence among the world's leading countries. The Asia-Pacific, or the now- Indo-Pacific region, is strategically important due to the shift of centre of power from the Atlantic countries to the countries along the Indian and Pacific Oceans. The Indo-Pacific region comprises world's leading economies, such as China, Japan, India, etc., and is a prominent hub of world trade. Besides economic importance, this region presents highly diversified natural and anthropogenic interactions. The concerned region is projected to emerge as the world's most dynamic and growth centre by 2025 [1]. The Indo-Pacific region is one of the most dynamic and vibrant economic regions globally, accounting for a significant portion of global GDP and trade. The region is home to several of the world's largest economies, including China, Japan, India, Australia, and many emerging economies. Many megacities, such as Mumbai, Tokyo, Delhi, Shanghai, etc., are in the Indo-Pacific region. The region is also a significant destination and source of foreign direct investment (FDI) due to its sizeable consumer market, strategic location, and competitive labour costs. The region's trade links and integration have been boosted by various initiatives, including the Regional Comprehensive Economic Partnership (RCEP), the Comprehensive and Progressive Agreement for Trans-Pacific Partnership (CPTPP), and the Belt and Road Initiative (BRI) [2]. Indo-Pacific nations account for 65% of the global population, generate 62% of the global GDP, 46% of the global foreign international trade, and approximately half of the global maritime transport [3].

The Indo-Pacific region has experienced exceptional economic growth over the previous four decades, along with its diversity, which is a strength. Several nations, such as China, India, etc., are going through fast changes in their economies, urbanization and population expansion, social upheaval, and technological advancement, and these nations possess shared economic stimuli. Megacities are the ones where changes are more prominent. In this context, Mumbai, Delhi, Shanghai, Beijing, etc., is no longer an exception. These megacities are the global economic powerhouse now. The changes in various human-centric spectrums of megacities have resulted in massive environmental degradation, including rapid climate change, which in turn is increasing the need for sustainable urban development while creating obstacles to attaining them.

India perceives the area as open, balanced, integrated, and inclusive in the Indo-Pacific region. India underlines the strategic connections, shared difficulties, and opportunities between the Indian Ocean and the Pacific constantly. India intends to work jointly to create a regional security and peace structure. The nations must develop a shared, rules-based system for the area through conversation for everyone to enjoy prosperity and security. India encourages an Indo-Pacific region marked by rule-based, open, fair, and stable trade that pulls all countries up on a wave of investment and trade [4].

Mumbai has been crucial to India's place in the world. Mumbai's business leaders have represented India as an investment destination and internationally projected its goods, services, entrepreneurs, and prospects. They have also acted as the country's economic envoys. Economic growth, booming industry, art, culture, etc., contribute to Mumbai's positioning as a global megacity. Mumbai is especially attracting global attention as a 'foci' of the Indo-Pacific region with a wide range of urban functioning [37].

Mumbai, India's financial capital, is the country's most populous and seventh largest populous city in the world. Mumbai, or the Greater Mumbai, consists of Mumbai City and Mumbai Suburban and comes under the Mumbai Metropolitan Region (MMR). The Greater Mumbai lies south of the Ulhas River and is drained into the Arabian Sea by Mithi, Oshiwara, Poisar, and Dahisar rivers. Along its 146 km coastline, it comprises several creeks and bays. The eastern coast is covered with extensive mangrove swamps, whereas the west coast is sandy and rocky [5, 6].

Mumbai city comprises 38% residential, 33.43% recreational spaces and transport, 8% industrial, 1% office, and 3% commercial land uses. According to the Disaster Management Department of BMC (2019), Mumbai's GDP is predicted to be $310 billion, or more than 6% of the country's overall GDP. Mumbai has an estimated GDP of $310 billion, contributing to more than 6% of the nation's GDP. The key sectors contributing to the city's economy are finance, IT & ITES, textiles, entertainment, gems, jewellery, and leather processing [7].

In contrast to the economic growth, 42% of the city's population lives in slum areas with unhygienic, congested, and poor housing conditions [7]. Dharavi is one of the largest slums in the world. In Mumbai, vast disparity in socio-economic conditions exists between slum and non-slum areas. Along with its rapid urbanization and economic development, the city suffers from an increasing crime rate, unemployment issues, and severe climatic and human-induced hazards. The recurrent climate disasters, such as floods and droughts, have a tremendous impact on livelihoods, forcing people to migrate to other cities and work in the informal sector [7]. In 2022, Mumbai was the first Indian city to release a Mumbai Climate Action Plan (MCAP). Mumbai is a unique city in the Indo-Pacific region regarding its Climate Action Plan and its linked strategies to combat transport, waste, and other issues [7]. Megacities are the prominent contributors to GHG emissions and other anthropogenic phenomena resulting in climate change. Hence, assessing, formulating, implementing, and monitoring the climate with changing urban functioning is necessary. The MCAP pioneers other Indian megacities in this respect while standing in line with other leading cities of influential countries in the Indo-Pacific region.

Excellent public transport system is the lifeline of Mumbai city; airways and highways connect Mumbai megacity with the other parts of the country and the world. Major National Highways such as NH-4, NH-8, NH-66, etc., connect the city with significant parts of the country. The Mumbai suburban railway mainly consists of six lines, viz. Western, Central, Harbour, Trans-harbour, Nerul-Ural, and Vasai Road-Roha lines. Mumbai is connected with other megacities via different national and international airlines through Chhatrapati Shivaji International Airport. Apart from the existing airport, the Navi Mumbai airport is under construction to tackle the congestion in air transport. Much international trade is conducted through Nhava Sheva Port (the second largest container port in India) of Mumbai.

Being an Indo-Pacific centre, Mumbai is a unique megacity worldwide. Mumbai, as well as India, has emerged as an essential participant in the Indo-Pacific system. Still, the Global Positioning of Mumbai can be further accelerated through strategic planning and implementation. In the formulation of plans, the significant parameters of the city should be discovered and analysed, around which the city's ecology functions daily. In this chapter, intra- and inter-regional economy and transport have been discussed to understand the potentiality of Mumbai and its interconnectedness with other megacities, particularly within Indo-Pacific nations.

2 Economic Profile of Mumbai

Mumbai is India's largest city (by population) and the country's financial and commercial capital, generating approximately 6.16% of the total GDP. It serves as an economic hub of India, contributing 10% of factory employment, 25% of industrial output, 33% of income tax collections, and 60% of customs duty collections. The economic powerhouse of India not only contributes the highest GDP of $209 billion and is responsible for 25% of industrial output, 70% of maritime trade in India, and 70% of capital transactions to the nation's economy. The city, best known as India's industrial centre, is home to numerous corporations and operates in various industries, including textiles, petrochemicals, gems, leather, etc. Over 20% of Maharashtra's GDP comes from Mumbai and its immediate surrounding regions, popularly known as Metropolitan Regions (MMRs). Mumbai's per capita income ofroughly three times higher than India's. According to UN projections, Mumbai's population is expected to reach around 27 million by 2020. Mumbai's GDP per capita at purchasing power parity (PPP) is anticipated to reach $23,000 by the 2020–2021 fiscal year. The main port of India on the Arabian Sea is Mumbai. It handles over 60% of India's total sea trade [8].

2.1 *Emerging Data Centre Industry*

Mumbai, India's thriving financial centre, has reached a critical milestone in the technology sector. According to a recent Knight Frank report, the city has emerged as the Asia-Pacific region's third largest data centre market, demonstrating enormous development and potential. Mumbai's data centre business is prospering and attracting domestic and international investments, with a total capacity of 2,337 MW and surpassing the 2GW milestone in Q1 [9]. This initiative emphasizes the city's growing importance in the global digital landscape and represents a fundamental shift in India's technological infrastructure. Data centres are the backbone of today's digital economy, providing the infrastructure for processing, storing, and distributing massive amounts of data. The exponential development in data consumption, fuelled by e-commerce, cloud computing, and digital services, has increased the need for reliable and scalable data centre facilities. Mumbai has positioned itself as a suitable destination for data centre investments in the world due to its strategic location, talented workforce, and extensive connectivity.

Mumbai's data centre sector has grown dramatically this decade. The operational capacity is 270 MW, accounting for approximately 11.55% of the total capacity. Furthermore, more than half of the total capacity, 1,272 MW, is in the early phases of development. This indicates a robust pipeline of upcoming data centre projects, showing investors' rising confidence in the business. NTT-Netmagic and CtrlS, two leading data centre providers, dominate the Mumbai market regarding live capacity. The rise of new companies, such as Stonepeak-backed Digital Edge, shows a variety of offers and greater competition. These players are leveraging the city's potential, expanding the city's data centre landscape, and pushing additional innovation. STT GDC India, India's largest data centre provider, is a significant data centre colocation service provider in Mumbai and Navi Mumbai [10].

According to Knight Frank's research, Mumbai is one of the top data centre markets in the Asia-Pacific region. While Shanghai (2,692 MW) and Tokyo (2,575 MW) are top and second, Mumbai's phenomenal growth has moved it to third place with its 2,337 MW capacity. This ranking affirms Mumbai's worldwide competitiveness and strengthens its position as a regional technology powerhouse [11].

India's data centre industry has witnessed steady growth and is expected to increase the supply capacity to 1400 MW by 2025. Mumbai Data Centre Market Leaders are Equinix, STT Telemedia, CtrlS, NTT Data, and Nxtra Data Limited. Mumbai's data centre capacity is predicted to increase by 2.65 times by 2023, according to a report by JLL; the market will be driven by the transition from 4 to 5G technology and an annual increase in the number of mobile subscribers [12].

2.2 *Finance and Stock Exchange Centres*

Mumbai is one of the world's top centres of commerce in terms of financial flow. The city contributed to several innovations in the financial market of India and is the centre of stock market culture, financial market regulation, and other derivatives trading, electronic and online trading. Mumbai continuously attracts industrial investments from domestic as well as foreign institutions. Share markets in the city transact almost 70% of the country's stocks. The National Stock Exchange (NSE) is one of India's largest stock exchanges in Mumbai (the oldest in Asia). Important companies listed on the NSE include Reliance Industries, HDFC Bank, SBI, and ICICI Bank. The stock market capitalization in the city during 2005–2010 tripled from $553 billion to $1631 billion, and turnover of the stock market rose from $314 billion to $801 billion. The stock markets also cater to nearly 1700 foreign institutional investors. There are other stock exchanges like BSE and MCX-SX as well. Global investors purchase shares from these stock exchanges and earn a considerable profit. The government has increased its focus on Mumbai, dubbed India's financial capital for various reasons. It houses an estimated population of over 22 million people and is home to prestigious financial institutions, banks, and companies recognized nationwide [13, 14].

2.3 Industry

Mumbai is a diverse manufacturing hub contributing one-fourth of India's industrial production. Mumbai also has a strong and growing information technology (IT) sector. The availability of excellent infrastructure with investment-friendly policies, a conducive business environment, and a skilled workforce in the city has attracted industries in various sectors such as Automobiles and its Components, Electronic Systems Design and Manufacturing, Gems and Jewellery, IT and Information Technology Enabled Services (ITES), Pharmaceuticals and Chemicals, Engineering, Textile, etc. Industrial parks, IT parks, logistic parks, mega textile hubs, smart industrial cities, plug-and-play infrastructure, connectivity to industrial clusters, incubation centres for startups, etc., have been developed in the city. The city is home to some of the leading industries in India, including IT, electronics, chemicals, pharmaceuticals, engineering products, cement, and diamond cutting and polishing. Numerous companies like Tata Motors, Reliance Industries, Mahindra & Mahindra have offices here. The banking industry contributes to the economy of Mumbai and is home to some of the largest financial institutions in India. The following key sectors play a vital role in the city's economy.

2.3.1 Gems and Jewellery

Mumbai is the hub of India's jewellery industry, receiving most of the country's gold and rough diamond imports (two major jewellery segments in India). It is also the largest exporter of gems and jewellery, accounting for US$28.32 billion or 69% of the total export of gems and jewellery from India. Although most diamond processing is undertaken in Surat, Bhavnagar, Ahmadabad, and Bhuj (cities neighbouring Gujarat), Mumbai has many semi-automatic factories and laser-cutting units. Most of these are located in the city's special economic zone, which in itself accounts for 50% of India's gem and jewellery exports. In March 2019, Maharashtra's chief minister announced the development of an INR 14,467 crore (US$2 billion) ultra-modern and high-tech gems and jewellery park. The 21-acre complex will be set up at Navi Mumbai on the outskirt of Mumbai. It will have manufacturing units, residences for industrial workers, commercial areas, and commercial support services. The Gem and Jewellery Export Promotion Council expects the park to generate an annual turnover of INR 41,467 crore (US$5.8 billion) [14].

2.3.2 Automobiles

The automotive industry has significantly aided the industrialization of Mumbai. The city has a representation for every automotive, including two- and three-wheelers, passenger cars, and commercial vehicles. Mumbai is home to the corporate offices of many significant automakers, including Tata Motors, the largest automaker in India. The city's established manufacturers, robust engineering sector, and extensive trained labour base are its main competitive advantages in the automobile and auto component industries. Hindustan Petroleum Gas Agency, Audi India Ltd., FCA India Automobiles Pvt. Ltd., Hinduja Group Ltd., Mahindra & Mahindra Ltd., Tata Motors Ltd., etc., are Mumbai's leading service providers. Other significant auto hubs near Mumbai include Nagpur, Nashik, Aurangabad, and Pune.

2.3.3 IT and ITES Sectors Industry

India has earned a well-deserved position in the international market for information technology. The share of IT products (mainly software and engineering services) in India's total exports has increased steadily. Mumbai, the country's financial and commercial capital, provided the first mover and shaker in the information technology industry. Despite competition from Bangalore, Mumbai has carved out a niche in India's IT industry, with many multinational corporations and small software firms located here. IT companies can use excellent facilities at the Santacruz Electronic Export Processing Zone (SEEPZ) and the International Infotech Park in Vashi, Navi Mumbai.

2.3.4 Cotton and Textile Industry

Mumbai is known as the 'cottonpolis' of India as it is the most important centre of cotton production. Mumbai's cotton textile industry proliferated because of the city's warm, humid climate, ideal for cotton growth. Mumbai has a port that aids in product import and export. The cotton textile industry in the region expanded quickly. The Indian textile and clothing sector is predicted to develop at a 10% CAGR from 2019 to 2025–2026, reaching US$190 billion. India accounts for 4% of the global textile and clothing trade. During the cotton season 2021–2022, the estimated production was 362.18 lakh bales. Domestic consumption for the cotton season 2021–2022 is expected to reach 338 lakh bales. Cotton production in India is expected to reach 7.2 million tonnes (43 million bales of 170 kg each) by 2030, owing to rising consumer demand. Exports of readymade garments (RMG) cotton, including accessories, total $7.68 billion in FY23 and are expected to reach $7.68 billion in January 2023 [15].

2.3.5 Tourism Industry

Tourism is one of the fastest-growing sectors in the State, generating substantial foreign exchange and also creating significant employment prospects. The State is one of India's most popular tourist destinations, offering diverse experiences from beaches, wildlife sanctuaries, hill stations, natural caves, waterfalls, forts, colourful festivals, ancient pilgrimage shrines, museums, and historical monuments. Maharashtra has always attracted tourists from different states and countries. According to the 'India Tourism Statistics, 2022' report, the State ranked fifth in domestic tourist visits and second in foreign tourist visits during 2021 [16].

2.4 Mumbai: The Hub of Indian Cinema

Mumbai is the birthplace of the Indian cinema industry, which began in 1896 with the first film screening at the Watson Hotel in Mumbai. Dadasaheb Phalke created India's first silent feature film, Raja Harishchandra, and by 1930, the cinema industry was producing more than 200 films each year. Today, Bollywood has expanded its market beyond its traditional territories in South Asia, bringing unprecedented scholarly interest and media attention to the world's most prominent producer of films. The Indian film industry has constantly fared well over the years and has emerged as the world's largest in the number of films produced annually. There are around 400 production firms with 32 corporate houses, and this industry employs and provides a living for over 60 lakh people [17]. The industry turnover crossed $1.2 billion in 2010, hovered around $2 billion in 2015, and is projected to grow steadily to $2.8 billion by 2020 [18]. India's media and entertainment industry has grown from 728.4 billion INR in 2011 to 1262.1 billion INR by 2016, with the largest share of television followed by print media [19]. A recent report highlighted that the Indian media and entertainment industry had reached INR1.61 trillion in 2021 with a growth rate of 16.4%, and the same is expected to reach INR2.32 trillion by 2024 [20] (Fig. 1).

In recent years, the Indian film industry, particularly Bollywood, has occupied a prominent place on the 'global' map of cinema and films. Because of its global effect on movies, music, dance, and other creative forms, Bollywood has developed a potent brand in global cultural events.

The Bollywood industry in India is suitable for studying the emergence and internationalization of a local company in a developing country. The Indian film cluster is increasing at 16% per year, making it India's second fastest-growing sector. In recent years, Bollywood has seen a 60% growth in exports and is well on its way to incorporating into the global market, contributing to Indian economic success.

Bollywood, in particular, also has tourism benefits, with Bollywood locations boosting tourist visitor numbers, i.e. an indirect channel through which the Indian film industry contributes to gross domestic product (GDP).

The power of Bollywood is driving India as a 'soft power' that is not only promoting Indian culture, its songs, dances, values, and beliefs but is also drawing

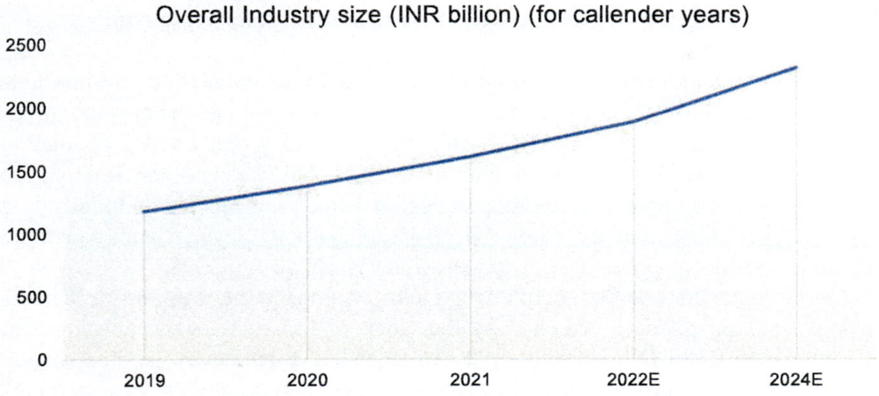

Fig. 1 Growth of Indian media and entertainment industry (2019-2024E). *Source* [20]

huge revenues into the country. In 2019, Indian film box office revenues exceeded US$2.5 billion, representing over half of the Indian GDP. Mumbai's Bollywood has become a global emblem of Indian culture and modernity, with a growing viewership in South Asian, Arab, and Southeast Asian countries, influencing numerous cultural practices in these regions. In the Middle East, Central Asia, Africa, and Latin America, Bollywood films are well-known. However, the United States and Europe now account for 60% of all overseas revenues. It has recently been announced that Mumbai plans to develop a 'sister city square' at Bandra Kurla Complex to honour the relationship between Mumbai and its 15 'sister cities' worldwide. It will be a central global cultural exchange zone and a tourist destination [20].

3 Transport

India is the world's seventh largest country in land area, and the most populous country globally. India has a very good transportation network. According to World Bank estimates, roadways carry around 85 per cent and 60 per cent of India's passenger and freight traffic, respectively [21]. As of 2020, India's rail network delivered 8.086 million originating passengers and 1208.41 million tonnes of freight traffic [22]. Aviation in India is roughly broken between military and civil aviation, with the latter being the world's fastest-growing aviation sector [23]. India has the world's eighth biggest waterway network, consisting of rivers, canals, backwaters, and creeks. Waterways are underutilized for goods movement in India, with inland waterways moving just 0.1% of total inland traffic (The Economic Times, 2016). According to the 2011 Indian census, over 21% of families have two-wheelers, whereas 4.7% of households have automobiles or vans [24]. India's car sector is now expanding quickly, with an annual production of about 4.7 million automobiles

[25] with a 10.5% yearly growth rate [24], and in the future, vehicle utilization is anticipated to skyrocket.

Mumbai, commonly known as Bombay, was the official name of the Indian state of Maharashtra until 1995. Mumbai is India's financial centre and the most populous metropolis, with a city-proper population of 12.5 million people. With a population of over 23 million, Mumbai is the core of the Mumbai Metropolitan Region, the world's sixth most populous metropolitan region. Mumbai is located on India's west coast on the Konkan coast and boasts a deep natural harbour. Seven islands constitute Mumbai, and Mumbai was designated an Alpha World city in 2008.

The Golden Quadrilateral, connecting Delhi, Mumbai, Chennai, and Kolkata, is an ambitious programme launched by the Govt. of India. Also, the development of Mumbai to Delhi Dedicated Freight Corridors and Privatization and expansion of the Mumbai airports are enhancing transport sector capacity and improving its efficiencies [21].

3.1 Air Transport of Mumbai

Mumbai has one international airport, Chhatrapati Shivaji Maharaj International Airport, which serves Mumbai and the metropolitan region of Mumbai. The airport was opened in 1942 and is mainly owned by the Airports Authority of India. The airport is operated by Mumbai International Airport Limited (MIAL), a joint venture between the Airports Authority of India and a consortium led by GVK Industries Ltd., which was hired in February 2006 to modernize the airport. In 2019, it was the 14th busiest airport in Asia and the 41st busiest airport in the world by passenger traffic, and it was the second busiest airport in the country in terms of total and international passenger traffic after Delhi. In 2018, it had around 49.8 million passengers.

Regarding freight traffic, it is also the second busiest airport [26]. In March 2017, the airport surpassed London's Gatwick Airport as the busiest single-runway airport in the world. Gatwick Airport eventually exceeded this at the end of 2019 when passenger counts in Mumbai dropped [27].

The airport's IATA (International Air Transport Association) code, BOM, is related to the city's old legal name, 'Bombay'. It has two operational terminals distributed across 750 hectares (1,850 acres) and manages around 950 aircraft movements daily. Apart from Mumbai's Chhatrapati Shivaji Maharaj International Airport, the proposed international airport is under construction at Navi Mumbai (Fig. 2).

There are two runways of Mumbai's Chhatrapati Shivaji Maharaj International Airport which intersects:

a. Runway 09/27 has a length of 3,660 m (12,008 ft). It has a northeast-southwest orientation. The runway's identifier '09/27' denotes that it may be used for takeoffs and landings in both directions: 09 (to the northeast) and 27 (to the southwest).

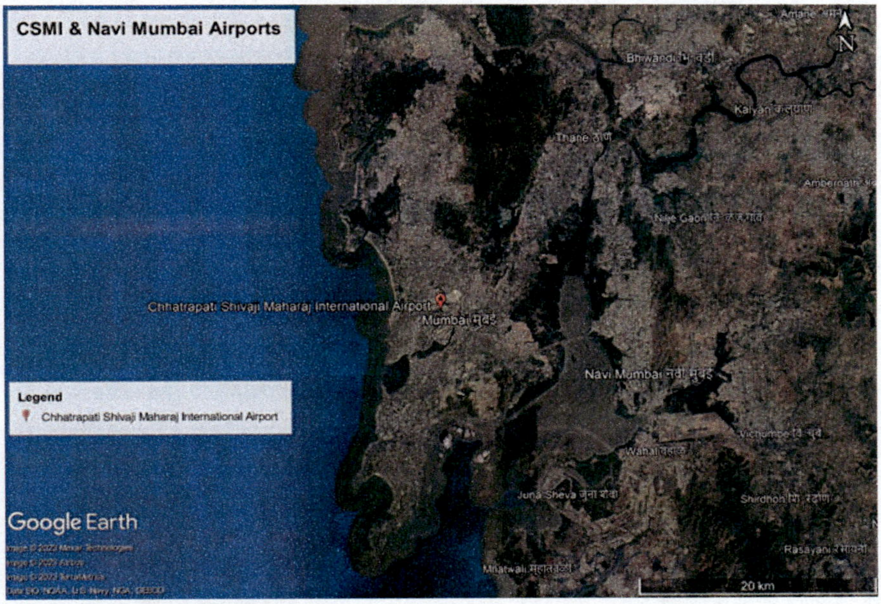

Fig. 2 Location of Chhatrapati Shivaji Maharaj international airport. *Source* Google Earth

b. Runway 14/32 has a length of 3,445 m (11,302 ft). It is positioned north to south. It, too, has a dual designation, permitting operations in both directions: 14 (towards the north) and 32 (towards the south) (Wikipedia, n.d.).

3.1.1 Terminals of Airport

The airport has terminals: Terminal 1 in Santacruz for domestic flights and Terminal 2 in Sahar for international and domestic flights. While both terminals use the same airside amenities, they are geographically separated on the city side and must be reached by a 15–20-min landside journey. A designated General Aviation Terminal serves passengers utilizing private and non-scheduled flight operations.

Terminal 1, known as Santacruz Airport, is used mainly for domestic flights. This was the old Santacruz building, which was initially integrated but was renamed Terminal 1 following the inauguration of the Sahar building for international operations and became a domestic operations terminal. Terminal 1 has two portions, 1A and 1B. Domestic flights, mainly low-cost airlines, are operated from Terminal 1.

Terminal 2 is mainly used for both domestic and international flights. It incorporates extra taxiways and apron spaces for aircraft parking, with a capacity of 40 million people annually. It also services some domestic flights operated by full-service airlines such as Air India. Terminal 2 has been completely renovated and improved to provide world-class facilities and services. It comprises additional taxiways and apron spaces for aircraft parking.

Private and non-scheduled flight operators mostly use the General Aviation Terminal. In April 2011, the terminal was certified for international operations. Passengers departing and arriving on private planes and corporate jets can use the facility [28].

3.1.2 Passenger as Well as Cargo Complex of Mumbai in the Domestic and International Levels

Mumbai is well connected by air, both domestically and internationally. Multiple airlines are operating, making Mumbai a desirable location regarding air accessibility. Consequently, many individuals may quickly move from one part of the country to another. The passenger flow was not so high until 2013, from 2014, the flow of passengers kept increasing at both domestic and international levels. The foreign passenger airlines are shown below in Table 1.

The Air Cargo Complex, located west of International Terminal 2, has been in service since 1977. The cargo apron can accommodate five wide-body planes. The Central Board of Excise and Customs in Mumbai has selected Mumbai International Airport Limited (MIAL) as cargo custodians. The freight terminal is an elevated terminal structure in which all inbound domestic freight is managed from the basement level, and outgoing cargo is managed from a higher level. The Air Cargo Complex of Mumbai's Chhatrapati Shivaji Maharaj International Airport handles domestic and international cargo [28]. Table 2 shows airlines which handles international cargo.

3.1.3 Passenger and Cargo Flow

According to Table 3 there was a 13.5% growth in the passenger flow in Mumbai airport in 2010–2011, a 5.8% increase in 2011–2012, but there was a sudden decline in passenger growth in 2012–2013, which was −1.7% compared to the previous year. The growth of passengers increased to 6.7% and the growth percentage of passengers increased to 13.7% from 2014 to 2016, but then the growth percentage kept declining from 2016 to 2017 and 2017 to 2018 which are 8.4 and 7.4 percentages, respectively.

Table 4 indicates a 15% growth in the passenger flow in Mumbai airport in 2010–2011, then a decline in the growth of 1.9% in 2011–2012 and 3.4% in 2012–2013. There was an increase in the growth of cargo complexes in the airport of Mumbai from 2013–2018. Before 2010, the cargo complex was lower in the international airport of Mumbai.

According to the Airports Authority of India (AAI), Mumbai's international airport has registered the highest passenger traffic among all metro airports in the country. In the financial year 2022–2023, Mumbai's domestic and international passenger traffic combined was 102%. On the other hand, at the Air Cargo Complex

Table 1 International passenger airlines of Mumbai

Airlines	Destinations
Air Canada	Seasonal: London–Heathrow, Toronto–Pearson
Air France	Paris–Charles de Gaulle
Air India	Abu Dhabi, Bangkok–Suvarnabhumi, Doha, Dubai–International,, Kuwait City, London–Heathrow, Newark, New York–JFK, San Francisco, Singapore
IndiGo	Abu Dhabi, Bangkok–Suvarnabhumi Dhaka, Doha, Dubai–International, Istanbul, Kuwait City, Nairobi–Jomo Kenyatta, Phuket, Singapore
Singapore Airlines	Singapore
SpiceJet	Dubai–International, Jeddah
Vistara	Abu Dhabi, Bangkok–Suvarnabhumi, London–Heathrow, Mauritius, Singapore,
Air Mauritius	Mauritius
All Nippon Airways	Tokyo–Narita
Azerbaijan Airlines	Seasonal: Baku
Batik Air	Malaysia
British Airways	London–Heathrow
Cathay Pacific	Hong Kong
EgyptAir	Cairo
Emirates	Dubai–International
Gulf Air	Bahrain
Iraqi Airways	Baghdad, Najaf
Kenya Airways	Nairobi–Jomo Kenyatta
Malaysia Airlines	Kuala Lumpur–International
Rwanda Air	Kigali, Mombasa SaudiaJeddah, Riyadh Seasonal: Medina
Singapore Airlines	Singapore
Sri Lankan Airlines	Colombo–Bandaranaike
Qatar Airways	Doha
Swiss International Air Lines	Zürich
Thai Lion Air	Bangkok–Don Mueang

(continued)

Table 1 (continued)

Airlines	Destinations
VietJet Air	Hanoi, Ho Chi Minh City Seasonal: Da Nang, Phu Quoc
Vietnam Airlines	Hanoi, Ho Chi Minh City
Turkish Airlines	Istanbul
Virgin Atlantic	London–Heathrow
Lufthansa	Frankfurt, Munich
Oman Air	Muscat
Yemeni	Aden

Source [28]

in Mumbai during 2019–2020, the growth of the cargo complex was not too high; due to the outbreak of COVID-19 pandemic, the growth of the cargo complex seemed to be much lower, according to the Air Cargo Complex Mumbai Customs. However, the Air Cargo Complex in Mumbai is the largest in India in terms of cargo volume, cargo value, paperwork filed, and revenue collected. The income from customs for the fiscal year 2021–2022 was Rs. 31,556 crore [30].

3.1.4 Future Developments of the Air Transport in Mumbai

Mumbai's air transport will likely expand significantly in the coming years. The projected Navi Mumbai International Airport (NMIA) is being built in the neighbouring city of Navi Mumbai to relieve congestion at Chhatrapati Shivaji Maharaj International Airport and to meet the region's expanding air traffic demand. Once operational, NMIA will boost capacity and improve connection to Mumbai's air transport system.

Thus, Chhatrapati Shivaji Maharaj International Airport is the hub of Mumbai's air transportation system. The airport is a vital link between the city and local and international destinations. Mumbai's air transport enables travel and trade for the city and the surrounding area because of its modern infrastructure, huge airline network, and continuing growth attempts.

3.2 Water Transport of Mumbai

The Nhava Sheva Seaport, popularly known as the Jawaharlal Nehru Port, is a large cargo port in Mumbai, Maharashtra, India. It is located in the Mumbai Harbour's Nhava Sheva neighbourhood on the eastern edge of the Arabian Sea, Raigad district,

Table 2 International cargo airlines of Mumbai

Airlines	Destinations
Aerologic	Frankfurt, Hanoi
Aero trans cargo	Hong Kong, Munich
Cathay Cargo	Amsterdam, Bangkok–Suvarnabhumi, Frankfurt, Hong Kong, London–Heathrow, Milan–Malpensa, Paris–Charles de Gaulle
China Airlines Cargo	Amsterdam, Taipei–Taoyuan
CMA CGM Air Cargo	Paris–Charles de Gaulle
DHL Aviation	Frankfurt
Emirates SkyCargo	Bangkok–Suvarnabhumi, Dubai–Al Maktoum
Ethiopian Airlines Cargo	Addis Ababa, Guangzhou, Xiamen
FedEx Express	Bangkok–Suvarnabhumi, Dubai–International, Guangzhou, Hahn, Hong Kong, Milan–Malpensa, Memphis, Paris–Charles de Gaulle, Tokyo–Narita
Georgian Airlines	Baku
IndiGo CarGo	Ras Al Khaimah, Sharjah
Lufthansa Cargo	Almaty, Bangkok–Suvarnabhumi, Cologne/Bonn, Frankfurt, Hanoi, Hong Kong, Krasnoyarsk
MASkargo	Kuala Lumpur–International
Qatar Airways Cargo	Doha, Macau
Saudia Cargo	Dammam, Jeddah, Riyadh
Sichuan Airlines	Chengdu–Shuang Liu
Silk Way West Airlines	Baku
SpiceXpress	Ras Al Khaimah, Singapore
Turkish Cargo	Doha, Dubai–Al Maktoum, Hanoi, Istanbul
YTO Cargo Airlines	Kunming, Nanning

Source [28]

Maharashtra, and is operated by the Jawaharlal Nehru Port Trust Authority (JNPTA). This port is accessible by Thane Creek, a node of Navi Mumbai (Fig. 3). After Mumbai Port, it is the principal port of the Mumbai Metropolitan Region, Maharashtra and Western India. Its popular name is derived from the names of the nearby villages of Nhava and Sheva. It is also the Western Dedicated Freight Corridor terminal.

The Nhava Sheva Seaport is India's largest container port, handling the majority of container traffic for Mumbai as well as the surrounding areas and hinterland. It

Table 3 Domestic and international passengers handled by the international airport of Mumbai

Passengers (in million)	2010–2011	2011–2012	2012–2013	2013–2014	2014–2015	2015–2016	2016–2017	2017–2018
Domestic	20.0	21.0	20.3	21.9	25.2	30.0	32.7	34.8
International	9.1	9.7	9.9	10.3	11.4	11.6	12.4	13.6
Total Passengers	29.1	30.7	30.2	32.2	36.6	41.7	45.2	48.5
Growth (%)	13.5	5.8	−1.7	6.7	13.7	13.7	8.4	7.4

Source [29]

Table 4 Domestic and international cargos handled by the international airport of Mumbai

Cargo (in '000' tonnes)	2010–2011	2011–2012	2012–2013	2013–2014	2014–2015	2015–2016	2016–2017	2017–2018
Domestic	199.8	190.3	182.4	181.1	207.7	209.0	234.9	258.4
International	470.4	467.2	452.7	467.6	486.5	496.2	547.4	648.0
Total Cargo	670.2	657.5	635.2	648.7	694.3	705.2	782.3	906.3
Growth (%)	15.0	−1.9	−3.4	2.1	7.0	1.6	10.9	15.9

Source [29]

Fig. 3 Location of Nhava Sheva international seaport of Mumbai *Source* Google Earth

is an integral part of India's foreign trade and serves as a significant entry and exit point for goods entering and exiting the country.

The port was built to relieve congestion at the Mumbai Port, which has restrictions owing to its location and capacity. The Nhava Sheva Seaport was built beginning in 1982 and completed in 1989. The Jawaharlal Nehru Port Trust (JNPT), a government-owned organization, manages the port.

Jawaharlal Nehru Port is India's second largest container handling port after Mundra Port, having handled 4 million TEUs (twenty-foot equivalent units), in containers between 2016 and 2021. Jawaharlal Nehru Port has a full-service Customs House, 30 Container Freight Stations, and access to 52 Inland Container Depots nationwide [31].

The Nhava Sheva Seaport has a well-developed infrastructure for handling various cargo kinds. The Jawaharlal Nehru Port Container Terminal (JNPCT), Gateway Terminals India Private Limited (GTIPL), Nhava Sheva International Container Terminal (NSICT), and Bharat Mumbai Container Terminals (BMCT) are among the container terminals in the city.

It has a cargo capacity of 62.15 million tonnes. The Nhava Sheva International Container Terminal (NSICT) was the country's first privately run container terminal. GTIPL (Gateway Terminals India Private Limited), APM Terminals' third container terminal with a capacity of 1.3 million TEUs (twenty-foot equivalent units), was commissioned in 2006. NSICT is a new freestanding container terminal with a quay length of 330 m and a capacity of 12.5 million tonnes. The first phase, with a capacity of 2.4 million TEUs, was completed in December 2017. By the end of Phase 2, the terminal will have a total capacity of 4.8 million TEUs per year and a quay length of 2,000 m [32].

As a major seaport, Nhava Sheva contributes significantly to the Indian economy. Automobiles, equipment, electronics, chemicals, textiles, and food goods are among the commodities handled. Because of the port's efficient operations and connection, several industries and enterprises have established operations in and around the port area, contributing to regional growth and job possibilities.

To accommodate the increasing demand, the Nhava Sheva Seaport has been expanding. The fourth container terminal, Bharat Mumbai Container Terminals (BMCT), opened in 2018, is the most noteworthy development. This new terminal enhanced the port's capacity and efficiency dramatically.

The Nhava Sheva Seaport is India's crucial containerized freight hub, carrying much of the country's foreign commerce. Due to its modern infrastructure, efficient operations, and strategic position, it enables commerce and fosters economic progress in Mumbai.

While the historical water transport methods played an essential part in Mumbai's history, the city's transport environment has changed significantly with the introduction of contemporary transit modes such as highways, railroads, and airports. These advances have resulted in a change in emphasis from water to land-based transportation networks. Water transport, however, retains cultural and historical relevance in Mumbai. Nevertheless, the Nhava Sheva seaport of Mumbai is India's most prominent international container seaport.

3.3 International North-South Transport Corridor

The International North-South Transport Corridor (INSTC) is a multimodal trade route connecting India, Iran, Russia, and other Central Asian, Caucasus, and European nations. The corridor is intended to improve connection and expedite the transit of products between these regions.

The INSTC comprises mainly rail, road, and maritime transportation networks. It begins on the western coast of India, travels through Iran, and then continues northward via the Caspian Sea to Russia. The corridor extends to neighbouring nations such as Armenia, Azerbaijan, Kazakhstan, and Turkmenistan (Fig. 4).

India's involvement in the expansion of the INSTC is demonstrated by its 2.1 billion investments, which include building the port of Chabahar in Iran and a 500-km railway route between Chabahar and Zahedan. Chabahar can now process ultra-large container ships.

The INSTC allows India to trade with Iran and Central Asia without going via Pakistan. Naturally, this has ramifications for reaching out to Afghanistan and Central Asia, both essential parts of our extended neighbourhood. The corridor will also give

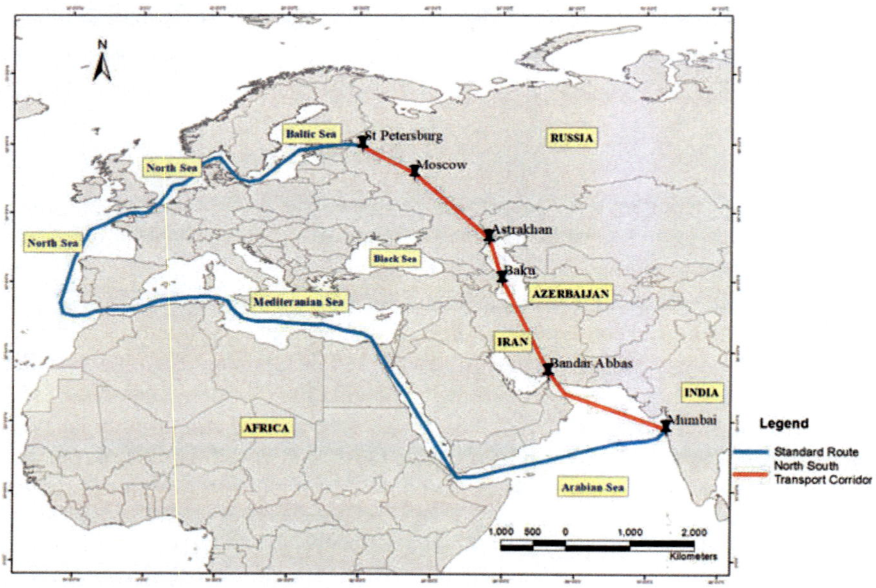

Fig. 4 International North-South Transport Corridor. *Source* Prepared from The Wire [33]

access to prospective markets throughout Eurasia. Aside from the fact that this North-South route saves roughly two weeks of journey time, positioning the INSTC as an alternative to the traditional deep sea Suez Canal route is the focus of geostrategic and economic diplomacy, mainly for India, Russia, and Iran.

India, Russia, and Iran created the legislative foundation for the INSTC multi-modal freight network comprising ship, rail, and road on September 12, 2000, during the Euro-Asian Conference on Transport in St. Petersburg. Other members include Turkey, Oman, Syria, Belarus, and Ukraine; Tajikistan, Kyrgyzstan, and Kazakhstan in Central Asia; and Armenia and Azerbaijan in the Caucasus. Pakistan, Turkmenistan, and Afghanistan are not signatories to the INSTC pact but are interested in using the corridor. The transport corridor is significant as an essential alternative economic growth corridor and a response to the European Union's (EU) economic and political dominance. The objectives of INSTC are as follows:

- To reduce transportation costs.
- For increasing trade volumes and promoting economic cooperation among the participating countries.
- To provide an alternative route to traditional sea routes and allow faster goods movement between Asia and Europe.
- For enhancing connectivity, the corridor has the potential to boost economic development by facilitating cultural exchange and strengthening political ties between the countries involved.

The North-South Transport Corridor is a major infrastructure project with potential regional integration and trade diversification benefits. Along its path, it can open up new markets, enhance access to resources and drive economic growth.

The International North-South Transport Corridor is vital to the transportation of Mumbai, one of India's major cities. This corridor is a transport infrastructure network that connects Mumbai's northern and southern areas and facilitates the flow of products and people. It consists of roadways, railroads, and ports.

a. **Road Transport:** The North-South Transport Corridor is comprised of well-developed road networks that allow for the seamless flow of traffic between Mumbai's northern and southern portions. The Western Express Highway and the Eastern Express Highway function as arterial roadways, linking numerous suburbs, business regions, and industrial zones. These roads reduce travel time and congestion by providing effective connections for commuters, commercial vehicles, and logistical activities.
b. **Rail Transport:** Mumbai's railway network is critical to the North-South Transport Corridor. The Western Railway and Central Railway lines are important transportation corridors that connect the northern and southern areas. Local rail services, such as the Western Line and Central Line, constitute the city's lifeline, supporting millions of inhabitants' daily commutes. These railways also facilitate the transit of products via designated freight lanes, allowing for efficient freight transportation across Mumbai.
c. **Port Connectivity:** Mumbai has essential ports like the Jawaharlal Nehru Port Trust (JNPT). The North-South Transport Corridor offers vital connectivity between these ports and Mumbai's northern and southern districts. This connection promotes commerce and economic progress by ensuring the smooth movement of products, including imports and exports. Efficient transport linkages between ports and industrial districts improve logistical operations and help Mumbai maintain its prominence as a commercial hub.
d. **Intermodal connection:** The North-South Transport Corridor also facilitates intermodal connection, allowing for the efficient transfer of commodities between modes of transportation. Road, rail, and port infrastructure integration promotes goods transfer from one mode to another. Well-designed logistics parks and container depots, for example, placed along the corridor, provide consolidation, warehousing, and distribution facilities, improving the transportation system's overall efficiency.
e. **Economic Development:** The North-South Transport Corridor is critical to Mumbai's economic development and recruiting investors. Efficient transportation infrastructure enhances communication between business centres, industrial zones, and residential regions. The interconnectedness promotes corporate growth, increases trade activity, and supports the creation of industrial plants and logistical centres throughout the corridor. The corridor's accessibility and efficient movement of commodities and people boost economic output, job development, and living standards.

Compared to China's Belt and Road Initiative (BRI), the INSTC is still in its early implementation phases. Nonetheless, it provides a geostrategic opposition to the New Silk Road's expanding network. With that as a strategic alternative, there is a solid foundation for collaboration between the China-Central Asia-West Asia Economic Corridor and the INSTC, becoming a primary transport and logistic centre. The Caspian area has gradually and quietly grown transportation and multimodal networks. The corridor can be expanded to the Baltic, Nordic, and Arctic areas.

Once the early kinks are worked out, there is expected to be a significant increase in INSTC freight traffic, which is anticipated to reach 14.6–24.7 million tonnes annually. By 2030, incremental freight traffic between them and South Asian and Persian Gulf nations might equal (4.4–9.0 million tonnes), or around 75% of total potential container traffic. The INSTC has significant relevance for India in diversifying energy import destinations, Iran as the primary transit hub on the North–South and East-West pathways, and Russia as the Eurasian trade and connectivity leader [33–36].

Overall, in Mumbai, the International North-South Transport Corridor is a vital transportation backbone, combining road, rail, and port facilities to flow products and people efficiently. It is crucial in promoting economic growth, strengthening connectivity, and improving the city's transportation environment.

4 Conclusion

This paper focused on the economic and transport opportunities offered by Mumbai's mega-urban agglomeration nationally and globally. The economy had led Mumbai to emerge as a megacity of global importance. Mumbai is the financial capital of India and accounts for 6.6% of the total GDP share of the country. It is the third-largest emerging data centre in the world. The tertiary sector contributes approximately 66% of the country's economy. All these factors contribute to Mumbai's emergence as an essential hub of the economy globally. The enlarging economy attracts people worldwide through FDI and stock exchange centres (NSE, BSE, etc.). The employment opportunity of the city acts as a pull factor, giving the term '*Mayanagri*' (city of dreams) to the city.

The city's economic development would only be possible in the transport sector. The city is well connected with the other parts of the country and the outer world through road, rails, air, and water. In the financial year 2022–2023, Mumbai's domestic and international passenger traffic growth was combined to a percentage of 102. The tremendous growth in Mumbai's air transport highlights the city's connectedness with the world. The recent development of the Delhi -Mumbai Industrial Corridor and North-South Transport Corridor serves as a backbone for international trade.

In recent years, the Indo-Pacific region has attracted the attention of the world as the most populous area, with high regional diversity and higher climate risks. Mumbai is the first Indian city to release a Climate Action Plan (CAP) on 13 March 2022. Mumbai comes, along with cities like Beijing, Shanghai, Guangzhou (China),

Tokyo, Osaka (Japan), Bangkok and Singapore (SE Asian cities), which are taking scientific measures to combat climate change in the process of building the cities sustainable. Mumbai is pioneering other Indian Megacities in framing their climate action plan. It is worth mentioning that in the Indo-Pacific region, Mumbai holds a special place in this regard, addressing the climate and environment of the city as an essential function along with its economic, cultural, and social aspects.

With proper scientific and strategic planning and implementation, Mumbai can achieve higher global importance, the process of which has already started. India is rapidly developing in the world economy; as an influential player, Mumbai is helping the country in overall development. It can also serve as a model to other potential Indian cities and the Indo-Pacific region in catering to economic prosperity and diversity and protecting the climate for humans' sake.

References

1. McKinsey Global Institute (2019). Asia's future is now. https://www.mckinsey.com/~/media/mckinsey/featured%20insights/asia%20pacific/asias%20future%20is%20now/asias-future-is-now-final.pdf. Accessed 27 June 2023.
2. Studyiq (2023). Editorial of the day: geopolitics related to Indo Pacific. https://www.studyiq.com/articles/geopolitics-related-to-indo-pacific/. Accessed 5 April 2023
3. The Government of the Republic of Korea (2022). Strategy for a free, peaceful, and prosperous Indo-Pacific Region. https://www.mofa.go.kr/eng/brd/m_5676/view.do?seq=322133. Accessed 27 June, 2023
4. Ministry of Defence (2022). India stands for free & rules-based Indo-Pacific crucial for economic development of the region & wider global community, says Raksha Mantri at Indo-Pacific Regional Dialogue in New Delhi. PIB Delhi. https://pib.gov.in/PressReleaseIframePage.aspx?PRID=1878750. Accessed 25 June, 2023.
5. MCGM-Municipal Corporation of Greater Mumbai (2012) Mumbai revised city development plan (2005–2025). Municipal Corporation of Greater Mumbai, Government of Maharashtra
6. MMRDA-Mumbai Metropolitan Region Development Authority (2016) Draft Mumbai metropolitan regional plan (2016–2036). Mumbai Metropolitan Region Development Authority, Government of Maharashtra. http://mmrda.maharashtra.gov.in:8080/documents/10180/8037279/1.%20Draft%20MMR%20Plan%20Report,%202016-36%20Colour/d107b724-c039-4ea1-840f-58a92bb9daf2?version=1.3. Accessed 5 June 2019
7. BMC—Brihanmumbai Municipal Corporation (2022) Mumbai climate action plan. https://mcap.mcgm.gov.in/. Accessed 10 June, 2023.
8. FICCI (2020). Why Mumbai? https://www.indiachem.in/why-mumbai.php Accessed 20 June, 2023.
9. The Economic Times (2023a) Mumbai 3rd biggest data centre market in Asia-Pacific, breaches 2GW mark. May 23, 2023 https://cio.economictimes.indiatimes.com/news/corporate-news/mumbai-3rd-biggest-data-centre-market-in-asia-pacific-breaches-2gw-mark/100435040. Accessed 10 June 2023
10. Newsband (2023). Mumbai's data centre market surges to third largest in Asia-Pacific region. May 29, 2023. https://newsband.in/?p=41289. Accessed 25 June, 2023.
11. IBEF (2023). Mumbai has emerged third on the list of top data centre markets in the Asia-Pacific region: Report. May 23, 2023. https://www.ibef.org/news/mumbai-has-emerged-third-on-the-list-of-top-data-centre-markets-in-the-asia-pacific-region-report#:~:text=As%20per%20a%20Knight%20Frank,by%20Tokyo%20(2%2C575%20MW). Accessed 18 June, 2023.

12. The Economic Times (2023b). Data centre industry to boost realty demand. Apr 14, 2023. https://economictimes.indiatimes.com/news/industry/data-centre-industry-to-boost-realty-demand/articleshow/99501928.cms?utm_source=contentofinterest&utm_medium=text&utm_campaign=cppst. Accessed 10 June 2023
13. Santana (n.d.). 7 reasons why Mumbai is the financial capital of India. https://mumbai7.com/why-mumbai-is-the-financial-capital-of-india/. Accessed 5 June 2023.
14. India Briefing (2019) The economy of Mumbai—India's commercial capital. https://www.india-briefing.com/news/economy-mumbai-india-commercial-hub-6704.html/#:~:text=Mumbai%20is%20the%20hub%20of,gems%20and%20jewelry%20from%20India. Accessed 5 June, 2023.
15. Mordor Intelligence (n.d.). India data center market size & share analysis - growth trends & forecasts (2023—2028). https://www.mordorintelligence.com/industry-reports/india-data-center-market. Accessed 28 June 2023.
16. Ministry of Tourism (2022). India tourism of statistics, 2022. Government of India. https://tourism.gov.in/sites/default/files/2022-09/India%20Tourism%20Statistics%202022%20%28English%29.pdf. Accessed 5 June, 2023.
17. Dungarpur SS (2014) Indian cinema-a vanishing legacy. J Film Preser 91:27–34
18. Muthuvelu S, Verma H, Jain N (2020) Emergence of Indian film industry in the international markets: facilitators and impeders. Int J Asian Bus Inform Manag 11(3):32–44. https:// doi: https://doi.org/10.4018/IJABIM.2020070103
19. KPMG India-FICCI (2017). India's media and entertainment industry report, 2017. http://producersguildindia.com/Pdf/2017.pdf. Accessed 28 June 2023.
20. E&Y-FICCI (2022). Tuning into consumer. https://assets.ey.com/content/dam/ey-sites/ey-com/en_in/topics/media-and-entertainment/2022/ey-ficci-m-and-e-report-tuning-into-consumer_v3.pdf. Accessed 28 June, 2023.
21. The World Bank (2011). India transportation. https://www.worldbank.org/en/news/feature/2011/09/23/india-transportation. Accessed 28 June, 2023.
22. Ministry of Railways (2021). Indian Railways annual report & accounts 2019–20. https://indianrailways.gov.in/railwayboard/uploads/directorate/stat_econ/Annual-Reports-2019-2020/Indian-Railways-Annual%20-Report-Accounts%20-2019-20-English.pdf. Accessed 25 June 2023.
23. The Telegraph (2016). India crowned world's fastest growing aviation market in 2015 as economy takes off. Bloomberg, 1 January. Retrieved from https://www.telegraph.co.uk/finance/economics/12077311/India-crowned-worlds-fastest-growing-aviation-market-in-2015-as-economy-takes-off.html. Accessed 5 June 2023.
24. Ministry of Statistics and Programme Implementation (2016). Motor vehicle writeup—Statistical year book India 2015. http://www.mospi.nic.in/mospi_new/upload/SYB2015/CH-20-MOTOR%20VEHICLES/MOTOR%20VEHICLE-WRITEUP.pdf. Accessed 15 June 2023.
25. OICA (2017). 2017 production statistics. International Organization of Motor Vehicle Manufacturers. https://www.oica.net/category/production-statistics/. Accessed 15 June 2023
26. Indiaonline (n.d). Chhatrapati Shivaji Maharaj international airport. https://www.indiaonline.in/about/tourism/mumbai-airport. Accessed 15 July 2023
27. ATC GUILD India (2022). Mumbai automation issue-regard—ATC GUILD India. https://atcguild.in/login/logged/ISSUES_TAKEN/2022/Mumbai%20Automation.pdf Accessed 4 July 2023.
28. Wikipedia (n.d.). Chhatrapati Shivaji Maharaj International Airport. https://en.wikipedia.org/wiki/Chhatrapati_Shivaji_Maharaj_International_Airport Accessed 15 July 2023
29. Mumbai International Airport Limited (n.d.). Statistics: Mumbai international airport Limited. Association of Private Airport Operators. https://www.apaoindia.com/?page_id=923. Accessed 15 July 2023
30. Times of India (2023). At 102%, Mumbai airport saw maximum growth in passengers in. April 28. Retrieved from https://timesofindia.indiatimes.com/city/mumbai/at-102-city-airport-saw-max-growth-in-passengers-in-22-23/articleshow/99830372.cms Accessed 28 June, 2023.

31. The Hindu. (2019). JNPT crosses 5 mn TEUs of container handling in 2018. https://www.thehindubusinessline.com/economy/logistics/jnpt-crosses-5-mn-teus-of-container-handling-in-2018/article25910908.ece/amp/. Accessed 12 May 2023
32. Alchetron (2022). Jawaharlal Nehru port. https://alchetron.com/Jawaharlal-Nehru-Port. Accessed 12 May 2023
33. The Wire (2022). The political economics of the international North-South transport corridor. June 30. https://thewire.in/world/political-economics-international-north-south-transport-corridor-india-iran-russia. Accessed 12 June 2023
34. Silk Road Briefing (2023). The expansion of the international north-south transport corridor: Geopolitical updates. April 4. https://www.silkroadbriefing.com/news/2023/04/04/the-expansion-of-the-international-north-south-transport-corridor-geopolitical-updates/. Accessed 10 June 2023
35. Vinokurov Evgeny (2022). Russia and Middle east need international north-south transport corridor. https://vinokurov.info/russia-and-middle-east-need-international-north-south-transport-corridor/. Accessed 5 June 2023
36. Wikipedia (2020). International north–south transport corridor. https://en.wikipedia.org/wiki/International_North%E2%80%93South_Transport_Corridor#:~:text=The%20route%20primarily%20involves%20moving,Astrakhan%2C%20Bandar%20Anzali%2C%20etc. Accessed 5 June 2023
37. Akshay M (2021). Relocating Mumbai in the geoeconomics of the Indo-Pacific. ORF Foundation. https://www.orfonline.org/expert-speak/relocating-mumbai-in-geoeconomics-of-indo-pacific/. Accessed 20 June, 2023.

Dr. Amit Chatterjee has a combined experience of seventeen years in teaching, research and industry and is presently an associate professor in the Department of Geography at Visva-Bharati, Santiniketan. Prior to join Visva-Bharati, he worked as an associate professor at Sidho-Kanho-Birsha University and prior to that as an assistant professor and program coordinator (urban and regional planning) at the School of Planning and Architecture (SPA), Bhopal from 2011 to 2021. He wrote his doctoral dissertation on future cities and urban sustainability models. He served the industry in various capacities. He had the experience for working for mega planning projects, including Navi Mumbai Special Economic Zone, Dholera town planning scheme, besides formulating the twin city development plan, highway projects in PPP mode, slum-free city planning under RAY across various cities in India. He has completed successfully a number of collaborative research and consultancy projects, including those on urban co-benefits (UNU, Japan), Urban Missions in India-targets, performance and linkages to UN-SDGs (GIZ), Politics of Care in Pandemic Time (UCL's Global Engagement Funds 2020/21), urban biodiversity (UNU, Japan), Shelter for All under Design Innovation Centre, National Rurban Mission of Madhya Pradesh, Development Plan of Saugor Cantonment, and currently working on Restoration and conservation of wetlands near urban areas (ICSSR, India and JSPS, Japan). He had recently published a co-authored book titled "Satellite towns in Neo-Metropolitan Development in India-Lessons from Selected Cities" by Springer and presently working as a co-editor of a book focused on urban water ecosystem (Routledge). He received appreciation from

UN-Habitat for Sustainable Cities Integrated Approach Pilot Projects (SCIAP) in India. He had presented research papers in ISOCARP, RSAI, RGS, AUC, etc. He contributes papers to reputed international journals in planning. His research articles appeared in publications like Routledge, Springer, ISOCARP, etc.

Mitali Mondal is a sincere and enthusiastic student. She has recently pursued a Masters' degree in Geography from Visva-Bharati, Santiniketan. Her entire student life until post-graduation has been in Visva-Bharati. Her interest in geography has grown since her childhood, when she was first introduced to maps by her mother. As a student of Visva-Bharati, she has an interest in various fields of art and culture and is very good at multi-tasking. She attended a cultural exchange programme for 22 days at Tagore Gymnasium in Berlin, Germany. She was an NSS volunteer for two years. She holds immense passion for music, and she achieved the National Scholar Award at the Ministry of Culture, Govt. of India, in vocal music. She is eager to learn new things and loves to search for bookish knowledge in the real world. She has learned QGIS and is willing to continue her studies in the Remote Sensing and GIS fields. Her dissertation work in her Masters' studies was based on Solid Waste Management in urban areas. She is a believer of "neo-determinism" and wants to serve the world for a sustainable future through her career.

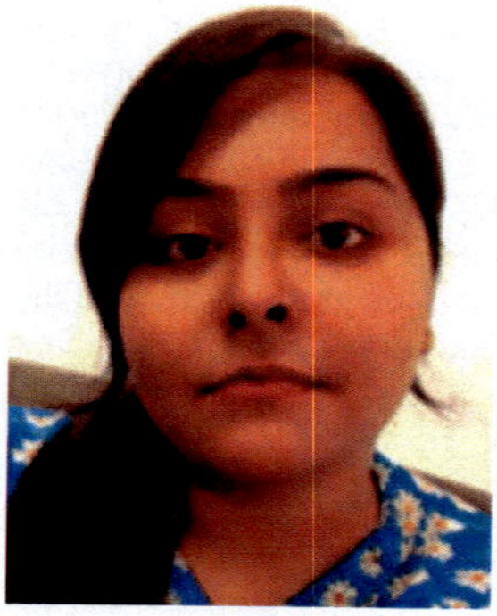

Jilik Dhar is an enthusiastic and hardworking student who recently pursued a Masters' degree in Geography from Visva-Bharati University. She also holds a Bachelor of Science Geography degree from Calcutta University, from which she graduated with honors. Jilik has completed a variety of certified and non-certified internships from well-known organizations such as GSITI, NIDM and Bhuvan. She is also proficient in HTML as well as tools like QGIS. Recognizing the value of practical experience, she has actively sought chances to deal with real-world geographical difficulties. She has taken part in fieldwork missions to remote areas, where she conducted environmental evaluations, surveyed topography and gathered significant data. She likes enjoying the outdoors, trekking in hilly areas and immersing herself in new cultures on her travels in addition to her academic interests. She is enthusiastic to continue her academic path, pushing the frontiers of geographical knowledge and contributing to a more sustainable and fairer planet, motivated by her passion of geography and desire to make a good effect.

Md. Imran Nazir has recently been completed post-graduate in Geography from Visva-Bharati University. He stood First rank in Pre- Degree & Graduation in Visva-Bharati. As he grew up in a village and spending his entire childhood in the lap of Santiniketan, he has a keen observation on environmental phenomena. He has interested in several branches of Geography like as—Population Geography, Urban Geography, Environmental Geography, etc. He also has a deep passion in tour and traveling to know the unknown facts and mystery of India and the World through the vision of a Geographer. He has a dreamt of writing books and wants to contribute in his field through optimistic vision.

Conclusion

International Collaborative Research: Indo-Pacific Smart Megacity System: Emerging Architecture and Megacity Studies and Conclusions

T. M. Vinod Kumar

Abstract This chapter has two parts. The first part discusses the goals and organizational details of the international collaborative research project "Indo-Pacific Smart Megacity System: Emerging Architecture." In the second part, in consultation with the team leaders of the Area studies, including the city study, their general conclusions of the area study are presented.

Keywords Study organization of study · Results

1 Smart City Research and Indo-Pacific Smart Megacity System: Emerging Architecture and Megacity Studies

The International Collaborative Research Project on Indo-Pacific Smart Megacity System: Emerging Architecture is the fifteenth in a series of books that Professor T.M. Vinod Kumar conceived, coordinated, implemented, and edited about articulating the various aspects of smart cities and the roles of Smart People in Smart Cities. These books discuss how smart people create Indo-Pacific Megacity System. The Indo-Pacific Smart Megacities is an international collaborative research project documented in two part and published by Springer-Nature. The first part is "Indo-Pacific System of Megacities: Emerging Architecture." The book documents research studies on a few megacities in the Indo-Pacific. While innumerable academic writers hailing from the East and the West postulated with their logic the rationale of Indo-Pacific, they are irrelevant when compared to a program development region, the conceptualization of a political leader the then Prime Minister Shinzo Abe of Japan delivered the famous address to the Indian parliament famously remembered as the Indo-Pacific the confluence of two great ocean Indian Ocean and the Pacific Ocean that covers an assortment of rich, middle-income and low-income countries and tiny

T. M. Vinod Kumar (✉)
School of Planning and Architecture, New Delhi, India
e-mail: tmvinod@gmail.com

T. M. Vinod Kumar (ed.), *Indo-Pacific Smart Megacity System*, Advances in 21st Century Human Settlements, https://doi.org/10.1007/978-981-99-6218-1_8

island nations to giant islands and small populated sovereign nations to giant populated Nations like India and China both from the east and the West. It was in the aftermath of the Tsunami in 2004 that Japan pooled its effort with that of the United States, Australia, and India along with numerous countries in the region, jointly saving the Indo-Pacific devastation to the ecology and human life, the compassionate joint action of many like-minded countries moved Prime Minister Shinzo.

The first book in this series, entitled "Geographic Information System for Smart Cities" in 2014 [1], was aimed at creating a comprehensive and spatial self-awareness of City functioning every second and every day in real-time, which is the foundation of any Smart City, or Smart data city. Geospatial technologies, sensors, and analytics can augment awareness and use it in real-time for several types of use by Smart People. The book is about utilizing GIS for various urban issues commonly seen globally.

These Smart People thereby progress toward their self-directed goals, which nobody must tell them, such as they demand Smart City that connects more people for intelligent community action and Smart Economic Development that changes their living to prosperity. They aspire to the highest quality of enhanced life in "smart city living," which they can very well afford by expanding many folds the economic development opportunities they can identify without any help to satisfy the higher income and fulfilling employment needs to sustain Smart People. No intelligent person in a city is an island or isolated elite. Still, they share a common destiny and common urban space, the urban realm, and shared social and physical infrastructure. As the regulator, the Government ensures all citizens are provided services irrespective of their income level and social status or whether they are above or below the poverty level. A hierarchy of Government exists in any city. Still, its Governance needs to be for a Smart City that is fully aware of itself every millisecond and against the Government comes to know about the issue when a case is filed in the court, which takes many decades to get a final judgment. The existing governance systems are obsolete, a product of sixteenth-century or earlier designed by colonial rulers and not for intelligent inhabitants, built on the model of East India Company's administration requirements in India or elsewhere. Local Governments shall avoid such approaches for Smart City economic development and intelligent global megacity creation and functioning. However, those who aspire to live in Smart Cities are in the twenty-first century and no more part of an exploitative colonial empire under the iron hand of a colonial administrator with a robust iron framework who cannot look beyond the interest of the colonial empire but effectively conceal it. Therefore, twenty-first-century Smart Cities require a Smart City e-governance system which was the subject of the second book entitled "E-Governance for Smart Cities" [2]. There are three parts, State Art Surveys, Domain Studies, and Tools based on E-Governance in Smart Cities in E-Governance in Smart Cities book.

The third book in this series is "Smart Economy in Smart Cities" [3]. This book explores possibilities for rapid change in the income level and employment opportunities of those Smart People below or above the poverty level in a Smart City and to consistently make the NDP growth rate to a desired higher level for the next many decades. Then, the current urban local economic development trend shall progress toward Smart City Economic Development. A 10% NDP growth rate envisaged for the next three decades in India, which was distant in the past and many other countries, can only be realized through Smart City Economic Development. The projection that the Indian GDP in PPP will be the third largest in the World after China can be realized through an intelligent metropolitan or megacity economy. Smart Cities and the related conceptualization boast of the Smart Economy, but not much has been systematically researched or documented about it so far. This calls for a study of many cities worldwide to demonstrate what constitutes a Smart Economy. There are two groups of cities being studied in this book. Some of them have been chosen as Smart Cities by learned societies, but others are not but aspire to be Smart Cities. These call for different approaches to research design and studies. It was seen from case studies that both these cases in other countries emphasize different methods, proving that there are no cookbook solutions. The cities studied in this book were spread across several major continents and regions, including North America, Europe, Africa, the Indian subcontinent, and East Asia. They are Ottawa in Canada; Stuttgart in Germany; Bologna in Italy; Dakar in Senegal; Lagos in Nigeria; Nairobi in Kenya; Cape Town in South Africa; New Delhi, Varanasi, Vijayawada, and Kozhikode in India; Hong Kong in China, Cape Town, Dakar, Nairobi, and Lagos in Africa.

The fourth book in this series is "E-Democracy for Smart Cities" [4]. The World over, participatory democracy is worshiped and preached, but what is practiced is representative democracy at the city level and beyond, which is not participatory democracy. It is believed that in meta cities, megacities, and metropolitan cities, only representative democracy with elected representatives will work, and so also in large democracies with billions of populations like India. However, democracy was practiced in small cities like Athens in Greece and the Licchavi Republic in India in ancient times much before Athens, which is well documented in Buddhist scriptures and adopted by Buddha in his Sangha. A face-to-face democracy existed in ancient times in many parts of the world. In these cities, everyone in the City sat together and jointly decided on all aspects of the city/country during war or peacetime. Today it is there in Indonesian village democracy and many places and countries in the Indo-Pacific. Citizens not only took part in decision-making but acted together as one Government and selected or elected a few among them to work even as temporary administrators for a task and as a regulator. There were no permanent administrators then, which colonial Governance and post-colonial republics propagated. With the advent of ICTs in Smart Cities of the twenty-first century, it is possible to go back to face-to-face democracy in meta city with a 20 million-plus population, a megacity with a ten million-plus population, and metropolises with a one million-plus population that, by any measure, is much superior to representative democracy which is not participatory at all. We do not want middlemen, the citizen's representatives, to make legislation. However, all citizens using electronic media demonstrate the creation of regulations based on their ability to think and act under the Constitution. The fourth book is all about E-Democracy in Smart Cities in action. It is divided into three parts, State the Art Surveys, Domain Studies, and Tools, and the Issue of E-Democracy in Smart Cities.

The fifth book in the series was "Smart Metropolitan Regional Development: Economic and Spatial Design Strategies" [5]. However significant these cities may be, these cities need to be converted to smart metropolises using the specific design of economic and spatial strategies and not by buying on-the-shelf smart technologies alone. The city studies for the "Smart Metropolitan Regional Development" result in many insights on many smart spatial and economic strategies using the Internet of Things, Internet of Democracy, and Internet of Governance oriented to the specific issue of a town and its potential; taking into consideration that the Smart metropolitan City is an integrated six systems in which Smart Economy and smart people are dominant components. The smart Economy can relate to Smart Mobility or Smart Environment. Based on the elaboration of the Smart metropolitan city System, if one must develop any metro region, then a country-and city-specific economic and spatial design strategies for a Smart metropolitan city must be designed based on a local ecological and cultural system of the City and not a type of universal design based not on a mechanical input–output model. Location-specific, exceptionally robust ecologically and culturally acceptable economic and spatial strategies can be locally evolved by people and not their representatives, who may have other agendas like making money by corrupt means while in power, governed, and managed. This is the only way local culture will find expression in the Smart metropolitan City using specific economic and spatial strategies by using local, creative talents of intelligent people in many institutions in Smart Cities. Sixteen cities were studied in this project, namely Pittsburgh in the United States, Stuttgart in Germany and Naples in Italy, Dakar in Senegal, Conakry in Guinea, Abuja in Nigeria, Johannesburg in South Africa, and Nairobi in Kenya, Ahmedabad-Gandhi Nagar, Bengaluru, Chandigarh, Jaipur, Kozhikode, New Delhi, Surat in India, and Hong Kong and greater Pearl River Delta Region from China.

The sixth book, published in mid-2019 by Springer-Nature, is entitled "Smart Environment for Smart Cities" [6] as a product of international collaborative research. This book aims to develop the Design Protocol and Practice of Smart Environmental Resources Management for Smart Cities. Environmental Resources are common proprieties where an active role of Government and People is needed. Hence, its management is a joint and synchronous effort of E-Democracy, E-Governance, and IoT system in a 24-h, 7-day framework on any resource in any Smart City. Smart environmental resources management is a practice that uses information and communication technologies, the Internet of Things, Internet of Governance (E-Governance), and Internet of People (E-Democracy), along with conventional resource management tools to realize coordinated, effective, and efficient management, development, and conservation that equitably improves ecological and economic welfare without compromising the sustainability of development ecosystems and stakeholders. This book presents many cities' case studies Hong Kong in China, Ahmedabad, Gandhi Nagar, Chandigarh, Kozhikode, New Delhi, Patna, Surat, Yokohama in Japan, Nairobi in Kenya, and Dubai in UAE that are centered on one or all environmental resource each in a city.

The seventh and eighth book in two volumes published by Springer-Nature in 2020 are on smart living for smart cities [7, 8]. The first volume focuses on city case studies, and the second is on community studies and ways and means. Based on international collaborative research, this book aims to develop the state-of-the-art design of "Smart Living" for metropolises, megacities, and meta cities as well as a community and neighborhood level. Smart living is one of six components of Smart Cities: smart people, smart Economy, smart environment, smart mobility, and Smart Governance. Smart living in any Smart City can only be designed and executed by active roles of Smart People and Smart City Government and is a joint and synchronous effort of E-Democracy, E-Governance, and ICT-IoT system in a 24-h, 7-day framework on all activities. In addition to the uses of information and communication technologies, the Internet of Things, Internet of Governance (E-Governance), and Internet of People (E-Democracy), the design of intelligent living utilize domain-specific tools for many aspects of living by age cohorts to realize the coordinated, effective and efficient management, development, and conservation that improves ecological, social, biophysical, psychological, and economic wellbeing equitably without compromising the sustainability of development ecosystems and stakeholders. This book presents many case studies covering cities such as Hong Kong, Dehradun, Kozhikode Metropolitan area, and Shimla centered on one or two domain-specific innovative living components.

The ninth and tenth book is on Smart Global Megacities [9, 10]. The editor and coordinator of the book series T.M. Vinod Kumar and many authors who participated in the earlier eight books felt a gap in knowledge about the Smart global megacity. Like all earlier books, funding for such a collaborative research project was another issue, so international travel and consultations were ruled out. Universities and research centers dominated in collaboration with these—smart megacity city research projects. We also found that along with universities, some not-for-profit national and international networks and institutions, city governments, and regional governments in certain countries also came forward to take part in this collaborative research program. The editor and coordinator of the project again felt that this international project should not seek any external funding other than the internal resource mobilization from within the participating universities. This book, the first volume, highlights Tokyo, Mumbai, New York, Hong Kong-Shenzhen, Kolkata, Chennai, and Kochi-Kannur out of about thirty-three megacities in the World with ten million-plus inhabitants.

The eleventh and twelfth books were entitled Smarter Master Planning: Case Studies of Digital Innovations and Domain Innovations [11, 12], respectively. Master Planning is practiced all over the World even though this well-tested instrument of Master Planning carries with it numerous defects apparent to the consumer citizen, planners, and other regulators of urban affairs. The two-volume books try to answer some of the shortcomings of Master Planning by investigating innovative approaches. These books research the Smarter Master Planning of Cities, discussing innovations using city case studies. Cities are urban agglomerations, the accurate representation of urbanized geographic space as per the census of India and UN-HABITAT and not Municipal boundary, or any arbitrary planning area or towns administrative boundary fixed arbitrarily and there starts the not so satisfactory Master Planning of cities which needs to be innovated since there is no replacement. Master Planning approach differs in countries as per the governing system followed, whether it is a Communist, Capitalist, Secular, Theocratic, or Democratic Government, besides based on city-specific urban issues and long-range vision of the City. The Constitution decides the governing system and approaches to the City Master Plan by several regulatory legislation encompassing all aspects of city life. This differs from country to country and federal state to state. The Constitution is a living document that makes the government system work.

Based on international collaborative research, this book presents a state-of-the-art design for "Smart Master Planning" for all metropolises, meta cities, and megacities and at sub-city zonal, community, and neighborhood levels. Smart Master Planning accepts that all cities are smart cities in a limited way as far as the six components of smart cities, namely smart people, smart Economy, smart environment, smart mobility, and smart Governance, are concerned. Smart Master Planning in any city can only be designed and executed by active roles of smart people and Smart City government and is a joint and synchronous effort of e-democracy, e-governance, and ICT-IoT system in a 24-h, 7-day framework for all activities. In addition to the use of information and communication technologies and remote sensing, the design of Smart Master Planning utilizes domain-specific tools for many aspects of a city to realize the coordinated, effective, and efficient planning, management, development, and conservation that improve ecological, social, biophysical, psychological, and economic wellbeing equitably without compromising the sustainability of development ecosystems and stakeholders. This book will present twelve case studies covering more than twelve cities or more cities centered on domain-specific smart planning components.

Case studies of Digital Innovation Master Planning include the Application of Artificial Neural networks in Master Planning for cities, a 3-D GIS Planning support system, and Digital Spatial Planning Incorporating Machine to machine Automation for Smart economic communities and M2M digital integration.

Case studies of Domain innovation include Urban Land Management, Master Planning for Water Management, Comprehensive master Planning Innovations, Smart use of Master Plan Basics, Integrated master Planning, and Citizen centric Master Planning [11, 12].

The following thirteenth and fourteenth books are about containment, life, work, and restarting cities and regions affected by COVID-19, using selected empirical case studies. This book presents the spread of Coronavirus spatially and temporally, analyzes containment strategies used, and includes recommended strategies. Further, it analyzes how life and work transform during the lockdown and gradual opening up and presents the future of work and life in cities impacted by COVID-19. This book discusses the concept of smart life and works in cities post-COVID-19 such that they do not reduce the quality of work and life and cannot create adverse economic and living consequences called the restart of a city after COVID-19.

Using selected empirical studies, these two books (urban and regional case) propose to model Coronavirus spread spatially and temporally by GIS Modeling in study areas, then analyze all containment strategies so far applied and come out with the authors' recommended strategies which may not be that was used by city Government or Union Government. Selected Cities of particular interest are studied. Special interest because the state of Kerala and Maharashtra in India the worst affected in India by the COVID-19 pandemic, and the book focuses on that with many studies [13, 14]. These thirteenth and fourteenth books were COVID-19, Containment, Life, Work, and Restart: Urban and Regional Studies.

This book is the sixteenth book "Indo-Pacific Smart Megacity System: Emerging Architecture and Megacity Studies" using selected empirical urban studies. While the fourteenth book focuses on regional case studies through themes such as delineation, Profiles. Regional Economy, Interconnections, International trade, mobility, Climate Change, Disaster Management, Security, Institutions, and Governance, with an introduction and conclusions. The book proposes to model Indo-Pacific Megacity System spatially and temporally by modeling in study areas around the World, then analyzing all strategies so far applied and coming out with the authors' recommended strategy. Then design smart life and work in cities post-COVID-19 such that life and work do not reduce the quality of work and life and cannot create adverse economic and living consequences. The first book in this project concentrates on sectoral studies within the region.

Part 1, Indo-Pacific System of Smart Megacity System, explores the Indo-Pacific Region, and the Megacity study in Part 2 studies individual megacities as part of the megacity system, such as Andamans, Chennai, and Kochi-Kannur. Kolkata, Mumbai, and Osaka.

The study teams from respective countries studied individual megacities. In contrast, the region is studied by study teams located within Indo-Pacific or outside, such as Italy, Slovakia, and the Czech Republic within the European Union, some of which have no individual country-specific strategies of engagement such, for example, Slovakia with Indo-Pacific but others such as France, Germany, the Netherlands within European Union do have the country for countries engagement policies. This editor hoped to achieve a balanced view of sectoral studies not influenced by the country's policy of Indo-Pacific.

2 Design of the Collaborative Research Program

Research Collaborations worked out are purely voluntary and without any financial support that appears to bind a project together. We could achieve this without money. Since collaborators are universities, governments, research institutions, professional networks, and not-for-profit associations as a general pattern from several countries, complete independence for pursuing the research was free of the baggage of ideologies of granting organization or their Government. They need not accept existing Smart City policies of study cities in their research or even their Government's policies for engagement with the Indo-Pacific region. This was never an issue in project execution and timely completion. The coordinator Editor of the project has no financial or administrative control over any institution taking part in the project since he was not in receipt of any grants and did not distribute it. Typologies of the institutions involved in this international project, all these autonomous institutions are guided by the highest standard of scholarship and prompt completion of research and publication within the milestone dates.

3 General Research Questions on Smart City Research and Indo-Pacific Projects.

The collaboration in this international research project requires that all participating institutions formulate their research questions and research the helpful methodology to the country, federal state, or Indo-Pacific where these study cities/regions are found. The approaches must differ Depending on the type of City; some lead to Smart cities, and some do not. However, the lack of empirical evidence on the Smart City opens a new area of research: What strategy intervention brings about a Smart City? This is the central focus of the book series.

Do smart cities awaken social, cultural development, and ecological (environmental) management through Smart City development? This question lies at the heart of the proposed international collaborative research program on Indo-Pacific, and unpacking it gives us four interrelated research questions, as follows:

I. What constitutes a Smart City beyond its official definition by UN-HABITAT? This will need to identify the key ingredients and their role in making Smart cities and Spatial changes in cities.

II. What changes the Smart City bring to social development, cultural preservation, heritage conservation, and ecological management? This calls for understanding the interlinkages between a Smart City on the one hand, with social development, cultural preservation, heritage conservation, and ecological management on the other.

III. How and what processes help the changes to the Smart City? Do cities bring to social development, cultural preservation, heritage conservation, and ecological management? These may include:

   a) innovation–diffusion (by ICTs and other modes),
   b) spatial planning,
   c) sectoral planning (including economic, social development, cultural preservation, and ecological management),
   d) Heritage conservation and management plan and
   e) institutional and Governance processes, among others.

IV. How and what changes can be brought to improve the processes to achieve enhanced/best results? These changes relate to the various processes, as Research Question iii mentions.

V. A deeper understanding of changes in the social, cultural, and ecological system of the Smart City with the advent of the Smart City and Smart People for innovative city development is the focus of the study. This research program and the institutions selected as academic collaborators are an effort to address this research gap.

4 Scope of Research

The following outlines the areas that may be covered when researching the "Smart City" program. This is a sign only, left to the team to decide what is appropriate.

I. A time series study of changes in the urban parameters and finding distinctive features of evolving to the Smart City.
II. Study of theories of the Smart City interventions at the Smart City level and modeling for study city.
III. The concept of accessibility in the Master Plan and its changes to the increasing use of ICT in Smart Cities for Smart City development.
IV. Changing the role of the hierarchy of service areas or watersheds in a Smart City is influenced by the increasing use of ICT to make the City smart.
V. Evolving structure of megacity urban agglomeration and changes needed in a Smart City when transformed.
VI. Evolving structure of cities in an urban agglomeration and changes are needed because of the increase in the use of ICT.
VII. Change of spatial standards in a Smart metropolitan city.
VIII. Changes required in zonal policies and plans.
IX. Study Town and Country Planning legislation and suggest changes per the particular Smart City requirement.
X. Change of role of community-based organizations (for example, Residential Associations in India) in a Smart City with an increase in the use of ICT.
XI. Change of role of Ward Committee in a Smart City with an increase in the use of ICT.
XII. Change of role of the Municipal Council in a Smart City with an increase in the use of ICT.
XIII. Change of role of the Planning system in a Smart City with an increase in the use of ICT.

Note: The scope of research can be further elaborated by the collaborating institutions but need not be uniform for all study cities. Each university department taking part in this research program shall incorporate relevant Smart City Development features appropriate to the goals of each department. The coordinator of this project does not intend to dictate the direction of the research and has a diverse group of collaborating universities. They should orient their study strictly based on the academic goals of their department.

5 Study Cities

The study city will be selected as a study area by each of the collaborating universities independently, which will be the place for the one-year and two-semester combined effort to conduct this research. Universities taking part in this program adopted diverse types of collaboration. Some universities used their doctorate and post-doctorate students, while others used students at master's and first professional degree levels. A post-doctoral student in the department can work on a narrow subject area in the study as individual work. While graduate and undergraduate students can work on design solutions for Smart City, Research institutions can charter their strategic research areas.

6 Project Details

One City will be selected as a study area by each collaborating university independently, which will be the place for the one-year and two-semester combined effort to conduct this research. Universities taking part in this program adopted diverse types of collaboration.

The project details of the study city are given in Table 1.

7 Way of Working the Program

7.1 Integrating Smart City Research with Academic Programs

As tabulated above, this international collaborative research program and study cities were conducted by many diverse university departments, research institutions, and others, as shown in the table above.

7.2 Role of Students

This international collaborative research program is meant for students since they are the future and are part of an internal academic program of the university. We consider them to be the leading actor and shall be given a vital role in this program. Many in that age group will live in the Smart Cities than their older faculty. Under the direction of the faculty, new concepts were introduced in the studio, and empirical studies were conducted around these concepts.

Table 1 Project details

Table of Contents Indo-Pacific Project			
Chapter	**Title**	**Author**	**Association**
1	Indo-Pacific Smart Megacity System: Emerging Architecture	T M Vinod Kumar	School of Planning and Architecture, New Delhi
2	Delineation of a system of Megacity in Indo-Pacific. Here the region's core is the Indo-Pacific Ocean, and the periphery is the landmass containing the megacities	Ashikha Raoof, Asha Devadas	Independent Consultant,
3	Profiles of megacities and current problems and prospects of individual cities and the collective system for economic integration	Ashikha Raoof, Asha Devadas	Independent Consultant
4	Prospects of Systems of Megacities and individual megacities for the Regional Economy	Vitálišová, Katarína; Borseková, Kamila; Vaňová, Anna	Department of Public Economics and Regional Development; Research and Innovation Center, Faculty of Economics, Matej Bel University in Banská Bystrica
5	Existing interconnections through rail, air, and ocean of megacity systems, their capacity and performance time series analysis, and emerging issue	Natalia Ivanova	Petersburg State Transport University, St Petersburg, Russia
6	International trade among the megacity systems/countries in its issues and barriers	Vitálišová, Katarína; Lacová, Žaneta; Považanová, Marianna	Department of Public Economics and Regional Development, Faculty of Economics, Matej Bel University in Banská Bystrica, Slovakia
7	Mobility of money, goods, and services among the systems of megacities time series analysis	Natalia Ivanova	Petersburg State Transport University, St Petersburg, Russia
8	Climate Change in the Indo-Pacific	Suruchi Bhadwal	TERI New Delhi
9	Disaster Management in the Indo-Pacific	Ramesh Srikonda	SPA Vijayawada
10	Security of the Indo-Pacific region and rule-based diplomacy and other options emerging	Gajula Sai Niranja, Vivek Noble, Santhosh Mathew	Centre for South Asia Study, Pondichery University

(continued)

Table 1 (continued)

Chapter	Title	Author	Association
	Table of Contents Indo-Pacific Project		
11	Institutions in Indo-Pacific, treaties, and covenants a critical study for reinforcing the megacity system	Antonella Contin	Department of Architecture and Urban Studies (DAStU), Politecnico di Milano, Milan, Italy
12'	The architecture for megacity Systems governance	Antonella Contin	Department of Architecture and Urban Studies (DAStU), Politecnico di Milano, Milan, Italy
13	Conclusion: Regional Architecture of Indo-Pacific Smart Megacity Systems	T.M. Vinod Kumar	School of Planning and Architecture, New Delhi

7.3 *Role of Faculty*

The faculty is the program's designer within the framework of existing curricula in each participating university's design studios and theory courses.

I. The project duration is one academic year or two semesters.
II. They guide and monitor student work as usual in the academic program.
III. They check students' input to the monthly progress report.
IV. They rewrite the output of the project for a book to be published by an international

Publishers give due credit to their work.

7.4 *Co-Design and Co-Production of Knowledge*

This international collaborative research program is founded on the principles of co-design and co-production of knowledge. In today's interconnected world, such collaboration is physically and intellectually possible—thanks to the Internet and ICTs. The collaborative aspect of the research program will be actualized in the form of:

I. Co-design the program with the partner academic institutions.
II. Co-production of knowledge through an interactive process of sharing, reviewing, and completing research findings.
III. Within each partnering institution, co-design and co-production of knowledge can be implemented through design studio/laboratory work between faculty and students.

7.5 Research Output

The key output of the "Smart City" research program will be these books edited by the coordinator Professor T.M. Vinod Kumar, to be published by Springer-Nature, an internationally reputed publisher, in 2023.

8 Summary of Conclusions and City Case Studies

8.1 The Smart Megacity System of Indo-Pacific: Emerging Architecture and Megacities Studies

Configuration of the Megacity System in the Indo-Pacific.

Based on the detailed study of many aspects, the authors conclude the following.

a. Peace and Prosperity in the Indo-Pacific and various milestone events on regional economic development in Indo-Pacific region.
b. Megacity system
c. The Indo-Pacific Region
d. How Megacity System Works in the Indo-Pacific
e. Evolution of QUAD and QUAD Working Groups conclusions
f. The First QUAD Summit March 12, 20,219
g. The Second QUAD Summit September 24, 2021
h. The Third QUAD Summit March 3, 2022
i. The fourth summit on May 24, 2022, in Tokyo
j. The fifth summit in Hiroshima 2023.
k. Indo-Pacific Policies, Strategies, and Programs of Outside Nations.
l. Indo-Pacific Strategy of France
m. Policy guidelines for the Indo-Pacific region Germany—Europe—Asia: shaping the twenty-first century together.
n. Indo-Pacific: Guidelines for strengthening Dutch and EU cooperation with partners in Asia.
o. The US Indo-Pacific Strategy February 2022
p. Joint Statement of the Leaders of India, Israel, United Arab Emirates, and the United States (I2U2)., July 14, 2022 [5].
q. The joint statement of Summit leaders in Hiroshima

8.1.1 Conclusions Configuration of the Megacity System in the Indo-Pacific

The G7, G8, and G20 Countries are the largest economies in the World and the strongest in terms of military and defense capabilities. Both aspects are equally crucial for regional development. They are the countries that develop the latest and

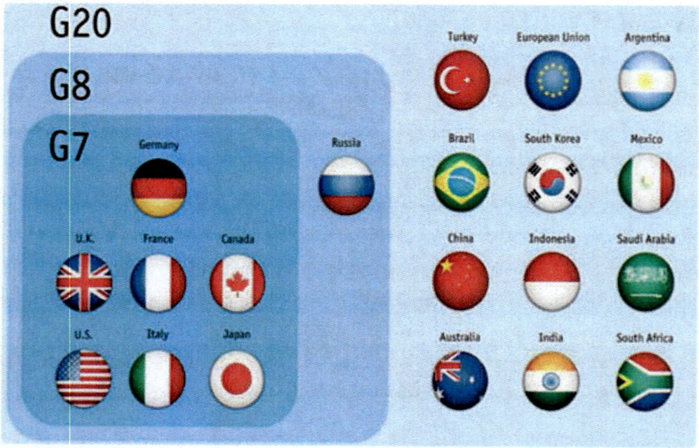

Fig. 1 G7,G8 and G20 country grouping

emerging innovative technologies with the potential to transform work and life and can become their market leaders. They are also a depository of the World Megacity system of thirty-plus megacities and Indo-Pacific Megacity system of twenty-one megacities discussed earlier. They are represented in the diagram below by their national flags (Fig. 1).

France, Germany, the United Kingdom, Japan, and the United States are either members of the QUAD group of countries or have independent policies, programs, and proposals in the Indo-Pacific with sizeable and recently strengthened diplomatic presence there in the Indo-Pacific and fully ready for collaborative action. Canada has set up a working group to formulate Government of Canada policies to be part of QUAD and published it. Italy, although not active now, is open to East Asia. Among the G20 Countries, India, Australia, Indonesia, China, South Korea, and the European Union are present in the Indo-Pacific either as QUAD members or with independent country policies, programs, and investment proposals and presence. All world systems of megacities and Indo-Pacific Megacity systems are part of these countries articulating their economic potential. Argentina, Brazil, and South Africa can be a potential grouping of countries with the United States or India. It can form the fourth QUAD after the UAE, Israel, United States, and India QUAD 2 and the United States, Japan, Australia, and India the QUAD 1 or QUAD.

It can be observed that no attempt is made to form an independent QUAD organization with headquarters and branches, as in the case of the United Nations or SARC. The QUAD uses individual participating country bureaucracy efficiently and with no duplication of effort and expert working groups independently or jointly as it develops for QUAD programs formulation of Implementation and uses the available bureaucracy and private parties as deemed most suitable and efficient for speedy and efficient Implementation, for the Governance of QUAD based on their programs. The programs are made in consultations with QUAD core member countries and mutually agreed upon by governments who wish to join with independence to fully join or partially join, depending upon the country's orientation to a particular component of the program of QUAD. There is no compulsion to join. The countries' sovereignty is respected, and governments can join fully or partially in any QUAD action program.

Once the country joins the QUAD programs, the barriers between people and distant international and multi-country QUAD must be broken. This can be executed if research and development are conducted and arrive at a detailed design of such Smart Global Economic Communities in the Indo-Pacific, which is the subject matter of two books to be published in 2024 by an international team of authors.

8.1.2 Megacity System is in the Indo-Pacific Region

The Indo-Pacific region is modeled (which is a simplification of reality) as a system of 21 plus megacities as today. This system sustains by constant flows of humans, goods and services, and communication; money flows exhibiting mutual dependencies due to interactions. This system also includes all dependent countries, megacity sub-regions, and other dependent city systems hierarchically below the megacity system. Administratively, Megacities spread to one or more Federal states, such as National Capital Region in Delhi, or districts as many as five sections in Kochi Kannur megacity in Kerala, which is seldom a Union Territory. Megacities have their sub-regions and zones of influence based on daily interaction and interdependencies within the national locations of these megacities. The landlocked countries outside megacities use the megacities' gateway infrastructure and related institutions for their economic activities.

Megacities deceptively seem to have grown historically as standalone and independent giant conurbations as an economic superpower among a hierarchy of cities, forgetting other megacities in the system. They are neither independent nor standalone cities. It has a local and international economy based on the types of economic interactions. Megacities themselves are interlinked with each other megacities functionally for a variety of higher-order economic, social, cultural, and ecological interactions with wide-ranging spatial impact beyond these cities, creating an integrated Indo-Pacific region increasingly. It had impacted the life and working of the inhabitants, contributed to more outstanding human accomplishments, and valued destinations for immigration.

As such megacities, the generator of economic superpower demands an uninterrupted daily flow of non-polluted air and water, assured protection of their territories and loss of valuable urban land from the rise of the sea due to climate change, potable water, perennial and reliable energy supply, perishable and nonperishable food supplies, a perennial source of increasing income, and expenditure to households, all types of utilities and services of highest standards, and essential other supplies to survive. It should provide work to earn an adequate income to survive honorably in a competitive environment affording a high standard of living within a high land value megacity, commensurate with a higher income than other cities.

Suppose one computes the total expenditure to sustain the megacities. In that case, one can find private households and non-government-owned entities in megacities spend a disproportionately higher share of the cost than the Government. Still, Governance is rightfully given a higher set of overriding power by the country's Constitution to protect the rights of all.

Because of all the above factors, Megacities are overly complex. In many cases, external factors beyond the national boundary and their external interaction had a more significant say in megacity development. At the same time, internal economic activities make them sustainable by providing a large market of ten million or above population in a limited geographic area for easy market access due to higher megacity densities.

A megacity can be considered a complex of overlapping spatial systems and involves many systems. For example, energy, water and sewerage, food, transport, health, education, biodiversity, and side-by-side ecological, economic, social, and cultural systems connect one area to another. This complex network of systems, interconnections, and flows can be described as a megacity system of systems. Many of these system components originate from outside the hierarchy of administrative regions. The City cannot politically and administratively influence the system environment's unpredictability, such as war, natural disasters, terrorism, climate change, economic recession, or other primary medium or minor ecological collapses. These eventualities point toward a crucial role of integrated megacity system planning, management, and development to design transcending national boundaries. As events unfold, the megacity system demands higher adaptability and resilience for sustaining a significant population of ten million and above.

Indo-Pacific came up in 2004 spontaneously as a response to the Tsunami by a group of countries who helped each other and themselves in making all systems stable and sustainable in as limited time as possible in many countries sharing what they best have to the situation. The World has faced many uncertainties in the past. The Indo-Pacific megacity system consisting of twenty-one megacities and the connected region as the highest repository of human capabilities, financial powers, superior infrastructure, knowledge, skills, and related higher-level institutions, has an essential and confident role in finding remedies for maladies for these uncertainties of natural disasters like tsunamis, earthquakes, floods, and the effects of climatic changes and even economic recession manufactured by sanctions or collapse of National Economy like that of Sri Lanka though autocratic president "overrules." Megacity systems collectively have a role in helping those affected by these calamities. There are also manufactured calamities like war which can be avoided by superior dialogue and diplomacy. The spread effect of economic sanctions affects more people in more countries in a connected world who may not be directly involved in the conflict and drives them to food and energy scarcity or scarcity of medicines and equipment, for example, faced during the pandemic COVID-19 or natural disasters or artificial Ukraine war. The solution is to identify many alternatives before the event that impacts supply chains so that these alternatives can substitute for the scarcity and even recessions that can come in. These megacities could identify scenarios of issues and alternate remedies as a joint effort of money, skills, and institutional capabilities of many megacities systems together transcending nations.

8.1.3 The Economic Strength of Indo-Pacific Megacity System

The two goals of the Indo-Pacific are peace and prosperity. Peace is an absence of war and conflict with the commitment of all countries to a rule-based order. A peaceful and stable megacity system ensures prosperity with no downward and unpredictable fluctuations in income and expenditure. Like many countries in the Indo-Pacific, India, with around one-third population living in cities a GDP urban share is about seventy percent and is now much above. With the increase in the size of the cities, GDP share increases, and hence megacity share of GDP is always higher than small cities, including metropolises. With Dhaka, as an example, the only megacity in Bangladesh, its share in GDP is as high as 60% of the National GDP, and Cairo about 50% on the higher side. Still, in India, with eight to nine megacities. the individual share is much lesser and evenly distributed to less than 10 to 16% but unevenly spatially distributed with most of them in the West and the South part of India while in China, it is primarily eastern part leaving the western part less developed in China and the east part in India. It looks like sub-regions of megacities are more developed than other regions where access to Megacities is limited, so economic opportunities and development of these sub-regions. Once the GDP of all twenty-one megacities in Indo-Pacific is combined, the percentage of the GDP of Indo-Pacific megacity system could be much higher than the rest of the region. It is also seen that a higher GDP share of the megacity means more significant opportunities for higher achievements, creating a situation in which other indicators of human development tend to be higher side by side, as illustrated by Bangladesh, where the secret of high achievements in social indicators is hidden in megacity Dhaka.

For aspirations of good living in megacities, high-paying work must be available continuously. Enhanced living is demanded with increasing income, and the megacity shall provide it. A perennial generation and flow of new knowledge for all economic, social, and cultural activities are the basis of the megacity system's sustenance. Such knowledge can be shared if the megacity collaborates, transcending national boundaries. Megacities can self-organize through self-organized supply chains for all of the above.

8.1.4 Sharing Economy, Platform Eeconomy, and Indo-Pacific Megacity System

A megacity system is a group of megacities that, in their evolution, have become interdependent because of the many relationship networks that interconnect them to a greater extent than other cities elsewhere. The air, sea, rail, road, broadband, and internet communication connection enables money flow to connect all megacities for economic interaction. E-Commerce, for example, Amazon, a global exporting entity for many countries, makes small sellers with minimal capital and income who can never be conventional shopkeepers become international sellers exporting their wares, using the megacity system and other services such as logistics, accounting, and legal services and sharing at the same time the technologically advanced e-commerce infrastructure platform of Amazon.

So, sharing the common digital platform becomes critical to megacity system economic development such as e-commerce. Just like an e-commerce platform with all its specialized algorithms and routines that move products to consumers' destinations from different origins selected in such a way that minimizes the time of delivery from purchase time, makes payment and makes refunds, and connects megacity for commercial activities. Machine to Machine (M2M) systems enabled platforms for supply chain component productions for many small and medium industries sharing common electronic platforms for unified and synchronous production. Machine to Machine (M2M) systems in the megacity system can be made into self-organized platforms to develop complex manufacturing supply chains and related algorithms in many megacities subjected to resilience under shocks. This can be for defense supplies or civil supplies. A similar platform can be there for service economies of diverse types in a megacity system and logistics. We must integrate e-commerce and e-production using M2M, generally a platform in a megacity system in an integration spatially executed with logical algorithm run decision-making any time on any day. Once it is achieved, e-commerce and M2M production systems are well integrated with a megacity system is the basis of megacity systems as engines of economic growth and economic, social, and ecological development. This, in short, is the platform economy of the Megacity System.

As indicated, twenty-one megacities exist in the Indo-Pacific out of thirty-three in the universe. We have an integrated megacity system once these megacities are interconnected along with e-commerce and spatially scattered M2M small and medium industries network for production with the 21-megacity system. Suppose one brings all of them into one electronic platform and generates different scenarios for the Indo-Pacific policies of countries such as Germany, France, the European Union, Australia, Japan, and India. In that case, we have a megacity development framework for the Indo-Pacific. The same system can answer supply chains of production and commerce when unanticipated events like sanctions follow the Ukraine wars. This is the challenge of the megacity system before the knowledge community.

Heterogeneity of megacity system Governance is given, and the author is a proud Hindu who believes in extreme and uncompromised tolerance to all. So, we have autocracy as a governance mechanism for Chinese megacities in Indo-Pacific and democracy for many others in the Indo-Pacific such as India, Japan, Indonesia, and many others. There can be a mixture of the two with overriding power with the President in a democracy such as Sri Lanka, which may be discarded eventually. The rights of citizens vary as per the country's Constitution, and there is no uncontrolled freedom in any democracy. Platform economy works differently in these systems followed.

The primary beneficiary of components of the platform economy in the megacity system is people below the poverty level, which effectively connects the Government to People by direct money transfer to their bank accounts when they are distressed or face calamities, making them easy to get bank accounts from Aadhar cards which is another internet platform. A man with one taxi car shares it and works with Uber on another mobility platform. A stay-at-home spouse from a lower middle class becomes an entrepreneur globally in E-commerce with Flipkart or Amazon. So also, a small-scale industrial establishment, however small it may be, participating in a feature-rich platform economy and benefits. These are all examples of the sharing economy of intelligent cities. Without a platform economy and high megacity infrastructure, the rapid transition of the low-income group to the middle-income group and the rich to the super-rich in the megacity system is not possible for several reasons. The land-use configuration I suggested makes access to employment opportunities in the megacity historically highest in the history of human civilization.

This is the age of sanctions. In addition to the United Nations, which gives out binding sanctions, we have non-binding, legally optional sanctions from the United States, European Union, NATO countries, and many others. Like autocracy and democracy co-exist in the megacity system, the UN and other countries' sanctions co-exist with others. While urban systems under Autocracy and Democracy do not change instantly, the sanctions triggered by war or trade war can change urban systems. The platform technology behind e-commerce and M2M production systems of supply chains to megacity systems must be fine-tuned as such sanction event unfolds. A set of predesigned different supply chains are enabled to circumvent food, medicine, and energy scarcity that results and to avoid recession from sanctions and to avoid creating more people below the poverty level in the third country that is not involved in sanctions or war. Only a platform-based megacity system can perform and incorporate unexpected events and calamities since it is part of Governing responsibilities. Although it is yet to be worked out by the megacity system of Indo-Pacific, it is a feasible option for Think Tanks to work out. Here, for example, Government and private e-commerce and M2M production systems shall come under one megacity system to overcome the impact of such unexpected and shocking eventualities.

Here system theory is the basis. Systems theory, the base of the platform economy, is the interdisciplinary study of systems, i.e., cohesive groups of interrelated, interdependent parts that can be natural or human-made. Every design is bounded by space and time, influenced by its environment, defined by its structure and purpose, and expressed through its functioning.

One of the noteworthy aspects of Indo-Pacific is that there are four leading countries, the United States, Japan, Australia, and India, and their willingness to collaborate with all countries of the Indo-Pacific as well as outsiders such as the European Union, UK, Germany, Netherlands, and France, all of them having their Indo-Pacific strategies and willing to be partners in the prosperity of the Indo-Pacific. They all have a genuine interest economically in these regions as a legacy of long historical connections of these countries in the area.

The Indo-Pacific Megacity system starts with the concept put forward as an interdependent region with common goals arising out of shared issues and a unified search for effective and outcome-oriented solutions with the participation of countries as per their capabilities using advanced technologies. The Indo-Pacific was, earlier at its inception, a spontaneous mental creation for disaster management when countries collaborated as if they were one family. Now, these need to be extended to all eventualities in the region, the impact of war, the aftermath of a fight against opponents, and sanctions and countersanctions. The additional dimension is that the Indo-Pacific region can be a focal area of the twenty-first century for regional economic development, which shall be based on integrating the megacity system and emerging platform technologies. Just like countries have the infrastructure and protocol of action once a Tsunami happens and other calamities or the Pandemic COVID-19 happen, how can alternatively supply chains disruptions in a variety of situations having economic and social impacts be designed and preparations for necessary action be put in place jointly by megacity system which can be immediate or run into many years and months can be created. While the e-commerce system and M2M system looks at only fulfilling the demands for the commercial product, the megacity system looks at holistically all aspect of living and working and its environmental impact in all unexpected eventualities to make the megacity system sustainable and within a brief time overcome difficulties with collaborative country actions.

8.1.5 How Indo-Pacific Integrated Regional Development Through Megacity System is Functioning

We know how these systems we discussed above can work through various case studies on the Smart City series we jointly researched and authored by academics from Africa, Asia, Europe, the Middle East, and North America, such as,

I. Smart Economy in smart cities [3]
II. Smart Metropolitan Regional Development [5]
III. E-Democracy for Smart Cities [4]
IV. E-Governance for Smart Cities [2]
V. Smart Global Megacities Two volume [9, 10]
VI. Smart Living for Smart Cities Two books [7, 8]
VII. Smart Environment for Smart Cities [6]
VIII. Smart Master Planning Case Studies of Domain and Digital Innovations Two Volume [11, 12]

In addition, we know the following.

a. There is a movement of money, information, people, and goods and services between all twenty-one and above megacities in the Indo-Pacific, and this system functions every millisecond.
b. This system exists because of thousands of island chains and about forty countries in East Africa, Asia, the Middle East, ASEAN, and SARC countries. Twenty-one or more megacities solely depend on Integrated Regional Development and Integrated Disaster management and one joint fight against environmental and

pandemic calamities such as COVID-19, Tsunami., earthquakes, and disappearance of island chains due to sea rise of Indo-Pacific for climate change. These are all existential threats where an integrated regional solution alone is possible. Individual countries cannot solve these issues independently.

c. Smart City technologies are the building blocks of the Indo-Pacific regional development; all cities are the smart City in evolution and smart global megacity systems, such as with the use of ICT, IoT, Machine 2 Machine Systems, Industry 4, 5G, and 6G applications in Industry 4, Intelligent cloud, and the blockchain. Fintech, E-Commerce, and E-Governance are there. Some are widely used, for example, for direct debit of cash to Bank accounts during distress and disasters circumventing the cumbersome and corruption-ridden bureaucracy, which amounts to several Billion dollars in India, and easiness of people experiencing poverty to open bank accounts in India using Aadhar cards and widespread use of Unified Payment Interface UPI of India in Nepal, Singapore, UAE, Bhutan, Australia, and more countries to follow.
d. We know that existing urban land-use planning starting from the land-use classification that exists today is out of date and irrelevant when ICT, IoT, and M2M enabled Smart Cities to evolve in the Indo-Pacific because its past foundation is monoculture land use which needs to be mixed land use totally and everywhere with limited dependence on even emerging environment friendly and individually owned electric Cars since workplaces are distributed in all mixed land use starting from home and then inexpensive local community shared faculties in walking distance. Some land uses and establishments such as bank branches or administrative buildings, for example, can be eliminated by being replaced by UPI or its evolutionary all-compassing systems of future Banking under experimentation once the development of E-Governance, E-Democracy, Smart Mobility, E-Commerce, M2 M production system, and Home-based work culture a product of the post-COVID-19 era has also made these redundant. COVID-19, with a compulsory work@home culture of Intel and Apple computer industries, was the most productive year for Intel and Apple when the 13th generation chip came up much faster than usual. Apple computers came out with M1, M3, and M2 Pro chips and moved toward M2, and high-powered, compact, and less energy-consuming Apple studio-type computers came up. Brick and Mortar Bank is irrelevant for Fintech end-users; Mall is outside for E-commerce users, large capital-intensive and less labor-intensive robotically run factories are irrelevant for M2M users in production and twentieth-century organization that houses in one location in many buildings are also out of date for Indo-Pacific which is based on emerging technologies. This land-use classification and zonal regulations must change.
e. The Megacity System future in the Indo-Pacific is emerging technologically dependent and unpredictable. So if there is a Management and Plan for the development of the Indo-Pacific, it must be a three- to 5-year long-range perspective plan and annual plans and planning. The implementation time delay may be reduced to milliseconds using intelligent technologies such as IoT and M2M systems.

f. Smartphones are enabled with fast progressing full ownership of smartphones in megacity systems with high bandwidth and 5 G or emerging 6 G. Internet-based Platforms for planning, management, and Implementation as practiced in the Fintech, E-Commerce, E-Democracy, E-Governance, M2m in Industry 4, Uber-type Taxi service shall be the backbone of Indo-Pacific Megacity System for Regional Development.
g. How we work, live, and govern in the Indo-Pacific must change once they accept advanced technologies.

8.1.6 What Are the Ways and Means for Achieving Indo-Pacific Integrated Regional Development Through Megacity System?

a. Accept that every institution, Government and Non-Government, goes through a life cycle of birth, growth, and death like all living entities, just like industrial products, the nature of everchanging services and commerce sectors. This is an opportunity to regenerate and reform using emerging technologies. Then rebuild institutions by accepting the reshaping of these institutions using ICT, IoT, M2M,5G,6G, Industry 4, and all emerging technologies.
b. Organize smart internet-based spatially and highly specialized and non-spatial communities for work involving primary, secondary, and tertiary sectors using emerging technologies with consequent enhanced smart living and appreciable enhancement with E-Democracy in living fully enabled.
c. Smart communities shall embrace Sharing Economy in all walks of work and life instead of a capitalist-owning economy.
d. Analyze and redesign all economic activities and reconfigure them using broadband internet and advanced emerging technologies to increase value addition and internal rate of return.
e. Renewables shall be part of life, starting with energy and combining production with renewable can reach production and service delivery with a near-zero marginal cost for long-range sustainability.
f. Increase production and distribution of relevant, high-quality knowledge workers from academic institutions to meet the development needs and make it easy to access their work opportunities in the Indo-Pacific, removing barriers.
g. Achieve an exceedingly elevated level of easiness to access Government services and continuously evaluate Governance by satisfaction scores of services by consumers and reform through e-democracy combined with e-governance and not only simply conforming administrative rules of law.
h. The spatial implications of all the above shall be part of the megacity system's integrated development and the Indo-Pacific's regional development.
i. Develop continuously Smart and technologically Up to Date Production systems for end-user supply chains in all the Megacity System's primary, secondary, and

tertiary economic activities. The fundamental steps of a supply chain in order are Sourcing raw materials, refining those materials into essential parts, combining those basic parts to create a product, order fulfillment for sales, product delivery, customer support, and return services. These can be dispersed spatially. The time it takes for any of these processes from start to completion is known as lead time, and using algorithms reduces it. The spatial and infrastructure requirements shall be accounted for.

8.2 Prospects of Systems of Megacities and Individual Megacities to the Regional Economy

The importance of cities and megacities is one of the defining features of the twenty-first century. The chapter provides a clear and comprehensive overview of current thinking within the academic and scientific sector and daily practice from two megacities in Europe—London and Paris since they are the only nodes which can interact with Indo-Pacific in their economic, defense, and strategic interests resulted from their multi-dimensional physical, institutional, economic, and social linkages in this region.

The Indo-Pacific region is important for the United Kingdom and France due to preserving the free flow of goods and services, defending the countries' sovereignty, and maintaining international stability. Partnership with Indo-Pacific countries, bilaterally or through the EU, is the crucial point of the policy of these countries, strongly supported by diplomatic power. Presented examples of megacities can be, in many ways, an inspiration for megacities in the Indo-Pacific and the development of megacities in Europe, despite historical, political, cultural, and religious differences.

Paris and London are the gateways for sea, air, and communication, institutional and academic linkages with the Indo-Pacific. These megacities have to face various challenges on the one hand, but on the other hand, there are new possibilities and solutions for their "sustainable" development.

The chapter identified new paradigms for urban space development, the role and importance of current development concepts and innovation systems, and the role and importance of integrated, open, multi-level, cross-sectorial planning and management. An integrated approach to spatial, financial, sectoral, intersectoral, and institutional development, particularly in terms of the participation of the relevant actors in development, is becoming a prerequisite for the megacity's development. Awareness of the direction of urban development and an assessment of current practice approaches also create the preconditions for transforming megacities into sustainable, smart, and resilient cities of the twenty-first century in the context of the 2030 Agenda.

8.3 International Trade Among the Megacity Systems/ Countries in Its Issues and Barriers EU 27 and Selected Indo-Pacific Countries

Global development is strongly correlated with the changes at a local and regional level. The researches show that megacities, as one of the urbanization trends, contribute significantly to world economic growth and global GDP. In the chapter, we focused on analyzing international trade among European Union countries as a unit and selected Indo-Pacific countries where megacities are important trade partners for the European Union. EU27 is one of the giant trading blocks in the world economy. Since the standard trade policy falls within the exclusive competencies of the European Union, the EU acts as one unit in trade relations with third countries. Although mutual trade was fully liberalized only between member countries of the EU, due to the importance of third countries as EU trade and investment partners, the EU signed with them various preferential agreements. The growing importance of the Indo-Pacific region in the world economy and the EU external relations is reflected by improving economic relations between the EU and these countries. In 2021, EU adopted Strategy for Cooperation in the Indo-Pacific. The deeper analysis is devoted to countries—India, Indonesia, Japan, Thailand, Philippines, and Bangladesh.

We discovered that trade and investment relations with India and Indonesia have deepened recently. An analogical tendency can be observed in relations with Japan, too. Negotiations with India for EU-India Free Trade Agreement (FTA), Investment Protection Agreement (IPA), and Geographical Indications Agreement (GIs) are currently ongoing. The EU is negotiating a Free Trade Agreement (FTA) with Indonesia. The most developed relations are between the EU and Japan, catalyzed particularly by Economic Partnership Agreement (EPA).

Thailand, the Philippines, and Bangladesh are not the key trade partners for the European Union, but the trade relationships are developed gradually because of their location in the ASEAN region. Even though countries are not so economically developed, they represent the market potential in the secure future. The importance is also a great effort by the European Union to support these countries' social and environmental development, fostering human rights and solving poverty. The unstable political situation influenced the cooperation with Thailand in the country, but after reconsideration by the EU, the ambition is to establish free trade agreement. A similar situation is in the relationships with the Philippines. The trade relationship with Bangladesh European Union has developed more intensively from the COVID-19 pandemic. The cooperation aims to support social development, respect for labor rights, and a responsible approach to business activities.

Despite some issues, gradual liberalization between EU and Indo-Pacific countries is assumed to increase mutual trade, contribute to production and employment growth, and promote sustainable development.

8.4 Metropolitan Architecture and Sustainable Habitat in the Indo-Pacific Region to Reinforce the Megacity System with Urban Rural Pattern

The over-reliance on technology and efficiency in the name of national or racial identity has led to a loss of uniqueness and division of society into elitist groups. This results in increased inequalities and vulnerabilities in both physical and social dimensions, particularly in the countryside. To fully embrace a new alternative model of civilization, it is crucial to incorporate digital technologies and balance the values of rural civilization with science and technology. Enhancing prosperity, equity, and climate neutrality is essential for a sustainable future in the Indo-Pacific region. Policy guidelines must act as a source of learning and catalyst for sustainable change, addressing potential disparities in regional socioeconomic and policy-making capacities shaped by the Twin Transitions.

Interstitial actors must accentuate their role as mediators in urban change, addressing the transformative role of intermediaries in urban change. The skills and resources required for policy learning and changes depend on personal identity and community goals. The goal must be to significantly transform cities and territories and explore various interpretations and partnerships for a biopolitical project, leading to a new understanding of life.

Mitigating the risk of losing traditional skills, culture, local expertise, and markets in the Indo-Pacific Region requires creating a new informational ecology that produces new abilities and skills for a new metropolitan lifestyle, regenerative agriculture actions, and civic Economy. The concept of instability and threats in the Indo-Pacific landscape highlights the challenges of modern world complexity to traditional linear territorial development. Twin Transitions require new forms of Governance and compelling service access issues. Understanding megacities trends requires dealing with different cycles at various scales and paces, considering factors such as real estate dynamics, urban development strategies, demographic trends, and economic cycles. The global population is expected to increase by 2 billion by 2030, with 4.9 billion people living in cities.

However, shrinking cities are also growing, with decreasing areas escalating faster than boomtowns. Demographic trends and prospects indicate that most urban systems in Africa, Asia, and the Americas are shrinking.

Tokyo and Delhi are two examples of how overlapping megacity cycles can simultaneously determine different stages. Tokyo has been one of the world's largest megacities since half of the last century, but Delhi is expected to replace it soon. Demographic trends will significantly influence urban drifting trends, increasing the demographic gap between urban and rural areas. Urban shrinkage in China is a significant concern, with 90 (18.1%) shrinking cities and 118 (23.7%) locally shrinking cities. This mainly affects smaller cities near larger urban areas in the Yunnan province, which is prone to high dependence on imported food and energy.

The 2020 Philippine census revealed a decline in population growth, with Metro Manila experiencing a decrease from 1.58% in 2010 to 0.97% in 2020. Metropolitan

Architecture Projects aim to recognize interphase zones among urban landscapes and design linkage patterns to create enjoyable urban spaces. This approach will create a common temporal space between cities and the countryside using technology to connect old and new places, networks, and landscapes.

The proposal addresses societal problems, territory, and social disparities and promotes sustainable practices in response to the environmental crisis and lack of welfare. It focuses on enhancing urban-rural linkages through green and digital transformations, applying the twin transition sustainably. The objectives include mapping vulnerability and threats, demonstrating green infrastructure territorial intelligence, developing sustainable agriculture and food production, developing decision-making tools, focusing on education and skill proficiency, and developing a preventive approach to environmental emergencies. These objectives aim to create a cultural and intellectual transition, reflecting on biopolitics and promoting a cultural and intellectual transition. The global population is proliferating, and Asia is expected to shrink in the latter half of this century, resulting in extended megalopolises.

Global climate change is causing a significant decrease in population growth, with Europe, Latin America, and Asia being the most affected. Asia is the most impacted due to its large population. Population shrinkage can have positive impacts, such as strengthening human rights and reducing environmental impacts. Urban planning has primarily focused on growth, but there is a need to develop a shrinkage model for the megalopolises of Asia, such as Japan, which is expected to experience a significant population decline. The relationship between depopulation, spatial issues, and planning in addressing population decline is crucial, as ignoring shrinkage could lead to a sustainable thinning of the urban fabric with extended infrastructure.

Conventional studies on urban shrinkage have mainly focused on Western cities' post-industrial, post-socialist, and suburbanization processes. The twenty-first century is already witnessing a parallel phenomenon of growth and shrinkage, with the first significant population shrinkage occurring in Europe, Latin America, and Asia. The consequences of population shrinkage on an unprecedented scale are unknown, but developing a visionary model for Japan is relevant, allowing for further discussion and consideration of the implications of this unprecedented phenomenon.

The proposed approach utilizes open-source data and the Metropolitan Cartography Methodology to address potential vulnerabilities and threats in implementing the twin transition. It aims to improve prosperity, community, and quality of life while balancing efficiency, equity, and civil Economy. The methodology emphasizes designing cities considering the natural environment, preserving heritage, and involving communities in planning. It also promotes continuous feedback between inhabitants and decision-makers, promoting a more informed understanding of risks and the use of place-based technologies. Case studies are selected to apply the methodology in specific productive and socioeconomic contexts.

Metropolitan systems rapidly develop, often disregarding essential planning guidelines, particularly regarding spatial distribution. Correct spatial distribution rules significantly influence the perception of value in urban and metropolitan systems. Experts in urban and metropolitan development management sciences must be able to rely on a large set of skills to be equitable and efficient. The analysis and assessment process is based on the choice of elements to be evaluated and the procedures to be adopted, inspired by problem-solving principles. Value assessment involves estimating metropolitan assets in quantitative and qualitative terms, considering the economic value and cultural implications.

Soil degradation is a pressing issue due to anthropic pressure and urbanization, driven by disadvantaged households using low-quality housing, mass public production lowering quality standards, localization factors, immigration, high-tech infrastructure needs, and innovative covenants. Risks in territorial and environmental factors are vast, affecting various disciplines. The proposal aims to qualify urban risks, identify risk drivers over time, and understand the temporal range of adverse effects. It also addresses the lack of social perception, which can lead to the fragmentation of agricultural land after urbanization processes. A taxonomy of territorial risk can be outlined to identify territorial strategies to reduce urban risks, with two main types: natural origin risks, which are directly related to the natural system, and anthropic interference due to localization activities.

The study aims to develop a strategic vision of territorial shrinkage for Japan, a leading shrinking country in Asia, by analyzing spatial analysis. The research focuses on the Japanese population decline and its impact on the region's sustainability and identity. The study identifies three sub-research questions: what could be the shrinkage model for an Asian megalopolis that can enhance territorial sustainability and identity, the Japanese population shrinkage mechanism, the characteristics of the 'in-between' territories of Japanese and Asian megalopolises, and how to maintain shrinking agricultural regions faster than cities.

The primary cause of population decline in Japan is demographic change, but spatial distribution and economic activity also contribute. Spatial planning strategies can address these issues by directly confronting population shrinkage consequences. The study proposes a hypothetical shrinkage model for Asian megalopolitan shrinkage, focusing on the Tone River desakota in the Kanto region of Japan. This strategy contributes to infrastructural shrinkage, agriculture, optimization of high-density urban settlements, and increased communication within the geographic region. The post-megalopolitan procedure involves strategically manipulating territorial shrinkage through capital concentration in the 'in-between' desakota regions, focusing on the proximity between agricultural supply, urban activities, and food production. This model can be adopted in other future shrinking megalopolises in Asia.

This thesis explores the desakota shrinkage model and its potential applications in Asian megalopolises. It highlights the need for further research on the feasibility of this model, the environmental impact of urban and agricultural mixtures, and practical methods for urban weight shift. The chapter also discusses the changing patterns of Indo-Pacific urbanization, the Desakota model in the Tokyo region, and the Philippines' economic growth trend until 2034. It highlights the importance of sustainable development strategies, such as monitoring river flows and groundwater sources, to address food security and climate change threats. The chapter also presents a strategic planning example for the shrinkage of Chinese rural heritage, focusing on the case of Wengding village.

The metropolitan architecture project aims to reconnect Wengding village with its rural heritage and create a sustainable future for the community by incorporating regional architecture techniques sensitive to local culture and environment. The study focuses on the role of culture in urban development, particularly for small and intermediate cities, and identifies practical knowledge-to-action proposals for urban development.

The study explores the principle of harmony in Chinese culture, focusing on integrating harmony into an operating culture encompassing anthropological, social, and economic aspects of cooperation with nature. Nature is crucial in shaping the relationship between humans and technology, and technology must align with nature's intentions. Culture serves as the soul that integrates technology and nature, providing a framework for understanding the relationship between humans and nature. China's significance lies in its technological advancements and its potential to help overcome its deep-seated tendencies toward particularism and discover new ways of understanding its unique forms of unity. Overcoming particularism involves more than rule by law or rule by people, but also urbanity, which can be seen as culture as an innovation operator for a global city in the post-COVID-19 situation.

Chinese culture can teach us how a community can develop skills in a moment of crisis, especially concerning the relationship between human beings' nature and the unknown. The development of the Chinese City as an "instant city" highlights the need to highlight a Chinese approach to "eventuality," such as being present and not succumbing to the past or the future. This is a fundamental challenge that China must grapple with in the twenty-first century.

Chinese societies have demonstrated innovative behavior in public spaces, such as bricolage. However, harnessing and transforming these energies into productive innovation remains a challenge. The proposed issues aim to promote innovation by channeling and utilizing these energies effectively. The Desakota model is essential in defining the metropolitan City of the second modernity, as it reconnects with the natural life cycle and embodies the temporal complexity between spaces due to technology. This raises questions about our position relative to nature and the value and sustainability of our actions.

The Desakota model also raises the question of giving back to rivers, as the ancestral Indo-Pacific world used water as a cultural resource. This geographic Economy is built on a chain of values that determines supply chains and choices that prioritize activities in specific places.

8.5 *Japanese Methods for Promoting Harmonious Regionalization in Asia*

Harmonious regionalization and economic development have long been both an ideal and a source of political controversy in Japan. There can be no doubt that harmony and development have been a significant source of Japanese internationalism in the postwar era, but the debate has remained on exactly how to carry this out. While there is epoch-defining change across Eurasia regarding international trade—including new systems for international settlement and the construction of new trade corridors—Japan has struggled to flesh out a proactive role within this new, emerging international movement for economic development. Barring a brief stint toward the end of the second Abe administration (2013-–2020), where, from 2016, Prime Minister Abe made a diplomatic about-face and sought to normalize and strengthen relations with regional partners, Japanese politics has struggled to adapt to this emerging reality. On the other hand, Japanese manufacturers and businesses must respond pragmatically to new challenges and opportunities within a complex international political economy.

A clear example is among lithium-ion battery manufacturers in Japan, who have sought to survive international competition through several means. Nonetheless, there remains a high degree of risk to Japanese firms—small-to-medium-sized enterprises, in particular—as the structural complexity of the international Economy limits their sense of global risks and, thus, their adaptation strategies for various contingencies. Manufacturers generally respond to tensions and change within the IPE by determining access to markets and risks to supply chains, seeking to implement diversification strategies as insurance against uncertainty. Among other concerns, manufacturers in growth industries like lithium-ion batteries rely heavily on technological advancements and will seek to identify collaboration areas for innovation. For this reason, concerns will remain over the guarantee of intellectual property protection and general geopolitical considerations when considering the wisdom of making long-term investments in joint projects in any region.

However, given that these are concerns of most manufacturers in the region, there is, despite the risk, ample opportunity to realize an economic regionalization that mutually benefits existing partners. It is likely that because of the complexity of the regional economy, manufacturers may adopt a long-term outlook that allows them to navigate through periods of tension and plan for sustainable growth in the industry. It is not unreasonable to suggest this could contribute to the formation of a harmony of interest in the Asian region, not least because cooperation and collaboration are of paramount importance to achieve developmental goals during periods of high uncertainty, such as when the world economy undergoes such change as is taking place in the present day. Japanese manufacturers would likely seek to maintain existing partners for as long as possible. Hence, the principles of peaceful coexistence that have formed the bedrock of Chinese diplomacy and economic relations between China and Japan in the past comprise a framework by which cooperation might be maintained in uncertain times.

8.6 Positioning of Mumbai in the Indo-Pacific Megacity System: Economic and Transport Opportunities

Mumbai, the financial capital of India, is one of the prime megacities in the Indo-Pacific region, attracting attention as the most populous area with high regional diversity and economic prosperity. In recent years, Mumbai has emerged as a prime center in geopolitics, global economy, international trade, and inter-regional transport, art, and culture (Hindi film industry). Mumbai has an estimated GDP of $310 billion, contributing 6% of the nation's GDP. The key sectors contributing to the City's Economy are finance, IT and ITES, textiles, entertainment, gems, jewelry, and leather processing. Mumbai's data center sector has grown dramatically in the last few decades. According to Knight Frank's research, Mumbai is one of the top data center markets in the Asia-Pacific region (IBEF, 2023). The National Stock Exchange (NSE) is one of India's largest stock exchanges in Mumbai (the oldest in Asia). The Mumbai film industry consists of around 400 production firms with 32 corporate houses, and this industry employs and provides a living for over 60 lakh people (Dungarpur, 2014). A recent report highlighted that the Indian media and entertainment industry had reached INR1.61 trillion in 2021, with a growth rate of 16.4%, expected to reach INR2.32 trillion by 2024.

Mumbai has an excellent public transport system that is the lifeline of Mumbai city; airways, highways, and waterways connect Mumbai megacity with the other parts of the country and the world. Mumbai connects to other megacities via different national and international airlines through Chhatrapati Shivaji International Airport. Apart from the existing airport, the Navi Mumbai airport is under construction to tackle the congestion in air transport. Nhava Sheva Port of Mumbai, India's second-largest container port, is a prime hub of international cargo movement. The Golden Quadrilateral, connecting Delhi, Mumbai, Chennai, and Kolkata, is an ambitious program launched by the Govt. of India. Also, the development of Mumbai to Delhi Dedicated Freight Corridors and the privatization and expansion of the Mumba airports are enhancing transport sector capacity and improving its efficiencies (The World Bank, 2011). The International North-South Transport Corridor (INSTC) comprises mainly rail, road, and maritime transportation networks; connecting Mumbai (India) will help increase trade volumes and promote economic cooperation among the participating countries.

In 2022, Mumbai was the first Indian City to release a Climate Action Plan (CAP). Mumbai and other megacities in Indo-Pacific nations such as Beijing, Shanghai, Tokyo, Bangkok, and Singapore are taking scientific measures to combat climate change by building sustainable cities.

8.7 *Is Kolkata Ready to Be the Gateway to the Indo-Pacific Region?*

The Indo-Pacific region, pivotal to defining the twenty-first century, is the world's fastest-growing region. It offers innumerable opportunities in the trade, research, manufacturing, and education sectors. The stability of this region is an assurance of the development of the global Economy and international security. India has actively advocated the concept of free and open trade in the Indo-Pacific, introduced by Prime Minister Shinzo Abe in 2016 [15]. Taking this stance of India as the basis, this research delves into the plausibility of the Kolkata megacity as being the gateway to the Indo-Pacific for India.

The medieval city of Kolkata, having a continuous industrial and port history for over 300 years, has emerged as a megacity in the eastern part of the Indian subcontinent. This mega-agglomeration is the center of economic, political, and business activities in East India and has held geopolitical importance since colonial times. It has been bereft of its centrality as a political, industrial, and leading economic hub with the shifting of the capital by the British and post-independence industrial and economic policies at the national level.

Despite these ramifications, Kolkata has been a popular livelihood destination for migrant labor from the neighboring states of Uttar Pradesh, Bihar, Odisha, Sikkim, and seven North Eastern states, which, along with cross-India migration, has probably pushed Kolkata into becoming a premature megacity.

The port of Kolkata, developed by the British and the key trading post with Southeast Asia and Europe, started to lose its gleam after Independence. This influx of migrants put considerable strain on the resources of Kolkata and the state of West Bengal, aggravating social problems and causing severe overcrowding. The increasing population pressure, dense urban environment, increase in slums and squatters, and substandard infrastructural facilities in terms of water supply, roads, electricity, sewerage, health, and education raises the question of readiness of Kolkata to be the gateway to the Indo-Pacific region.

Post-independence state policies of industrialization, decades of prolonged labor unrest, and the political milieu, too became a deterrent in the continuation of Kolkata being the country's leading manufacturing and port city. As discussed by the authors in Chapter 5, decentralization of manufacturing industries and regional development are the mantras that can bring the Kolkata region to the forefront of the industrial and economic environment, especially in light of the agricultural and mineral-rich hinterland.

References

1. Vinod Kumar TM, Associates, (eds) (2014) Geographic information systems for smart cities. Copal Publishing Group, New Delhi
2. Vinod Kumar TM (ed) (2015) E-Governance for smart cities. Springer, Singapore

3. Vinod Kumar TM (ed) (2016) "Smart economy in smart cities" international collaborative research: Ottawa, St Louis, Stuttgart, Bologna, Cape Town, Nairobi, Dakar, Lagos, New Delhi, Varanasi, Vijayawada, Kozhikode, Hong Kong" Springer, Singapore
4. Vinod Kumar TM (ed) (2017) E-Democracy for smart cities. Springer, Singapore
5. Vinod Kumar TM (ed) (2018) "Smart metropolitan regional development: economic and spatial design strategies" Springer-Nature, Singapore
6. Vinod Kumar TM (ed) (2019) "Smart environment for smart cities" Springer-Nature, Singapore
7. Vinod Kumar TM (ed) (2020) "Smart living for smart cities: case studies" Springer -Nature Singapore
8. Vinod Kumar TM (ed) (2020) "Smart living for smart cities: community study, ways, and means" Springer-Nature Singapore
9. Vinod Kumar TM (ed) (2021) "Smart global megacities: TOKYO, MUMBAI, NEW YORK, HONG KONG-SHENZHEN, KOLKATA New York," Springer-Nature Singapore
10. Vinod Kumar TM (ed) (2021) "Smart global megacities: CHENNAI, KOCHI-KANNUR" Springer -Nature Singapore
11. Vinod Kumar TM (ed) (2022) "Smarter master planning: case studies of digital innovations" Springer-Nature Singapore
12. Vinod Kumar TM (ed) (2022) "Smarter master planning: case studies of domain innovations" Springer-Nature Singapore
13. Vinod Kumar TM (ed) (2022) "COVID-19, containment, life, work, and restart: city studies" Springer-Nature Singapore
14. Vinod Kumar TM (ed) (2022) "COVID-19, containment, life, work, and restart: regional studies" Springer-Nature Singapore

Professor T. M. Vinod Kumar had 50 years of experience in Urban Planning, as a teacher, researcher, and adviser/consultant and worked in India, China, Bhutan, Nepal, Malaysia, Indonesia, and Hawaii, USA. He was Dean of Studies, Head of the Department of Urban Planning, Head Centre for Systems Studies and Analysis, Head of, the Centre for GIS and Remote Sensing, and Centre for Urban Studies of the School of Planning and Architecture, New Delhi; Visiting Professor National Institute of Technology, Calicut, and Institute of Technology Bandung, Indonesia and Professional Associates, East-West Resources Systems Institute Honolulu, Hawaii, Fellow Centre for the Study of Developing Societies, Delhi, Project Manager in Council for Social Development, New Delhi, Regional Program Coordinator at the International Centre for Integrated Mountain Development (ICIMOD) and Planner-Engineer at the Ford Foundation. He is the author of many books and journal articles. and edited "Geographic Information System for Smart Cities" (Copal:2014), "E-Governance for Smart Cities" (Springer:2015) "Smart Economy in Smart Cities" (Springer:2016) and "E-Democracy for Smart Cities" (Springer:2017), "Smart Metropolitan Regional Development: Economic and Spatial Design Strategies" (Springer:2018)," Smart Environment for Smart Cities" (Springer:2019), "Smart Living for Smart Cities Case studies" (Springer:2020)" Smart Living for Smart Cities Community study and Ways and Means" (Springer:2020), "Smart Global Megacities Collaborative Research-Tokyo, Mumbai, New York, Hong Kong-Shenzhen, Kolkata" (Springer 2021) and Smart

Global Megacities Collaborative Research-Chennai and Kochi-Kannur" (Springer:2021), Smart Master Planning: Case Studies of Digital Innovation" (Springer 2022), Smart Master Planning: Case Studies on Domain Innovations" (Springer:2022). COVID-19, Containment, Life, Work, and Restart, Urban Studies (Springer:2022) and. COVID-19, Containment, Life, Work, and Restart, Regional Studies (Springer:2022).